PRAISE FOR *CONGO*

"[A] detailed and well-researched biography, thoroughly rooted in the lived experience of the Congolese. . . . It is clear that the author is not your typical historian dryly publishing his findings, but a literary artist with a pen almost as sharp as Lumumba's tongue."   —ThinkAfricaPress.com

"This sweeping history of Congo begins during the precolonial era and brings readers all the way up to the current era of warlords and civil war. Van Reybrouck's carefully researched and elegantly written book takes in the reader with compelling portraits of ordinary people that enrich what would otherwise be a fairly conventional historical narrative."
                                                              —*Foreign Affairs*

"Beyond the retelling of slave and ivory trading, Belgian colonialism, and unstable independence, Van Reybrouck offers the perspective of ordinary Congolese caught in the broad sweep of that nation's turbulent history. The usual historical figures are here, from the adventurer Henry Morton Stanley to Belgian King Leopold II, from liberator Patrice Lumumba to the brutal ruler Mobotu Sese Seko, later overthrown by Laurent Kabila. But also present are elders, some in their hundreds or nineties, recalling their everyday lives in the midst of malaria outbreaks, missionaries, racial designations by colonial whites that exacerbated tribal differences, violence and oppression, economic instability and political upheaval, even the joy of hosting the fight between Muhammad Ali and George Foreman. Van Reybrouck draws on interviews and anthropological research to offer dense detail of dress, custom, diet, beliefs—all the ingredients of everyday life. This is a compelling mixture of literary and oral history that delivers an authentic story of how European colonialism, African resistance, and the endless exploitation of natural resources affected the lives of the Congolese."   —Vanessa Bush for *Booklist*

"Van Reybrouck's extensive account reveals the depth and breadth of exploitation, particularly under Belgian colonial rule, and how Congo's story is one fraught with the toxic cycle of 'desire, frustration, revenge.'"
                    —*Publishers Weekly*

"A mixture of multiple and passionate private testimonies, history, reportage, literature and even poetry. This is Gide's *Voyage to Congo* transposed to the present day, reread by Claude Lévi-Strauss, annotated by Fernand Braudel and remixed by Studs Terkel, Howard Zinn, or any other expert in oral history!"   —*Le Monde*

"A masterpiece."   —*L'Espresso*

"Van Reybrouck makes a good case for the importance of Congo to world history and its ongoing centrality in a time of resurgent economic colonialism, this time on the part of China."   —*Kirkus Reviews*

"A monumental history . . . more exciting than any novel."
                    —*NRC Handelsblad*

"An unbelievable tour de force."   —*HUMO*

"An absolute masterpiece!"   —VPRO Radio

"Breathtaking."   —*Trouw*

"If you are looking to read one book on Congo this year, this is it. David van Reybrouck combines deep historical investigation with extensive ethnography. The result is an illuminating narrative."
                    —Mahmood Mamdani, director of the Makerere Institute of
                    Social Research and author of *Good Muslim, Bad Muslim*

"A well-documented and passionate narrative, it reads like a novel. In a first-person rendering, Van Reybrouck retraces the stories and confrontations that bring together Congolese postcolonial clamors of the twentieth century and the late-nineteenth-century colonial adventures that explain them. As an eye, a judge, and a witness, a talented writer testifies."

—V. Y. Mudimbe, author of *The Invention of Africa*

"The English-speaking world has been impatiently awaiting this translation. *Congo* is a remarkable piece of work. Van Reybrouck pulls off the tricky feat of keeping a panoramic history of a vast and complex nation accessible, intimate and particular. He does this by talking to the Congolese, who know their history better than anyone else."

—Michela Wrong, author of *In the Footsteps of Mr. Kurtz*

"A blockbuster of a book—richly documented, crisply written, wide-ranging in scope and sources. No other book that I can think of does a better job of capturing the human dimension of the Congo tragedy."

—René Lemarchand, professor emeritus, University of Florida

"A triumph of writing 'history from below,' of letting people tell stories of survival with dignity and purpose that challenge established narratives. I am sure this translation will, as did the original, find a wide and enthusiastic readership."

—Johannes Fabian, professor emeritus, University of Amsterdam

"The Congolese are looking for their identity. They really need to delve into their own past. This book is so well written that it will help to increase our awareness." —Dr. Denis Mukwege

Also by David Van Reybrouck

CONGO

The Epic History of a People

DAVID VAN REYBROUCK

Translated from the Dutch
by Sam Garrett

An Imprint of HarperCollins*Publishers*

Flemish
Literature
Fund

The translation of this book is funded by the Flemish Literature
Fund (Vlaams Fonds voor de Letteren, www.flemishliterature.be).

A hardcover edition of this book was published in 2014 by Ecco,
an imprint of HarperCollins.

FIRST ECCO PAPERBACK EDITION PUBLISHED IN 2015.

Designed by Suet Yee Chong
Maps by Jan de Jong

Library of Congress Cataloging-in-Publication Data has been
applied for.

ISBN 978-0-06-220012-9

HB 12.12.2023

À la mémoire d'Étienne Nkasi (1882?–2010), en reconnaissance profonde de son témoignage exceptionnel et de la poignée de bananes, qu'il m'a offerte lors de notre première rencontre.

Et pour le petit David, né en 2008, fils de Ruffin Luliba, enfant-soldat démobilisé, et de son épouse Laura, qui ont bien voulu donner mon nom à leur premier enfant.

(To the memory of Étienne Nkasi [1882?–2010], deeply grateful for his exceptional testimony and for the handful of bananas that he offered me when we first met.

And for little David, who was born in 2008, the son of the demobilized child-soldier Ruffin Luliba and his former wife, Laura, who were so kind as to give my name to their first child.)

Le Rêve et l'Ombre étaient de très grands camarades.
(The dream and the shadow were the best of comrades)
—Badibanga, *L'élephant qui marche sur des œufs* (Brussels, 1931)

CONTENTS

LIST OF ILLUSTRATIONS

MAP 1: GEOGRAPHY

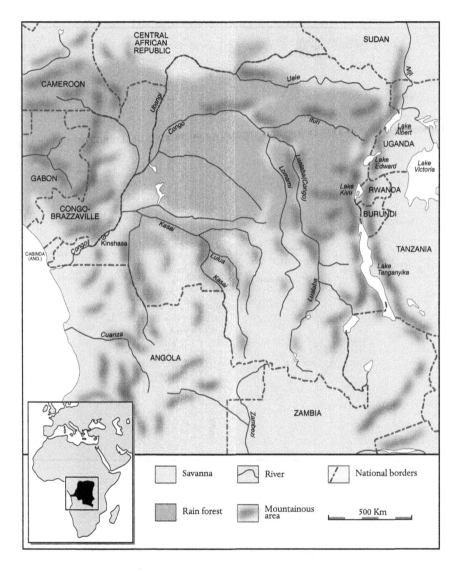

Savanna	River	National borders
Rain forest	Mountainous area	500 Km

MAP 2: POPULATION, ADMINISTRATION, AND RAW MATERIALS

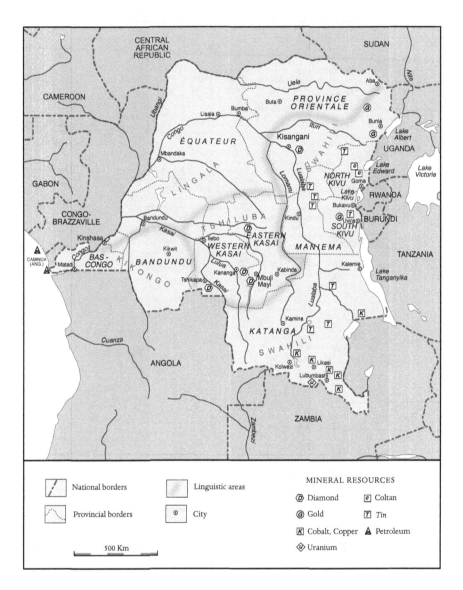

INTRODUCTION

I T IS STILL THE SEA, OBVIOUSLY, BUT YOU CAN SEE THAT SOME-
thing has changed, something about the color. The low, broad roll-
ers rock the ship as benevolently as ever; there is still nothing but ocean,
yet the blue is gradually becoming tainted with yellow. And that produces
not green, the way you might remember from your lessons in color theory, but murkiness. The glimmering azure has vanished. There is no more
turquoise billow beneath the noonday sun. The boundless cobalt from
which the sun arose, the ultramarine of twilight, the leaden grayness of
the night: gone.

From here on, all is broth.

Yellowish, ochre, rusty broth. You are still hundreds of nautical miles
from the coast, but you know: this is where the land starts. The force with
which the Congo River empties into the Atlantic is so great that it changes
the color of the seawater for hundreds of kilometers around.

Once, aboard the old packet boats, this discoloration made the first-
time traveler to Congo think he was almost there. But the crew and old
hands soon made it clear to the greenhorn that it was still a two-day sail
from here, days during which the newcomer would see the water grow
ever browner, ever dirtier. Standing at the stern he could see the growing
contrast with the blue ocean water that the propeller continued to lift up
from deeper layers. After a time, clumps of grass would begin drifting by,
chunks of sod, little islands that the river had spit out and that were now
bobbing about dazedly at sea. Through the porthole of his cabin he per-
ceived dismal shapes in the water, "chunks of wood and uprooted trees,
pulled up long ago from darkened jungles, for the black trunks were leaf-
less and the bare stumps of thick branches sometimes roiled at the surface
for a moment, then dove again."[1]

In satellite images, one sees it clearly: a brownish stain that stretches out up to eight hundred kilometers (about five hundred miles) westward at the high point of the rainy season. It looks as though the dry land is leaking. Oceanographers speak of the "Congo fan" or the "Congo plume." The first time I saw aerial photos of it, I couldn't help but think of someone who slashes his wrists and holds them under water—but then eternally. The water of the Congo, the second longest river in Africa, actually sprays into the ocean. The rocky substrate keeps its mouth relatively narrow.[2] Unlike the Nile, no peaceful maritime delta has arisen here; the enormous mass of water is forced out through a keyhole.

The ocher hue comes from the silt that the Congo collects during its 4,700-kilometer-long (about 2,900-mile) journey: from the high springs in the extreme south of the country, through the arid savanna and the weed-choked swamps of Katanga, past the endless equatorial forest that covers almost the entire northern half of the country to the rugged landscapes of Bas-Congo and the spectral stands of mangrove at the river's mouth. But the color also comes from the hundreds of rivers and tributaries that together form the drainage basin of the Congo, an area of some 3.7 million square kilometers (about 1.4 million square miles), more than a tenth of all Africa, coinciding largely with the republic of the same name.

And all those tiny bits of earth, all those torn-off particles of clay and mud and sand go floating along, downstream, to wider waters. Sometimes they hang suspended in place and glide on imperceptibly, then roil in a wild raging that mixes the daylight with darkness and foam. Sometimes they get stuck. Against a rock. An embankment. Against a rusty wreck that howls silently up at the clouds and around which a sandbank has formed. Sometimes they encounter nothing, nothing at all, nothing but water, different water all the time, first fresh, then bracken, finally salt.

That is how a country begins: far before the coastline, thinned down with lots and lots of seawater.

BUT WHERE DOES THE HISTORY BEGIN? Also much further away than you might expect. In 2003, when I first considered writing a book about the country's turbulent history—not only the postcolonial period, but also the colonial and a part of precolonial times—I decided it would only be worth doing if I were able to include as many Congolese voices as possible. To at least challenge the Eurocentrism that I would doubtlessly find on my path, it seemed to me that I would have to go systematically in search of the local perspective or, better yet, of the diversity of local perspectives, for there is of course no single Congolese version of history, just as there is no single Belgian, European, or simply "white man's" version. Congolese voices, in other words, as much as possible.

The only problem was: how does one set about doing that in a country where the average life expectancy during the last decade has never risen above forty-five? The country itself was turning fifty, but its inhabitants no longer were. There were, of course, the voices that came bubbling up from forgotten or nearly forgotten colonial sources. Missionaries and ethnographers had documented marvelous stories and songs. Numerous texts had been written by the Congolese themselves—to my amazement, I would even come across a native "ego document" from the late nineteenth century. But I was also looking for living witnesses, for people who would share their life stories with me, even the trivia. I was looking for what rarely ends up on the page, because history is so much more than that which is written down. That applies everywhere and always, but certainly in areas where only a tiny upper crust has access to the written word. Because of my training as an archaeologist, I attach great importance to nontextual information, which often provides a fuller, more tangible picture than textual information does. I wanted to be able to interview people, not necessarily the big decision makers, but everyday people whose lives had been marked by the broader scope of history. I wanted to ask people what they had eaten during this or that period. I was curious about the clothes they'd worn, what their house looked like when they were a child, whether they went to church.

It is, of course, always risky to extrapolate to the past from what people tell one today: nothing is so contemporary as our memories. But while opinion can be extremely malleable—informants sometimes sang the praises of colonialization: was that because things were really so good for them back then? Or was it because things were so bad for them now? Or was it because I'm a Belgian?—the memories of commonplace objects or actions often exhibit greater permanence. In 1950 you either had a bicycle or you did not. You spoke Kikongo with your mother as a child or you did not. You played soccer at the mission post or you did not. Memory does not discolor at the same rate everywhere. The trivial details of a human life retain their color longer.

So I wanted to interview ordinary Congolese people about their ordinary lives, although I don't like that word *ordinary*, for often the stories I was told were truly extraordinary. Time is a machine that crushes human lives to bits, I learned during the writing of this book, but occasionally there are also people who crush time.

Yet still: how was I to get started? I'd hoped to be able to speak on occasion to someone who might still remember the final years of the colonial period. I unquestioningly assumed that there would be no eyewitnesses to the period before World War II. I would have been very pleased to think that some older informant could still tell me something about his parents or grandparents in the period between the wars. For earlier periods I would have to navigate by the shaky compass of written sources. It took a while, however, before I realized that the average life expectancy in Congo is not so low because there are so few old people, but because so many children die. It is the country's hideous infant mortality rate that undercuts the average. During my ten journeys to Congo I soon met people of seventy, eighty, and even ninety years of age. One time, a blind old man of almost ninety told me a great deal about his father's life: in that way I was able to descend indirectly to the year 1890, a dizzying depth. But that was nothing compared to what Nkasi told me.

FROM THE AIR Kinshasa resembles a termite queen, swollen to grotesquery and shuddering with commotion, always active, always expanding. In the sweltering heat it stretches out along the river's left bank. On the far shore lies its twin sister, Brazzaville, smaller, fresher, shinier. The office towers there have mirrored windows. This is the only place on earth where two capital cities can view each other, but in Brazzaville, Kinshasa sees only its own, shabby reflection.

Kinshasa's palette is varied, but they are not the intact pigments of other sun-drenched cities. Nowhere will you find the saturated hues of Casablanca, the warm coloration of Havana, the deep-red tints of Varanasi. In Kinshasa every lick of paint fades so quickly that the people seemed to have given up on it: pallid colors have become an aesthetic of their own. Pastels, the missionaries' favorite hues, are dominant. From the tiniest boutique selling soap or prepaid mobile-phone refill cards up to and including the exuberant volume of a newly built Pentecostal church, the walls are always a faded yellow, faded green, or faded blue. As though illuminated day and night by neon lighting. The crates of Coca-Cola piled to form huge bulwarks in the yard at the Bralima brewery are not scarlet, but a dull red. The shirts of the traffic policemen are not a bright yellow, but urine colored. And in the brightest sunlight even the colors of the national flag flap rather wanly.

No, Kinshasa is not a colorful city. The soil there is not red, as in other parts of Africa, but black. Beneath the layers of pastel paint the walls are consistently drab. When masons along the Boulevard Lumumba lay their stones in the sun to dry, you see a color fan of grays: wet, dark-gray blocks beside mouse-gray ones that are still brittle, and ash-gray blocks beside those. The only color that really stands out here is the white of dried manioc, also known as cassava, the tuber that forms the staple for large parts of Central Africa. The plastic tubs of ground meal beside which the female merchants squat glisten so brightly that the women are forced to squint. Beside them lie piles of manioc roots, hefty, bright-white stumps that remind one of sawed-off tusks. Seeing those untidy piles from the air it looks as though the subsoil is baring its teeth, angry and fearful as a baboon. A grimace. The crooked ivories of a drab city. But pearly white, indeed. Impeccably white.

Imagine you could skim over this town like an ibis. A chessboard of rusty corrugated-iron roofs is what you would see, parcels of dark-green foliage. The grisaille of the *cités*, too, the poor districts of Kinshasa rolling on and on. We would circle above neighborhoods with leaden names like Makala, Bumbu, and Ngiri Ngiri, and down toward Kasavubu, one of the oldest neighborhoods for "inlanders," as the Congolese were called in colonial times. We would see Avenue Lubumbashi, a straight stretch of arterial with countless smaller streets and alleys emptying into it, but which has never been paved. It is the rainy season, the street is covered in puddles the size of swimming pools. Even the most skilled cabby becomes bogged down here. The inky-black mud spatters from beneath his screeching tires and sullies the flanks of his rattling, but newly washed, Nissan or Mazda.

We would leave him behind, cursing, and soar on to Avenue Faradje. In the courtyard of number 66, past the concrete wall topped with shards of glass, past the heavy metal gates, something white is glistening. We zoom in. It is not manioc or ivory. It is plastic. Hard, white, extruded plastic. It is a potty. A child is sitting on it, a darling little one-year-old girl. Her coiffure: a plantation of young palm trees bound together close to the crown with yellow and red elastic bands. Her yellow dress with the floral pattern is draped over her rear end. Around her ankles there are no panties: she doesn't have those. But she is doing what all one-year-olds all over the world do when they don't understand exactly why that potty is so damned important: she is screaming, furiously and heartrendingly.

I SAW HER SITTING THERE on Thursday, November 6, 2008. Her name was Keitsha. It was a traumatic afternoon for her. Not only was she being denied the joy of spontaneous defecation, but she was also facing the most terrifying thing she had ever seen in her short life: a white person, something she knew about only from her worn-out, handicapped Barbie doll, but then big, and alive, and with two legs.

Keitsha would remain on her guard all afternoon. While the members of her family sat talking to this peculiar visitor and even sharing bananas and peanuts with him, she remained at a safe distance, staring for minutes at how he dug *his* hand too into the crackling bag of nuts.

Fortunately I had not come for her, but for her forefather, Nkasi. I left the courtyard with the howling child behind and slid aside the thin sheet covering the doorway. The room was almost completely dark. As my eyes tried to grow accustomed, I heard the roof cracking in the heat. Corrugated iron, of course. A faded blue wall, like everywhere else. "Christ est dieu" was written on it in chalk. Beside that, in charcoal, someone had scribbled a list of cell-phone numbers. The house as address book; for years, paper in Kinshasa has been prohibitively expensive.

Nkasi was sitting on the edge of his bed. His head hung down. With his old fingers he was trying to do up the final buttons of his shirt. He had only just awoken. I approached and greeted him. He looked up. His glasses were attached to his head with a rubber band. Behind the thick and badly scratched lenses I made out a pair of watery eyes. He let go of his shirt and took my hand in both of his. A striking amount of strength still in those fingers.

"Mundele," he murmured, "mundele!" He sounded moved, as though we hadn't seen each other in years. "White man." His voice was like a rusty gear slowly creaking into motion. A Belgian in his home . . . after all these years . . . That he would live to see this.

"Papa Nkasi," I spoke into the semidarkness, "I am very honored to meet you." He was still holding onto my hand, but gestured to me to sit down. I located a plastic garden chair. "How are you?"

"Aaah," he moaned from behind his lenses, so scored with scratches that you could hardly see his eyes, "I'm afraid my *demi-vieillesse* is acting up again." Beside the bed was a little bowl that obviously served as spittoon. On the grimy mattress lay an enema syringe. Its rubber bulb looked chapped and brittle. Here and there I saw a piece of foil of the kind used to package pills. Then he had to laugh at his own joke.

So how old was that anyway, that middle-old age? He definitely looked like the oldest Congolese I had ever met.

He didn't have to think about it long. "Je suis né en mille-huit cent quatre-vingt-deux."

Eighteen eighty-two? Dates are a relative thing in Congo. I have had informants tell me, when I asked how long ago something had happened: "A long time ago, yes, a long, long time ago, at least six years, or no, wait, let's say: eighteen months ago." My desire to provide a Congolese perspective would never meet with complete success: I myself am much too fond of dates. And some informants are fonder of an answer than they are of a correct answer. On the other hand, though, I had often been struck by the precision with which they were able to recall facts from their own lives. In addition to the year, they were also frequently able to name the month and the day. "I moved to Kinshasa on April 12, 1963." Or: "On March 24, 1943, the ship set sail." It has taught me, above all, to be very careful with dates.

Eighteen eighty-two? Let's see, that would mean we were talking about Henry Morton Stanley's day, the establishment of Congo Free State, the arrival of the first missionaries. That was even before the Berlin Conference, the famous meeting in 1885 during which the European powers determined the future of Africa. Could I really be face-to-face with someone who not only remembered colonialism, but was in fact born in the precolonial era? Someone born in the same year as James Joyce, Igor Stravinsky, and Virginia Woolf? It was almost impossible to believe. That would mean the man was 126 years old! And that would make him not only the oldest man in the world, but also one of the oldest people ever. In Congo, no less. Three times the country's average life expectancy!

And so I did what I would have done in any other situation: check and double check. In this case, that meant digging up the past, little by little, with endless patience. Sometimes that worked promptly, at other moments not at all. Never before had I spoken like this with such a distant past, never before had it felt so fragile. Often, I was unable to understand him. Often, he began a sentence and stopped halfway, with the surprised look of someone who goes to fetch something from the cupboard and suddenly no longer knows what he was looking for. It was a struggle against forgetfulness, but Nkasi not only forgot the past, he also forgot to forget. The gaps that arose healed over immediately. He was unaware of having lost anything. I, on the other hand, was doing my best to bail out an ocean steamer with a tin can.

Finally, however, I came to the conclusion that his year of birth just very well might be correct. He talked about events in the eighties and nineties of the nineteenth century that he could only have known about firsthand. Nkasi had not attended school, but he knew historical facts of which other elderly Congolese from his region were entirely ignorant. He came from Bas-Congo, the area between Kinshasa and the Atlantic Ocean where the Western presence had first made itself known. If the map of Congo looks like a balloon, Bas-Congo is the neck through which everything passes. His memories, therefore, I could check against well-documented events. He spoke with great precision about the first missionaries, Anglo-Saxon Protestants who had settled in his homeland. They had, indeed, begun their evangelism around 1880. He mentioned the names of missionaries who, as it turned out, had come to the area in the 1890s and had moved to a nearby mission post around 1900. He spoke of Simon Kimbangu, a man from a neighboring village who we know was born in 1889 and started his own religion in the 1920s. And he talked above all about how he, as a child, had watched them build the railway between Matadi and Kinshasa. That took place between 1890 and 1898. The construction in his part of Congo began in 1895. "I was twelve, fifteen at the time," he said.

"Papa Nkasi . . ."

"*Oui?*" Whenever I addressed him he would look up, slightly distracted, as though he had forgotten there was a visitor in the room. He made no effort to convince me of his advanced age. He talked about what he still knew, and seemed amazed at my amazement. He was clearly less impressed by his age than I, who wrote down an entire notebook full.

"How is it that you know the year when you were born, anyway? There was no registrar's office back then, was there?"

"Joseph Zinga told me about it."

"Who?"

"Joseph Zinga. My father's youngest brother." And from that there followed the story of the uncle who had gone with a British missionary to the mission post at Palabala and attended catechism classes himself, during which he learned about the Christian calendar. "He told me I was born in 1882."

"But then, did you know Stanley?" Never, in all my life, had I thought I would ask someone that question in earnest.

"*Stanlei?*" he said. He spoke the name in the French way. "No, I never met him, but I heard about him. He came to Lukunga first, and then to Kintambo." The chronology, in any case, jibed with the journey Stanley had made between 1879 and 1884. "I did know Lutunu, though, one of his boys. He was from Gombe-Matadi, not far from us. He never wore trousers."

The name Lutunu rang a bell. I remembered that he was one of the first Congolese to serve the white men as a "boy." Later, the colonizer would make him an inland chief. But he had lived until the 1950s: Nkasi could have met him much later as well. That, however, definitely did not apply to Simon Kimbangu.

"I knew Kimbangu back in the 1800s," he said emphatically. It was the only time, with the exception of his year of birth, that he referred directly to the nineteenth century. They had lived in neighboring villages. And, he added: "We were more or less the same age. Simon Kimbangu was greater than me in *pouvoir de Dieu,* but I was greater in years." During later visits as well, he confirmed time and again that he was a few years older than Kimbangu, a man born in 1889.

IN THE WEEKS AFTER my initial visit, I went by to see Nkasi several times. At the house where I was staying in Kinshasa I would run back through my notes, put together the pieces of the puzzle and search for gaps in his story. Each visit lasted no more than a couple of hours. Nkasi indicated when he was growing tired or when his memory was failing him. The conversations always took place in his bedroom. Sometimes he would sit on the edge of his bed, sometimes on the only other piece of furniture: a worn-out car seat that stood on the floor. Once I was able to talk with him while he was shaving. Without a mirror, without shaving cream, without water, only a disposable razor that he never disposed of. He ran his fingers over his chin, made a whole host of strange faces and scraped the white plastic razor across his weathered skin. After a few hesitant scrapes he would knock out the tiny hairs against the edge of his bedstead. The white stubble floated to the darkened floor.

In one corner of the room was a pile of odds and ends: what remained of his belongings. A broken Singer sewing machine, a pile of rags, a big can of Milgro powdered milk, a gym bag, and a linen bundle. The latter item had caught my eye during my first visit. It looked like it contained something round. "What's in that package?" I asked him one time. "*Ah, ça!*" He reached for the bundle. Slowly, he unwrapped it and held out a beautiful pith helmet. A black one. I didn't even know they existed. Without my asking, he put it on and smiled broadly. "Ah, monsieur David, I lived my entire life in the white man's grasp. But within two or three days I am going to die."

Moving about was very difficult for him. He used the handle and stick of an old umbrella as a cane, but preferred to rely on the support of a few of his daughters. Nkasi had had five wives. Or six. Or seven. Accounts differed. He himself had lost count. There were always a few family members outside in the courtyard. Estimates varied concerning the number of his children. Thirty-four was the figure heard most often. In any case, four pairs of twins, everyone seemed in agreement on that. Grandchildren? Definitely more than seventy.

I was also introduced to his two younger brothers, Augustin and Marcel, ninety and one hundred years old, respectively. Marcel did not live in Kinshasa, but in Nkamba. I spoke with Augustin's son, a smart, sensible man who had not yet reached middle age. Or so I thought. Until he told me that he was sixty. It was almost too much for me to believe: he truly looked no older than forty-five. What an extremely resilient family, it occurred to me, what a wild quirk of nature. Three ancient brothers, all of them still alive. There had been two sisters as well, but they had died recently. Also well into their nineties.

Fourteen people lived here in three little adjacent rooms, but there was family visiting all day. Nkasi shared his room with Nickel and Platini, two boys in their twenties. One of them had a sweater that read *Miami Champs*. As the eldest, Nkasi got the bed each night, a foregone conclusion; the young people slept on the floor on woven banana-leaf mats. During the day they sometimes took a nap on their grandfather's thin mattress.

Nkasi ate manioc, rice, beans, sometimes a little bread. The family couldn't afford meat. After one particularly long session he realized I must be hungry and, using his umbrella stick, slid over to me a bunch of little bananas and a bag of peanuts. "I can tell. The head is closed, but the belly is open. Take it, eat." There was no sense in refusing. Every time I visited I brought something along and would buy a supply of soft drinks. The family, like countless others in the *cité*, had a modest beverage dépôt, where they sold drinks from the Bralima brewery by the bottle. But they had no money to buy the cola and orangeade themselves. One time I watched as Nkasi, sitting in his car seat, poured a bit of Coca-Cola into a plastic mug. Bloodcurdlingly slow, he held the cup out to Keitsha. It was a poignant scene: the man who had apparently been born before the Berlin Conference (and before the invention of Coca-Cola) was handing a drink to his granddaughter, born after the Congolese general elections of 2006.

The first time I met Nkasi was on November 6, 2008. The day before had been an auspicious one in world history. At a certain point, Nkasi reversed our roles. Would I mind if he asked me a question? There were more things to talk about than just the past. He had heard a rumor and could hardly believe it. "Is it true that a black man has been elected president of the United States?"

NKASI'S LIFE RAN PARALLEL to the history of Congo. In 1885 the region fell into the hands of King Leopold II of Belgium. Leopold named it the État Indépendant du Congo (Independent State of Congo), commonly referred to in the Dutch language as Congo-Vrijstaat. In 1908, in the face of virulent criticism at home and abroad, he transferred his holdings to the Belgian state. It would continue to be called the Belgian Congo until 1960, when it became an independent country, the Republic of Congo. In 1965 Joseph-Désiré Mobutu carried out a coup that kept him in power for thirty-two years. During that period the country received a new name, Zaïre. In 1997, when Mobutu was dethroned by Laurent-Désiré Kabila, it was renamed the Democratic Republic of Congo. The "democratic" part required some patience, however, for it was only in 2006 that the first free elections were held in more than forty years. Joseph Kabila, son of Laurent-Désiré, was elected president. Without having moved about much himself, Nkasi had lived in five different countries, or at least in a country with five different names.

Although the country as conceived by Leopold II in no way corresponded with any existing political reality, it did exhibit a striking geographical cohesion: it coincided to a great degree with the drainage basin of the Congo River. Each stream, each watercourse empties at some point into that single, powerful river and theoretically contributes to that brown spot in the ocean. That fact is a purely cartographic one: in actual practice, that hydrographic system was not seen as a unit. But ever since then, Congo—a country of 2.3 million square kilometers (about 900,000 square miles), the size of Western

Europe, two-thirds the size of the Indian subcontinent and the only country in Africa covering two time zones—has been the country of that one river. Despite the many name changes, it has always borne the name of the mother of all currents (the Congo, the Zaïre). Today's inhabitants speak of it in French as *le fleuve*, the stream, just as the inhabitants of the Low Countries speak of "the sea" when they mean the North Sea.

The Congo is no straightforward river; its course describes three-quarters of a circle and runs counterclockwise, as though one were turning back the hands of an analogue watch forty-five minutes. That big curve has to do with the even and relatively flat topography of the Central African interior. The Congo, in fact, makes one huge meander through an area of gently rolling hills that is mostly only several hundred meters above sea level. During its thousand-kilometer-long journey the river descends less than fifteen hundred meters (about 4,900 feet). Areas above two thousand meters (6,500 feet) are found only in the farthest eastern part of the country; the country's highest point lies directly on the border with Uganda: Mount Stanley, 5,109 meters (16,604 feet), the second highest peak in Africa, with a permanent layer of snow and a (dwindling) glacier. The eastern mountains, along with a chain of elongated lakes (the four so-called Great Lakes, of which Lake Tanganyika is the largest), are the result of major tectonic activity, as witnessed by the area's still-active volcanoes. This serrated eastern edge of Congo is a part of the Rift, the great fault line cleaving Africa from north to south. Climatologically, this mountainous area can be relatively chilly: a city like Butembo, for example, close to the Ugandan border, has an average annual temperature of only seventeen degrees Celsius (about sixty-three degrees Fahrenheit), while Matadi, not far from the Atlantic Ocean, has an average of twenty-seven (about eighty-one degrees Fahrenheit). Elsewhere, the equatorial setting produces a tropical climate with high temperatures and great humidity, although regional differences are considerable. In the equatorial forest, afternoon temperatures vary from thirty to thirty-five degrees (about eighty-six to ninety-five degrees Fahrenheit), while to the extreme south there may be frost on the ground during the dry season. The duration of the dry season and the time it commences also vary.

Two-thirds of the country is covered by dense equatorial forest, with its 1.45 million square kilometers (565,500 square miles) the largest tropical rain forest outside the Amazon Basin. From the air it resembles one huge and endless head of broccoli, occupying an area three times the size of Spain. To the north and south, the woods (*la forêt*, as the Congolese call it) gradually changes to savanna. Not an endless, National Geographic sea of yellow waving grass but a woodland savanna that gradually fades into brush savanna as one travels away from the equator. The country's biodiversity is spectacular, but increasingly threatened. Three of the most important zoological discoveries of the twentieth century were made in Congo: the Congo peacock, the okapi, and the bonobo. The discovery of a new primate in the twentieth century was something of a miracle in itself. Congo is the only country in the world where three of the four great apes are to be found (only the orangutan is absent): but the chimpanzee and particularly the mountain gorilla are highly endangered species as well.

Twentieth-century ethnographers distinguished some four hundred ethnic groups in the interior, each of them a society with its own customs, social structure, artistic traditions, and often its own language or dialect. These groups are usually indicated in the plural form, to be recognized by the prefix *ba-* or *wa-*. The Bakongo (sometimes rendered as baKongo) belong to the Congo people, the Baluba (or baLuba) to the Luba people, the Watutsi (or waTutsi, sometimes even waTuzi) to the Tutsis. In the chapters that follow I will use the standard expression as it is found in English. I will therefore speak of both the Bakongo and the Tutsis which, although not very consistent, is all the more convenient. The singular form (Mukongo or muKongo) is one I have avoided as much as possible. *Kongo* with a *K* refers to the ethnic group living close to the mouth of the Congo river; *Congo* with a *C* to the country and the river. The languages of these groups usually start with the prefix *ki-* or *tschi*: Kikongo, Tshiluba, Kiswahili, Kinyarwanda. Here too I have given precedence to common usage. Therefore: *Swahili* rather than *Kiswahili*, *Kinyarwanda* rather than *Rwandan*. Lingala is the exception to the rule, although languages in Lingala also start with *ki*. Once I even heard someone speak of "*kiChinois.*" And *Kiflama* is the language of the Baflama, the Flemish (derived from *les flamands*): Dutch, in other words.

The massive anthropological wealth of Congo must not blind one to the country's great linguistic and cultural homogeneity. Almost all the languages are Bantu languages and exhibit structural similarities. (*Bantu* is the plural form of *munt'u,* meaning "the people.") This is not to say that Nkasi will automatically understand someone from the other side of the country, only that his language will resemble that other one just as the Indo-European languages resemble one author. Only in the extreme north of Congo are languages spoken that are fundamentally different and belong to the Sudanese linguistic group. Everywhere else, Bantu languages became common with the spread of agriculture from the northwest. Even the Pygmies, the original hunter-gatherers of the jungle, made the switch to Bantu languages.

In Congo, ethnic awareness is a relative concept. Almost all Congolese can tell you with a certain precision to which ethnic group they and their parents belong, but the extent to which they identify with that group varies widely in accordance with age, place of residence, education, and, more crucial than all the rest, living conditions. Groups become more tightly knit in proportion to the extent to which they are threatened. At various moments in one's life one may attach greater or less importance to ethnic background. If the turbulent history of Congo makes anything clear, it is the elasticity of what was once referred to as "tribal awareness." It is a fluid category, and one I shall refer to more often.

Although the names of the provinces and their number have changed often, still there are several regional designations used invariably by the inhabitants to delineate the parts of this enormous country. Bas-Congo, as mentioned, is the neck of the balloon. Matadi, its administrative center, is a seaport some sixty miles inland where container vessels, tacking against the Congo's powerful current, can load and unload. Farther upstream, rapids render the river unnavigable. Kinshasa, a city of an estimated eight million inhabitants, who call themselves Kinois, is located precisely at the spot where the balloon widens. From this point the river once again becomes navigable, until deep into the interior. To the east of Kinshasa one finds Badundu, an area between forest and savanna that includes Kikwit and the historically important region of Kwilu. Beside it, in the country's heart, lies Kasai, the diamond country. Its principal city is Mbuji-Mayi, which has grown in recent years to become the country's third, perhaps even second largest city, due to the rush for diamonds. Farther east one arrives in the area once known as the Kivu, but which has now been divided into three provinces: North Kivu, South Kivu, and Maniema. Both Kivus form the vulnerable top of the balloon, with Goma and Bukavu as major centers directly on the border with Rwanda. This is a heavily populated farming area. Due to its altitude, sleeping sickness does not occur here and it is possible to raise cattle; the soil and climate, furthermore, are well suited to high-grade agriculture (coffee, tea, quinine).

To the north of the line Bandundu-Kasai-Kivu stretches the largest part of the rain forest, administratively categorized under two megaprovinces, Équateur and Orientale province, both long earmarked for subdivision. Mbandaka and Kisangani respectively are their capitals. Kisangani in particular has played a key role throughout the history of Congo. To the south of this central east-west axis lies another megaprovince, Katanga, with Lubumbashi as capital. This mining region forms the economic heart of the country. Katanga has a territorial peninsula sticking out to the southeast, as though a clown had given a twist to the balloon that is Congo: it is, in fact, the product of a late-nineteenth-century border dispute with England. While Katanga enjoys a great wealth of copper and cobalt, and Kasai relies on its diamonds, the earth of Kivu contains tin and coltan, and that of Orientale province, gold.

The country's four major cities, therefore, are Kinshasa, Lubumbashi, Kisangani, and, more recently, Mbuji-Mayi. At the time of this writing they are connected by neither rail nor paved road. At the start of this third millennium, Congo has less than one thousand kilometers of asphalt roads (and most of those run to the outside world: from Kinshasa to the port at Matadi and from Lubumbashi to the Zambian border, to facilitate the import of goods and the export of mineral ores). Rail service these days is almost nonexistent. The boats from Kinshasa to Kisangani take weeks to reach their destination. Anyone hoping to travel from one city to the other takes a plane. Or takes a great deal of time. One rule of thumb says that a journey that took one hour during the colonial period now corresponds to a full day's travel.

Kinshasa is and always has been the country's navel, the knot in the balloon. More than 13 percent of the country's sixty-nine million inhabitants lives in one of the capital city's twenty-four districts, but the bulk of the Congolese population still lives in the countryside. Bas-Congo, Kasai, and the area around the Great Lakes are particularly densely populated. French is the language of the government and higher education, but Lingala is the language of the army and the ubiquitous pop music. Four native languages are officially recognized as official languages: Kikongo, Tshiluba, Lingala, and Swahili. While the first two truly constitute ethnic languages (Kikongo is spoken by the Bakongo in Bas-Congo and Bandundu, Tshiluba by the Baluba in Kasai), the latter two are trade languages of a far greater range. Swahili arose on the African east coast and is spoken not only in all of the eastern Congo, but also in Tanzania and Kenya. Lingala arose in the province of

Équateur and made its way down the Congo River to Kinshasa. Today it is the country's fastest-growing language, and is also spoken in neighboring Congo-Brazzaville.

And, speaking of neighboring countries: Congo has no less than nine of them. Starting at the Atlantic and proceeding clockwise, they are: the Angolan enclave of Cabinda, Congo-Brazzaville, the Central African Republic, South Sudan, Uganda, Rwanda, Burundi, Tanzania, Zambia, and Angola proper. Only Brazil, Russia, and China have more neighbors—ranging from ten to fourteen. This results in diplomatic complexities, and Congo was and is no different, both in colonial times and in the present. For the last century and a half, border disputes and territorial conflicts have been a constant, just as some stretches between Russia and China have long been a bone of contention.

WHERE DOES THE HISTORY BEGIN? A long way out to sea, a long way from the coast, even a long time before Nkasi was born. A tendentious urge exists to let the history of Congo begin with the arrival of Stanley in 1870, as though the inhabitants of Central Africa wandered sadly before that time through an eternal, immutable present and had to wait for a white man to come through and free them from the wolf trap of prehistoric listlessness. Central Africa, it is true, did gain major historical momentum between 1870 and 1885, but that absolutely does not mean that the inhabitants before that were suspended in a solidified state of nature. They were not living fossils.

Central Africa was a region without writing, but not without a history. Hundreds, yea thousands of years of human history preceded the arrival of the Europeans. If a heart of darkness existed back then, it was sooner to be found in the ignorance with which white explorers viewed the area than in the area itself. Darkness, too, is in the eye of the beholder.

I would like to illustrate that far-distant history with the use of five virtual slides, five snapshots. And I would like to wonder aloud what the life of, say, a twelve-year-old boy looked like at each of those five moments. The first snapshot was made about ninety thousand years ago. The date is taken fairly at random, but it happens to be the only reliable date we have for the oldest archaeological remains in Congo.

How bizarre to place Congo's history in the hands of a European. How Eurocentric can one be? It was in Africa, between five and seven million years ago, that the line of humans split off from that of the primates. It was in Africa, four million years ago, that humans began walking upright. It was in Africa, almost two million years ago, that the first purpose-driven stone tools were made. And it was in Africa one hundred thousand years ago that the complex, prehistoric behavior of our species arose, a behavior characterized by long-distance trading networks, advanced tools of stone and bone, the use of ochre as a dye, early systems for counting, and other forms of symbolism. Congo lay a little too far to the west to join in that evolution from the very start, but extremely primitive and undoubtedly extremely ancient tools have been found at many locations, most of them, unfortunately, poorly dated. The area also produced a number of the world's most impressive prehistoric celts, skillfully crafted stone axes of up to forty centimeters long.

Ninety thousand years ago, let us say. Let's imagine the shore of one of the four Great Lakes to the east, the one known today as Lake Edward. Our twelve-year-old boy could have been sitting there, at the spot where the Semliki River flows out of the lake. Perhaps he was part of that little group of prehistoric people whose remains were meticulously exhumed in the 1990s. Once a year, a group of hunter-gatherers assembled at this spot, always when the catfish were spawning. This tasty, slow-moving fish with its grim feelers can easily grow to over seventy centimeters in length and weigh more than ten kilos. Normally, it lives at the bottom of the lake, beyond the reach of people. But at the start of the rainy season it moves to shallow water to spawn. To this end, the fish is even equipped with a special respiratory organ. Useful, but risky: as early as ninety thousand years ago, people along that lake were already carving harpoons from bone, the oldest known such artifacts—in other parts of the world, that took place only twenty thousand years ago. Using a rib or a femur, a spearhead was fashioned with deadly notches and barbs. A twelve-year-old boy like ours may very well have learned to impale a fish like that, or perhaps one of the many smaller species. Very possible indeed that he also dug up lungfish, eel-like animals that nestle into a shallow recess at the start of the dry season and remain there for the eight months of summer. Paleontological research has shown that the area was much drier then. Elephants lived there, along with zebras and warthogs, animals typical of an open landscape. But the proximity of water meant there were also hippopotami, crocodiles, marsh antelope, and otters. The wind blew over the water, the bushes rustled, a fish writhing in pain beat its tail wildly and impotently against the wet rocks. And leaning over it on his harpoon, a boy's voice might have been heard: excited, grim, and exultant. A snapshot, no more than that.

A second slide: it is twenty-five hundred years before the start of our calendar. Our twelve-year-old boy was then a Pygmy in the dense rain forest. Agriculture did not yet exist, not by a long shot, but he would certainly have tasted the nuts of the wild oil palm. The remains of such early inhabitants have been found beneath rocky overhangs in the Ituri forest. Scattered amid a few crude stone implements there were the pits of prehistoric palm nuts. Did these forest dwellers live there? Or did they come there only sporadically? No one knows. The tools, in any case, were made from quartz and river stones that had been gathered locally. The twelve-year-old boy may have belonged to a small, extremely mobile group of hunter-gatherers, who must have possessed a very thorough knowledge of their everyday environment. They hunted apes, antelopes, and porcupines, they picked nuts and fruit, dug for tubers, and knew which plants were medicinal and which were mind expanding.

This too, however, was not a closed world. Even then there was contact with the outside world. Flint and obsidian were traded over great distances, sometimes up to three hundred kilometers (185 miles) away. Perhaps our boy was the first Congolese of whom we have a written record. Perhaps he was captured as a slave and abducted from the jungle, taken across the savanna and the desert, a journey that took months, to a river and a seemingly endless journey downstream: the Nile. His escort was delighted with his catch: a Pygmy, the rarest and most costly thing imaginable. His divine master in the north would send him a special summons: a letter he would later have carved in stone: "Come and bring me the dwarf, the dwarf you have brought me from the land of the spirits, alive, safe and well, to dance the holy dances to the entertainment and felicity of Pharaoh Neferkare. Be careful he does not fall in the water."[3] The hieroglyphs were chiseled into the rock at Aswan, at the grave of the expedition's leader, twenty-five hundred years before our era began. "The land of the spirits": it was the first time Congo was mentioned in writing.

Next slide, the third. We find ourselves around AD 500. In Europe, the Western Roman Empire has just collapsed. A twelve-year-old Congolese boy in those days led a completely different life from his predecessor. The wandering was over, from now on he would be more or less sedentary: he no longer pulled up stakes several times a year, but only a few times in his life.

Approximately two thousand years before our Christian era began, agriculture was practiced for the first time in what is now Cameroon. This new source of nutrition resulted in a growing population. And because it was extensive agriculture, new fields had to be cultivated each year. Slowly but surely, the farming lifestyle spread across Africa. It was the start of the Bantu migration. That migration should not be seen as a great trek on the part of farmers who one day packed their bags and arrived a thousand kilometers further along with the statement "This is it!" It was a slow but steady shift southward (to the north, after all, lay the Sahara). In the course of three millennia, agriculture took over all of central and southern Africa. The hundreds of languages spoken in that huge area are, as noted, related even today. The forest in Congo formed no real obstacle for the Bantu-speaking farmers. Along the rivers and elephant trails, they pushed their way into the area. There they came in contact with the local forest dwellers, the Pygmies. By the year 1000, the entire region had been settled.

The great innovation around AD 500 was the cultivation of plantain, the cooking banana, a crop with an origin as unclear as its flavor is delicious. Our twelve-year-old boy was in luck: during the centuries before, the principle staple had been the yam, a nutritious tuber rich in starch, but rather bland to the taste. For his mother, who tended the fields, the plantain had major advantages: unlike the yam, it did not draw the malaria mosquito. The yield was ten times that of the tuber, the plantain required less care and was much less taxing for the soil. At that time, his father must also have climbed palm trees to harvest palm oil. Perhaps they had a few chickens and goats, maybe even a dog. In addition, they still did a great deal of foraging, fishing, and hunting. The son must have gathered termites, caterpillars, grubs, slugs, mushrooms and wild honey. With his father and other men from the village he hunted antelope and bush pigs. To catch fish he set out weirs or dammed streams. He had, in other words, an extremely varied diet. Agriculture still accounted for only 40 percent of his intake.

In the year 500, our boy's father probably had several iron tools. That was another innovation of the day: the earliest known metallurgy in the area arose during the first centuries of our era. Before that, people had used exclusively stone tools. His mother almost certainly owned pots of baked clay. Ceramics had come along centuries earlier. Pots and bowls and metal were luxury goods that his parents obtained by barter and exchange, just like costly animal skins and rare pigments.

The family lived in a smallish village with a few other families, but between villages new forms of cooperation were arising. Agriculture's ongoing march resulted in family ties across a larger area. Perhaps even then each village would have had its own gong or slit drum, a hollow log on which one could produce two tones, one high and one low; this allowed them to send messages across a great distance. Not vague alarm signals, but extremely precise messages, entire sentences, bits of news and stories. When someone died, his name, nickname, and calls of condolence were ruffled around. When a hut burned down, when a prey was killed or a family member came to visit, the villagers drummed the news from place to place. Early in the morning or late in the evening, when the air was cold, you could hear the beats for miles around. Distant villages passed the information along to even more distant villages. The peoples of Central Africa never developed a system of writing, but their *langage tambouriné* (drummed language) was ingenious. Information was not stored for the future, but broadcast immediately across field and forest and shared with the community. Nineteenth-century explorers were amazed to find that the villages where they came ashore had been expecting them for a long time. When they learned that a drummed message could travel up to six hundred kilometers (370 miles) in a single day, they spoke laughingly of the *télégraphe de brousse* (bush telegraph). They didn't know that this form of communication easily preceded Morse's invention by a millennium and a half.

Next slide, more than one thousand years later. Let's say: 1560. Italy is caught up in the Renaissance. Jan Breughel the Elder is painting his masterpieces. The first tulip appears in Holland. What did the life of a twelve-year-old Congolese look like? If he was born in the forest, he now almost certainly lived in a larger village, a village with dozens of houses and hundreds of inhabitants, ruled over by a chieftain. The chief's power was based on his name, fame, honor, wealth, and charisma. Only he was allowed to don the skin and teeth of the leopard. He had to rule like a father, never placing his own interests before those of the community. A number of such villages together formed a kind of circle that helped prevent conflicts over farmland and ward off intruders.

If our young fellow was born on the savanna, he would have noticed that this system had developed a step further there. A number of circles together formed a district, in some cases even a kingdom. The first real states, like those of the Kongo, the Lunda, the Luba, and the Kuba, appeared on the savanna south of the equatorial forest from the fourteenth century on. The expansion was made possible by agricultural surpluses. Some of the states were the size of Ireland. They were feudal, hierarchical societies. At the top stood a king, a village chieftain to the nth degree, the father of his people, protector and benefactor of his subjects. He cared for the community, consulted the elders, and settled disagreements. The result of this political construct is not hard to imagine: a great deal indeed depended on the personality of the king. One could be lucky or unlucky. When power becomes so personalized, history becomes manic-depressive. This certainly applied to the kingdoms of the savanna. Periods of prosperity alternated rapidly with periods of decline. The changing of the throne led almost invariably to civil war.

If our imaginary boy grew up along the lower reaches of the Congo, he was a subject of the Kongo Empire, the most famous of all these feudal principalities. Its capital was called Mbanza-Kongo, today a place in Angola, just south of Matadi. In 1482 the coastal inhabitants of that empire had seen something extremely remarkable: huge huts looming up out of the sea, huts with flapping cloths. When those sailing ships anchored off the coast, the people along the shore saw that there were white people in them. These had to be ancestors who lived at the bottom of the sea, a kind of water spirit. The whites wore clothes, lots more clothes than they did, which seemed to be made from the skins of strange sea creatures. All highly peculiar. The inexhaustible quantities of cloth the strangers had with them made the people think they probably spent most of their time weaving, there below the ocean.[4]

But these were Portuguese sailors who, in addition to linen, also came bearing the consecrated wafer. The king of the Bakongo, Nzinga Kuwu, allowed them to leave four missionaries behind in his empire and sent four dignitaries with them in exchange. When the latter returned a few years later with weird and wonderful stories about that distant Portugal, the king burned with the desire to learn the Europeans' secrets and, in 1491, let them christen him Don João. Several years later though, disappointed, he returned to his polygamy and divinations. His son, Prince Nzinga Mvemba, however, was to become a deeply devout Christian and to rule over the Kongo Empire for four decades (1506–43), under his Christian name of Afonso I. It was a period of great prosperity and consolidation, during which the king's power was founded on trade with the Portuguese. When those Portuguese asked for slaves, he had raids carried out in neighboring districts. It was an ancient practice—slavery was an indigenous phenomenon, anyone with power also had people—but his cooperation created so much goodwill with the Portuguese that Afonso was allowed to send one of his sons to Europe to attend seminary. In Lisbon the son in question, eleven-year-old Henrique, learned Portuguese and Latin and then moved to Rome, where he was enthroned as bishop—the first black Catholic bishop in history—before returning home. But Henrique was of weak constitution and died a few years later.

The Christianizing of Congo was therefore undertaken by Portuguese Jesuits and later by Italian Capuchins as well. These activities in no way resembled the missionizing of the nineteenth century; here the church made its appeal expressly to the upper reaches of society. The church stood for power and affluence, and that appealed to no little extent to the top of the Kongo Empire. The wealthy had themselves baptized and assumed noble Portuguese titles. Some of them even learned to read and write, although a sheet of paper at that time cost as much as a chicken, and a missal cost as much as a slave.[5] Yet churches were built and cult objects (*fétiches* in French) burned. Where sorcery was found, Christianity was obliged to triumph. A cathedral arose in the capital, Mbanza-Kongo, and governors in the provinces had churches built as well. The population at large viewed the new religion with interest. While the Christian priests hoped to bring them the true faith, the people saw them as their best protection against sorcery. Many had themselves baptized, not because they had abandoned witchcraft, but precisely because they believed in it so fervently! The crucifix became highly popular as the most powerful of all cult objects to ward off evil spirits.

In 1560, after Afonso's death, the Kongo Empire went through a deep crisis. Chances are that our twelve-year-old boy wore around his neck a crucifix, a rosary, or a medallion, perhaps an amulet his mother had made. Christianity did not oust an older belief, but fused with it. Years later, in 1704, when the cathedral at Mbanza-Kongo had already fallen to ruin, a local black mystic would live amid the ruins and claim that Christ and the Madonna were members of the Kongo tribe.[6] When missionaries traversed the lower reaches of the Congo in the mid-nineteenth century, they still met with people with names like Ndodioko (from Don Diogo), Ndoluvualu (from Don Alvaro) and Ndonzwau (from Don João). They also saw rituals being performed before crucifixes three centuries old, but now decked out with shells and stones and roundly claimed by all to be indigenous.

Around 1560, in addition to an amulet, our boy also adopted different eating patterns. The Atlantic trade brought new crops to his district.[7] From the moment the Portuguese established their colony on the coast close to Luanda, the change came quickly. In much the same way that the potato reached ascendancy in Europe, corn and manioc quickly conquered all of Central Africa. Corn grew from Peru to Mexico, manioc came from Brazil. In 1560 our boy of twelve would have primarily eaten porridge made from sorghum, a native grain. From 1580 on, however, he began eating corn and manioc. Sorghum could be harvested only once a year, corn twice and manioc the whole year through. While corn did well on the drier savanna, manioc flourished in the more humid forest. It was more nutritious and easier to cultivate than plantain or yams. The tubers rarely rotted. All one had to do was clear a new plot each year; it was during this period that slash-and-burn agriculture originated.[8] If he was lucky, the boy's bowl also featured sweet potatoes, peanuts, and beans—regular ingredients even today in the Congolese kitchen. Within a few decades the diet of Central Africa had been radically transformed, thanks to globalization on the part of the Portuguese.

Congo, in other words, did not have to wait for Stanley in order to enter the flow of history. The area was not untouched, and time there had not come to a standstill. From 1500 it took part in international trade. And although most of the forest's inhabitants would not have known it, each day they ate plants that came from another part of the world.

Fifth slide. Final snapshot: we have arrived in the year 1780. If our boy was born then, there is a sizeable chance that he became merchandise for the European slave drivers and ended up on the sugar plantations of Brazil, the Caribbean, or in the south of what would later be the United States. The Atlantic slave trade lasted roughly from 1500 to 1850. The entire west coast of Africa was involved in it, but the area around the mouth of the Congo most intensively of all. From a strip of coastline some four hundred kilometers (250 miles) long, an estimated four million people were put on transport, equaling almost a third of the entire Atlantic slave trade. No less than one in every four slaves on the cotton and tobacco plantations of the American South came from equatorial Africa.[9] The Portuguese, the British, the French, and the Dutch were the major traders, but that does not mean that they themselves penetrated far into the African interior.

Beginning in 1780 greater demand for slaves in the United States resulted in a major upscaling of the trade. From 1700 onward, between four and six thousand slaves were shipped annually from the Loango Coast north of the Congo; by 1780 that number had risen to fifteen thousand annually.[10] This increase was felt far into the equatorial forest. If our boy was abducted during a raid, or sold by his parents in times of famine, he would have ended up with one of the important traders along the river. He would have been forced to sit in an enormous dugout, perhaps twenty meters in length, which could carry between forty and seventy passengers. He may have been chained. In addition to dozens of slaves, the canoe would also have carried ivory, the rain forest's other luxury good. A Pygmy who had killed an elephant would not, after all, have gone himself to the coast to sell the tusks to an Englishman or a Dutchman. Trade went by way of a middleman. In the opposite direction as well: a keg of gunpowder could easily take five years to make its way from the Atlantic coast to a village in the interior.[11]

And then the journey began, downstream. For months the captives floated down the broad, brown river through the jungle, until they arrived at the section that was no longer navigable. There arose the huge and supremely important market of Kinshasa. People gathered there from all over. One heard the bleating of goats, dried fish hung on racks, manioc loaves were piled beside textiles from Europe. You could even buy salt there! The air was filled with shouts, prayers, laughter, and argument. There was as yet no city, but the activity was in full swing. Here the trader from the interior would sell his slaves and ivory to a caravan leader, who would take his goods overland to the coast, three hundred kilometers (185 miles) farther. Only there would our twelve-year-old boy see a white man, for the first time in his life. He would be haggled over for days.

We do not know how his crossing to the New World went. But a rare eyewitness account by a West African slave who was shipped to Brazil in 1840 provides a bit of a picture:

> We were thrown naked into the ship's hold, the men close together on one side, women on the other; the ceiling of the hold was so low that we could not stand up straight, but were forced to squat or sit on the floor; day and night were the same to us, the close quarters made it impossible to sleep, and we grew desperate with suffering and fatigue. . . . The only food we were given during the journey was grain that had been soaked and boiled. . . . We suffered greatly from a lack of water. Our rations were one half liter a day, no more than

that; and a great many slaves died during the crossing. . . . If one of them became defiant, his flesh was cut with a knife and pepper and vinegar were rubbed into the wound.[12]

The international slave trade had an enormous impact on Central Africa. Regions were torn apart, lives destroyed, horizons shifted. But it also brought with it an extremely intensive network of regional commerce along the river. If you had to go down the Congo River anyway with a shipment of slaves and tusks, you might as well fill your dugout with less luxurious goods to sell along the way. And so fish, manioc, cane sugar, palm oil, palm wine, sugarcane wine, beer, tobacco, raffia, baskets, ceramics, and iron were taken along as well. Each day, some forty metric tons (forty-four U.S. tons) of manioc were transported along the Congo, over distances of no more than 250 kilometers (155 miles).[13] Usually this was in the form of manioc loaves, *chikwangue*: boiled manioc gruel cleverly packaged in banana leaves. A hefty meal in itself, leaden on the stomach, but not perishable and easy to transport.

The importance of this regional trade should not be underestimated. In a world of fishermen, farmers, and hunters, a new professional category arose: that of merchants. People who had traditionally lived by tossing their nets discovered that a greater catch could be obtained by plying the river. Fishermen became merchants, and fishing villages marketplaces. Trading had always been carried out on a modest scale, but now commerce became a trade in itself. Many were none the worse for it. Some came to possess dugouts, wives, slaves, and muskets, and therefore power. Anyone possessing gunpowder had influence. And so the traditional authority of the tribal chieftains was shaken to the foundations. Centuries-old social forms were eroded. Anarchy reared its head. Political ties based on village and family were being elbowed out by new economic alliances between traders. Even the once so powerful Kongo Empire became dissolute.[14] A gigantic political vacuum arose. International trade was flourishing, but it resulted in total chaos far into the African interior.

Ninety thousand years of human history, ninety thousand years of society . . . such vitality! No timeless state of nature occupied by noble savages or bloodthirsty barbarians. It was what it was: history, movement, attempts to contain the misery, attempts that sometimes brought new misery, for the dream and the shadow are the closest of friends. There had never been anything like standing still; the major changes followed each other with ever-increasing momentum. As history moved faster, the horizon expanded. Hunter-gatherers had lived in groups of perhaps fifty individuals, but the earliest farmers already had communities of five hundred. When those societies expanded to become organized states, the individual was absorbed into contexts of thousands or even tens of thousands of people. At its zenith, the Kongo Empire had as many as five hundred thousand subjects. But the slave trade annihilated those broader ties. And in the rain forest, far from the river, people still lived in small, closed societies. Even in 1870.

IN MARCH 2010, as I was putting the finishing touches to this manuscript, I booked a flight to Kinshasa. I wanted to visit Nkasi again, this time accompanied by a cameraman. I resolved to take him a nice silk shirt, for poverty cannot be combated with powdered milk alone. Regularly, during the long months of work on this book, I had called his nephew to ask how Nkasi was getting along. "Il se porte toujours bien!" (He's still doing fine) was always the cheerful announcement from the other end. Less than one week before my deadline, five days before my departure, I called again. That was when I heard that he had just died. His family had left Kinshasa with the body, to bury him at Ntimansi, the village in Bas-Congo where he had been born an eternity ago.

I looked out the window. Brussels was going through the final days of a winter that knew no respite. And as I stood there like that, I could not help thinking about the bananas he had slid over to me during our first meeting. "Take it, eat." Such a warm gesture, in a country that makes the news so much more often for its corruption than for its generosity.

And I had to think about that afternoon in December 2008. After a long talk Nkasi had needed a rest, and I entered into conversation with Marcel, one of his great-nephews. We were sitting in the courtyard. Long lines of wash had been hung out to dry and a few women were sorting dried beans. Marcel was wearing a baseball cap with the visor turned to the back and was leaning back comfortably in a plastic garden chair. He started talking about his life. Although he had been good at school, he had now been relegated to the *marché ambulant* (walking marketplace). He was one of the thousands and thousands of young people who spent all day crossing the city with a few articles to sell—a pair of trousers, two baskets, four belts, a map. Sometimes he would sell only two baskets a day, a turnover of less than four dollars. Marcel sighed. "All I want is for my three children to be able to go to school," he said. "I liked school so much myself, especially literature." And to prove that, in a deep voice he began reciting "Le soufflé des ancêtres," the long poem by the Senegalese poet Birago Diop. He knew great chunks of it by heart.

> Listen more often
> To things rather than people
> You can hear the voice of the fire,
> Hear too the voice of the water.
> Listen to the bush
> Sobbing in the wind:
> It is the breath of the dead.
>
> Those who died never went away:
> They are in the shadow that lights up
> And also in the shadow that folds in upon itself.
> The dead are not beneath the ground:
> They are in the leaves that rustle,
> They are in the wood that groans,
> They are in the water that rushes,
> They are in the water at rest,
> They are with the people, they are in the hut.
> The dead are not dead.[15]

Winter on the rooftops of Brussels. The news I just received. His voice that I can still hear. "Take it, eat."

MAP 3: CENTRAL AFRICA IN THE MID-NINETEENTH CENTURY

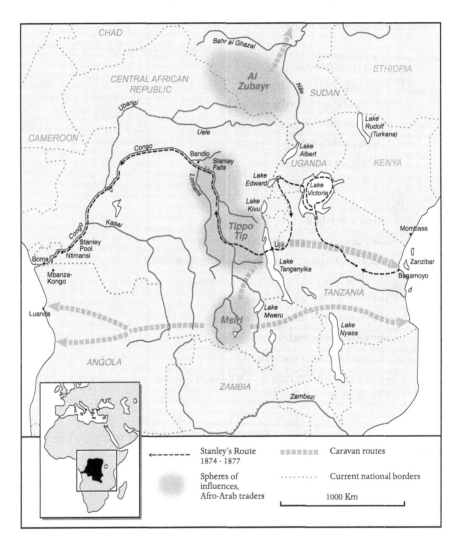

← - - - - Stanley's Route 1874 - 1877	▨▨▨▨▨▨▨ Caravan routes
Spheres of influences, Afro-Arab traders	· · · · · · · Current national borders
	├────────┤ 1000 Km

CHAPTER 1

NEW SPIRITS

Central Africa Draws the Attention of East and West

1870–1885

N O ONE KNOWS EXACTLY WHEN DISASI MAKULO WAS BORN. But then neither did he. "I was born in the days when the white man had still not arrived in our area," he told his children many years later. "We didn't know then that there were people in the world with skin of a different color."[1] It must have been around 1870–72. He died in 1941. Not long before, he had dictated his life's story to one of his sons. It would appear in print only in the 1980s; twice in fact, in Kinshasa and again in Kisangani, but Zaïre, as Congo was called in those days, was as good as bankrupt. The publications were sober, with limited print runs and distribution. And that is unfortunate, because the life story of Disasi Makulo is above all a fantastic adventure. To understand the last quarter of the nineteenth century in Central Africa there is no better guide than Makulo.

Where Disasi was born, however, he knew very well: in the village of Bandio. He was the son of Asalo and Boheheli, a Turumbu tribesman. Bandio lay in the district of Basoko, now Orientale province. The heart of the equatorial forest, in other words. Aboard the boat from Kinshasa to Kisangani, a few weeks' journey upstream along the Congo, one passes on the port side a few days before arrival the large village of Basoko. It is on the northern bank, at the confluence with the Aruwimi, one of the Congo's larger tributaries. Bandio is to the east of Basoko, a ways back from the river itself.

His parents were not fishing folk; they lived in the jungle. His mother raised manioc. With her hoe or digging tool she would chop at the earth to pry loose the thick tubers. She lined them up to dry in the sun and, a few days later, ground them to flour. His father worked with palm oil. Climbing high into the trees with his machete, he chopped off the bunches of greasy nuts. Then he would press them until the lovely juice ran out, a deep orange, a sort of liquid copper that has added to the region's wealth since time immemorial. That palm oil could be used to trade with the fishermen along the river. Commercial ties had existed for centuries between the riverine inhabitants, who had fish in abundance, and the people of the forest with their surpluses of palm oil, manioc, or plantains. The result was a balanced diet: the protein-rich fish was taken to the rain forest, the starchy crops and vegetable oil were left on the banks.

Bandio was a relatively insular world. The radius of activity covered in a human lifetime was limited to a few dozen kilometers. People sometimes visited another village to attend a wedding or arrange an inheritance, but most of them left their region seldom or never. They died where they were born. When Disasi Makulo entered the world with a shriek, the villagers of Bandio knew nothing of the outside world. They knew nothing of the permanent presence of the Portuguese a thousand kilometers to the west, along the Atlantic, nor in fact of the existence of an ocean. The Portuguese colony of Angola had lost much of its splendor, as had Portugal itself, but—for Africans as well—Portuguese remained the major trading language along the coast south of the mouth of the Congo. Nor did Disasi's people know that, since the eighteenth century, the British had taken over the trade of the Portuguese along the Congo's lower reaches and embouchure. That the Dutch and the French had settled there as well: they could never have guessed that, for none of those Europeans ever made their way inland. They remained on the coast and the area immediately behind it, waiting till the caravans led by African traders reached them with their goods from the interior: ivory in particular, but also palm oil, peanuts, coffee, baobab bark, and pigments such as orchil and copal. Not to mention slaves. Although the trade in human beings had been abolished throughout the Western world in those years, it went on in secret for quite some time. The Westerners paid with precious cloth, bits of copper, gunpowder, muskets, and red or blue pearls or rare seashells. This latter commodity was no act of clever Western fraud. As with official coinage, those shells were piece goods of great value that could be transported easily and were impossible to counterfeit. But Bandio was too far away to see much of that. If such a white, gleaming shell or bead necklace actually happened to make it to their area, no one knew exactly where it came from.

Newborn Disasi's fellow villagers may have known nothing about the Europeans on the west coast, but they were even less informed about the great upheavals taking place more than a thousand kilometers to the east and north. Beginning in 1850, the Central African rainforest had also attracted the attention of merchants from the island of Zanzibar, as well as from the African east coast (present-day Tanzania) and even from two thousand kilometers away in Egypt. Their interest was prompted by a natural raw material that had been valued around the world for centuries as a luxury good for the manufacture of princely Chinese tablets, Indian figurines, and medieval reliquaries. That material was ivory. High-grade ivory was found in huge quantities in the African interior. The tusks of the African elephant comprised the largest and purest pieces of ivory in the world, weighing up to seventy kilos and more. Unlike the Asian elephant, already rare by that time, the female of the African species bore tusks as well. In the mid-nineteenth century this seemingly inexhaustible treasure trove was the subject of increasingly close perusal.

In the northeast of what would later become Congo, where the rain forest meets the savanna, traders from the Nile Valley were active: Sudanese, Nubians, and even Egyptian Copts. Their clientele lived as far away as Cairo. The traders traveled to the south by way of Darfur or Khartoum. Slaves and ivory were the major export products, razzias and hunting parties the principle form of acquisition. By 1856 the entire trade had gradually entered the hands of a single individual: al-Zubayr, a powerful trader whose empire in 1880 extended from Northern Congo to Darfur. Officially, his trading zone was a province of Egypt; in practice, it comprised an empire unto itself. The Arab influence spread all the way to southern Sudan.

But it was above all Zanzibar, an unsightly island in the Indian Ocean off the coast of present-day Tanzania, that played a crucial role. When the sultan of Oman settled there in 1832 to control the flow of trade in the Indian Ocean, the move had far-reaching consequences for all of eastern Africa. Zanzibar, itself rich only in coconuts and cloves, became the global transfer table for slaves. The island exported to the Arabian Peninsula, the Mideast, the Indian subcontinent, and China.

In 1870 the villagers of Bandio noticed none of this. But the Zanzibar traders possessed excellent firearms and so they themselves moved farther and farther into the interior, farther than the Europeans to the west ever had. Some of them were pure Arabs, others were of mixed African blood. Often they included African converts to Islam, whom we refer to as Afro-

Arab or Swahilo-Arab traders; in the nineteenth century, however, they were called *les arabisés*. Swahili, a Bantu language with many Arabic loan words, spread all over Eastern Africa. Starting at Zanzibar and the town of Bagamoyo on the coast, huge caravans began heading inland from 1850 on, until they reached the shores of Lake Tanganyika, eight hundred kilometers to the west. The settlement of Ujiji, where Stanley would "find" Livingstone in 1871, became a major trading post. From the lake's far shore the caravans moved even farther inland, into the area now known as Congo. As with the trading empire of al-Zubayr, one saw spheres of commercial influence solidify into political entities. In southeastern Katanga, Msiri, a trader from the African east coast, took over an existing realm: the ancient, but by-then mordant Lunda Empire. From 1856 to 1891 he was lord and master over this region rich in copper and controlled all trade routes to the east. His interests, at first purely commercial, in this way took on political form.

A bit farther to the north, the notorious ivory and slave trader Tippo Tip reigned supreme. As son of an Afro-Arab family from Zanzibar he answered directly to the Sultan, but soon he became the most powerful man in all of eastern Congo. His authority was felt in the area that stretched between the Great Lakes to the east and the headwaters of the Congo (also referred to there as Lualaba), three hundred kilometers (185 miles) to the west. Tippo Tip's power was founded not only on his exceptional business sense, but also on violence. At first he had acquired his luxury goods—slaves and ivory—in a friendly fashion: like other Zanzibaris, he established pacts with local leaders for the purposes of bartering. A number of those leaders became vassals of the Afro-Arab traders. Yet, from 1870 on, all that changed. As more and more tons of ivory began flowing eastward, traders like Tippo Tip grew in power and wealth. In the final account, the sacking and pillaging of entire villages proved more cost effective than bartering for a few tusks and adolescents. Why spend days chattering with the local village chieftain, refusing lukewarm palm wine that your religion forbade you to drink anyway, when you could just as easily torch his village? In addition to ivory, this new approach also produced additional slaves to carry that ivory. Raiding became more important than trading; firearms tipped the scales. The name Tippo Tip sent shivers down the spines of those inhabiting an area half the size of Europe. In fact, it wasn't even his real name (that was Hamed ben Mohammed al-Murjebi), but probably an onomatopoeiac form derived from the sound of his rifle.

At Disasi Makulo's home in Bandio, however, no one had ever heard of

Tippo Tip. The stage was still empty, the world still a verdant green. In the wings, to the left and right, foreign traders—European Christians and Afro-Arab Muslims—stood awaiting their cues, ready to push on into the heart of Central Africa. It was only because the region's power structures were already in a wretched state, due in part to the European slave trade carried out in the centuries before, that their offensive was even possible. Not much was left in those days of the once so-powerful native kingdoms, and social structures in the jungle had always been less complex than those on the savanna. The political vacuum in the interior, therefore, offered new economic opportunities for foreigners. That is putting it nicely. In reality, the period to come was one of administrative anarchy, rapaciousness, and violence. But not yet. Little Disasi still lay slumbering, tied to his mother's back, his cheek pressed to her shoulder blade. The wind rustled in the treetops. After a thunderstorm, the rain forest went on dripping for hours.

"ONE DAY, a few people from the riverside came to visit my parents." Thus begins Disasi Makulo's earliest recollection. He must have been five or six at the time. The strangers brought with them a very peculiar story. "They said they had seen something bizarre on the river, a spirit perhaps. 'We saw a huge, mysterious canoe,' they said, 'that rowed itself. In that canoe is a man, white from head to toe, like an albino, covered completely in garments, you could see only his head and his arms. He had a few black men with him.'"[2]

Besides fish and palm oil, the peoples of the river and the jungle also traded information. The river people, of course, had a tendency to come up with weird news anyway—you could hardly imagine the crazy things they heard from fishermen and traders further along!—but this report sounded particularly strange. What's more, it was no secondhand account. The clothed albino they had seen was no one less than Henry Morton Stanley. The little group of black men were his bearers and helpers from Zanzibar. That huge, mysterious canoe was the *Lady Alice*, his eight-meter-long steel boat. After he had found the presumably lost physician, missionary, and explorer David Livingstone on the shores of Lake Tanganyika in 1871, the *New York Herald* and the *Daily Telegraph* of London had commissioned Stanley to carry out, from 1874 to 1877, what would become the mother of all exploratory expeditions: the crossing of Central Africa from east to west, a staggering journey through festering swamps, hostile tribal territories, and murderous rapids.

It was around the middle of that same century that Europe had come down with the fever of discovery. Newspapers and geographical societies challenged adventurers to explore mountain ranges, chart rivers, and map jungles. A sort of mythical fascination arose for "the sources" of streams and rivers, in particular that of the Nile. Shortly before his meeting with Stanley, the Scotsman Livingstone had found the Lualaba, a broad but unnavigable river in eastern Congo that flowed north, and which he thought could very well constitute the headwaters of the Nile. In 1875 the Englishman Lovett Cameron stood on the banks of that same river. Cameron, however, realized that a bend to the west later on was all it would take to make this, in fact, the Congo, the mouth of which was already known thousands of kilometers away on the Atlantic coast. Neither of them succeeded in following that river. Stanley did.

He left Zanzibar with his caravan in 1874 and, just to be sure, took his own ship along with him. The *Lady Alice* could be taken apart and portaged like a set of Tinker Toys. What a strange sight that must have been: a long caravan threading its way across the boiling hot savanna of Eastern Africa, hundreds of kilometers from any navigable current, with at the back a group of twenty-four porters bearing the man-size, glistening sections of an otherworldly steel hull.

Stanley subjected Lake Victoria and Lake Tanganyika to a very close inspection. Then, setting out to the west in 1876, he entered the territory of the much-feared but, upon closer acquaintance, also gallant Tippo Tip, with whom he made a deal. In return for a generous reward, Tippo Tip and his men would accompany Stanley a long way to the north along the Lualaba. It was what we today might call a "win-win" situation: Stanley was protected by Tippo Tip, and Tippo Tip could expand into the new territories he discovered along with Stanley.

It worked, although the presence in Stanley's entourage of the most notorious slave driver of all did generate great hostility among the local population. No one knew what an explorer was; Stanley was seen as just another trader. Spears and poisoned arrows came raining down on more than one occasion, and more than once, there were casualties. Although in his writings Stanley tended to exaggerate the number of such clashes (which did his reputation no good), their frequency indicates how much the Arab slave trade had disrupted the area. After passing a series of cataracts, the river became navigable and turned off to the west. Stanley named the spot Stan-

ley Falls (later Stanleyville, today's Kisangani). Bidding farewell to Tippo
Tip and accompanied by several native canoes, he moved on alone into the
area where no European or Afro-Arab trader had ever been before.

On February 1, 1877, at two in the afternoon, his ship passed the area
where the friends of Disasi Makulo's parents lived. Drums had warned the
inhabitants along the banks of his approach, and they had prepared themselves well.[3] A war party of forty-five large dugouts carrying a hundred men
each headed for Stanley's little flotilla. He noted: "In these savage areas our
mere presence awakens the most furious passions of hatred and murder, as
a low-lying ship in shallow water stirs up muddy sediment." It was, indeed,
one of the most impressive military confrontations on his journey. Hundreds of sinewy arms paddled in unison. The canoes approached the *Lady
Alice* on waves of foam. At their bows, warriors with colorful feather head-
dresses were standing ready with their spears. At the stern sat the village
elders. There was a deafening sound of drums and horns. "This is a bloodthirsty world," Stanley wrote, "and for the first time we feel that we hate the
filthy, rapacious ghouls who live here."[4] As soon as the first cloud of spears
came raining down, musket fire rang out. Stanley shot his way to shore.
Once on land he found piles of tusks, and in the villages he saw human skulls
mounted on poles. By five that afternoon he was gone.

It seemed like a one-off incident, a terrible apparition, an inexplicable
epiphany. Peace and quiet returned, or at least so the villagers thought. But
that afternoon passage would change their lives, and especially that of Disasi
Makulo.

One week later, for the umpteenth time, Stanley asked a native what this
river was called. For the first time he was told: "Ikuti ya Congo" (This is the
Congo).[5] A simple answer, but one which filled him with joy: now he knew
for sure that he would not end up at the pyramids of Giza, but at the Atlantic.
He soon began seeing the first Portuguese muskets as well. The attacks from
the riverbanks tapered off, but malnutrition, heat, illness, fever, and rapids
continued to take their toll on this historic crossing of Central Africa.

On August 9, 1877, more than six months after passing through Disasi's
homeland, to the extreme west of that vast area, near the sleepy trading
post of Boma close to the Atlantic, an exhausted and emaciated white man
dropped his things. No one knew that this bundle of starvation and misery was the first European to have followed the entire course of the Congo.
Of the four white men who had left the east coast with him, Stanley was

the only one who survived. Of the 224 members of the expedition, only
ninety-two reached the west coast of Africa. It was a heroic journey, and
one with far-reaching consequences: within the space of three years, from
1874 to 1877, Stanley had circumnavigated and mapped two gigantic lakes,
Lake Tanganyika and Lake Victoria; he had unraveled the complex hydrol-
ogy of the Nile and the Congo and charted the watersheds of Africa's two
largest rivers; and he had carefully documented the course of the Congo
and blazed a trail through equatorial Africa.[6] The world would never be the
same. Today, Stanley's name is associated sooner with that one, awkward
sentence—"Dr. Livingstone, I presume?"—with which he tried to maintain Victorian decorum in the tropics, than with his much more impressive
achievement, which would change forever the lives of hundreds of thousands in Central Africa.

THE PEOPLE in Disasi Makulo's region thought they had seen a ghost. How
could they know that many thousands of kilometers to the north there was
a cold and rainy continent where, in the course of the last century, something as mundane as boiling water had changed history? They knew nothing of the industrial revolution that had altered the face of Europe. The
existence of a society, largely agrarian as their own, which had suddenly
acquired coal mines, smokestacks, stream locomotives, suburbs, incandes-
cent lighting, and socialists, was beyond their ken. In Europe it was raining
inventions and discoveries, but none of that had trickled down to Central
Africa. It would have taken the large part of an afternoon to explain to
them what a train was.

The forest inhabitants could not have dreamed that the industrialization
set in motion by the power of steam would change not only Europe, but
the whole world. More industry meant greater production, more goods, and
so more competition for markets and natural resources. The circles within
which a European factory did its buying and selling were expanding all the
time. Regional became national, national became global. World trade was
growing like never before. Around 1885 steamships replaced sailing ships
on the long-distance routes. The tea drunk by a rich Liverpool family came
from Ceylon. In Worcester, a sauce was made on an industrial scale using
ingredients from India. Dutch ships carried printing presses to Java. And in
South Africa, special ostriches were being raised so that women in Paris, London, and New York could wear large, bobbing feathers on their bonnets. The

world was growing smaller and smaller, time was going faster and faster. And the nervous heartbeat of this new era could be heard everywhere in offices, train stations and border posts in the hectic tapping of the telegraph.

Industrialization definitely served the European powers' expansive urge. In faraway places one found inexpensive raw materials and, with a bit of luck, even new customers. But that did not immediately lead to colonization. No one out to maximize operational profits would thinking of founding an expensive colony. Anyone swearing by the principles of free trade (and every industrialist did so in those days) would be loath to turn to anything as protectionist as an overseas territory. Industrialization alone, therefore, cannot explain the rise of colonialism. In purely commercial terms, a colony was not even necessary. In Central Africa, one could have gone on for a time trading bales of cotton for tusks. No, there was another element needed to make colonial fever break out, and that was nationalism.

It was the rivalry between European nation-states that caused them, from 1850, to pounce so promptly upon the rest of the world. Patriotism led to a craving for power, and that craving, in turn, to territorial gluttony. Italy and Germany had only recently become distinct, united nations and they found overseas territories something that befit their newly acquired status. France had been shamefully whipped by the Prussians in 1870 and attempted to remove the blot on its reputation with colonial adventures abroad, particularly in Asia and Western Africa. England derived great pride from its navy, which had ruled the world's waves for decades, and from its empire, which stretched across the globe, from the West Indies to New Zealand. Proud tsarist Russia was interested in expansion as well, and set its sights on the Balkans, Persia, Afghanistan, Manchuria, and Korea.

The bitter struggle manifested itself in Asia before reaching Africa. Europeans had been familiar with that region much longer and knew that lucrative dealings lay in wait there. (They were still less sure about Africa.) By the time Disasi saw his first white man, in the person of Stanley, the British already controlled the entire Indian subcontinent with offshoots to Baluchistan in the west and Burma to the east. To the southeast the French were busy acquiring Indochina, which included the present-day Laos, Vietnam, and Cambodia. The Dutch still ruled over the enormous island group that would later be called Indonesia, and had done so for more than two centuries. The Philippines were in Spanish hands, but would soon become American territory: the United States, a cluster of former British

and French colonies, had itself become a colonial power. China and Japan
resisted on all sides the pressure of Western colonizers, but were forced
with great reluctance to sign treaties concerning trade tariffs, concessions,
spheres of influence and protectorates. From 1850 the globalization that
had begun in the sixteenth century entered a period of decisive accelera-
tion. And it was the heady mix of industrialization with nationalism that
would lead to the colonialism typical of the nineteenth century.

   That applied most definitely to Central Africa too. At first the European
interest in the region had been largely commercial. Until 1880 the players did
not feel much of an urge to transform their economic activities into political ones. Colonies were not really called for. Without the rise of national
rivalries in Europe, large parts of Central Africa would most probably have
fallen under the political sway of Egypt and Zanzibar.[7] That process was
already under way. In the east, Tippo Tip and Msiri ruled over empires that
owed allegiance to the sultan of Zanzibar. Farther north, al-Zubayr ran a
huge area that was officially a province of the khedive of Egypt. In other
words, the creation of an entity like "Congo" was anything but inevitable.
Things could have gone very differently. The region was not predestined to
become a single country. That Disasi would ever become a countryman of
Nkasi, the old man I had met in Kinshasa, was not written in the stars. The
two boys may have differed less than ten years in age, but one of them lived
in the equatorial jungle and the other along the lower reaches of the Congo,
some twelve hundred kilometers (about 750 miles) away. They spoke different languages, had different customs and knew absolutely nothing about
each other's culture. That they became compatriots was not due to them or
to their parents, but was the result of envy in that mad, northern part of the
world of which they had no cognizance.

   No, the contemporaries of those two children could not have known
that envy cropped up more often in Europe. And that it was precisely for
that reason that the major nations had agreed in 1830 to the formation of a
new, minuscule country. Belgium, as that ministate was called, had turned
its back on the the United Kingdom of the Netherlands after a fifteen-year
mariage de raison, and could still serve as a buffer between ambitious Prus-
sia, powerful France, and proud England. It could, perhaps, even temper the
mutual envy between those countries. That was how it had been viewed in
1815, too, after the Battle of Waterloo. For centuries, the region had served
as a battlefield for the armies of Europe, and now it was to be a neutral zone

for the promotion of peace. In 1830 it declared its independence. One big step for the Belgians, one small step for mankind. No one in Central Africa lost any sleep over it.

No one at all, in fact, had ever heard of Belgium. No one could imagine that the first king of that little country would soon sire a son who would bring to bear a most disturbing ambition. The father, a melancholic prince who became a widower at an early age, was satisfied enough with his kingship. But his overweening son, the future Leopold II, seemed to bridle at the limited territory over which he ruled. *"Il faut à la Belgique une colonie"* (Belgium must have a colony), he had engraved on a paperweight destined for the desk of his finance minister when he was only twenty-four. Precisely where that colony should be, that was less important. Even before assuming the throne, he had cast a wistful glance at Dutch Limburg, Constantinople, Borneo, Sumatra, Formosa (Taiwan), Tonkin (Vietnam), parts of China or Japan, the Philippines, a few islands in the Pacific or, if need be, a few islands in the Mediterranean (Rhodes, Cyprus). But from 1875 he fell under the spell of Central Africa. He devoured the reports sent back by explorers, licked his chops at the prospect of a glorious adventure, and daydreamed about a heroic enterprise. It was not merely personal ambition or megalomania, as is often claimed. No, Leopold believed with all his heart that an invigorating involvement abroad, wherever that might be, would benefit both the finances and the morale of the young Belgian nation. Whatever else may be said about him, he did it not only for himself, but also for people and fatherland. Fully in tune with his times, the young king effortlessly reconciled warm-blooded patriotism with coolly calculating commercialism.

In 1876 the impetuous young ruler brought together thirty-five explorers, geographers, and entrepreneurs from all over Europe to discuss the status of Central Africa. Officially, his intention was to halt the Afro-Arab slave trade and promote science, but those closest to him knew that he himself desired a healthy slice of "ce magnifique gateau africain" (this magnificent African cake).[8] His outrage about the slave trade was, for that matter, selective: that Westerners had also dealt in human cargo until quite recently, and that some of them still continued with that even in his day, were things about which he remained silent. The meeting was to be an illustrious one. For four days, adventurers from all over Europe, men more commonly found poking about the tropics in sweat-soaked shirts, were his guests at the royal palace.

They dined with the king and his wife and were driven through the streets of Brussels in chic coaches. Lovett Cameron was there, the man who had crossed Central Africa from east to west via the savanna south of the equatorial forest, as was Georg Schweinfurth, the maker of important discoveries on the savannas north of the jungle, and Samuel Baker, who had approached the region from the upper reaches of the Nile. During the previous few decades, awesome progress had been made in the exploration of Africa.

Until somewhere around 1800, the continent that lay closest to Europe was also the one most unfamiliar to Europeans. Since the sixteenth century, Portuguese, Dutch, and British merchants on their way to India had become more or less familiar with its coastlines, but for centuries the African interior remained terra incognita. The West's presence went no further than a few European outposts on the west coast. At the start of the nineteenth century, Africa comprised one of the two blank spots on the map of the known world; the other was Antarctica. By then, the Amazon Basin had been largely charted.

Three-quarters of a century later, however, European cartographers knew with fair precision where the oases, caravan routes, and wadis of the Sahara lay. They had accurately localized the volcanoes, mountains and rivers in the savanna of Southern Africa. The sketches on their drawing tables filled rapidly with exotic place names and the descriptions of peoples. But the map pored over by the conferees in Brussels in 1876 contained one large, white spot in the middle. All of them had circled around it at some point. It looked like a nameless plain without words or color, a yawning chasm covering no less than one-eighth of the continent. At most, it contained the occasional, hesitant squiggle or dotted line. That spot, that was the equatorial jungle. That forlorn dotted line, that was the Congo River.

While the delegates in Brussels were talking and attending plays at the king's expense, Stanley was making his crossing of Central Africa. On September 14, 1876, the day that Leopold officially closed the conference, Stanley left the western shore of Lake Tanganyika and advanced on the upper reaches of the Congo. If there was one day on which the political fate of the region was, if not sealed then certainly determined to a great extent, it was that one. It must have been the least of Stanley's worries at that point (he was more concerned about the rain forest, the natives, and the slave drivers), but starting in on that stage of the journey would ultimately lead him to the mysterious river that guided him through the ostensibly impenetrable forest of Central

Africa. That day in Brussels, the decision was made to set up an international association, the Association Internationale Africaine (AIA), in order to open up the area scientifically by means of establishing a number of outposts. The association had national committees, but its leader was Leopold.

In Europe, the news of Stanley's crossing came as a bombshell. King Leopold understood immediately that Stanley was the man who could help him realize his colonial ambitions. He immediately sent two emissaries to Marseille to welcome him back to Europe in January 1878 and to invite him to the royal palace at Laeken. As an Englishman, however, Stanley first tried to interest Britain in his adventure, but when he was turned down in London he decided to accept Leopold's invitation. The two men discussed plans at length. The king became so caught up in his enterprise that the queen began to wonder what would become of him "should he ruin himself with chasing after shadows." The first secretary of the AIA complained to the queen: "Madame, let us stop this—I am unable to do anything else, all I do is argue with His Majesty, but he works behind my back with rogues. It is driving me mad! And the King is bringing himself to ruin, but then completely."⁹ It was to no avail. The king had his way: in 1879, Stanley left again for Central Africa, now at Leopold's expense, for a period of five years. This time the explorer was going to travel in the opposite direction, from west to east, upstream. But that was not the only difference. Stanley's journey from 1879 to 1884 was fundamentally different from that from 1874 to 1877. On that first occasion he was commissioned by a newspaper; now he had been hired by Leopold's international association. The first time he had set out to cross Africa as quickly as possible; this time he was charged with establishing outposts here and there along the way—a time-consuming business. He had to parley with local chieftains and man the stations as well. The first time he had been an adventurer and a journalist, now he was a diplomat and official.

DISASI MAKULO TURNED TEN, then twelve, and began hearing more and more about a new tribe, the "Batambatamba." The older children and adults spoke of them in fear and horror. Batambatamba was no ethnic name, but an echoism that designated the Afro-Arab traders. They had arrived in his region now, the farthest west they would ever come. In his village he heard the stories: "We have seen people who walk to and fro; they carry a kind of hollow stick, when you strike it you hear a sound, *Bam Bam*, and grains come out of it that wound people and kill them. Terrible!"¹⁰

Still, it all seemed very far away, as bizarre as that story about the albino and his boat without rowers. One day, Disasi Makulo's parents let him go off with his aunt and uncle.[11] It was 1883, but the years still had no numbers.

It was very hot that day. When we came to a river called Lohulu, between Makoto and Bandio, my uncle and I decided to bathe. My Aunt Inangbelema waited for us a little further along. While we were swimming and splashing each other cheerfully, the Batambatamba heard us and surrounded us. My aunt was singing lullabies to soothe her crying baby. None of us were thinking of possible danger.

Suddenly there was a scream. "Help! Help! Brother Akambu, the warriors are attacking me. . . ."

We jumped out of the river and saw that my aunt had already been seized by our enemies. One of the attackers pulled the baby out of her hands and laid it on a red ants' nest. We were so shocked that none of us could get to him. Uncle Akambu and my little cousin ran away and hid in the bushes. I remained at a distance, to see what they would do to my aunt. Unfortunately, one of the men spotted me. He ran after me and caught me. Then my Uncle Akambu and my cousin were captured as well.

Until that hideous day, Disasi's life had taken place in his village and a few nearby settlements. Now he was brutally torn away from those familiar surroundings. Stanley's journey, and his deal with Tippo Tip in particular, had opened up the equatorial forest to Afro-Arab slave hunters. That resulted in a wave of violence. The Batambatamba plundered villages and put them to the torch, they murdered, and they took prisoners. The local inhabitants in turn painted their faces and attacked the foreigners' camps at night, slaughtering the intruders with their spears amid loud war cries.

Disasi's assailants were probably slaves themselves, plundering on their master's behalf. Disasi would soon meet that master, a man who traveled through the jungle in a spotless white robe: Tippo Tip! He probably also saw Salum ben Mohammed, Tippo's cousin and close associate.[12] The fresh slaves were assembled at the village of Yamokanda.

Here one could buy back prisoners. Many prisoners were let go because their parents brought ivory. My father came with a few tusks as well, but Tippo Tip told him that it was not enough for four people. He let my Uncle Akambu, my Aunt Inangbelema and my little cousin go. Concerning me, he told them: "Go home and come back with two more tusks." I remained behind, along with the other prisoners who had not been bought back.

The slave driver in question, however, decided not to wait and left that very same day. The adult prisoners were shackled, the children were not. Huge canoes were waiting along the banks of the Aruwimi. "All you could hear during that ghastly cruise was the sound of weeping and sobbing." Disasi knew he was leaving his home ground and could no longer be bought back. Later he heard that his father had returned to the camp with the ivory, as demanded, but that the caravan had already left.

The trip eastward was not a fortunate one. "For us, that journey down-river was nothing but a departure towards death, although they told us that they wanted to protect us and make us like them." That latter statement was not meant cynically. The slaves of the Afro-Arab traders were not sent to huge cotton or sugar cane plantations, as in America. Some of them would go to gather cloves in Zanzibar, but most of them would serve as domestic slaves to wealthy Muslims, in places as far away as India. Many converted to Islam and climbed the social ladder. A start was made on their conversion during the journey itself.

One day something strange happened to us. While our *mwalimu* [teacher] was teaching us to read the Koran, we saw downstream something like very large canoes coming in our direction. There were three of them. Everyone, both we and the locals, were startled, because we believed that these were new assailants coming upstream to murder and plunder as well. The locals fled in their canoes, to hide on the little islands in the river, others of them disappeared into the forest immediately. We remained where we were, our gazes fixed on those strange canoes. Before long they moored along the banks. We saw white men and black men getting off: it was Stanley with a few whites, on his way to establish a post in Kisangani [Stanleyville]. Stanley was no stranger to the people along the banks. The Lokele called him "Bosongo," meaning "albino."

Stanley was indeed traveling with three steamboats. He was carrying out King Leopold's orders to establish stations here and there and negotiate with local chieftains. It was during this journey that he noted that his crossing had opened up the interior not only to Western trade and civilization, but also to the slave drivers from the east, who were moving farther downriver all the time. It was then that he realized that the Arab traders could very well beat him to the punch and arrive at the river's lower reaches in no time. They had now come to just below Stanley Falls (Kisangani); soon they could be at Stanley Pool (Kinshasa). If that happened, Leopold's plans could be relegated to the rubbish bin. It was during this journey that he realized what he was up against: the slave traders had dozens of canoes and a few thousand troops. He had three little boats and a few dozen helpers.[13]

In Disasi's area, Stanley saw along the banks only burned villages and charred huts, "the remains of once- populous settlements, scorched banana plantations and felled palms . . . all bearing equal testimony to merciless destructiveness." Further along he saw the slave camps beside the river. In late November 1883 he arrived at the camp where Disasi was being held:

> The first general impressions are that the camp is much too densely peopled for comfort. There are rows upon rows of dark nakedness, relieved here and there by the white dresses of the captors. There are lines or groups of naked forms, standing or moving about listlessly; naked bodies are stretched under the sheds in all positions; naked legs innumerable are seen in the positions of prostrate sleepers; there are countless naked children, many mere infants, forms of boyhood and girlhood, and occasionally a drove of absolutely naked old women bending under a basket of fuel, or cassava tubers, or bananas, who are driven through the moving groups by two or three musketeers.[14]

First he went to establish a post at Stanley Falls, but on December 10, 1883, he returned to the slave camp. Little Disasi witnessed a remarkable scene. "Tippo Tip went to meet Stanley. After a long talk in an incomprehensible language, Tippo Tip called out to our overseer. He gathered us together and brought us over to the two gentlemen." Disasi had no idea what was going on. Once the discussion was over, Stanley's men fetched two rolls of cloth and a few bags of salt from the ship's hold. His Koran teacher told him, with pain in his heart, that this white man wanted to buy him and his companions. Stanley took eighteen children with him.[15] Militarily, he was too weak to take any action against the Batambatamba. The only thing left was for him to take the fate of a few children to heart. He bought them away.

A new phase in Disasi's life began. The atmosphere on board was cheerful. "We shout, we laugh, we tell stories. No one has a rope around his neck and we are not treated like animals, as we were when we were with the Arabs." But it would be too simple to state that Stanley had freed them from slavery. Traditionally, slavery in Central Africa was seen principally as a matter not of robbing you of your freedom, but of uprooting you from your social setting.[16] It was gruesome, to be sure, but for reasons other than commonly assumed. In a society so characterized by social feeling, "the autonomy of the individual" did not equal liberty at all, as Europeans had been proclaiming since the Renaissance, but loneliness and desperation. You are who you know; if no one knows you, you are nothing. Slavery was not being subjugated, it was being separated, from home. Disasi had been uprooted from his surroundings and would remain uprooted. He valued Stanley therefore not so much as his liberator, but as a new and better master.

Never was that clearer than on the next day, when he sailed past his home ground again. Disasi thought Stanley would return him to his parents, but to his surprise the boat did not slow. "That's where we live! That's where we live!" he shouted. "Take me back to my father!" But Stanley spoke, as Disasi would recall a lifetime further along:

> My children, do not be afraid. I did not buy you in order to harm you, but in order that you might know true happiness and prosperity. You have all seen how the Arabs treat your parents and even little children. I cannot let you return home, because I do not want you to become like them, cruel savages who do not know the True Lord. Do not mourn the loss of your parents. I will find other parents for you who will treat you well and teach you many good things; later you will be like us.

Having said that, Stanley immediate cut a roll of cloth into pieces and gave each child a loincloth, so that they would be decently clothed. "That present pleased us," Disasi recounted, "and his goodness made us feel his fatherly love already."[17]

Meeting Stanley constituted a drastic turn in Disasi Makulo's life. For many of his contemporaries, however, there were very few changes at all.

The men continued to burn off their plots, the women planted corn and manioc, fishermen mended their nets, old people talked in the shade, and children caught grasshoppers. Everything seemed to go on the way it always had.

Yet that was only the surface. Those who had actually seen those peculiar Europeans were often deeply impressed. These shabby men showed up to buy a few chickens and spent the afternoon talking to the village chieftain, but they did all they could to make an impression on the local population. Mirrors, magnifying glasses, sextants, compasses, timepieces, and theodolites were produced intentionally, for effect. That did not always result in enthusiasm. In some villages, people believed that the death by natural causes of some inhabitants could be blamed on the strange thermometers and barometers demonstrated by the white men.[18] Awe was mingled with suspicion. Only later would this lead to large-scale violence, when the local population was subjected to European authority by force of arms.

There was often doubt about whether these Europeans were actually common mortals. The shoes they wore made it seem as though they had no toes. And because white, in large parts of sub-Saharan Africa, was the color of death (the color of human bones, of termites, of tusks), they almost had to come from the land of the dead. They were seen as white ghosts with magical powers over life and death, men who popped open umbrellas, and could bring down an animal at a hundred yards. The Bangala referred to Stanley as *Midjidji*, the spirit; the Bakongo called him *Bula matari*, the stonebreaker, because he could blow up rocks with dynamite. Later, the term *Bula matari* would also be used to refer to the colonial regime. In Disasi Makulo's village too, he was seen as a phantom. E. J. Glave, one of Stanley's helpers, was first referred to as *Barimu*, ghost, and later as *Makula*, arrows. The Bangala gave Herbert Ward, another helper, the nickname *Nkumbe*, black hawk, because he was such a skilled hunter.

And the way these white people moved from place to place was so peculiar as well. By steamboat! The Bangala who lived along the river in the interior thought these travelers ruled over the water and that their boats were drawn by huge fish or hippos. After a parley, when they saw the white man disappear into the hold to fetch pearls, cloth, or copper bars, they thought he had a special door in the ship's hull through which he could descend to the bottom of the river and collect these means of payment.[19]

A first wave of evangelization followed immediately in the wake of exploration. It was carried out by Anglo-Saxon and Scandinavian Protestants who had started on the west coast right after Stanley's crossing. The Livingstone Inland Mission began its proselytization in 1878, starting from the mouth of the Congo. In 1879 the Baptist Missionary Society set out from its base at the Portuguese colony to the south, the Svenska Mission Förbunet began in 1881, and the American Baptists and Methodists followed in 1884 and 1886. Two French Roman Catholic congregations were also active from 1880: the Missionaries of the Holy Spirit in the west and the White Fathers (Society of Missionaries of Africa) in the east. Such undertakings were anything but without risk. Anyone setting out for Central Africa in those days knew that it could be the death of them. Sleeping sickness and malaria took a heavy toll. The British Baptist Thomas Comber lost his wife only a few weeks after they arrived in Central Africa. He himself would later die of a tropical illness as well, as would his two brothers, his sister, and his sister-in-law: six members of a single family. A third of all Baptist missionaries sent out between 1879 and 1900 died in the tropics.[20] With no prospects of financial gain or worldly power, the first missionaries were truly deeply devout people who saw it as their duty to let others share in the truth that possessed them so completely.

When it came to impressing people, the early missionaries had their own bag of tricks. This was advisable, particularly in those areas that had been in contact with the white man for some time. The ivory trade had had more consequences than prosperity alone. In 1878, when the British Baptists George Grenfell and Thomas Comber headed north as the first white missionaries from the Portuguese colony, they stumbled upon the town of Makuta, halfway between Mbanza-Kongo in Angola and the Congo River. The local chieftain didn't like the newcomers' looks.

Ah, so they haven't come to buy ivory! Well then, what do they want? To teach us about God! About dying, more likely. We already have more than enough of that: the deaths in my city go on and on. They must not come here. If we allow the white man in, that will be the end of us. It's bad enough that they are on the coast. The ivory traders already take far too many spirits away in the tusks, and they sell them; we are dying too quickly. It would have been better if the whites had not come to cast a spell over me.[21]

Although one of the two men would later suffer a gunshot wound at Makuta, the Protestant evangelists—thanks in part to the miracle of technology—succeeded elsewhere in winning the hearts and minds of the local population. To the chief of the Bakongo, British Baptists displayed a number of mechanical toys. In addition to a wind-up mouse, they also showed him what they called a "dancing nigger," a mechanical doll that played the fiddle and hopped about.[22] Merriment and awe were guaranteed. Music boxes were another fine example. But the cleverest of all were the slide shows, depicting scenes from the Bible, that some missionaries projected at night with the help of magic lanterns. For the native population, such things must have seemed absolutely out of this world.[23]

Talking with Nkasi in his stifling room about those first pioneers was a mindboggling experience. The conversation went in fits and starts; all I received were wisps of memory, but the fact that more than a century later he still recalled the arrival of white missionaries indicated how very special those wisps were. In reference to the British Baptists he had spoken quite precisely of "English Protestants who came to Congo from Mbanza-Kongo in Angola." He mentioned the mission posts at Palabala and Lukunga, both founded by the Livingstone Inland Mission and transferred to the American Baptist Missionary Union in 1884. He also remembered "Mister Ben," as I jotted down phonetically in my notepad. Later I discovered that this must have been Alexander L. Bain, an American Baptist particularly active in the area from 1893 on.[24] But most of all he talked about "Mister Wells" or "Welsh," mister and not *monsieur,* for French was not yet spoken in Congo. "I saw him at the Protestant mission at Lukunga. He was an English missionary who gave us lessons. He lived with his wife in Palabala, close to Matadi."

For a long time, I wondered who that man might have been. Was it the American Welch, a follower of the energetic American Methodist bishop William Taylor, who established three missions in the area in 1886 (although not at Palabala or Lukunga)?[25] Or was "Mister Welsh" the nickname of William Hughes, a British Baptist who had manned the Bayneston mission post in the same area from 1882 to 1885?[26] Finally I arrived at Ernest T. Welles, an American Baptist who had sailed for Congo in 1896 and who had translated Bible passages into Kikongo as early as 1898. He had to be the one. He was a direct colleague of Mister Bain and turned out to have been associated with the Lukunga mission for a time. In his letters home he wrote about native assistants who had helped him print his Bible translations.[27] That was interesting. Nkasi, after all, had told me that his father's youngest brother had worked for that missionary. Those first evangelists, in any case, made an indelible impression on the young Nkasi. What he still remembered best was their simplicity and friendliness. "Mister Wells," he mused during one of our talks, "went everywhere on foot, he was extremely kind."

JANUARY 1884. Stanley had been preparing his journey home for weeks. The eighteen children he had with him he distributed among the stations established on his way upriver, such as Wangata and Lukolela. Disasi Makulo and one of his young comrades were the last on board, and wondered what was going to happen to them. Finally they arrived at the "pool," where the river widened and Stanley had established the Kinshasa station. He had left the running of that station in the hands of his faithful friend Anthony Swinburne, a young man of twenty-six who had traveled with him for a decade. It was to Swinburne's care that Disasi and his friend were entrusted. Saying farewell to Stanley was hard: "From the first day of our liberation to the moment of farewell, he had been a father to us, full of benevolence," Disasi wrote. In our day Stanley is often criticized as an archracist, a reputation he owes to his hyperbolic writing style and his association with Leopold II. In fact, however, his attitude was much more nuanced.[28] He had great admiration for many Africans, maintained deep and sincere friendships with a number of them, and was greatly loved by many. His combination of kidnapping and bargain hunting was, of course, highly idiosyncratic, but he seems to have been sincerely concerned with the welfare of the children he had bought out of slavery. Disasi recounted:

> Mister Swinburne received us with open arms. What Stanley had predicted proved true. Here we found ourselves in a situation that in no way differed from what a good father and a good mother offer their children. We were fed well and clothed well. During his free hours, he taught us to read and write.[29]

That Swinburne had any free hours whatsoever is little short of a miracle. Within only a few years he had developed Kinshasa into the best of all stations along the Congo. It lay close to the river, among the baobabs. He had bananas, plantain, pineapple, and guava planted nearby, as well as rice and European vegetables. He kept cows, sheep, goats, and poultry. The air was fresh and healthy. The station was known as the Paradise of the Pool.[30] His clay house had a grass roof and three bedrooms. The verandah around it was a place where people came to eat and read. Behind Swinburne's house were the huts of his Zanzibaris. The stations of that day were often no more than a simple dwelling inhabited by a white man. It served to assist travelers, promote scientific research, disseminate civilization, and, if at all possible, do away with slavery. In practice, it was actually a sort of minicolony aimed at exercising a certain authority over the surrounding region. Little islands of Europe. The Zanzibaris made up its standing army. There was, as yet, nothing like a general occupation of the interior.

Behind Swinburne's station began a huge plain, bordered on the horizon by hills. Today this is the site of one Africa's biggest cities; in the nineteenth century it was a marshy area full of buffalo, antelopes, ducks, partridges, and quail. On the drier stretches the villagers raised manioc, peanuts, and sweet potatoes. Their villages were a few kilometers away. Swinburne was on very good terms with the local population. His patience and tact made him not just respected, but loved. He spoke their language and they called him the "father of the river." Yet he was not, when he deemed it necessary, shy about intervening. When a local chieftain would die, for example, he would regularly do his best to prevent the man's slaves and wives from being killed and buried along with him. This greatly amazed the villagers: how could anyone worthy of the name of chieftain be allowed to arrive all alone in the kingdom of the dead?

In order to set up a station, Stanley and his helpers first had to establish contracts with the local chiefs. That was the way European traders at the Congo's mouth had been doing it for centuries. They rented plots of land in exchange for a periodic payment. Swinburne, too, had closed several such contracts, often after palavering for days. Starting in 1882, however, Leopold grew impatient. His international philanthropic association had meanwhile been transformed into a private trading company with international stake-holders: the Comité d'Études du Haut-Congo (CEHC). The king ordered his agents to obtain larger concessions, within a much shorter period, and preferably for perpetuity. Rather than carry out lengthy negotiations to rent a plot of land, they now had to quickly buy up entire areas. And even that was not enough: Leopold wanted to purchase not only the ground, but also all rights to that ground. His commercial initiative had become a clearly political project: Leopold dreamed of a confederation of native rulers fully dependent on him. In a letter to one of his employees, he made his aims perfectly clear: "The text of the treaties Stanley has signed with the chieftains does not please me. It should at least contain an article stating that they relinquish their sovereign rights to those territories. . . . This effort is important and urgent. The treaties must be as brief as possible and, in the space of one or two articles, assign all rights to us."[31]

As a result, Stanley's helpers entered into real treaty-making campaigns. They went from one village chieftain to the next, armed with Leopold's marching orders and terse contracts. Some of them lost no time in doing so. During the first six weeks of 1884, Francis Vetch, a British army major, established no less than thirty-one treaties. Belgian agents like Van Kerckhoven and Delcommune both signed nine such contracts in a single day. Within less than four years, four hundred treaties were established. They were written without exception in French or English, languages the chieftains did not understand. Within an oral tradition in which important agreements were sealed with blood brotherhood, the chiefs often did not understand the import of the cross they made at the bottom of a page filled with strange squiggles. And even if they had been able to read the texts, they would not have been familiar with concepts of European property and constitutional law like "sovereignty," "exclusivity," and "perpetuity." They probably thought they were confirming ties of friendship. But those treaties did very much indeed stipulate that they, as chieftain, surrendered all their territory, along with all subordinate rights to paths, fishing, toll keeping, and trade. In exchange for that cross the chieftains received from their new white friends bales of cloth, crates of gin, military coats, caps, knives, a livery uniform, or a coral necklace. From now on the banner of Leopold's association would fly over their village: a blue field with a yellow star. The blue referred to the darkness in which they wandered, the yellow to the light of civilization that was now coming their way. Those are the dominant colors in the Congolese flag even today.

The reason behind Leopold's sudden haste could be traced, once again, to rivalry among the European states. He was afraid that others would beat him to the punch. And some of them did. To the south, the Portuguese were still asserting their rights to their old colony. And to the north, Savorgnan de Brazza had begun in 1880 to establish similar treaties with local chieftains. Brazza was an Italian officer in the service of the French army, officially charged with setting up two scientific stations on the right bank of the Congo. France itself took part in the Association Internationale Africaine chaired by Leopold, and those two stations were the French contribution to the king's initiative. But Brazza was also a fanatical French patriot who, at the behest of no nation whatsoever, was busy establishing a colony for his beloved France: it would later become the republic of Congo-Brazzaville.[32] By 1882 people in Europe had begun to realize that someone was independently buying up large sections of Central Africa. That led to great consternation. Leopold had no choice but to act.

An Italian personally buying pieces of Africa for France, and an Englishman, Stanley, buying others for the Belgian king: it was called diplomacy, but it was a gold rush. In May 1884 Brazza crossed the Congo with four canoes in an attempt to win Kinshasa for himself. But there he ran into Swinburne, the agent with whom Disasi had now been living for the last four months. Brazza tried to make the local village chieftain a higher bid and so nullify the earlier agreement, but that resulted in an unholy row. There was a brusque discussion with Swinburne, followed by a scuffle with the chieftain's two sons and Brazza's hasty departure. For Leopold's enterprise, the loss of Kinshasa would have been disastrous. It was not only the best but also the most important of his stations; it was located at a crossing of the trade routes, a place where boats moored and caravans left, where the interior communicated with the coast. The import of the incident with Brazza was almost certainly lost on Disasi, but for generations to come it remained vitally important: the area to the north and west of the river would become a French colony, known as French Congo; the area to the south would remain in Leopold's hands.

Yet still, this episode highlighted a major weakness. In military terms, Stanley could easily deal with someone like Brazza—he had troops and Krupp cannons, while Brazza traveled virtually alone—but as long as Stanley's outposts were not recognized by the other European powers, not a round could be fired.[33] Leopold knew that too. Starting in 1884 he devoted himself to a diplomatic offensive unparalleled in the history of the Belgian monarchy: the drive for international recognition for his private initiative in Central Africa.

Leopold cast about in search of a masterstroke. And found it.

Central Africa was at that point exciting the ambitions of many parties. Portugal and England were quibbling over who was allowed to settle where on the coast. The Swahilo-Arab traders were advancing from the east. A recently unified Germany hankered after colonial territory in Africa (and would ultimately acquire what was later Cameroon, Namibia, and Tanzania). But Leopold's biggest rival was still France, that much was clear. That country, in the face of all expectations, had finally been rash enough to accept Brazza's personal annexations, even though it had not asked for them in the first place. Brazza had gone too far. Leopold could have turned his back on France in anger, but instead the king decided to calmly seize the bull by the horns. His proposal: would France allow him to go about his business in the area recently opened up by Stanley, on condition that—in the event of an eventual debacle—it be granted the *droit de préemption* (right of first refusal) over his holdings? It was an offer too good for the French to refuse. The chance that Leopold would fail, after all, was quite real. It was as though a young man had discovered an abandoned castle and set about restoring it with his own hands. To the neighbor he says: if it becomes too costly for me, you'll have the first option! The neighbor is all too pleased to hear that. It was a brilliant *coup de poker,* and one that would also impact other parts of Europe. The agreement took Portugal down a notch or two; obstructing Leopold might mean it would suddenly find itself with mighty France as its African neighbor. The British, on the other hand, were quite charmed by the guarantee of free trade that Leopold presented so casually.

The mounting competition over Africa between European states called for a new set of rules. That was why Otto von Bismarck, master of the youngest but also the most powerful state in continental Europe, summoned the superpowers of that day to meet in Berlin. The Berlin Conference ran from November 15, 1884, to February 26, 1885. Tradition has it that Africa was divided then and there, and that Leopold had Congo tossed in his lap. Nothing, however, could be further from the truth. The conference was not the place where courtly gentlemen with compass and straightedge convivially divided African among themselves. Their aim, in fact, was the complete opposite: to open Africa up to free trade and civilization. To do that, new international agreements were needed. The drawn-out conflict between Portugal and England concerning the mouth of the Congo made that clear enough. Two important principles were established: first, a country's claims to a territory could be based only on effective occupation (discovering an area and leaving it to lie fallow, as Portugal had done for centuries, was no longer sufficient); second, all newly acquired areas must remain open to free international trade (no country was to be allowed to impose trade barriers, transit charges, or import or export duties). In practical terms, as Leopold would soon notice, this meant that colonization became very expensive. In order to allow free access to merchants from other countries, one had to invest a great deal in one's effective occupation. But although the criteria

of effective occupation did speed up the "scramble for Africa," no definitive divvying up of the continent took place as of yet. The delegates to the
conference met no more than ten times over a period of more than three
months; Leopold himself, in fact, never made it to Berlin.

In the corridors and backrooms, however, any number of arrangements
were made. Multilateral diplomacy was practiced during the plenary meetings, but bilateral diplomacy set the tone during coffee breaks. Before the
conference even started, the United States recognized Leopold's Central
African claim. It accepted his flag and his authority over the newly acquired
territory. Yet that sounds more impressive than it actually was. The America
of that day was not the international heavyweight it would become during
the twentieth century, and it had no interests whatsoever in Africa. Of much
greater importance was the German stance. Bismarck considered Leopold's
plan quite insane. The Belgian king was laying claims to an area as large as
Western Europe, but he held only a handful of stations along the river. It
was a string of beads, with very few beads and a lot of string, to say nothing of the enormous blank spots to the left and right of it. Could this be
called an "effective occupation"? But, oh well, as ruler of a little country
Leopold hardly posed much of a risk. Besides, he was anything but impe-
cunious and he *was* terribly enthusiastic. What's more, he guaranteed free
trade (something you could never be sure of with the French and Portuguese) and pledged to extend his protection to German traders in the area.
After all, Bismarck figured, perhaps the territory was indeed an ideal buffer zone between the Portuguese, French, and British claims to the region.
Rather like Belgium itself in 1830, in other words, but then on a much larger
scale. It might make for a bit of peace and quiet. He signed.

The other countries at the conference could do little but follow the
host's lead. Their recognition was not granted at a formal moment during
the plenary session itself, but throughout the course of the conference. With
the exception of Turkey, all fourteen states agreed: that included England,
which had no desire to cross Germany on the eve of an important agreement concerning the Niger. Later, more or less accidentally, the conference
even agreed to the vast boundaries of which Leopold had been dreaming.
And so Leopold's latest association, the Association Internationale du Congo
(AIC), was internationally recognized as holding sovereign authority over
an enormous section of Central Africa. The AIA had been strictly scientific-
philanthropic in nature, and the CEHC commercial, but the AIC was overtly

political. It possessed a tiny, but crucial, stretch of Atlantic coastline (the mouth of the Congo), a narrow corridor to the interior bordered by French and Portuguese colonies, and then an area that expanded like a funnel, thousands of kilometers to the north and south, coming to a halt only fifteen hundred kilometers (over nine hundred miles) to the east, beside the Great Lakes. It resembled a trumpet with a very short lead pipe and a very large bell. The result was a gigantic holding that was in no way in keeping with Leopold's actual presence. The great Belgian historian Jean Stengers said: "With a bit of imagination one could compare the establishment of the state of Congo with a situation in which an individual or association would set up a number of stations along the Rhine, from Rotterdam to Basel, and thereby obtain sovereignty over all of Western Europe."[34]

At the close of the Berlin Conference, when Bismarck "contentedly hailed" Leopold's work and extended his best wishes "for a speedy development and for the achievement of the illustrious founder's noble ambitions," the audience rose to its feet and cheered for the Belgian ruler. With that applause, they celebrated the creation of the Congo Free State.

Shortly after gaining control over Congo, Leopold received a visit at his palace from a British missionary who brought with him nine black children, boys and girls of twelve or thirteen, all contemporaries of Disasi. They came from his brand-new colony and wore European clothing: dress shoes, red gloves, and a beret—their nakedness had to be covered. They were, however, allowed to sing and dance, the way they did during canoe trips. The king, his legs crossed, watched from his throne. When they were finished singing he gave each child a gold coin and paid for their journey back to London.[35]

Meanwhile, ignorant of all this, Disasi Makulo was sitting on Swinburne's veranda in Kinshasa, practicing his alphabet. The weather was lovely and cool. A slight breeze blew across the water. He saw steamboats and canoes glide across the Pool. On the far shore lay the settlement of Brazzaville, by then part of a different colony that would, from 1891 on, be called the French Congo. How his life had changed, in only eighteen months! First a child, then a slave, now a boy. No one had experienced the great course of history firsthand the way he had. He had been uprooted and borne along on the current of world politics, like a young tree by a powerful river. And it was not nearly over yet.

MAP 4: CONGO FREE STATE, 1885–1908

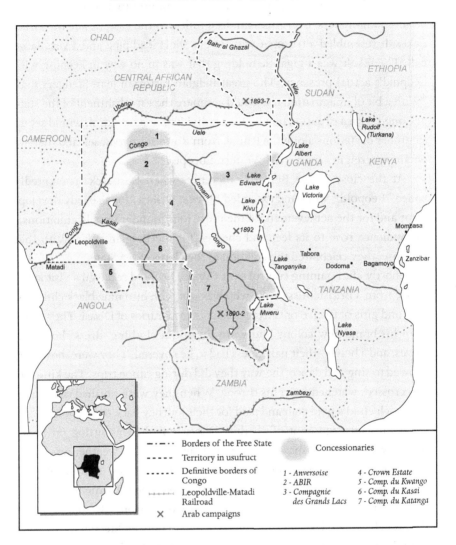

Borders of the Free State

Territory in usufruct

Definitive borders of
Congo

Leopoldville-Matadi
Railroad

✕ Arab campaigns

Concessionaries

1 - Anversoise
2 - ABIR
3 - Compagnie
    des Grands Lacs

4 - Crown Estate
5 - Comp. du Kwango
6 - Comp. du Kasai
7 - Comp. du Katanga

CHAPTER 2

"DIABOLICAL FILTH"

Congo Under Leopold II

1885–1908

O N JUNE 1, 1885, KING LEOPOLD II AWOKE IN HIS PALACE AT Laeken a different man: in addition to being king of Belgium, from that day on he was also sovereign of a new state, the Congo Free State. That latter entity would continue to exist for precisely twenty-three years, five months, and fifteen days: on November 15, 1908, it was transformed into a Belgian colony. Congo began, in other words, not as a colony, but as a state, and one of the most peculiar ever seen in sub-Saharan Africa.

To start with, its head of state lived more than six thousand kilometers (about 3,700 miles) to the north, a four-week journey by ship from his empire—a journey that he himself, by the way, never undertook. From his investiture in 1885 to his death in 1909, Leopold II never set foot in his Congo. In view of the inherent risks to personal health engendered by such a journey at the time, that is hardly surprising. The heads of state of other European powers did not travel to their recently acquired holdings in Central Africa either. The more curious fact is that the Belgian king, unlike his colleagues, was the complete and absolute ruler over his overseas territory. Kaiser William I, Queen Victoria, and Jules Grévy, president of France's Third Republic, also ruled over vast stretches of Africa in 1885, but none of them owned those areas personally. Their colonial policies were not a private matter but a government affair, watched over by parliament (chamber of deputies) and cabinet. But the Belgian king ruled over the new state in a personal capacity.

Officially, the Kingdom of Belgium at that point still had nothing to do with Congo; it only happened to share a head of state with that remote tropical backwater. In Belgium, Leopold was a constitutional monarch with limited powers; in Congo he was an absolute ruler. This extremely personalized regime made him more closely resemble a fifteenth-century king of the Kongo Empire than a modern European monarch. And he acted as though he truly did *own* this empire of his.

Leopold's acquisition of so much power, incidentally, took place almost by sleight of hand. The European superpowers had not recognized him, but his Association Internationale du Congo, as sovereign administrator over the Congo basin. Yet when he abandoned that paper tiger for what it was after the Berlin Conference and began behaving ostentatiously as ruler of the Congo Free State, no one seemed to protest. People saw him as a great philanthropist with a great many ideals and even more means at his disposal.

On the ground, however, things went quite differently. His ideals turned out to be rather pecuniary, his means often extremely shaky. At first, the Congo Free State existed only on paper. Even by the end of the nineteenth century, Leopold had no more than fifty stations, each of which ruled—in theory, at least—over a territory the size of the Netherlands. In actual practice, large parts of the territory eluded his effective occupation. Katanga was still largely in the hands of Msiri, Tippo Tip was still lord and master to the east, and various native leaders refused to bow to his authority. Until the very end of the Free State itself, the number of representatives of his government remained limited. By 1906 there were only fifteen hundred European state officials among a total of three thousand whites (the rest were missionaries and traders) in the country.[1]

Indicative of the sketchy state of affairs was that no one knew exactly where the borders of Leopold's empire lay. Least of all Leopold himself. When it came to those borders, he had a tendency to change his mind. Before the Berlin Conference that, of course, was understandable: nothing had as yet been fixed. On August 7, 1884, he, along with Stanley, had drawn up a preliminary sketch of the future territory at the royal villa in Oostende. Stanley unfolded the very tenuous map he had made after his African crossing, a large blank roll showing only the Congo River and its hundreds of shoreline villages. It was to this sheet of paper that the king and Stanley added a few hastily penciled lines. It could almost not have been more arbitrary. There was no natural entity, no historical inevitability, no metaphysical fate that predestined the inhabitants of this area to become compatriots. There were only two white men, one with a mustache, the other with a beard, meeting on a summer afternoon somewhere along the North Sea coast to connect in red pencil a few lines on a big piece of paper. Nevertheless, it was that map that Bismarck would approve a few weeks later and that would set in motion the process of international recognition.

On December 24, 1884, the king pulled out his pencil once again. He was on the verge of losing to the French the area to the north of the Congo's mouth, a region for which he had entertained great hopes and that he would surrender only with pain in his heart. As compensation, on that dark day before Christmas, he set about annexing another area: Katanga. Quite literally, annexation in this case meant poring over a map and thinking, like that mythical first landowner of Jean-Jacques Rousseau's: "Ceci est à moi" (This belongs to me). Not a single soldier was involved. It was a game of Risk, not of Blitz. So Katanga it was, Katanga it would have to be. Leopold was not particularly delighted. Katanga consisted of savanna, with less ivory to be found than in the rain forest. Only decades later would it become clear that the earth there abounded in ores and minerals. But Leopold simply doodled it into the picture.

In 1885 France and England approved the new borders. Which is not at all to say, however, that they would be incontestable from then on. During the twenty years that followed, a great many territorial disputes would arise: with France about Ubangi, with England about Katanga, and with Portugal about Luanda, the area that bordered on Angola. And as though that were not enough, during the first years of the Free State, Leopold tried to press on to the headwaters of the Zambezi River, to Lake Malawi, Lake Victoria, and the headwaters of the Nile, in fact to the whole area to the east of his holdings. His lust for land was insatiable. Why all the hurry? His African state was still extremely shaky. Wouldn't it have been better for him to clean up his own internal backyard before thinking about moving on to something else? After all, his means were considerable but they were not inexhaustible, were they? All true enough, but Leopold realized that soon there would no more opportunities for new acquisitions in Central Africa. An understandable concern. As easily as he had swept together hundreds of thousands of square kilometers before 1885, as ploddingly did that go afterward. Until 1900 he kept alive the hope of further expansion, but none of his plans succeeded. He had his sights set on the Nile in particular, and even made a grab for the Sudan, where he apparently hoped to become a new-fangled pharaoh. But Uganda and Eritrea attracted him as well. And mean-while, outside of Africa, he also lay in wait to appropriate the Philippines or parts of China. . . .

Congo's definitive borders would be established only in 1910. But then what is definitive? In 1918 the map was altered anew when Belgium received Rwanda and Burundi (formerly part of German East Africa) as mandated territories. During World War I, the eastern border had already been tam-pered with. A piece of Katanga was added in 1927. And even as late as 2007, discussions were still going on concerning the exact border between Congo and Angola.

TODAY, THE CONGO FREE STATE is notorious not so much for its vague borders as for its crushing regime. And rightly so. Along with the turbu-lent years before and after 1960, the year of independence, and the decade between 1996 and 2006, that period is seen as the bloodiest in the nation's history. But during the first few years, there was none of that. From 1885 to 1890 history ran its course in relative calm. Europeans were still engaged primarily in trading in ivory, and made use of the stations established by Stanley beginning in 1879. The governing of the state itself remained a rather minimalist affair.

Yet things were not all sweetness and light. Some areas were marked by outspoken native resistance to the new authorities, but that resistance did not essentially differ from what had been seen in the past. Expeditions were attacked, local chieftains refused to fly the newcomers' flag, and they besieged government stations. It was hardly coincidental that these acts of resistance took place largely in areas on the periphery, such as Kwango in the southwestern Congo, parts of Katanga in the south, and Uélé to the north-east. There the traditional power structures had been less eroded by the tur-bulent events along the river, there one still had relatively robust empires. Which were, as that is called, then forcefully "pacified."[2]

Leopold II invested a great deal of his own money in expanding his state, particularly in the new outposts, which helped to extend his grasp on the territory. It constituted, however, an extremely light form of governance. He set up no bureaucratic state apparatus, but only created the minimum conditions needed to allow free trade to flourish. Costs were to be kept as low and profits as high as possible. His imperialism was based on decidedly economic motives. The revenues for which he hoped were not meant to develop the Free State, but to be funneled off to Brussels. Later that was often seen as avarice, and not entirely without good reason. Yet it is only part of the story. Leopold used one of his states, Congo, to provide the other, Belgium, with new élan. He dreamed of economic prosperity, social stability, political grandeur, and national pride. In Belgium, that is—near was his shirt, but nearer yet his skin. To reduce one's view of his enterprise to a case of unbridled self-enrichment would be to do injustice to the national and social motives for his imperialism. Belgium was still young and unstable; it had lost huge sections of its territory in Dutch-Limburg and Luxembourg, the Catholics and liberals of his day fought each other tooth and nail, and the proletariat was beginning to stir: altogether, this formed an explosive cocktail. The country was like a "boiler without a safety valve," Leopold thought.[3] Congo was to become that safety valve.

The place in Congo where the new state was most highly visible was, without a doubt, the town of Boma. In 1886 it became the country's first real capital. Today, time there seems to have stood still. There are few places in Africa where nineteenth-century colonialism has remained so visible. In 1926 it surrendered its status as capital to Léopoldville, and as port of call it was eventually outclassed by Matadi. To walk through Boma today is to wander through time. At the waterfront is an enormous baobab that has been poking its gnarly limbs at the sky for centuries. A little farther along one finds the old post office, dating from 1887. Like almost all buildings from that day it stands on cast-iron pilings, to prevent rotting and to ward off insects. Atop a little hill nearby is "the cathedral," a pompous name for an extremely modest chapel built entirely of iron. The walls, doors, and windows consist of prefabricated plates that were sent from Belgium in 1889 and assembled on the spot, as a sort of IKEA furnishing *avant la lettre*. But most impressive of all is the governor general's residence from 1908. That too was built on iron pilings and constructed of prefab metal plates; around them, however, was built a beautiful wooden façade featuring a spacious verandah, high-beamed rooms, plaster ceilings, and windows of skillfully cut glass. It was from here that the Free State was run: the governor general's instructions were given to his provincial governors, who passed them in turn to their district commissioners in the interior, and from there they went to the *chef de secteur* and, further down, to the *chef de poste*. At Boma, letters were postmarked, statistics compiled, and soldiers trained. Cases were judged and a regime was founded. It served, in truth, as the hinge between Congo and the outside world. And it was here, a few decades later, that the inhabitants, who had already grown accustomed to steamboats, printing presses, and marching bands, saw the strangest thing that had ever been seen: an automobile. A British industrialist had shipped in an eight-cylinder Mercedes with spoked wheels, followed a few years later by a LaSalle from the United States. "For his wife," the people of Boma will tell you today; the wrecks of those old-timers, the first two cars in Congo, still stand rusting beneath a lean-to just outside of town.

But it was not just the inhabitants of Boma who came in contact with the European way of life. Here and there around the country young Congolese were entering service as "boys." In that way they literally made their way into the white man's home, kitchen and bedroom. They saw that he did not sleep on a mat, but a mattress. They collected his sheets and his dirty laundry. They scrubbed sweat stains out of shirts and urine stains out of underpants. Hanging on the walls they saw photographs, of which they later told their friends: "When I was in the white man's house I saw people hanging straight up on the wall, but they couldn't speak, they remained silent. In fact, those were the dead. The white man had taken them prisoner."[4] It was an awkward acquaintanceship. Boys wondered why their boss swallowed pills every day and did not eat with his hands, why he became so angry about a spot on his glass, and why he always left the fish's head on his plate. (Wasn't that the tastiest part, after all? How wonderful to feel the little bones crack between your teeth and hear the eyes pop in your mouth.) In the evening they saw him writing beside a lamp, smoking a pipe or putting on a pair of spectacles. How peculiar, how peculiar it all was. The boy learned to cook in the Western fashion, he set the table, washed the dishes, and made the beds. He made sure that while doing the ironing—another bizarre habit!—no holes were burned in the linen. When the boss went on a journey he was often allowed to go along, and so found himself in places he would never have known otherwise. A good boy often received kudos, sometimes a beating, but rarely autonomy. Leopold had sworn to put an end to the Swahilo-Arab slave trade, but in essence there was no difference between the life of a Central African domestic slave on the Arab peninsula and a boy in the household of a European official in Congo.

And this was the life Disasi Makulo had led since Stanley entrusted him to Anthony Swinburne. He could have done worse, for Swinburne was patient and amiable and the station at Kinshasa comfortable and lively. Neither boy nor boss, however, however, could have guessed that their lives were about to be brutally upturned. But Leopold II had decided to do just that.

The country of Belgium may not have been directly involved in setting up the Free State, but the king increasingly began sending his subjects off to Congo. Belgian officers led expeditions, Belgian diplomats manned a consulate for him on Zanzibar, and the stations along the river were placed under the leadership of Belgian citizens. The British helpers appointed by Stanley began to be phased out. English as administrative language made way for French, although place-names such as Beach, Pool, and Falls remained. Words like *steamer* and *boy*, due in part to the influence of British and American missionaries, never disappeared. In Lingala, the language spoken along the river, a book was by then referred to as a *buku*, and the verb *beta* meant "to hit," a bastardization of "to beat."

Once the Berlin Conference was over, Leopold II had less and less use for the British. What's more, he had been forced to promise the French that Stanley—in their eyes the devil incarnate who had thwarted "their" Brazza—would never be given a senior post in the Free State.[5] In 1886 Leopold instated Camille Jansen as first Belgian governor general of Congo. The auspiciously inaugurated Association Internationale du Congo was gradually becoming an owner-run business with Belgian personnel. Among the three thousand whites who remained in Congo in 1900, seventeen hundred were Belgians.[6] They were well aware that one could easily lose one's life in this place, but they hoped above all to garner honor, fame, and money. This budding Belgian enthusiasm is not very well known. The lack of imperial zeal in the king's European homeland was not due to the fact that the monarch stood alone at the helm of his overseas enterprise. He may never have succeeded in galvanizing a broad cross-section of the Belgian people, but an urban elite of officers, diplomats, jurists, and journalists did warm to his plans. While in the provincial towns young men from the lower middle class dreamed of a life more heroic and glorious as a soldier, government agent, or missionary.

For a person like Anthony Swinburne, this Belgification was a particularly bitter pill to take: the man who had kept Kinshasa out of French hands and so hoped quietly for an appointment as provincial governor received a pat on the back and was then sent packing.[7] For his two boys, however, his dismissal was a chance in a million. Their master's employment was terminated in 1886; Swinburne headed back to England and took them with him. And so Disasi Makulo, once Tippo Tip's slave and destined to be shipped to Zanzibar and from there to the Arab peninsula or India, suddenly found himself in Europe.

> It was horrible to see the big boat and the sea for the first time. After we had lifted anchor to cross the sea, we felt ill and had to vomit. Despite all the care with which we were surrounded, we barely recovered during the entire crossing. After many days we arrived in England. Europe seemed to us like a dream, we could not believe that we were in the real world! The huge buildings, the streets that were paved so neatly, the cleanliness one found everywhere, the houses so well decorated inside. In the house where we stayed there was a sort of cupboard in which food could be kept for a long time without spoiling. The lives of the whites were truly very different from our own. Every day we were happy, the only thing from which we suffered was the cold. But they had us wear warm and heavy clothing.[8]

With that, Disasi became one of a handful of Congolese—a few hundred individuals at most—to arrive in Europe before 1900. Missionaries occasionally brought a few children back with them, to serve as teaching material during their lectures, and promotional material during their collection drives. To whet the young Africans' appetites for industry and diligence, they were taken along to shipyards, coal mines, and glass-blowing plants. A tiny group went to study at the Congo Institute in Wales. There, at Colwyn Bay, the British Baptist William Hughes had started a training institute for young Congolese with a calling: twelve of them left home for Europe between 1889 and 1908.[9] A group of around sixty boys and girls went in the year 1890 to the eastern Flemish village of Gijzegem, where they received schooling from the Reverend Father Van Impe. The boys boarded at the schools, the girls were spread over convents in Flanders. They wore blue and white sailor suits.[10] Others Congolese visitors ended up in ethnographic exhibitions; Pygmies in particular were a popular attraction at circuses and fairs. At the Antwerp World's Fair in 1885 one could view a "Negro village" with twelve Congolese. By 1894 their number had grown to 144. But the largest group of natives, some 267 of them, traveled to Tervuren in 1897 as exotic features in the colonial exhibition there. They built huts beside the park's pond and during the day played at being themselves, stared at by hundreds of thousands of Belgians who had come to see that for themselves: a Negro.

In addition to the wonders of the Western world, they were also regularly confronted with the inclemency of Europe's moderate climate. During the wet summer, seven of the delegation members to Tervuren died of influenza. Lutunu, a former slave who, like Disasi, had become a boy to a white agent, left for England with a few other children in the winter of 1884–85, along with the British Baptist Thomas Comber. Some of them developed earaches and sore throats, but refused to use Western medicines, which they believed caused one to go blind (true in any case of the quinine they had seen whites use to combat malaria in the tropics). Even though there was no respectable *féticheur* (traditional African healer) among them and no palm oil suitable for ritual use in all of Liverpool, they still succeeded in healing each other in the traditional manner.[11]

In 1895 a young man by the name of Butungu left for England with John Weeks, another Baptist. Butungu had received schooling at a mission post in the equatorial jungle along the river and could read and write. He too came home with a pile of tall tales about steamboats, seasickness, and salt water: stories about the sea, in other words. He recorded them in Boloki, his native language. It is the only known text by a Congolese from the nineteenth century.[12]

> And I saw so many things: sheep, goats, cows, you name it. There are all kinds of things in their country. If you don't believe me, just look at their cities, that's the way they are. And their villages are so clean. One day we went to a rifle show, with bullets fired in the air that exploded in light. . . . And when the cold arrived, I saw things like flakes, the flakes from the molondo tree. And I asked: "What is this?" The people told me: "That is snow." At our feet were hailstones, but hailstones are hard and this was soft. That was also the end of the year's circle. For six months there is only cold, and for the other six the sun shines. . . . So their country is not like ours at all. I did not see a single snake. The little animals they raise and that we have in our country as well do not live in the people's yards, although they too have cockroaches, rats, and cats. But they have built barriers around all the animals. If you go through

one of those barriers you can see all the animals, and even there the
people have built houses for the animals. Only the horse is allowed to
move about freely.

Butungu stayed in England for almost a year and a half. In addition to
farms, snow, and fireworks, he also saw London and "the many things the
people there have made." That was all he said about it. The homecoming to
his own village, however, he described most touchingly:

> I went to the Reverend's house and talked to myself. I looked around
> and saw my mother and I said: "That is really my mother." I went to her
> and called her, and she said: "Where is Butungu?" And I replied: "It is
> me." And she said: "So you have come back." I said: "Yes." We walked
> through the village and many came out to greet me.

Anyone who had traveled to mythical Europe had to tell his story a
hundred times over. Parents and children clung to his every word, family
members begged him for details. Only a tiny number of Congolese had been
there, but the whole village eavesdropped as he talked about his first train
trip: "The train went as fast as a fly, it was unbelievable!" Those who had
stayed at home saw the strangest objects up close. In addition to suits and
shirts, those who returned from Tervuren brought back bowlers, brooches,
walking sticks, pipes, watches, armbands, and necklaces as well as hammers, saws, planes, axes, fishhooks, coffee pots, funnels, and magnifying
glasses with which to light fires. Many of them had also bought a dog in the
village of Tervuren. Young Lutunu, after his journey to England, had even
sailed to New York, where he stayed with a missionary's sister. When he left,
she gave him an extremely peculiar present: a bicycle! Lutungu took it back
with him to Congo and so became Central Africa's first cyclist.

It was handy, many whites reasoned, to have your boys with you in
Europe. Not only did it draw a lot of attention, but it was also educational
for the young people themselves. Still, one had to be careful. Before you
knew it, a young man might learn too much during his journey. The British
Baptist George Grenfell traveled with a boy and a girl of nine to England,
but warned his hosts: "If we shower them with attention, we shall have
trouble relegating them to their former status once we return."[13] The Belgian socialist Edmond Picard mocked those colonials who paraded about

in their home country with their "model servants": "Often it does not take long before that luminous person drives to despair his incautious master, who has introduced him all too intimately to our refined civilization and our chambermaids."[14] The number of Congolese able to travel to Europe would always remain limited. Travel did not necessarily make a person more licentious, but it apparently made one less docile. That would manifest itself later on. Congolese veterans who returned from World War II in 1945 began to resent the colonial authorities. The intellectuals and journalists who returned in 1958 from the world's fair in Brussels began to dream of independence.

Disasi Makulo returned as well. Swinburne no longer worked for the Free State, but was still bound and determined to make his fortune as a trader in Congo. Along with Edward Glave, another Brit expelled by Leopold, he began buying up ivory. As soon as he arrived in Kinshasa, Congolese people began offering it to him. At a certain point there were no less than sixty tusks of ten to fifty kilos (22 to 110 pounds) each around his house. As soon as Swinburne was able to obtain a steamboat of his own, however, he sailed upriver; there he could buy up ivory for less than a third of the price.[15] And he was not the only one, not by a long shot. Riverine commerce, the exclusive domain of local carriers for almost four centuries, was now taken over entirely by Europeans. In a twinkling Leopold's free trade had devoured the old trading network. European trading posts and storehouses arose. Ocean steamers docked at Matadi and hoisted the ivory on board with cranes. In Antwerp there were warehouses packed full of tusks. In 1897, 245 metric tons (about 270 U.S. tons) of ivory were exported to Europe, almost half the world's production in that year. Antwerp soon outstripped Liverpool and London as the global distribution center for ivory.[16] Pianos and organs everywhere in the West were outfitted with keys of Congolese ivory; in smoky saloons the customers tapped billiard balls or arranged dominoes that were made from raw materials from the equatorial forest. The mantelpieces of middle-class homes sported statuettes made of "elfin wood" from Congo; on Sunday the people went out strolling with walking sticks and umbrellas whose handles had once been tusks. All this international free trade, however, stole bread from the mouth of local commerce.

It was primarily children and teenagers who became closely familiar with the European lifestyle. Young men got to know it as boys, the girls as

menagères. Despite the name, the *menagère* was less concerned with managing the household in the classic sense than with managing the hormones of
her employer. Because European women were considered unsuited to life in
the tropics, while at the same time it was recognized that an all-too-lengthy
period of sexual deprivation was bad for the white man's zest for work and
life in general, a great tolerance arose toward forms of concubinage with
a native woman. In 1900 there were eleven hundred white men in Congo
and only eighty-two white women, sixty-two of whom were nuns.[17] A great
many of the men therefore developed long-term, intimate relations with
one or more African women. Some of them spoke openly of their *menagère*
as "my wife," others developed a profusely libertine lifestyle. Often the
girls chosen were very young, twelve or thirteen; often the line between
affection and prostitution was unclear; often pure lust went hand in hand
with tenderness. Yet the relationships always remained asymmetrical. The
menagère slept under the same mosquito net as the white man, but she often,
voluntarily or not, did so on a mat on the floor.

The missionaries, of course, viewed this with dismay. Church attendance by Europeans in Congo, however, was many times lower than in
Europe itself: the minuscule cathedral at Boma was more than large enough
to accommodate the crowd on Sunday mornings. The Roman Catholic
rites were observed only at funerals. Disasi Makulo saw this with his own
eyes. In 1889, less than three years after his trip to Europe, his master Swinburne came down with gastric fever. Horrible sores covered his legs and he
declined visibly. Disasi and a friend fashioned a litter from hammocks and
started off with him for Boma. Along the way they stopped at the mission
post at Gombe, where the British Baptist George Grenfell attended to the
sick man for two weeks. When that did not help, they set off again on their
gruelingly long journey. At the Dutch trading post at Ndunga run by Anton
Greshoff, father of the writer Jan Greshoff, Swinburne died. He was only
thirty. "The whites we had met at the trading post hastened to prepare the
funeral. All of the whites in lovely suits and a crowd of blacks attended the
funeral," he noted. And he added: "That day we found the world the bitter-
est place of all, and our thoughts froze when we did not know whether our
lives would be subject to any further support."[18]

After the funeral Greshoff decided to bring the two boys back to Grenfell's mission post. Grenfell was a living legend who owed his reputation
to his remarkable combination of enthusiastic evangelization and a feverish

urge to explore. He had arrived in 1879 as one of the very first missionaries in Congo and died there in 1906, seemingly immune to all tropical illnesses. Beginning in 1884 he began piloting his steamboat *Peace* up countless, previously unexplored tributaries of the Congo. Within two years he covered twenty thousand kilometers (over twelve thousand miles) on the Congo, the Ubangi, the Kasai, the Kwango and other side rivers. He drew maps and set up posts. He was in fact, after Stanley and Livingstone, the third greatest among Congo explorers. Disasi Makulo had been enslaved by Tippo Tip, had been bought by Stanley, and had served as boy to Swinburne. Now, at around the age of eighteen, he and his friend became helpers to the most celebrated of all nineteenth-century missionaries in Congo.

> Grenfell received us as though he had known us for a long time. He took us along in his boat and, look, there we were on the river again. We made many trips on the river and the side rivers. At first we didn't understand the purpose of all that traveling back and forth. Only later did he explain to us that it was in order to explore the rivers and to study the various areas, so that they could set up mission posts there.[19]

The missionaries were dauntless. While many European civil servants were sowing their wild oats, the missionaries acted against what they saw as pernicious native customs such as human sacrifices, trial by poison, slavery, and polygamy. But all that, of course, was subjective. Many natives were not at all anxious to be Christianized. Disasi Makulo knew all about it:

> When the boat approached Bolobo, a huge crowd of villagers came and stood on the banks. They shouted and waved knives, spears, and weapons, because they thought we had come to wage war. To show them that we had not come to fight, Mrs. Bentley [the wife of another missionary] picked up her baby, held him in the air and showed him to the crowd. It was the first time the people had seen a white woman and a white baby. Curious now, they put down their weapons and came to us, whooping with enthusiasm, to admire these creatures. The boat landed quietly.[20]

Bolobo became the site of one of the most important missions. In the absence of white babies, the Protestants also availed themselves of native

children. Grenfell always took a few of "his" children along on his forays. They chopped wood for the steamboat, held the rudder, and served
as interpreters. As freed slaves they often spoke the language of their
native region, where the Christianizing had yet to begin. At Yakusu, the
missionaries' activities clearly benefited from the presence of a converted
native girl. The villagers recognized her tribal tattoos and knew that she
was one of them.[21] The spreading of the Word was therefore not simply a
matter of white versus black; black people too evangelized and played an
important role in the oncoming religious turnabout. Disasi Makulo also
became such an intermediary. Baptized in 1894, he helped with the Christianizing, and not without success. In one of his letters, Grenfell wrote:
"Disasi . . . worked well and created quite a favourable impression among
the natives."[22]

During his travels with Grenfell, Disasi returned to his homeland for
the first time. The reunion with his parents was gripping. The gong sounded
out news of the lost son's return. Relatives immediately slaughtered a few
goats and dogs, and proposed en passant that two slaves be sacrificed as
well. "When I saw that I was deeply indignant that such barbaric customs of
slavery and cannibalism continued to exist within my tribe." He protested
vehemently and even released the slaves, to the bewilderment of his former
neighbors. "Many of them wondered in amazement why I felt pity for these
slaves. Others accused me of having prevented them from eating the delicious flesh of a human. The dancing went on for two days, without a stop."[23]
Disasi Makulo was now a man caught between two cultures, loyal to his
tribe and to his new faith.

He was not the only one to find himself cut loose in a new moral universe. The first inhabitants of the missions were often children whom the
authorities of the Free State had taken away from areas marked by conflict.
They did not always come from the slave traders; some of them were victims
of tribal violence. Lungeni Dorcas, a girl from Kasai, was captured by warriors from the neighboring Basonge tribe. She had watched as her mother
and brothers were beaten and her youngest brother, still an infant, was
pounded against the ground till dead. Hers is one of the few female voices
known from that period:

> After a few days we heard that a white man would come to fight against
> our enemies and free us. Our captors, having heard that, began sell-

ing their prisoners. Then the white man came, he was an authority from the government and accompanied by a great many soldiers. He summoned the head of the village and said that he wanted to free all of the prisoners, including those of his subjects. He had them open a chest containing all kinds of beads, necklaces, *mitakos* [copper currency ingots], and textiles. We were struck by the beauty of those objects and were introduced to the Europeans. After he had freed us, he took us to Lusambo. That day a boat arrived there, steered by a white man. Our government official handed us over to him and it was he who took us to the Protestant mission at Kintambo. There we met many boys and girls from different tribes, all of them bought free like us.[24]

The importance of this account can hardly be overestimated, for it shows us in detail how mission posts acquired their first believers through the government's intervention and how that led to the establishment of the first interethnic communities. Young people totally unfamiliar with each other's language and culture suddenly lived together closely. The missionaries even went a step further. As the children grew up, the Europeans became multicultural matchmakers. Once again, Lungeni Dorcas: "To save us from all manner of complications in the future, the missionaries wanted us to marry only young Christians who had been raised by them as well." In her case, this meant marriage to an old acquaintance: "That was why they arranged for me to marry Disasi. And that is what came to pass."[25]

One generation earlier it would have been unthinkable for her to marry a man born eight hundred kilometers (five hundred miles) away; now she bore him six children—three boys and three girls. The mission deemphasized tribal ties, eased people away from their villages, and promoted the nuclear family as the alternative.

As a newlywed, Disasi remained deeply distressed with *la terrible barbarie* (the terrible barbarism) in his village.[26] He therefore proposed to Grenfell that he begin a mission post of his own. In 1902 he set up the mission at Yalemba, one of the first black-run posts in Congo. Grenfell came by to visit on occasion. After his many wanderings, Disasi had come home again:

The objective of my return was to help my own, to protect them and bring them the light of civilization. . . . I had decided that all the inhabitants of my village would come to live with me at the mission. I began with the members of my own family: my father, my mother, my sisters, my brothers, my nieces and nephews. At first, the other villagers did not want to leave their village. Only later, after great effort, was I able to persuade them to leave and settle down with me.[27]

Black catechists formed a bridgehead between two worlds. Old Nkasi had already told me something along those lines during our conversations. His father's youngest brother, Joseph Zinga, went after all with the Protestant missionary Mr. Welles to Palabala to become a catechist. That was how he had absorbed European ideas and knowledge and became familiar with our Christian calendar. "It was because of him that I know that I was born in 1882," Nkasi had said.[28]

MEANWHILE, THE CATHOLICS were also on the move. After early efforts by the Holy Ghost Fathers and the White Fathers, following the Berlin Conference, the Catholic mission work quickly gathered momentum. Now that Leopold II had withdrawn from his international association, he granted preference to Belgian missionaries, who were without exception Catholic. In 1886 Pope Leo XIII, who was on very good terms with Leopold, announced that the Congo Free State was to be evangelized by Belgians. The White Fathers, originally a Franco-Algerian congregation, now dispatched only Belgians. Young Scheutists and Jesuits, followed by Trappists, Franciscans, fathers of the Sacred Heart, and sisters of the Precious Blood, left from countless villages and towns throughout Belgium. They divided Congo's interior neatly among themselves. Protestant missionaries from England, America, and Sweden continued to be active, but lost some of their influence: they were forced to work in accordance with the dictates of the new country and to learn to live with the badgering of Catholic missionaries who absconded with their converts.

While the Protestants focused largely on the individual, based on their doctrine of the personal connection between Christ and the believer, the Catholics went in search of groups from the very start. For them, the collective experience of faith took pride of place. But how, for heaven's sake, did one go about finding groups like that? Once again, children provided the solution. Their first followers, like those of the Protestants, were often child slaves freed and entrusted to them by the state. At the mission post in Kimuenza, for example, the Jesuits began in 1893 with seventeen freed slaves, twelve workers from the Bangala tribe, two carpenters from the coast, two soldiers with their wives, and eighty-five children whom the state had "confiscated" from the Arabized slave traders. Together they formed *une colonie scolaire* (educational colony). Two years later, in April 1895, there were already four hundred boys and seventy girls, and even forty toddlers between the ages of two and three. In 1899 the mission had a church that could hold fifteen hundred worshipers, with three stained-glass windows and two bronze bells, one weighing two hundred kilos, the other six hundred (440 and 1,320 pounds). The bells had been cast in Belgium. One could hear them peal as far away as a two-and-a-half-hour walk from the mission post.[29]

Government help, therefore, was essential. But the intertwining of church and state went much further than the acquisition of new converts. When the Kimuenza mission was founded, an official called together the village chieftains to explain to them that the mission enjoyed the special protection of the state, and that they should not hesitate to sell to them chickens, manioc, and other provisions.[30] The state even saw to the maintenance of the school, on condition that four out of every five children completing their studies enroll in the Force Publique, the Free State's army! This much was clear: the Jesuits fought for Jesus, but also for Leopold. The school was run like a Belgian military academy.

> The little black boys had to salute and even march. . . . The daily routine is no different. At five-thirty, get up quickly to the bugle's call, wash quickly, then prayer: Pater, Ave, Credo in Fiote [Kikongo]. After prayers, breakfast. All gather on the square in front of the building that serves as refectory. The children line up. The sergeant screams: "Attenshun!" Silence immediately descends over the ranks. "Right, face!" The little column starts to move and lines up, ramrod straight and silent, at the tables. "Sit!" and all are seated. Then comes the order for which everyone has been waiting impatiently. "Eat!"[31]

After some time had passed, such *colonies scolaires* also proved to have their limitations: the influx of slave children was not endless, and, despite all the bell ringing, conversion among the neighboring "heathens" did not continue once all the alumni had vanished into the barracks. That was why the Jesuits followed up with the system of the *fermes-chapelles* (chapel farms). Close to an existing village they would establish a new settlement where local children learned to pray, read, and garden in relative isolation. The emphasis lay precisely on that relative isolation: they were to be kept away from their familiar culture long enough to prevent them from backsliding into "heathendom." "Civilizing these blacks, while leaving them in their own surroundings, is like reanimating a drowned man while at the same time holding his head under water," was the subtle simile applied.[32] At the same time, however, their new status as well-fed and well-dressed little catechists had to remain visible to the other, seminaked villagers: that, after all, provoked useful feelings of envy. The mission became a means to material welfare. For every child he brought to the chapel farm, the village chief received a gift. Little wonder then that one of them is known to have said: "White men, come to honor my village, build your house there, teach us to live like white men. We will give you our children and you shall make of them *mindele ndombe,* black white men."[33]

Mission posts became large-scale farms and display windows for a different way of life. The number of baptisms skyrocketed. Between 1893 and 1918 the Jesuits alone made some twelve thousand converts. At their Kisantu post in 1896 they had fifteen cows; by 1918 there were more than fifteen hundred. There was a carpenter's shop, a little hospital, and even a printing press.[34] Those who had finished their schooling stayed at the mission post and married. They worked as farmers, carpenters, or printers, and started families. Like the Protestant missions, these newly formed villages did not fall under the authority of a native chieftain. The traditional village with its countless contacts and variegated forms of solidarity lost ground to the monogamous family. Other religious orders adopted the chapel farm formula as well, but the system also met with fierce criticism. To inflate their baptismal registers, the missionaries were extremely prompt in labeling children as "orphans," even when there were still plenty of relatives to raise them in the African tradition. Whenever sleeping sickness broke out, children were plucked en masse from their villages. "The result was disastrous," one contemporary realized, "and it made the natives hate us."[35]

The missionaries' affability had its darker facets. As friendly as their smiles when dealing with the local population, just as underhanded were their methods at times. The Belgian missionary Gustaaf Van Acker explained how he, as White Father, dealt with the "talismans" of the native religion ("bones, hair, animal droppings, teeth, hundreds of filthy objects and more") that he found in "hovels" along the road":

So as not to annoy the people and to safeguard our studies, we could not act against all that diabolical filth; we had to smother our hatred and only on occasion, when we were alone, could we apply an enraged stomp and leave that mess in ruins. I hope that someday soon we may act more openly and in all of Oeroea, all its villages, along all its streets, replace these infernal signs and diabolical knickknacks with the True Cross. Oh my! So much work for so few champions of the Cross!"[36]

Some missionaries destroyed thousands of cult objects in this way.

In Boma I had the privilege of speaking with a few old inhabitants. Victor Masunda was eighty-seven and blind, but he remembered his father's stories with startling clarity.[37] "The first missionary my father saw," he said as we sat together drinking Fanta in his darkened living room, "was Père Natalis de Cleene, a huge man from Ghent, a Scheutist. He was the one who set up the *colonie scolaire* at Boma; it replaced the mission established by the Fathers of the Sacred Heart. Leopold had asked the pope for Belgian missionaries, and the Scheutists arrived."

He knew his history. What's more, the name of the missionary in question was completely accurate: I found it later in registers kept by the Scheutists. De Cleene was a famous missionary.

"Four or five years later that priest left town on horseback and set up the Kango mission in the Mayombe jungle. My father and mother lived in the jungle as well. Papa was fifteen. He was baptized in December 1901. He belonged to the second batch of students. His number was 36B. My mother was baptized in 1903. They married three years later. They left their village and settled at the mission's work camp."

I asked Masunda why they did that.

With an outburst of laughter still intended to cover his shame, he said: "In the jungle there were no chairs to sit on, like at the mission, the people there still sat on *tree trunks*! All they ate was bananas, yams, and beans. But one of the priests gave my father a rifle! Then he could hunt antelope, wild swine, and beavers!" More than a century later, he still sang the mission's praise: "In the jungle they wore rags and tatters, but at the mission my father was given a pair of short trousers and my mother a *boubou*, a little smock. He even learned to write a bit. There were children there from all over the place. His native tongue was Kiyombe, but that's how he also learned Lingala, Swahili, and Tshiluba."

A few days later, in the shade of a little mango tree, I spoke with seventy-three-year-old Camille Mananga. He was blind as well and came from May-ombe too. He told me not about his father, but about his grandfather. "He never wanted to be baptized. He climbed in his palm tree and made palm wine. He had four wives and a great many children. The missionaries felt that he should keep only one of them, but he felt responsible for all of them. And he never argued with them."[38] Converting adults was clearly a more difficult task.

The Protestant evangelists had looser ties with the state, but were not entirely independent of it either. In 1890, when the Free State requisitioned Grenfell's steamboat for the war against the Afro-Arab traders to the east, he protested vehemently. How could they think that his *Peace*—the name alone said enough—might be used to wage war! One year later, however, he was all too pleased to accept a personal commission from King Leopold: he was charged with surveying the border between the Free State and the Portuguese colony of Angola. That area was not only subject to international conflicts, but was also the site of the most violent uprisings against the new regime. So he, Grenfell, a British cleric, set out with an escort of four hundred soldiers from the Force Publique to chart and pacify the region. He was given a mandate to sign treaties and establish borders. Disasi Makulo accompanied him on that exhausting trip overland through hostile territory, "the most painful and dangerous journey we had ever made." He too noted the highly overt interweaving of mission and state: "The government provided our military equipment and porters." Disasi Makulo wore the uniform, the plus fours and fez, of the Force Publique.[39]

THE FINAL WAY in which young people came in contact with the Free State was through the armed forces. The Force Publique, a colonial army under the firm leadership of white officers, was set up in 1885. Most of those officers were Belgian, but there were also any number of Italians, Swiss, and Swedes. Without exception, the most prominent and highly valued foot soldiers were the Zanzibaris, men who had accompanied the explorers on their journeys, and then mercenaries from Nigeria and Liberia. These West Africans had a reputation as trustworthy and courageous soldiers. The first group of ten Congolese was conscripted in late 1885. They had been recruited in the rain forest by the Bangala, who took them to Boma. The Bangala themselves were known as a warlike tribe, and a great many of them would also be recruited in the years to come. As a result, their language, Lingala, began spreading rapidly: it would one day become the most important in the west of the country.

As the Free State's capital, Boma was also the country's first garrison town. It was there that young people, previously unable to tell time, learned to live by the minute. The recruits arose at six o'clock and went to bed at nine. The bugle's blare divided the day into drill, roll call, parade and rest. Military discipline was hammered in. The recruits learned to shoot, clean their rifles, march, and even play martial music. Yet even this stringent military discipline could not entirely disguise a large component of clumsiness. The cavalry had no horses, but donkeys—seventeen, to be exact. The artillery had a number of Krupp cannons, but no moving targets on which to practice. The soldiers had to make do with aiming at and firing upon herds of antelope.[40] Nevertheless, the Force Publique would become a factor to reckon with. During the first few years of the Free State, King Leopold dedicated half his available budget to the army. For many young men it formed the most direct and drastic acquaintance with the state. In the year 1889 there were fifteen hundred recruits; by 1904 that number had risen to seventeen thousand. During the final days of the Free State, the Force Publique had twenty-five thousand Albini rifles with bayonet, four million rounds of ammunition, 150 cannons, and nineteen Maxim machine guns, making it the largest standing army in Central Africa.[41] Unlike in Belgium, the young Congolese recruits were allowed to take their wives with them when they entered military service. The wives even received a modest stipend; an allowance was also paid for any child that might come along. In this way the army, like the missions, promoted monogamy and the nuclear family.[42] True families of military careerists arose.

In Kinshasa in 2008 I met Eugène Yoka, who had been an air force colonel for decades, back when the national armed forces still had planes. In Mobutu's day he had been part of the inner circle of pilots who flew French Mirage fighters in formation above the capital during national parades. His father, he told me, had been a professional military man as well and had experienced World War I. His grandfather, a Bangala tribesman from Équateur, was one of the first recruits in the Force Publique. Colonel Yoka had two sons, one of whom had joined the army and worked his way up to major.[43] Four generations of dedicated military men, serving the state for more than a century.

The Congo Free State's first five years were the mildest by far. The administration was still quite scanty and there was as yet no widespread terror. But during this period a growing group of natives, mostly children and young people, became directly acquainted with the European way of life in Congo. As boy, as *menagère*, as Christian, or as recruit, they entered houses unlike any they had ever seen, wore clothing unknown to them till then, and tasted unfamiliar foods. They learned French and adopted new ideas. A handful of them even witnessed firsthand how things went in Europe. And some of them actually propagated this new lifestyle, or at least their interpretation of it. Young catechists tried to convince their family members and fellow villagers of the fact that they led heathen lives. Young recruits showed off their uniforms and soldier's pay in their villages. Their wives went with them to the barracks, their children grew up there. A life outside the village began, just like with the chapel farms, where one no longer lived under the authority of the native chieftain, but under a strict European regime. The Free State brought about a deep change in the lives of many.

After 1890, though, things got grimmer. From then on, contact with the Free State no longer meant making acquaintance with another way of living, but a confrontation with violence, horror, and death. What's more, this new type of encounter took place on an exponentially greater scale. There where the Free State had at first impacted thousands or tens of thousands of Congolese, now millions were subjected to the (iron-fisted) presence of the state. To understand this radical turnabout, we must look again to the mastermind of the Free State himself, the one who devised, carried out, profited from and bore final responsibility for the whole enterprise: Leopold II.

The Belgian ruler had acquired his Congo in 1885 by dint of three promises. At the Berlin Conference he had promised both to safeguard free trade and to do away with the slave trade. To the Belgian state he had promised never to request funding for his personal project. And until 1890 he stuck to those promises: free trade flourished, the Belgian treasury was left untouched, and, although the fight against the slave trade had not yet been won, the missions did regularly receive "freed" children as a present. These were, quite literally, lavish promises. To facilitate free trade the king had to expand the necessary infrastructure and administration at his own cost. An expensive business, and one profited from largely by others. Leopold launched into all this in the hope of making serious profits of his own, but was ultimately disillusioned.

Between 1876 and 1885 he invested no less than 10 million Belgian francs, but the revenues in 1886 amounted to no more than seventy-five thousand francs.[44] By 1890 he had already spent 19 million francs on Congo. The huge fortune inherited from his father had gone up in smoke. The king was virtually bankrupt.

It was at that point that he decided to break two of his promises. He solicited money from the Belgian state and set about obstructing free trade with a vengeance. Despite a growing number of "Congo-philes" among an elite of bankers and industrialists, however, the Belgian parliament was not at all inclined to take part in a colonial adventure. Yet it could not simply look on passively as the head of state went bust. Reluctantly, therefore, a loan was arranged: by way of capital injection the king was given 25 million gold francs, later supplemented with an additional 7 million.[45] The country also invested heavily in the construction of a railroad. The agreement was that, in the event of continuing financial malaise, Belgium would take over Congo.

Much more dramatic for the situation in Congo itself was Leopold's unscrupulous series of decrees rendering all lands neither cultivated nor inhabited—including all raw materials to be found there—the immediate property of the Free State. For the European ivory merchants this constituted an enormous setback and for the locals it was an unmitigated disaster. At one fell swoop the king nationalized some 99 percent of the country. A Pygmy who shot an elephant and sold its tusks was no longer supporting himself in legitimate fashion, but stealing from the state. A British trader who bought the tusks was no longer participating in free trade, but receiving unlawfully obtained goods. On paper, free trade continued to exist—of course, it had to—but in practice it was dead as a doornail: after all, there was nothing left to trade, for the state now kept everything for itself.

In bookkeeping terms, Leopold's coup de théâtre was doubtlessly clever and sly, but in terms of culture and community it was a fiasco. He seemed to assume, for convenience's sake, that his subjects in the villages made use only of the spot where their huts and fields were located. In reality, however, the local communities needed areas many times that size. Extensive farming forced them to clear new fields each year, in the rain forest or on the savanna. What's more, entire villages often changed locations. And because no one lived from farming alone, they also made use of vast hunting and fishing grounds. Leopold's decision literally robbed the people of what was dear to them: their land. He had no idea of the extremely complex land-use rights in the region, let alone of local views on collective property ownership. He simply transplanted the Western European concept of private property to the tropics, and with that sowed the seeds of deep discontent within the Free State.

But what about his third promise, the fight against the slave trade? That was the only one he kept and actually focused on more intensely. That struggle, after all, provided him with the perfect cover for his expansionist ambitions. After a major antislavery conference was held in Brussels in 1890, the king stepped up his efforts. The battle was waged largely in three areas: ranging from south to north, those were Katanga, East Congo, and South Sudan. Those regions coincided with the historical spheres of influence of the three most important Afro-Arab slave traders: Msiri, Tippo Tip, and al-Zubayr.

Msiri's empire in Katanga was annexed between 1890 and 1892. Leopold wasted no time, in the knowledge that Cecil Rhodes was advancing on that same area from South Africa. Rhodes, a British imperialist whose megalomania was every bit Leopold's equal, was attempting to link up Britain's colonial holdings "from Cape to Cairo." But Katanga became Leopold's, and this time not merely on the map he had pored over on that Christmas Eve in 1884.

The struggle against the slave drivers in the eastern Congo was more problematic; they were well-armed and wealthy and had a great deal of experience with waging war in that area. In 1886 they had attacked the government post at Stanley Falls. To quiet things down, Stanley—with Leopold's permission—appointed Tippo Tip, the most powerful man in the region, as that post's new provincial governor. For Tippo Tip himself, this resulted in a conflict of loyalties. In a letter to King Leopold he wrote: "None of the Belgians in Congo like me and I see that they only wish me ill. I am starting to regret having entered the service of the Kingdom of Belgium. I see that they do not want me. And now I find myself at odds with all the Arabs as well. They are angry at me, because I deliver more ivory to Belgium than to them."[46] The economic interests of Europeans and Zanzibaris clashed so loudly that a confrontation could not be long in coming, especially as supplies of ivory began to dwindle. From 1891 to 1894, therefore, the Force Publique was sent out on the so-called Arab campaigns. Led by Lieutenant Francis Dhanis, those campaigns resulted

in 1892 in the destruction of Nyangwe and Kasongo, the two major trading centers for Swahili-speaking Muslims in eastern Congo. The power of
the Afro-Arab traders from Zanzibar was broken for good. Stronger both
militarily and economically, their empire was nevertheless too politically
divided. By that time, Tippo Tip had left Congo to retire on Zanzibar.
In Maniema and Kisangani, however, Islam remains in place as minority
religion to this day.

The hardest fighting took place in the north. For years Leopold per-
sisted in his dreams of annexing southern Sudan. Under the spell of Egyp-
tomania ever since since his honeymoon journey to Cairo in 1855, Leopold
was obsessed with the Nile. Snatching southern Sudan would also allow
him to seize the upper reaches of that legendary river. What's more, the
area was said to be rich in ivory. As early as 1886 he sent Stanley there
to free Emin Pasha, governor of the Egyptian province of Equatoria but
in fact a German physician from Silesia, from a hostile Mahdist army. In
truth, the mission was an early attempt to make southern Sudan a part
of Congo. In 1890 Leopold offered Stanley the staggering sum of 2.5 million gold francs to finish the job for him and even to capture the city of
Khartoum, but the explorer was no longer interested.[47] The king therefore
financed several expeditions of his own, led by Belgian officers: all of them
failed miserably. In 1894 the British granted him a portion of southern
Sudan in usufruct, but that did not satisfy him completely. One last time,
he assembled an expeditionary force. In 1896 the Force Publique moved to
northeastern Congo with the largest army Central Africa had ever seen,
intending to advance from there to the Nile. But they never got that far.
The soldiers mutinied en masse.

How could that have happened? Beginning in 1891, in the absence of
enough volunteers to maintain a substantial army, the Free State had set up a
draft system for the Force Publique. As they had for the mission posts, all village chieftains were now required to supply a few young men, one conscript
for every twenty-five huts. The period of military service was seven years.
It was an ideal way for village leaders to rid themselves of troublemakers,
agitators, and prisoners. The Force Publique, therefore, was able to grow by
virtue of the arrival of unruly characters with absolutely no desire to serve.
That manifested itself as well during the expedition to Sudan. Such forays
were no orderly march to the battlefield. Hundreds of women, children,
and elderly people traipsed along with the soldiers through the jungle; uni-

formed men carrying Albini rifles fought side by side with traditional warriors who whooped and waved their spears. This was no regular, national
army on the move, but a motley crew, a semi-organized gang reminiscent
more of a messy eighteenth-century band of brigands than any tight Napo-
leonic infantry square. And the chaos was not limited to the margins and
the camp followers, but extended into the very heart of the military apparatus. For such a huge group, victuals could not be carried along but had to
be obtained by improvisation. The local population was sometimes willing to sell provisions, but more frequently refused. And so the army took
what it needed. Plundering as they went, the troops blazed a trail toward
the promised Sudan. Brussels chose to see things differently, but there was
in fact little difference between the Force Publique and the Batambatamba,
the Afro-Arab gangs of slave drivers described by Disasi Makulo. Unrest was
inevitable.

In 1895 there had already been a barracks revolt in Kasai, with fatalities
on the European side as well: several hundred mutineers there had thrown
off the yoke of the state. But the fury unleashed by the troops on their
way to Sudan was unparalleled. Ten Belgian officers were murdered. More
than six thousand soldiers and auxiliaries turned on their commanders.
Led by the Batetela, the mutiny became a rebellion that lasted four years.
It was the first major, violent protest against the white presence in Congo.
Military historians have often pointed out the troops' low morale: ill and
underfed, large numbers of soldiers died; many of them received almost no
training; and the most recent additions were soldiers who had first fought
on the side of the Afro-Arab slave drivers and now had little interest in
doing battle for their conquerors. But the hardhandedness of the officers,
in combination with their extreme incompetence when it came to logistics
and strategy, also fed a deep-seated hatred. And that hatred rounded not
only on the officers themselves, and not merely on the Belgians, but on
whites in general.

A French missionary suffered a night of terror when he was taken prisoner by the mutineers. "All whites conspire together against the blacks,"
he heard one of his captors argue against letting him live. "All the whites
should be killed or chased away." The heated discussion was finally decided
in his favor, which was a stroke of good luck for historians as well. Later
he described his ordeal in a letter to his bishop, giving us today a fairly precise view of the motives behind the mutiny. One rebel leader told him: "For

three years now I have been choking back my anger against the Belgians, and especially against Fimbo Nyingi. Now we had the chance to avenge ourselves." Fimbo Nyingi was the nickname of Baron Dhanis, the expedition's commander, the same lieutenant who had led the troops in Eastern Congo. His nickname meant "many lashes." The missionary resolved to listen to their grievances. "They even became friendly and offered me coffee—very nice coffee, in fact. What they told me about the Belgians was indeed shocking: sometimes they had to work hard for months without pay, and the wages for arriving too late were a sound beating with the *kiboko*. They were hanged or shot for the most minor offences. At least forty of their leaders, they told me, had been killed for trifling matters, and the number of fatalities among the foot soldiers was beyond count." Belgian officers, they told him, sometimes had native chieftains buried alive. They cursed their troops and called them beasts and slaughtered them "as though they were goats."[48]

I had never thought I would still be able to pick up echoes from that dark and distant period in the early years of this third millennium. But one day, in the Kinshasa working-class neighborhood of Bandalungwa, I found myself at the home of Martin Kabuya. Martin was ninety-two years old, a former Force Publique soldier and a World War II veteran. He lived in the capital but his family came from Aba, the most northeasterly village in Congo, on the border with Sudan. His grandfather was a chieftain at the time of the Force Publique expeditions to southern Sudan. "His name was Lukudu, and he was extremely mean. That's why they buried him alive, with his head just above the ground," he recounted. A common practice, as it turned out. To break their resistance, recalcitrant chieftains were buried up to their necks, preferably in the hot sun and preferably close to an anthill. Some of them were forced to stare directly into the sun for hours. Their families, too, were destroyed: under the guise of "liberation," their children were taken away. "The Marist Brothers took all his children to the boarding school at Buta [six hundred kilometers, or about 370 miles, to the west]. Including my father. There he became a Catholic. He married at the mission post and had three children. I'm the youngest."[49]

While King Leopold's troops were combating the slave trade in the east, and trying out new methods of subjugation as they went along, things in the west of the country were not much better. There were no full-out wars there, but there were daily forms of coercion and terror. To circumvent the unnavigable stretch of the Congo, the railroad between Matadi and Stanley

Pool was built between 1890 and 1898. Without such a railroad, Stanley had
noted earlier, Congo was not worth a red cent. The system of porters was
simply too costly and too slow, especially now that the state was the prime
exporter. It took a caravan eighteen days to cover that route, a steam train—
even with frequent stops for water and wood—only two.⁵⁰ With great difficulty, Leopold was finally able to scrape together the funding for that project
(the money came from private investors and above all from the Belgian government): with even greater difficulty, the work commenced. During the
first two years only eight of the total of four hundred kilometers (about five
out of 250 miles) of rail were laid: the railroad had to wind its way through
the desolate, mountainous countryside east of Matadi. Three years later the
work was no further than kilometer marker 37 (milepost 23). Working conditions were extremely harsh. The crews were decimated by malaria, dysentery, beriberi, and smallpox. During the first eighteen months alone, nine
hundred African workers and forty-two whites died; another three hundred
whites had to be repatriated to Europe. Over the full nine years, the project
claimed the lives of some two thousand workers.

The organization had a military feel to it: at the top of the chain was a
Belgian elite, this time consisting of engineers, mining engineers, and geolo-
gists, led by Colonel Albert Thys, himself a military man and captain of
industry. Working under them were the manual laborers from Zanzibar and
West Africa, a crew of between two and eight thousand men. There were
also a few dozen Italian miners. But as fewer and fewer Africans proved willing to work in the hell called Congo, the organizers began recruiting workers from the Antilles and had hundreds of Chinese laborers shipped in from
Macao—almost all of whom succumbed to tropical diseases.

Just like in the army, the Congolese themselves barely took part at
first. The argument given was they were still indispensible as porters. Only
when the railroad had almost reached the halfway point at Tumba in 1895
were workers recruited from the local population. That was home ground
to Étienne Nkasi, the old man I had met in Kinshasa. "I was twelve or fifteen at the time," he told me during one of our conversations, "I was still a
child, incapable of hard labor. Kinshasa and Mbanza-Ngungu didn't exist
yet." Indeed, I reflected, Kinshasa was not yet a city at that time, at most only
a cluster of settlements; Mbanza-Ngungu (the former Thysville) had yet to
be built. It owed its existence to the railroad. At the highest point along the
track, precisely halfway between Matadi and Kinshasa, there had once been

a pleasantly cool and fertile hillside. It was there, between 1895 and 1898, that they built the town named after the project's chief engineer. Travelers spent the night there during their two-day journey. It was a fresh and verdant spot, where European crops were raised on a large scale. Today it is marked by rusty boxcars on rusty tracks, beside rusty Art Nouveau–style colonial homes.

"I was there when they built Thysville," Nkasi had remarked, amazing me for the umpteenth time with his memories of an incredibly distant past. "My father knew Albert Thys. He was the leader of a crew, my father was. Four black men pulled the white man's cart over the rails. The white man wore one of those white helmets. I saw that." He smiled, as though realizing only then how very long ago that was. "Papa worked at Tumba, at Mbanza-Ngungu, Kinshasa, Kintambo. I went with him everywhere." Those were indeed the posts along the route under construction. The job was finished in 1898. At the festive opening, white people trundled along from Matadi to Kinshasa, a nineteen-hour journey, in tuxedo and décolleté. Along the way there were fireworks and here and there blacks in uniform stood and saluted. At some of the stations the travelers were treated to hymns sung by the choirs from the local mission posts, accompanied by a rickety harmonium.[51] The celebrated narrow-gauge railway was in fact merely a tramway with open carriages, yet the opening of it constituted a milestone in the opening up of Congo. For Nkasi, however, the grand opening meant it was time to go home. He had been gone for three years. "When the work was finished, Papa went back to the village, back to my mother. To make more children. I was still the only one they had. Two babies had died after I was born. When he came back from the railroad, he made five more." I asked him about the trains back then. He remembered that too. "The engines, they ran on wood," he explained. "And when they started moving . . ." He sat up a little straighter on his bed, clenched his old fists, and began rocking his thin arms back and forth. "It went: toooot . . . tacka, tacka, tacka." Then he burst into noiseless laughter.[52]

Working on the railroad was not the worst that could overcome a Congolese, especially not after 1895. For at the moment when the first native laborers were taken on, a premium system came into effect. The white yard supervisor would agree with the black headman on a term within which a certain stretch of railroad was to be finished. If that goal was met, his team received a preestablished bonus. A company culture of incentives, *avant la lettre.* On top of his daily wages of fifteen centimes and his ration of rice, biscuits, and dried fish, the worker could in this way earn a little extra cash, valid albeit only at government shops, in the absence of a monetary economy in the rest of the country. Louis Goffin, the engineer who devised the premium system, spoke of "une cooperation du travail des noirs et du capital européen" (a working arrangement between black labor and European capital). The objective, according to him, was to instill the Congolese with enthusiasm for their work, purchasing power, and pride. The idea was "to create among the natives new needs, which would result in a love of work, a rapid development of commerce and, in that way, of civilization."[53] Once the railroad was finished a few Congolese remained employed as lathe operators in the machine shop, as stationmasters, or even as locomotive drivers. They were on the payroll, and therefore the first to be absorbed into the money economy. Each time I visited Nkasi, he spoke with great admiration of a man named Lema, his father's cousin. Lema had served as *boy-bateau* on the ships to Antwerp, but went to work on the railroad in 1900. "He became stationmaster at Lukala." "Where the cement factory is now?" "Yes, that's right. Stationmaster! He knew the white people."[54]

Elsewhere in the Free State, monetization was yet to arrive. Barter was still the norm. That, however, made it rather hard to collect taxes. The Free State needed revenues and considered it expedient that its subjects help pay for their country's development; no money, however, could be expected from those who had none. The beads, copper ingots, and bales of cotton of the past were of no interest to them. And so an arrangement had to be made for payment in kind: in goods or in labor. That, after all, was how it used to go, when a hunter presented a tusk or a portion of the spoils to the village chieftain. That had been a solid system in the past, but now it would lead to the total disjunction of Congolese society. The refusal to introduce a money economy to the interior, too, had painful consequences.

Leopold II played a rather dirty trick on free trade. As owner of almost all the ground in Congo, he himself was nonetheless unable to develop it. His solution was to grant huge concessions to a few commercial concerns: to the north of the Congo River, the Anversoise, a new company, was allowed to exploit an area of 160,000 square kilometers (about 62,400 square miles), approximately twice the size of Ireland. To the south of that same river, the ABIR (Anglo-Belgian Indian Rubber Company) received a permit for a comparable area. The king treated himself to an extremely generous chunk of jungle largely south of the equator: the Kroondomein (Domaine de la Couronne), covering 250,000 square kilometers (about 97,500 square miles), approximately ten times the size of Belgium. In order not to displease all the people all the time, Kasai remained a sort of natural reserve where free trade was allowed to muddle on for a while (it was later monopolized by the king anyway). The Compagnie du Katanga and the Compagnie des Grands Lacs also received enormous territories; their very names show that they were set up especially for the occasion. The major economic exploitation of the Congolese interior, therefore, lay in the hands of the king and a few privileged concessionaries. Yet these were not watertight worlds; Leopold himself was usually the main shareholder in the new companies or at the very least had a right to a substantial share in their profits. The concessionaries' management councils always included top administrative officials from the Free State. In Belgium, the king's financial adviser, Browne de Tiège, was not only chairman of the Anversoise and the Société Générale Africaine, but also a member of the board of ABIR, as well as the Société Internationale Forestière et Minière, the Société Belge de Crédit Maritime in Antwerp, and a handful of other partnerships.

Congo's economic potential now no longer revolved solely around ivory. In 1888 a Scottish veterinarian, John Boyd Dunlop, had come up with an invention that would not only considerably improve the comfort of thousands of travelers in Europe and America, but also rule, or even end, the lives of millions of Congolese—the inflatable rubber tire. In an age in which new inventions like the car and the bicycle still had to make do with ironclad wooden wheels, the rubber tire was a godsend. The worldwide demand for rubber took off. For Leopold, it was nothing short of a miraculous deliverance. Fewer and fewer elephants roamed the Free State, but rubber trees abounded. The timing could not have been better. His Congo was teetering on the verge of bankruptcy and Belgium was poised, however reluctantly, to take over. Suddenly, that turned out not to be necessary. In 1891 Congo was producing only a few hundred metric tons of rubber, but by 1896 that had grown to thirteen hundred tons (nearly 1,450 U.S. tons) and by 1901 to six thousand tons (6,600 U.S. tons).[55] From a moribund project in Central Africa, the Free State had suddenly become a true economic wonder. Leopold raked in millions and, after a very long wait and years of reckless entrepreneurship, finally received his return on investment. At last he could demonstrate what a colony was good for: economic boom, imperial fame, and national pride. With the revenues from Congo he was able to spruce up Belgium on a major scale. Brussels saw the construction of the Jubelpark Museum and a new royal palace, at Tervuren an immense colonial museum and park, inspired by Versailles, and at Ostend the celebrated Venetian esplanade.

The flip side was seen only in Congo itself, where—except for the tins of foie gras and bottles of champagne sent to government officials from Belgium—there was little glamour and glitter to be found. Not only did Leopold refuse to invest the proceeds from his rubber empire in Congo itself, he set about supervising the harvesting of that rubber in an extremely troubling fashion. There were nothing like plantations in Congo, only wild rubber. The harvesting of it was a long and arduous task that required the involvement of many manual laborers. The ideal form of taxation, therefore, had been found: the rubber itself. Natives had to go into the jungle to tap the rubber trees, collect the latex, and process it crudely into sticky clumps. Whereas taxes had formerly been collected in the form of manioc loaves or ivory, or by the impressments of porters, now the local population had to deliver baskets of rubber at prearranged intervals. The quota varied from region to region, but the principle remained the same. In the Crown Domain, the regional governor would draw up an estimate and the soldiers of the Force Publique would see to the collection of the rubber tax. In those areas where concessionaires operated, the collection was done by armed guards, the so-called sentries. In both cases it involved Africans with limited military training and little discipline.

The abuses to which this system led were, in fact, a foregone conclusion. The men paid to collect the rubber were paid for the quantity of rubber they collected. No rubber, no pay. They therefore did all they could to maximize yield. In actual practice that meant a universal reign of terror. Because they were armed, they were able to mercilessly terrorize the local population. The situation in the territories allotted to the concessionaires was appalling, but in the areas controlled by the Free State things were hardly better. Disasi Makulo witnessed it himself, at the mission post he had founded at Yalemba. Trouble awaited him not only from the heathen villagers of his region, he had to concede, but also from the Congolese from other parts of the country, now in the service of the Free State.

They often profited from the absence of their superiors. They abused, tortured and sometimes even murdered people. . . . At the mission post at Bandu was a man whose nickname was Alio [the eagle], because of his cruelty. He was the general overseer for the rubber deliveries. That man was terribly cruel. He killed a great many people! One day he and his crew crossed the river to go to the Turumbu, a tribe that lived on the right bank. As usual, he demanded goats, chickens, ivory, et cetera et cetera from every village he went to. This time he caused very serious problems. He even killed a man.

When I heard that he was on his way to my village of Bandio . . . I took a few boys from the mission along and went looking for him. When we arrived, we found him just as he was busy beating people and torturing them and plundering the village! Without wasting a minute I went up to him and said: "You are in the service of the state only for the purpose of overseeing the deliveries of rubber, not to abuse, rob, and murder. Give back immediately all that you have confiscated, or else I will report these facts to the authorities in Basoko."[56]

Disasi also witnessed the shooting of a girl from his village by the guardian of the rubber depot. His experiences were typical of all those who came in contact with the rubber policies of that day. The men were sent into the forest to collect rubber, the women were held hostage until enough rubber had been delivered. Human lives were not worth much, as several disturbing eyewitness reports show. "Two sentries, Bokombula and Bokusula, arrested my grandfather Iselunyako, because his rubber basket was not full enough. They put him in a well and trampled him underfoot. That is what killed him. When we showed his body to the white man, he said: 'Good. He was finished with rubber and therefore finished living.' "[57]

Eluo, a man from Esanga, related the following: "We had to supply fifty baskets of rubber. One day, during the administration of the white man Intamba [Meneer Dineur], we came back with only forty-nine, and they declared war on us. The sentry Lomboto came to our village with a few others. Along the way, as they passed a swamp, he saw my sister fishing. For no reason at all, Lomboto shot her with his rifle and killed her.' "[58]

Sexual violence took place in those days as well. A married woman recounted: "To punish me, the sentries Nkusu Lomboto and Itokwa removed my *pagne* and stuffed clay into my genitals. That was very painful."[59] Cruelty had a function.

> The village chieftain Isekifusa was killed in his hut. Two of his wives were murdered at the same time. A child was cut in two. One of the women was then disemboweled. . . . Boeringa's people, who had come along with the sentries, ate the bodies. Then they killed ten men who had fled into the forest. When they left Bolima, they left a part of Lombutu's behind, chopped into pieces and mixed with banana and manioc, in plain sight, to frighten the villagers. The child's intestines were hung up around the village huts. The child's body parts were impaled on sticks.[60]

HAD THE SYSTEM OF PREMIUMS applied during the construction of the railroad in Bas-Congo been introduced here as well, a very different set of dynamics would have been set in motion. People would have been rewarded for their efforts and motivated to continue producing. The Congolese, after all, were anxious for such rewards, but the authorities ignored this: "When we ask for *mitakos* [copper currency ingots], we get the *chicotte* [strop made of hippopotamus hide] instead," someone said.[61] The rubber had to flow freely to the state, at no cost. This was about taxation, not remuneration: in fact, what it boiled down to was pillaging.

The dirty work of collecting these revenues was left to subordinates with rifles. Because their white bosses wanted to be sure that they did not misuse their weapons to hunt for game, they had to account for every round of ammunition. At various places, therefore, there arose the custom of cutting off the right hand of those they had shot and taking it along as proof of what the bullet had been used for. To keep the hands from rotting they were smoked over an open fire, in the same way that food is preserved to this day. The tax collector, after all, saw his boss only once every few weeks. During the debriefing he was expected to present the hands as *pièces justificatives*, as "receipts" for expenses incurred.

Beginning in 1900 voices began to be raised in Europe against this Belgian ruler who had his employees cut off people's hands. A few photographs of Congolese with stumps for arms made their way around the world. This resulted in the widespread misconception that living persons were having their hands cut off in Congo on a major scale. That did happen, but much less systematically than most people thought. The greatest ignominy of Leopold's rubber policies was not that dead people's hands were cut off, but that the murdering took place so casually. The mutilation of corpses was a secondary effect. That does nothing, however, to detract from the fact that, in a number of cases, the atrocities truly knew no bounds. "When I was still a child," said Matuli, a fifteen-year-old female student at the Ikoko mission, "the sentries shot at the people in my village because of the rubber. My father was murdered: they tied him to a tree and shot and killed him, and when the sentries untied him they gave him to their boys, who ate him. My mother and I were taken prisoner. The sentries cut off my mother's hands while she was still alive. Two days later, they cut off her head. There were no white men present."[62]

By severing the limbs of living victims, the sentries not only saved on bullets, but were also able to steal the broad copper bracelets that women often had forged around their wrists or ankles. Boali's story is quite telling in that regard: "One day, when my husband was in the forest tapping rubber, the sentry Ikelonda came to my hut and asked me to give myself to him. I refused. Enraged, he shot me with his rifle; you can still see the wound. I fell to the ground and Ikelonda thought I was dead. To get the copper ring I wore around my ankle, he chopped off my right foot."[63] Had Boali shown any sign of life at that point, she would have been killed immediately.

But violence by Africans against other Africans was not the whole story: it was not only at the base of the pyramid of power that blood flowed. Many Belgians also took part in this. Physical violence was more widely tolerated in those days—Belgian cafés were the scene of weekly brawls, free-for-alls were a part of youth culture, corporal punishment was the standard at schools— yet some of the offences in Congo far exceeded the boundaries of custom. Floggings with the *chicotte* were an official disciplinary measure. The Belgian civil servant in charge established the number of lashes to be administered, his black aide-de-camp dealt them out during the morning or evening roll call, while the flag of the Free State waved over the proceedings. The strop had to be flat, the number of lashes was not to exceed fifty (to be administered in two series of twenty-five each), only the buttocks and lower back were to be lashed, and the whipping was to cease at the first show of blood. Some white people, however, did not abide so closely by the rules: they preferred a nonregulation strop, which was twisted and angular and therefore much more painful. They also included the stomach, loins, and sex organs in the flogging. Sometimes they prescribed punishments of up to four hundred lashes, and paid no heed to any bleeding or physical collapse. Pregnant woman, who officially were not to be punished in this way, still received a beating.[64]

Mokolo, a married woman, testified:

My husband's name was Wisu and every two weeks, along with Ebobondo and Ebote, he brought the rubber from our village to the trading post at Boyeka. We always supplied twenty baskets full, but then the whites began demanding twenty-five. Our people refused, and pointed out that our village was only a small one. But the next time they showed up with only twenty baskets, the white men became angry. One of them, Nkoi [the nickname used for Ablay], threw my husband to the ground and held onto his head. The other, Ekotolongo [the nickname of Félicien Molle], began beating him with nkekeles [canes], three of which even broke. Almost dead, Wisu was dragged by Ebobondo and Ebote to a dugout in which they traveled to Bokotola. But before they could go ashore there, he died. I saw Wisu's body, and you can still see the traces of my tears.[65]

The Free State administration contained out-and-out racists and sadists. Torture, abuses of power, and massacres occurred. A person like René de Permentier, an officer in the Force Publique, reveled in completely pointless bloodbaths. He had the *brousse* (jungle brush) cleared around his house so that he could shoot at passers-by from his veranda. Domestic personnel who made a mistake were slaughtered without mercy. Executions were a daily occurrence.[66] Léon Fiévez, a farmer's son from Wallonia who became a district commissioner in Équateur, indulged in bloody punitive expeditions. After only four months in public service, he had murdered 572 people.[67] During one of those expeditions, within a few days, he saw to the looting and torching of 162 villages, the destruction of the local fields, and the killing of 1,346 people. He was, however, also able to claim the greatest volume of incoming rubber in all the Free State.[68]

The lion's share of those Belgians who arrived to try their luck in Congo came from small provincial towns and the lower middle class. Many of them had served in the army and looked forward to adventure, fame, and fortune. But once in Congo they often found themselves alone at remote trading posts in a killing climate. The heat and humidity were relentless, the attacks of fever frequent. No one knew yet that malaria was transferred by mosquitoes. A young man like that, in the flower of his youth, might awake at night for no good reason, soaked in sweat, delirious, shivering, thinking about all those other white people who had suffered and died. He heard a jungle full of strange noises, recalled snippets of a brusque conversation with a village chieftain earlier that day, and thought back on the skittish glances of the people charged with collecting rubber, on the sinister hissing of their incomprehensible language. In his feverish visions between sleep and wakefulness he saw gleaming eyes full of suspicion, broad, shiny backs covered in tattoos, and the budding breasts of a young native girl who had smiled at him.

George Grenfell, the British Baptist who had cared for Disasi Makulo, was a keen observer of all this. For a long time he had been one of Leopold's fervent supporters, he had even taken upon himself the task of chairing the king's Commission for Native Welfare, in fact nothing but a paper attempt to spread oil on troubled waters. But Grenfeld's disaffection grew rapidly: "In view of the number of solitary posts manned by unmarried white men, with only a handful of native soldiers amid semi-docile and often cruel and superstitious peoples, it should come as no surprise if more madness comes to light. But it is the system that is to be condemned, more than the poor individual who, overpowered by fever and fear, loses control over himself and indulges in forms of intimidation in order to maintain his authority."[69] The Free State's administration prided itself on punctuality, state officials feigned a certain equanimity, the appearance of control was held high. But writhing beneath all this were feelings of fear, depression, melancholy, lethargy, despair, and total madness. People lost their heads.

The Free State condemned misconduct in word, but in deed it could not control its subordinates. There were almost no convictions. News of what was happening in Brussels reached Boma quicker than news from the rain forest. King Leopold too sounded dismayed when initial reports of atrocities began trickling in. He said: "The abuses must stop, or I shall withdraw from Congo. I will not let myself be sullied with blood or muck. This shameful behavior must stop."[70] That did not keep him, however, from reappointing infamous brutes like Fiévez, even though the king was well aware of the man's shameful deeds. Neither he nor his advisers nor his top officials in Boma wished to admit that the atrocities were inherent to the system employed. And yet, with profit maximization as the alpha and omega of the entire enterprise, people at all levels of the administration were pressured to collect more taxes, bring in more rubber, tighten the thumbscrews even further. The Free State system was a pyramid with Leopold II at its pinnacle, and under him the governor general at Boma and the various administrative levels, followed by the black soldiers of the Force Publique, and, at the very base, the native in his village. The physical violence may have been limited to the lowest rungs (on the part of rapacious soldiers and bugged-out officials in the interior, on the part of brutal sentries and completely deranged minds in the jungle), but the structural violence permeated all the way to the top, even unto the king's palace at Laeken. The official rule may have been that a native was to work no more than forty hours a month for the state, but as rubber became scarcer the natives had to go farther and farther into the jungle to collect their quota. No time was left for other forms of work. People became the state's bondsmen. Leopold II had, at least nominally, set out to eradicate Afro-Arabic slave trading, but had replaced it with an even more horrendous system. For while an owner took care of his slave (he had, after all, paid for him), Leopold's rubber policies by definition had no regard for the individual. One would be hard-pressed to choose between contracting the bubonic plague or cholera, but from a distance it would seem that the life of a Congolese domestic slave in Saudi Arabia or India was to be preferred to that of a rubber harvester in Équateur.

The consequences were horrendous. The fields lay fallow. Agriculture dwindled to the raising of only the most basic staples. Native commerce came to a standstill. Crafts in the process of refinement for centuries, such as iron smithing or woodcarving, were lost. The native population became listless, enfeebled, and malnourished. And so extremely susceptible to illnesses. Around the turn of the century, sleeping sickness became rampant. This illness, carried by the tsetse fly, had been known in the region for a long time, but the death rate had never been so high. It assumed truly pandemic proportions. In 1904 George Grenfell wrote: "In many districts, the current death rate is nothing less than alarming. Along the thousand miles of river (two thousand miles of shoreline) between Léopoldville and Stanleyville, after having counted the houses and made a rough estimate, I would strongly doubt whether if one hundred thousand souls still live in all of the town and villages along the way."[71] This, it should be remembered, was once the most populous stretch of the interior. In some villages, between 60 and

90 percent of population vanished. In 1891 Lukolela, one of the oldest trading posts along the banks of the Congo, had some six thousand inhabitants;
by 1903 there were fewer than four hundred.[72] It is impossible to say how
many people died as a direct or indirect result of Leopold's rubber policies.
There are simply no reliable figures. What's more, there was another reason for the depopulation; many people simply went away, away from the
river, away from the banks. They went to live deep in the jungle or crossed
the border to remain beyond reach of the state. They too became invisible.
A rare eyewitness to that first historic stream of refugees was interviewed
about this in 1903:

> How long ago was it that you left your houses? Was it when the big
> problems began, the ones you told us about?
> *Three years ago. This is the fourth year after we fled and came to live in this*
> *region.*
>
> How many days must one walk to reach your own country?
> *Six days of brisk walking. We ran away because we could no longer live*
> *with the things they did to us. Our village chieftains were hanged, we were*
> *murdered and starved. And we worked ourselves to death in order to find*
> *rubber.*[73]

It would be absurd in this context to speak of an act of "genocide" or a
"holocaust"; genocide implies the conscious, planned annihilation of a specific population, and that was never the intention here, or the result. And
the term *Holocaust* is reserved for the persecution and annihilation of the
Jews during World War II. But it was definitely a hecatomb, a slaughter on
a staggering scale that was not intentional, but that could have been recognized much earlier as the collateral damage of a perfidious, rapacious policy
of exploitation, a living sacrifice on the altar of the pathological pursuit of
profit. When sleeping sickness ravaged the population, Leopold II called
in the assistance of the Liverpool School of Tropical Medicine, the most
famous center for tropical medicine of its day. He would never have done
that if his intention had been to commit genocide. But that does not mean
that he immediately acknowledged his responsibility in the matter. That, in
fact, was something he never did.

The bloodstained rubber policy, nicknamed "red rubber," did not have the same impact everywhere. Équateur, Bandundu, and Kasai—the western part of the Congolese rain forest—were the hardest hit. That was where the exploitation was easiest to achieve, because of the big rivers. When I once asked old Nkasi, who came from Bas-Congo, about the days of the rubber quotas, he was unable to tell me anything. "That wasn't where we were," he said. "That was in the Mayombe." He may very well have been right. The Mayombe was a stretch of equatorial forest north of Boma, close to the ocean and the Portuguese enclave of Cabinda. It was one of the few places in Bas-Congo where rubber was harvested. Nkasi knew about it only by word of mouth. "The Portuguese there cut off people's hands," he added, but he wasn't completely sure about that. When I went on to ask whether he had witnessed the rapid spread of sleeping sickness, however, his nod was much more confident. "Yes, I saw that. Many young people died. A nasty disease." He repeated that final sentence again in his simple French. "C'est mauvaise maladie" (It's a bad disease).

FROM 1900 ON increasingly clear indications began coming in concerning the atrocities in the Free State. They were not immediately given credence. Protestant missionaries expressed their abhorrence in no uncertain terms, but in Belgium it was felt that they were simply frustrated about the influx of Catholic missionaries and the power they had lost. In Antwerp, Edward Morel, the employee of a British shipping company, began to realize that something was very wrong in Congo: he saw ships leave without cargo (except perhaps for guns and ammunition) and return full of rubber. That seemed more like plundering than bilateral trade, didn't it? His objections, however, were dismissed lightly, as the typical moaning of those Liverpudlian traders who never stopped whinging about the decline of free trade. Did little Belgium really stand to learn anything from the British? That's what they wanted to know. Weren't those imperial browbeaters in fact the worst malefactors of all, now that they had made mincemeat out of the defenseless Boers down in South Africa? The Boer War, after all, had met with disapproval in Belgium as well.

The tenor of the discussion changed a bit, however, after Roger Casement, the British consul in Boma, wrote a thoughtful but damning report in 1904. Casement was a highly respected diplomat. This was no British dockworker, but an official envoy of Great Britain, a man of great moral authority and long acquainted with the Congolese interior. His objections could no longer be dismissed; they led to major protests in the British House of Commons. Authors such as Arthur Conan Doyle, Joseph Conrad, and Mark Twain publicly voiced their disapproval. One year after the report appeared, King Leopold found himself compelled to send an international, independent investigative committee to Congo. Three magistrates, one Belgian, one Swiss, and an Italian, were allowed to travel around Congo for months and carry out interviews in his Free State. They would surely absolve him of all blame. But that is not how it went. The investigative committee went to work as a sort of Truth and Reconciliation Commission *avant la lettre*. They listened to hundreds of witnesses, compiled plaints, and wrote a down-to-earth report in which the Free State's policies were quite accurately dissected. It was a dry but devastating text, stating that the "taking hostage and abduction of women, the subjugation of chieftains to forced labor, the humiliations to which they are subjected, the *chicotte* used by harvest overseers, the violent actions on the part of blacks ostensibly occupied in 'guarding' the prisoners" were the rule rather than the exception.[74] The Brussels lawyer and professor Félicien Cattier followed the reasoning through to its most extreme conclusion: "The clearest and most incontrovertible truth arising from this report is that the state of Congo is no colonized state, barely a state at all, but a financial enterprise. . . . The colony is not administered in the best interests of the natives, nor even in the economic interests of Belgium: to provide the sovereign with the greatest possible financial gain, that was the motivation."[75]

The international pressure on King Leopold II was mounting. Something had to give, and the only option was for Leopold to part with his overseas territory and for Belgium to take over Congo. In December 1906 the knot was cut, but Leopold loitered over the modalities of the transfer for almost two more years. He wondered whether he might perhaps still be able to keep a piece of Congo for himself, the Crown Domain for example. It was with clear reluctance that he handed over his lifework. Shortly before the transfer he even ordered the Free State archives to be burned. But on November 15, 1908, the day finally arrived: on the occasion of the annual National Celebration of the Dynasty, the dynasty handed over Congo. The term *free state* itself had meanwhile become rather outmoded for a state without free trade, free employment, or free citizens. In its stead there had come a regime that revolved around a monopolistic economy, forced labor and bondage. From then on the region was to be called the Belgian Congo.

During the term of the Free State, the local population had had its first encounter with various aspects of the European presence. By the year 1908 some sixteen thousand children were attending missions schools, an estimated thirty thousand people had learned to read and write, sixty-six thousand had served in the army, and some two hundred thousand had been baptized.[76] Directly or indirectly, hundreds of thousands of locals had been effected by the rubber policies. Millions had been struck down by sleeping sickness and other infectious diseases.

And Disasi Makulo had seen it all from close by. He had been involved in the ivory trade when it was still free, he had served as boy to a celebrated British missionary, he had made countless journeys on the man's steamboat, he had literally experienced the interweaving of mission and state firsthand when—during one of his travels with Grenfell—he was conscripted into the Force Publique. He had been baptized, married a girl from a distant region, he swore by monogamy and the nuclear family, criticized traditional village life, and in the end became a catechist in order to Christianize his own region. And it was precisely at that point that he had personally witnessed the violence of the rubber policies.

But at his mission post, even more sorrow lay in store for him. His great mentor George Grenfell came to visit there in mid-1906. Grenfell looked like a man of eighty, but had still to turn fifty-seven. His years in the tropics had been long and grueling. He was worn out. Grenfell asked his former pupil and his converts to sing a hymn for him in their own language, Bobangi. Afterward he explicitly stated his desire to be buried at Yalemba, the mission post Disasi had founded himself. Disasi called him "he who remained our father until his death."[77]

TODAY, IN KINSHASA, one sees very little that hearkens back to those early years, but during my very first trip to Congo in December 2003 I was granted access to the former garage for city buses in the borough of Limete. Buses had been absent from the capital's streets for quite a few years already; the few broken-down examples that still existed had been converted into houses in which several families lived. The windshield wipers were used as clotheslines. The residents slept on the old seats, their arms draped over the aluminum bars. From the shadow of a hubcap or hood one heard the bleating of an unseen goat. It was an abandoned industrial estate, where nature was regaining its grip. After walking on a bit I saw, in the grass, a highly remarkable work of postmodern art. Never had I seen a more peculiar installation invested with such historical reminiscences. Lying on its stomach in a rusty steel boat was the bronze figure of a man at least twelve feet tall. I recognized the statue right away: it was the triumphant image of Stanley that had stood looking out boldly across the river for decades, atop the hill at Ngaliema. Designed and cast in Molenbeek, outside Brussels, it had been shipped here during the colonial period, but was pulled from its pedestal after independence. And now here he lay, old Stanley. The broad sweep of that arm with which he had once taken the Congo's measure was now pointing nowhere and at nothing. The fingers touched only the boat's rusty boiler. Power had become a cramp, courage something laughable. On the hull, close to the bow, I saw three letters: AIA, the abbreviation for Association Internationale Africaine. This was one of the three boats with which Stanley had braved the current between 1879 and 1884, for the purpose of establishing the odd trading post. It was in one of those boats that he taken aboard Disasi Makulo after buying him from a slave trader. Now Stanley lay felled in his own boat. The fleet with which he placed Congo under a new authority had become his mausoleum. Who knows what civil servant had come up with this ingenious bit of *bricolage*; it was probably a wrecking company that had improvised on the spot this scrap heap of history, but seldom had I seen a more ironic settling of accounts with colonialism than in Stanley's official monument lying flat on its belly in his own old tub.

The next day, in the tranquil, green district of Ngombe on the other side of town, I found the Baptists' old mission post. It lay along the riverbank in what is now the most exclusive neighborhood in Kinshasa. Their original building was still there, a simple construction on cast-iron stilts, just like in Boma. Around each individual stilt was a sort of vase: these had once been kept filled with petroleum, to ward off the termites. It must be, I think the oldest building in Kinshasa. I walked on a few steps to get a better view of the river. Kinshasa lies on one of the world's largest rivers, but with all the walls and barriers (it remains, after all, a national border) there are few places where one can truly see the water. On the slope down to the river, in the tall grass, there lay something that looked like a huge insect, or the ribcage of a bronze giant. It was the cooler of an enormous engine. Dozens of parallel brass tubes found their confluence in a sturdy steel rod. One of the Baptists' students told me that this was the engine of the *Peace*, the steamboat on which Grenfell had made all his journeys of discovery. When the ship itself was finally salvaged, this showpiece of industrial archaeology had been hoisted ashore. It seemed too good to be true. Not only do we possess the details of Disasi Makulo's formidable life story, but the two boats in which he plied the Congo are today still lying, rusting away, in the tall, silent grass of Kinshasa.

CHAPTER 3

THE BELGIANS SET US FREE

The Early Years of the Colonial Regime

1908–1921

L UTUNU LOOKED AT HIS WIFE. IT WAS BECOMING INCREAS-
ingly hard for her to walk. And she was still so young, he
thought. The lumps on the side of her neck were clearly visible now, like
a row of pebbles beneath the skin. He knew the signs, this was how it had
started with his children too. First a fever, headaches, and stiff joints, then
deathly fatigue and listlessness during the day, followed by sleeplessness.
He knew what she was in for. She would become increasingly groggy,
increasingly lethargic. Her eyes would roll up in her head, she would foam
at the mouth. Then she would lie down in a corner until it was all over.
What had he done to deserve this? All those who had this disease died. A
few years ago all his brothers and sisters had succumbed to smallpox, they
had dropped like flies. Then his two young boys, the first children she had
borne him, had died of sleeping sickness. And now her. Had she drunk
from a gourd used by someone else with the sickness? Had she eaten an
orange with brown spots on it? No one knew where the sickness came
from, no healer could offer a cult object or a medicine against it. Some
people said it was a punishment from the missionaries, that they spread
the sickness in indignation at those who still didn't accept their doctrines.[1]
Lutunu had no idea.

Around 1900 even his leader had died of it, Mfumu Makitu, the big chief
of Mbanza-Gomber. In 1884 he had been one of the first chieftains to sign
an agreement with Stanley. Back then their village had been along the caravan route from the coast into the interior, long before the railroad came.

Chief Makitu wanted nothing to do with the white newcomers at first, but he finally gave in. On March 26, 1884, along with several other chieftains, he had put his mark at the bottom of a sheet of paper which read: "We, the undersigned chiefs of Nzungi, agree to recognize the sovereignty of the Association Internationale Africaine, and in sign thereof adopt its flag (blue with a golden star). . . . We declare that from henceforth we and our successors shall abide by the decision of the representatives of the Association in all matters affecting our welfare or our possessions."[2] Lutunu remembered it as clearly as though it had happened only yesterday. Chief Makitu gave Stanley a generous welcome present, one of his youngest slaves: Lutunu himself. He was ten years old at the time. For his great display of loyalty, Makitu was rewarded in 1888 with a medal of honor: he become one of the country's first *chefs médaillés* (decorated chiefs). His prosperity continued to grow. Now, many years later, he had left behind sixty-four villages, forty wives, and hundreds of slaves.

Lutunu's life was as full of adventure as that of Disasi Makulo, so adventurous in fact that he is still remembered today. A street in Kinshasa was named after him, and old Nkasi, whose native village had been close to Lutunu's, had actually met him once in the distance past. "Lutunu, oh yes, I knew him!" he told me. It was the first time I had heard that name. "He came from my area, he was a little older than me. He was Stanley's boy. And he always refused to wear trousers. When the white man would call out 'Lutunu!,' he simply shouted back 'White man!' Just like that! White man!" Nkasi couldn't help laughing at the thought of it. Lutunu was special. A hotshot, on friendly terms with a lot of white people. When I returned to Belgium, I discovered that his story had been documented by a Belgian artist and writer.[3]

Like Disasi Makulo, Lutunu was a slave who fell into the hands of the Europeans. He served as boy to Lieutenant Alphonse Vangele, one of Stanley's earliest helpers. He too came in contact with the British Baptists: they set up one of their most important mission posts in his region and Lutunu later became boy to one of the missionaries, Thomas Comber. And that took him to Europe, just like Disasi. He was there when Comber went to England and Belgium; he was present when Comber was received by King Leopold II. He was one of the nine children who were allowed to sing a song for the king. He was the one who later sailed to America and, when he came home, achieved fame from Matadi to Stanley Pool, and received a host of breathless followers as the first cyclist in Congo. That Lutunu, in other words. And his madcap adventures were not nearly over yet. Perfectly unsuited for the patient translation of the Gospels into his native language, but all the more for the world at large, he sailed up the Congo with Grenfell and must have met Disasi Makulo. He became the guide and interpreter for the Belgian officers Nicolas Tobback and Francis Dhanis during their military campaigns. For a short while, he was even a soldier himself in the Force Publique. He went everywhere the white people did and knew the colonizers better than anyone else. "Lutunu!" "White man!" But he refused to wear their trousers. And he had no interest in being baptized.

But then his wife died and he was all alone. Children dead, family decimated. After all his wanderings, he had ended up back in his native village. He spoke with the Protestant missionaries there and was converted. He was already around thirty by that time. The dozens of slaves he had bought in the course of the years he now set free. He went to live at the mission post. Francis Lutunu-Smith, that was the new name they gave him.

When great chief Makitu died around the turn of the century, his successor according to local custom was an inexperienced sixteen-year-old boy. The missionaries suggested that Lutunu act as the boy's regent: that would be better for the village and better for the mission. It would allow them to exercise influence over the local authorities: Lutunu, after all, was one of their own. Just as Disasi Makulo was allowed to set up his own mission post, Lutunu was allowed to bear some administrative responsibility: thanks to the white man, the slave children of yore were acquiring a good deal of power.

Lutunu's life may have resembled Disasi's, but in piety he was no equal. Five years later he was suddenly expelled from the mission: he had taken an excessive liking to English stouts and lagers. The Congo Free State had dealt summarily with the endemic alcoholism of the local population. The consumption of palm wine was radically restricted; brandy, gin, and rum were banned. But Lutunu drank and danced. And although he continued to cherish his copy of the Bible, he suddenly turned out to be married to three different women, who bore him four, five, eight, twelve, seventeen children. Was the new religion really all that hard to reconcile with the old customs?

What did Congo's new status as a Belgian colony mean to him? Did he notice anything of the transformation from the Congo Free State to the Belgian Congo? Was 1908 a pivotal year for him and his family as well? Did the local population actually notice anything of that reshuffle?

Hard questions to answer.

The classic historical accounts often say: the atrocities of the Free State lasted until 1908, but as soon as Belgium took over the colony everything calmed down and Congo's history became *un long fleuve tranquille* (one long, smooth flow), which only much later, at the end of the 1950s, began to once again exhibit a few whitecaps.[4] From that perspective, the colonial period in the strictest sense, lasting from 1908 to 1960, was a long and tranquil intermezzo between two turbulent episodes. In present-day Belgium, people tend often to be more concerned about the atrocities of Leopold II and the murder of Lumumba—two moments, strictly speaking, that do not belong to the classic colonial period—than by the decades in which the Belgian parliament and therefore the Belgian people were directly accountable (or should have been) for what happened in Congo. That idea of peaceful stability is reinforced further by the lengthy tenure of a number of key figures. Between 1908 and 1960, Congo had only ten governor generals, some of whom remained in office for seven or even twelve years. The first two ministers of colonies, Jules Renkin and Louis Franck, were in service for ten and six years, respectively. A tranquil current with a few solid beacons, or so it seemed.

But perhaps those are only assumptions. There was, after all, no complete break with the years before 1908. The Belgian tricolor was raised over the capital city of Boma on November 15 of that year, as the flag of the Free State was lowered and folded up for good, but little change was seen afterward. Leopold's regime continued to cast a long, dark shadow over the colonial period. Furthermore, the half century of Belgian rule was anything but static. In fact it was characterized by a unique vitality—not only the oft-sung, unilinear dynamism of "progress," but also the multifaceted dynamism of a complex historical era marked by tensions, conflicts, and friction. A long, wide current that grew ever more powerful? No, more like a braided river with numerous side channels, rapids, and whirlpools.

There was certainly a great deal afoot in 1908, but at first that new dynamism was seen more in Brussels than in Congo. On paper, a new dawn had come. The Colonial Charter arranging the transfer of the Free State provided Congo for the first time with a sort of constitution. Very much aware of the misery suffered under the Free State, the Belgian ministers and secretaries laid out a completely new system of governance. Colonial policy was no longer based on the caprices of an obstinate ruler who could impose his will, but was established by the parliament, which was charged with ratifying laws concerning the colony's administration. In actual practice, such policy was largely conceived and implemented by the minister of colonies, a newly designed post with a rather absurd title. The plural form, copied from its foreign neighbors, was a misnomer: Belgium had only one colony. Parliament itself spoke out only rarely on "overseas" politics. On December 17, 1909, no more than thirteen months after his lifework was taken from him, Leopold died. His successor, King Albert I, adopted a much more discreet and less self-willed stance when it came to Congo. There was also the Colonial Council, a new government body designed to provide the minister with technical advice on a host of subjects. Of its fourteen members, eight were appointed by the king and six by parliament and the senate. And then there was the Permanent Commission for the Protection of Natives, an institution with noble aims but little influence. During the fifty years of its existence, the Permanent Commission met only ten times.[5] The financial arrangements changed as well: Leopold's shadowy arrangements—which allowed him to slush money back and forth between his own personal fortune and the civil list, the means put at his disposal by the nation itself—were gone for good. From now on, black-and-white transactions were the rule. Revenues from the colony were to go to the colony itself and no longer to building projects in Brussels; this also meant, however, that Congo was to support itself in times of crisis (although Belgium, in actual practice, sometimes footed the bill). The colony, in other words, was to bear the joys and burdens of having its own budget.

These were drastic administrative reforms. But a change was seen as well in the attitude with which the colony was governed. The adventuresome made way for the bureaucratic, foie gras for corned beef. After Leopold's antics, preference was given to a strict and sober approach. Belgium assumed its role as colonizer with more gravity than pride. The administration became highly officialized and in Belgian terms that meant extremely hierarchical and centralized. Its directives originated in Brussels and were given substance largely by people who had seldom or never been to Congo. This resulted on more than one occasion in conflicts with the European people in the colony itself. In Congo the governor general still reigned supreme, but his estimations of the situation in the colony were often at loggerheads with the orders handed down to him from Brussels. What's more, Belgian colonials had no say in colonial policy: they had no formal political power. They submitted and not always enthusiastically.

But if they felt passed over, how much worse must it have been for the Congolese themselves? The Belgian government's policies definitely had the natives' best interests at heart: that insight, after the experiences with red rubber, was quite firmly defined. But Belgium was not answerable to the people in the country. The government was not elected by them, nor did it consult them in any way. It took care of them, with loving kindness.

AS POORLY AS THE BELGIAN GOVERNMENT LISTENED to the people in Congo itself, just as carefully did it heed the words of science. The objective, as Albert Thys put it, was "une colonization scientifique."[6] No more ad hoc improvisation, but Cartesian methodicalness. Scientists were the embodiment of this new-fangled sobriety—impartial, businesslike, colorless, and reliable. Or so people assumed. For in actual practice, it was their supposed impartiality that allowed them to gain so much influence.

One of the first scientific groups to gain a say in this way was that of the physicians. Around the turn of the century, Ronald Ross, a British doctor born in India, discovered that malaria was not caused by breathing in "bad air" in swampy areas (mal aria in Italian; the disease was still common in the Po estuary at the time). It was the mosquitoes that lived and bred in the stagnant water there that transmitted the sickness. One of the great mysteries of the tropics, which had claimed the lives of countless missionaries and pioneers, had been solved. Ross received the Nobel Prize in 1902 for his discovery. But that was not all. Yellow fever and elephantiasis, the disease that caused such gruesome malformation of the limbs, also turned out to be spread by mosquitoes. The enigmatic sleeping sickness came from contact with tsetse flies. Leishmaniasis was carried by sand flies, typhoid fever by lice, bubonic plague by the fleas on rats. The bite of a tick could produce stubborn attacks of fever. A new field of study, tropical medicine, was born; it was to become a powerful tool in the service of colonialism. Leopold II had already invited scientists from Liverpool to the Congo to study sleeping sickness. In 1906, on the model of the Liverpool School of Tropical Medicine, he set up the École de Médecine Tropicale in Brussels, forerunner of the Antwerp Institute for Tropical Medicine.

For the inhabitants of Congo, this medicalization had major consequences. Even during Leopold's regime, field hospitals were set up here and
there in the Free State, where the victims of sleeping sickness were attended
to by nuns. These *lazarettes* were located on islands in the river or at remote
spots in the jungle and closely resembled leper colonies. Hospitalization
often took place under duress. The patients were subjected more to a sort
of quarantine than any form of nursing. No family, friends, or relatives
were allowed to visit. For many, therefore, referral to the *lazarette* felt like
the death sentence. The patients served as guinea pigs for all sorts of new
medicines, like atoxyl, a derivative of arsenic that produced blindness more
frequently than recovery. It was not always clear what was actually being
improved, the patient's health or the experimental medicine. Because the
aim was to isolate victims during the earliest stages of the sickness (when
it is most contagious but also most treatable), those who were quarantined
often felt perfectly healthy. Swollen lymph glands in the neck were often
their only complaint. The characteristic symptoms arose only during their
stay at the field hospital itself. The *lazarettes* therefore developed a bad reputation: people believed they were camps where colonial officials had one
intentionally infected with the sickness. Riots broke out and guards cracked
down, but many patients ran away and went back to their villages.

When Belgium took over Congo, for the first time in colonial history
a medical service was set up . . . in Brussels. The chain of command to the
local post administrators in the bush was extremely long, yet policy was
successfully tailored to fit the new situation. Field hospitals alone were not
enough. From now on, the movements of all Congolese had to be moni-
tored. A 1910 decree stated that all natives fell under a specific *chefferie* or
sous-chefferie (territory administered by a chief or smaller unit).[7] The boundaries of those areas were carefully demarcated, and existing territorial limits were taken into account. Anyone wishing to move over a distance of
more than thirty kilometers (about nineteen miles) for the duration of more
than one month, another decree from 1910 said, were obliged to carry with
them a medical passport that stated their region of birth, state of health,
and any treatments received to date. A passport could only be obtained
with the approval of the village chief or *sous-chef.* Those already infected
were kept under village arrest. Anyone traveling without the proper documents risked a fine.

It would be hard to overstate the importance of this measure, which had five far-reaching consequences. First, all Congolese, even those in perfect health, were no longer able to come and go as they pleased; their freedom of movement was severely curtailed. For a region with a permanently high degree of mobility, that took some getting used to. Second, each inhabitant was from now on pinned to a spot on the map, like a beetle to a piece of cardboard. The sense of belonging in the native communities had always been great, now it became absolute. One's identity was from then on chiseled in granite. Third, the local *chefs* became part of the local administration. That process had started already in Stanley's day (see the case of Makitu), but now it was formally confirmed. They constituted the bottom rung on the ladder of the official hierarchy and fulfilled an intermediary role between the state and its subjects. The colonial government, of course, preferred to work with docile local leaders. The officially appointed chief, therefore, was often a weak character with little moral authority; the real, traditional chieftain remained in the lee, in order to go on ruling as before.[8] Fourth, because the average *chefferie* comprised no more than a thousand inhabitants, larger ethnic entities dissolved.[9] The village fell directly under the authority of the state, and intermediate levels vanished. That too had an impact on tribal awareness: nostalgia arose for a former glory. And fifth, for many, the laws originating in faraway Brussels were their first immediate encounter with colonial bureaucracy. During the era of the Free State, hundreds of thousands of Congolese had been brought under the yoke of the distant oppressor; now, in principle, there was *no one* who remained out of reach. The number of Belgian colonials remained quite limited (a few thousand in 1920), but the colonial apparatus tightened its grip on the population and penetrated further and further into the life of the individual.

The state: in 1885 that had been a lone white man who asked the head of your village to fly a blue flag. The state: in 1895 that was an official who came to conscript you as a bearer or soldier. The state: in 1900 that was a black soldier who came to your village, roaring and shooting, all for a few baskets of rubber. But in 1910 the state was a black assistant nurse on the village square who felt your lymph glands and said that it was good.

The colonial administration hoped to get started quickly with a large-scale medical examination of the population; King Albert allocated more than a million Belgian francs to that end, but World War I delayed the process. Starting in 1918, however, teams of Belgian physicians and Congolese nurses began traveling from village to village, and many hundreds of thousands of villagers were tested. The state: that was the men with microscopes who frowned gravely as they looked at your blood. The state: that was the gleaming, sterile hypodermic needle that slid into your arm and injected some kind of mysterious poison. The state literally got under your skin. Not only was your countryside colonized, but so was your body and your self-image. The state: that was the medical pass that said who you were, where you came from, and where you were allowed to go.

Lutunu's life, in any case, became much more domestic as a result. The man who had been to every corner of his country, in addition to making trips to Europe and América, now remained in his village year after year. As assistant to an adolescent village chieftain, he probably advised the white supervisor about who was eligible for a travel pass and who was not. It need come as no surprise that this system left the door wide open for abuses. Passports were highly coveted, and some chiefs, clerks, and nurses were open to bribes. In the hope of obtaining a new, blank passport, villagers who wished to travel but had only recently been treated for sleeping sickness claimed to have lost their medical papers. Many people viewed the white man's medicine with profound suspicion. Atoxyl could blind you and the spinal taps used to treat the worst cases were very painful. Yet this is not to say that the people lived in irrational fear of everything wearing a white coat. Some treatments met with general approval, such as the operative removal of tumors caused by elephantiasis, but hypodermic needles were commonly seen as a way to spread diseases. The colonizer clearly underestimated the deep attachment to traditional medicine and rejected it out of hand as quackery and mumbo-jumbo. For many Africans, this made sleeping sickness the white man's disease, part and parcel of military domination, economic exploitation, and the political reshuffle.

And in all of this, physicians had power, a great deal of power. Doctors decided who could go where. They established the boundaries of those areas where travel was forbidden. They could force recalcitrant individuals to undergo treatment, and even punish them. They even had the authority to have entire villages moved, should there be sound public-health reasons for doing so. Village communities in areas where the tsetse fly flourished could be forced to pack up and move collectively. And should such a village refuse, doctors could call in the assistance of colonial officials and the Force Publique. More than healing individuals, this brand of medicine was intended to keep the colony on its feet.

After such mandatory migration, local communities often became disjointed. Bakongo tribespeople forced to leave their village behind sang with nostalgia and the blues: "Hey! Look at the village of our forefathers. / The shady village of palms that we were forced to leave. / Hey! The old folks? Hey! Hey! Hey! Our dead have vanished! / Hey! Look at our abandoned village! / Too bad!"[10]

Lutunu's village was allowed to stay put. In order to reduce the risk of sickness, however, he did something no one in his village had ever done before: he built a house of stone. From then on he no longer slept beneath a thatched roof, between walls of clay, but in a brick hut with galvanized iron for a roof. By then, after all, there were enough carpenters and masons to be found in nearby Thysville. They knew how to make bricks from earth and how to nail down corrugated iron. Sleeping sickness had destroyed Lutunu's family, but now he lived more or less like the white man did. Were his new brick walls hung, as a Belgian cabinet minister noted after a visit to a village chieftain in eastern Congo, with "extremely middling portraits of our rulers, as distributed everywhere by the colonial administration, and a few illustrations torn from magazines from Paris or London"? Did the occasional white visitor leave behind "a few pretty prints and a few tins of caramels" as a present?[11] No one knows. What we do know is that a few years later the colonial administration appointed him chief of the region, and that, a former slave, he was given authority over no less than fifty-two villages.

THE SECOND GROUP OF SCIENTISTS to turn their attention to the colony were the ethnographers. If the scandal of the Free State had made anything clear, it was the total lack of understanding of native culture. Félicien Cattier, the eminent Brussels scholar and a virulent critic of Leopold, had been quite explicit about that: "How can one hope to carry out useful work in the colonies, if one fails to first submit the native institutions, their customs, their psychology, the conditions of their economic existence and their societies to careful study?"[12] Some explorers and missionaries had shown interest in local customs, but many officers and agents of the Free State entertained rudimentary views, to say the least, about what they called "the Negro race." If any interest existed at all, it was focused primarily on the tangible aspects of the foreign culture: their baskets and masks, their canoes and drums, the shapes of their spears, the dimensions of their skulls.

But that, Cattier felt, was not enough. This was not about specific objects or individual traits. One had to develop an eye for the deeper layers of native society. And that called for serious study. "It would be fitting if there were to be set up in Congo, just as in the Dutch Indies or British India, a ministry or office of ethnological studies."[13]

And so it came about. With great to-do, the Bureau International d'Ethnographie was called into being, an institution manned by Belgian and foreign researchers whose goal it was to gather and analyze as much data as possible concerning the native population of Congo. What the École de Médicine Tropicale was for medicine, the Bureau International d'Ethnographie was for anthropology: an agency possessing expertise that became transformed into influence. Its members read travel diaries and mission reports and invested a great deal of time in drawing up exhaustive questionnaires, which were then sent to thousands of respondents in the colony: civil servants, traders, soldiers, and missionaries. They were asked to fill out all 202 sections, with themes ranging from marital customs to funeral practices and on to personal hygiene. The informants complied and the answers began flowing in. Within a four-year period, more than four hundred thousand bits of ethnographic data were processed.[14] This information ended up in a monumental series of books, the *Collection des monographies ethnographiques*, eleven volumes of which appeared between 1907 and 1914. Each volume dealt with a given population that was considered characteristic of a given geography: the Bangala for the riverside, the Basonge for the savanna, the Warega for the jungle . . . and attention was granted as well to the Mayombe, the Mangbetu, the Baluba, and the Baholoholo. A description was provided each time of all 202 sections of the questionnaire, adding up to more than six thousand pages of reading. It was the first attempt to systematically document native culture. The result was nothing less than an *encyclopédie des races noires* (encyclopedia of the black races).[15]

But the result was also that these "races" were suddenly seen as something absolute. The series sorted the inhabitants of Congo into clearly distinguishable blocs, each with its own identity, ethnic character, and customs. There was something to be said for it—there were, after all, undeniable differences—but the attempt to throw up a cultural wall around each of those groups was entirely artificial and served to obscure any interrelations. Yet that is precisely what the ethnographers did. At the outset of the project in 1908, the key collaborator, Edouard De Jonghe, resolved to examine "les peuplades une à une, en elles-mêmes" (the tribes one by one, as they are).[16] Methodologically, this step-by-step approach was quite understandable; it kept things orderly. But what was at first a guiding principle soon became an unshakeable conclusion. The "tribes" were eternal, free-standing, and immutable entities. After a few years the project's initiator, Cyrille van Overbergh, also a prominent Catholic politician, stated with certitude: "Generally speaking, the peoples of Congo have little in the way of mutual interrelations. . . . The tribes are independent of each other and retain their autonomy."[17] This roundly ignored the centuries-long, and by that time well-known, exchanges between various groups of the population. Pygmies lived alongside Bantu-speaking farmers; the Bobangi took their boats upriver and came in contact with dozens of other groups. Ethnically, the former savanna kingdoms of Bakongo or Baluba had been quite heterogeneous. Many natives were multilingual. The cultures of the various Bantu-speakers were closely related. But the early-twentieth-century anthropologist unraveled the population into specific racial threads, the same way the eighteenth-century taxonomist had once split up the animal kingdom into various species. Changeless throughout time, without admixtures.

Congo became like an old-fashioned typecase. The map of the colony from then on consisted of little compartments, each with its own tribe. A gigantic collection of ethnographic material was assembled in Tervuren, outside Brussels, all neatly categorized according to tribe. Because the physicians forced people to stay put, the anthropologists fell prey to the even more outspoken impression that the peoples they encountered "were tied to their respective territories," as the head of the Bureau International d'Ethnographie put it.[18] This "monographic vision" had major consequences. Not only did the white people in the colony begin to act accordingly, but the Congolese themselves began identifying with a tribe more and more. The genie of tribalism was out of the bottle.

Early anthropology was not at all seen as art for art's sake; it was intended to facilitate the colonizer's work. Recruiters for the Force Publique, for example, could profit from a description of the belligerence inherent to certain tribes. The medical services could learn about the hygienic conditions amid the peoples most heavily affected by sleeping sickness. The administrators in Brussels could adapt their legislation in accordance with what they read about traditional common law in the colony. And the mission congregations could adjust their tactics on the basis of which religion was prominent in which region. People acted on the insights gleaned from the *Collection des monographies ethnographiques*. Like the various European nationalities, the tribes were ascribed characters of their own. In Congo there arose the equivalent of the stingy Scotsman, the lazy Sicilian, the messy Spaniard, and the hardworking but humorless German.

The inhabitants of the colony, too, began seeing themselves and each other in that way. What about Lutunu, for example? He had seventeen children, thirteen of whom survived into adulthood. Beginning in 1910 they all fell under the same *chefferie*, had the same state-recognized village chieftain, and without medical permission were not allowed to leave the area—all ingredients that furthered a pronounced regional and ethnic awareness. In addition, they received schooling from the missionaries, for education in the colony was entirely in the hands of the church. In 1908 there were some five hundred missionaries in Congo, by 1920 around fifteen hundred. School attendance was not mandatory, but Lutunu with his bicycle and his brick house, must surely have encouraged them to learn to read and write like himself. He was, after all, one of the first literate inhabitants of Bas-Congo. His village lay within the sphere of influence of the British Protestants, but outside that area the power of the Belgian Catholics continued to grow.

And what were the Congolese taught, in those plain classrooms or in the shade of a tree? Reading and writing, of course. Arithmetic as well. Church history. Devotional stories. The provinces of Belgium. About the royal family, of course, but also a few lessons concerning their own country. About the slave trade, for example. "Tungalikuwa watumwa wa Wangwana / Wabeleji wakatukomboa," the children at Catholic mission posts in the interior sang. Literally: "The Arabized Africans were making us their slaves / The Belgians set us free." The melody was taken from the "Brabançonne," the Belgian national anthem. One of the oldest known school songs in Swahili provided a summary description of the colonization: "Once we were idiots / Sinning day by day / Sand fleas on our feet / Heads full of mould / Thank you, reverend fathers!"[19]

The songs and lessons of the Catholic priests and nuns were invariably taught in the local vernacular. Most of the missionaries came from Flanders and, in analogy to the Flemish struggle for linguistic recognition, considered one's own language to be a supreme good. That too served to bolster tribal pride. In a grammar used by the Missionaries of the Sacred Heart in Mbandaka in the 1930s, one finds the following reading exercise: "Our language is Lonkundo. . . . Although some people like to speak Lingala, we love our Lonkundo best. The language is very beautiful and has many shades of meaning. We are very fond of it. It is a language given to us by our ancestors."[20]

But ethnic identification also took place much more explicitly. Around that same time, pupils in Équateur read that "the people of Congo are divided in many groups. They stand out by reason of their dialects, their customs, and even their laws. Our true family is the tribe of the Nkundo."[21] This sounded like a literal echo of the *Collection des monographies ethnographiques*. The first textbooks used by the Marist Brothers (the oldest dates from around 1910) went even further. In Lingala, one read in a textbook:

The inhabitants of Congo are black. Their numbers have not yet been counted. They are estimated at up to sixteen million. They can be divided according to various tribes: Basorongo, Bakongo, Bateke, Bangala, Bapoto, Basoko, Babus, Bazande, Bakango, Bangbetu, Batikitiki, or Baka and many others.

The Basorongo live close to the ocean.

The Bakongo live upstream, close to Boma, Matadi, and Kisantu, on the river's left bank. They are dockworkers and heavy laborers.

The Bateke are found in Kitambo. They are specialized in buying and selling.

The Bangala live around Makanza, Mobeka, Lisala, and Bumba. They are large people. They wear tattoos on their faces and ears. They remove the lashes from their eyelids and file their teeth. They are not afraid of war. Are there not, after all, many Bangala in the government's army? They are intelligent.

The Bapoto and Basoko are brothers to the Bangala. They disfigure their faces with tattoos. They make big pestles and sound canoes, they forge spears and machetes. They kill lots of fish.[22]

AND SO IT GOES, on and on. Congo consisted of tribes, one was taught, each with its own territory and customs. Some were virtuous, others were not. The pupils, for example, were taught that the Azanda respected their chieftains, which was a good thing; the Babua did not and that was scandalous. The Bakongo killed elephants and were therefore very courageous. Mission schools were factories for tribal prejudice. Children who were not allowed to leave their villages were suddenly told that the Bakongo lived on the other side of their vast country and what they were to think of them. In many handbooks, Pygmies were depicted as bizarre aberrations. If you had never met one, you still knew what you were to think of them. "They excel in stealing other people's property," the pupils at Bongandanga read in the late 1920s, "they do not make friends with other people. . . . Most of the peoples of Central Africa are fond of keeping themselves clean and because there is plenty of water they bathe every day. But the Pygmies detest water and are very dirty. . . . In terms of ignorance, they stand head and shoulders above all other African peoples. They do not realize that living in a village with people of your own culture is better than moving around all the time."[23]

This is not to say that there had never been tribes—of course there had, there were major regional differences, different languages were spoken, different customs honored, different dances danced, different dietary patterns observed, and there had even been intertribal wars. But now the differences were being magnified and recorded for all time. It rained stereotypes. The tribes, in fact, were not communities that had been fixed in place for eons; their rigidity came only in the first decades of the twentieth century. More than ever before, people began identifying with one tribe as opposed to the other.

In the 1980s an old man from Lubumbashi recorded a few recollections of his childhood. The nascent mining operations then had brought people from various backgrounds together in the compounds: "In the olden days we didn't look at other people and say: 'That one there is from Kasai, that one is a Lamba, a Bemba or a Luba.' No. We were together." And, he added: "There was no difference. No one talked about the differences."[24]

The missions not only ran primary schools, but also set up seminaries to train talented pupils to become local priests. The first Congolese to be ordained was Stefano Kaoze, in 1917. He came from the Marungu mountains and was molded and made by the white fathers. In 1910, at the age of twenty-five, he had already come up with a first: his long essay "La psychologie des Bantu" appeared in *La Revue Congolaise*. This made him the first

Congolese to publish a text. And what do we read in the first paragraphs of this incontestable landmark document? What does a young Congolese intellectual write, one who has been saturated with Catholic mission schooling? Indeed, that tribal awareness in Africa was nurtured by European books: "When I had read a number of books about a number of tribes, I saw that most of the customs originate from the same background as those of the Beni-Marungu [Kaoze's own tribe]. Now that I realize this, I am going to tell who we are, we Beni-Marungu, and what we are not."[25] The books he read caused him to reflect on his own tribal identity. Is it any wonder that, later in life, he developed into a tribal nationalist, a champion of his own people and a defender of Congolese interests? "Potentially the most dangerous black man," a French nobleperson noted after a tour of the colony, "is he who has had a bit of education."[26]

MEANWHILE, NKASI'S LIFE DRIFTED ALONG CALMLY. When interviewing him, I was struck on a number of occasions by the fact that he had few memories of the early years of the Belgian Congo. When he spoke of the building of the railroad in the final decade of the nineteenth century, his eyes twinkled and the stories came of their own accord. But the decades that followed, which he spent back in his village, seemed to have been washed away. For a long time I wondered why, until I noticed that Lutunu's biographer was also rather laconic about that period in her subject's life. She too had noted blank spots in her conversations with her informant. Could that be a coincidence? I don't believe it is. I suspect that the legislation forcing people to remain in their villages also resulted in becalmed years with few spectacular events. World War I passed them by with barely a ripple, even for Lutunu who was by that time, after all, an assistant regent. When I asked Nkasi again whether he really could remember nothing of the Great War, he said: "I may have heard of it, but it didn't happen here."[27] His world had drawn in on itself once again. His youngest brother was born around that same time, yes, he remembered that. And in the end he had finally allowed himself to be baptized a Protestant. That was in 1916, at the Lukunga mission post. His Christian name became Étienne, but everyone continued to call him Nkasi.

For him, the big turnaround came in 1921: for the first time in a long time, he left his village again. To do that he first had to apply for a valid passport and *une feuille de route* (a travel pass), otherwise he would not be allowed to leave. Even today, a Congolese has trouble traveling through his country without an *ordre de mission*; Congo is one of the few countries in the world with a migrations service that also deals with *domestic* travel—due to the once-so-preponderant sleeping sickness. But Nkasi was in luck. His father's cousin worked for the railroad and so he was able to travel for free by train. He spent one whole day chugging across the grand landscape and arrived that evening in Kinshasa.

The place had changed unrecognizably since Swinburne had set up his post in the wilds there in 1885. Along the shores of Stanley Pool, some eighty companies had meanwhile built warehouses. Eight kilometers (about five miles) to the west lay the older military and administrative center, Léopoldville, where the British Baptists had once established their headquarters. In 1910 the two nuclei, Kinshasa and Léopoldville, were connected by a broad road. Today that is the Boulevard du 30 Juin, no longer a connecting road between two European settlements, but the city's hectic, smoking main arterial. When Nkasi arrived, however, there were no more than two hundred cars and trucks in Kinshasa. A thousand white people lived there, including one hundred and fifty women. The city numbered some four hundred houses built of durable materials.[28]

Nkasi found himself in a city under construction, a dusty flat full of building sites and avenues leading nowhere. To the south of the European district the colonizer had built a *cité indigène* (district for housing African workers), a three-by-four-kilometer (about five-square-mile) checkerboard neatly divided by straight lanes. Clay huts with thatched roofs stood on the tidy square plots. Around the houses, the inhabitants grew manioc and plantain. Here and there one saw a brick house with a corrugated iron roof. Children ran naked down the sandy alleyways. Women spent hours sitting the shade, combing each other's hair. Some of the house fronts were painted. It was there, he found out quickly, that one could buy rice, dried fish, and matches. This was a new world. Within only a few years, twenty thousand people had come to live here. Another twelve thousand settled in neighboring Léopoldville. They had arrived from all over the interior. They spoke languages he didn't understand and came from regions he had never heard of. Only four thousand of them were women. It was a man's world full of coarse shouting, roars of laughter, and homesickness. The *cité indigène* in no way resembled the traditional village; it was one huge camp of manual workers and tradesmen, but also of boys who made their way up to the white neighborhood each morning, and of vagabonds, the victims of sleeping sickness, thieves, and prostitutes.[29]

"I came to Kinshasa in 1921. I worked for Monsieur Martens," he told me. "He had sheds full of diamonds from Kasai. Diamonds came from the mines, but they were sorted in Kinshasa. My job was to fill sacks and empty them." To illustrate his words, he made a shoveling motion with his arms. "Fill them and empty them. I earned three francs a month."[30] To prevent thievery, the diamonds were not sorted at the mines themselves. The concentrate that came from the washing plant was instead taken to a central depot.

Nkasi's move to the big city that was soon to become the colony's capital was due to a twenty-milligram grain of glass that had been found years before at a spot many hundreds of kilometers to the east. In 1907 Narcisse Janot, a Belgian prospector traveling around Kasai with a geologist, found a chunk of crystal that did not look entirely unpromising. Because he did not have the instruments needed to carry out a petrological assay, he put it in a tube and took it back with him to Brussels. When he got home however, he forgot about it and the tiny stone remained among the many geological samples brought back by the expedition. It turned up again only years later. Further analysis proved that it was, indeed, a diamond.[31] A veritable rush ensued. Kasai turned out to be the source of high-grade diamonds fit for jewelry, but also of a rougher sort in great demand for industrial use.

At other spots as well, the colony's substratum proved to have highly welcome surprises in store. Back in 1892 the young geologist Jules Cornet had discovered extremely rich veins of copper in Katanga: areas such as Kambolove, Likasi, and Kipushi seemed particularly promising. That evening in his tent, he noted: "I would not dare to venture a figure concerning the enormous quantity of copper present at the sites I have recently examined: if I did, it would sound all too outrageous and unbelievable."[32] King Leopold II made him swear to keep his discovery a secret, so as not to rouse Britain's interest. Probably not unwise: the copper deposits of Katanga proved to be the richest in the world. Some areas of substrate contained up to 16 percent pure copper. In a few rivers in the hilly northeast of the country, close to the border with Uganda, two Australian prospectors found a number of unsightly chunks of metal that gleamed in the sun: gold. The sites at Kilo and Moto would develop into the most important gold mining area in Central Africa. And in 1915 another prospector in Katanga found a yellowish, extremely dense stone that reminded him of the work of Pierre and Marie Curie. Later analysis showed the stone to indeed be very rich in uranium. The place where it was found became the Shinkolobwe mine—for decades the world's major supplier of uranium ore.

Beneath its surface, Congo turned out to conceal a true "geological scandal," as Cornet put it. It was almost too good to be true. Until then, the economic exploitation of the area had been aimed exclusively at its biological riches—ivory and rubber—but now a far greater wealth was found to be lying a few meters under the ground. Katanga, the rather unpromising region that Leopold had annexed almost by accident in 1884, suddenly turned out to contain an improbably vast treasure trove. In addition to copper and uranium there were major deposits of zinc, cobalt, tin, gold, wolfram, manganese, tantalum, and anthracite coal. The discovery that the colony was sitting atop these immense mineral riches came, by the way, not a moment too soon. Revenues from rubber harvesting had begun sinking rapidly as from 1910. The world price for rubber was in free fall. In 1901 rubber had accounted for 87 percent of Congo's exports; by 1928 that was only 1 percent.[33] "These days," a traveler noted in 1922, "one no longer—or almost no longer—refers to rubber in Congo."[34]

It seemed like a historical déjà vu: in the same way that the rubber boom had arrived just in time to offset the dwindling ivory trade, mining began just in time to replace the ailing rubber industry. There is no other country in the world as fortunate as Congo in terms of its natural wealth. During the last century and a half, whenever acute demand has arisen on the international market for a given raw material—ivory in the Victorian era; rubber after the invention of the inflatable tire; copper during full-out industrial and military expansion; uranium during the Cold War; alternative electrical energy during the oil crisis of the 1970s; coltan in the age of portable telephonics—Congo has turned out to contain huge supplies of the coveted commodity. It has easily been able to meet demand. The economic history of Congo is one of improbably lucky breaks. But also of improbably great misery. As a rule, not a drop of the fabulous profits trickled down to the larger part of the population. That dichotomy, that is what we call tragedy. Nkasi, who once worked by the sweat of his brow to empty sacks of jewel-laden earth, profited very little indeed from the entire diamond business. Today he is poor as a pauper.

For the colonizer, however, these finds were extremely important. They signaled the start of the local mining industry, even today the most important branch of Congolese industry by far. But extracting and processing ore was not the same as buying tusks or commandeering baskets full of rubber. To achieve a profit here, one had to make huge investments. Crushers and rinsing installations had to be built, ovens, foundries, hoists, and rolling mills. What's more, the most important minerals came from regions far from the ocean. If Africa resembled a giant pear, then Katanga was "if not its heart, then certainly one [of] its best seeds."[35] That called for the construction of new railroads, harbors, telegraph lines, and roads.

All this was financed by the Belgian state and private capital. The goldmines of Kilo-Moto were at first entirely state owned; the enterprise went public in 1926. In other places one reverted to the system of concessionaries, the same arrangement that had made "red rubber" possible. Those companies operated with private capital, but there was also usually a lavish retombée (fall back, beneficial effect) for the colonial treasury. That took place not by means of direct taxation (before World War I, a tax on profits was still almost unheard of), but by the mandatory relinquishing of large packets of shares to the colonial government. That stock portfolio made it possible for the treasury of the Belgian Congo to fall back on what were often extremely ample dividends.

In 1906 three companies were set up that would play a crucial role in Congo's mining activities: the Union Minière de Haut-Katanga (UMHK), the Société Internationale Forestière et Minière du Congo (Forminière), and the Compagnie du Chemin de Fer du Bas-Congo au Katanga (BCK). Half the Union Minière's starting capital came from British investors, the other half from the Generale Maatschappij, the powerful Belgian holding company that had maintained a firm grip on the national economy ever since 1822. The company focused largely on Katanga. After the initial extraction activities had been carried out by a private company—the Compagnie du Katanga, run by Albert Thys, the same industrialist who had built the railroad in Bas-Congo—the Comité Spécial du Katanga (CSK) became involved. The CSK had a very special legal structure: it was not a classic enterprise but a semigovernmental organization run by the colonial state, a partnership sui generis, with public-private funding and unique privileges. It laid claim to exclusive mining rights for half of Katanga, and was also charged with the region's political administration. The CSK, although more a company than a government, even had its own police force. It was a state within the state. This odd situation continued even after the Union Minière came along in 1906. Economic and political interests remained tightly interwoven. As the supreme industrial colossus in Katanga, the CSK often had more say in the colonial administration than the colonial administration did in the company. The colonial government, for example, facilitated the recruitment of workers for the company. Katanga, in short, was subject to a form of administration unlike that in the rest of the country. It was that, among other things, which would later fuel the region's struggle for independence.

Forminière was set up with American capital. Because the diamond deposits were so widely scattered, the company was originally allotted a prospecting stake of no less than 100 million hectares (about 39,000 square miles), later reduced to 2 million hectares (about 7,700 square miles) with fifty mines in the area around Shikapa and Bakwanga. In 1913 Forminière extracted 15,000 carats in Katanga; by 1922 that had grown to 220,000 carats.[36]

BCK, finally, the third company set up in 1906, was a private railroad company founded with French-Belgian capital and charged with the construction of a rail connection between Katanga and Bas-Congo. It was along this line that ore was to reach the ocean without leaving the territory of the Belgian Congo. The only other alternative would have been to cross through Portuguese, German, or British colonies, thereby generating troublesome forms of dependence. The new railroad was finished in 1928. But BCK was involved in more than simply building railroads. The company also owned enormous mining rights, which would later serve it very well indeed. Its concession turned out to contain one of the world's largest deposits of industrial diamonds. The profits were spectacular and almost half of them flowed into the Congolese treasury.[37]

And Nkasi went on shoveling. The earliest mining activities, after all, called for manual labor, a great deal of manual labor. And who was going to supply that? The Belgians themselves? That seemed to be out of the question: "South of the Equator, a Belgian can carry out almost no other work than that of supervision. Continuous physical effort, every form of manual labor, which is difficult enough in itself, is more or less off limits to him."[38] For a time, in sparsely populated Katanga, consideration was given

to importing Chinese laborers; in view of the god-awful mortality rates
experienced during the building of the railroad, however, this idea was soon
abandoned. Anyone flying over Katanga today by helicopter, for example
from Kalemie to Lubumbashi, as I had the honor of doing in June 2007, can
learn a great deal about the region's social history. The UN aircraft in which
I was supposed to travel turned out, due to a shortage of passengers, to have
made way for a worn-out chopper with a Russian crew and Russian insignia.
Rather than a short, two-hour flight, it became a long and noisy six-hour
journey over an empty landscape. We flew at an altitude of no more than
three hundred meters (about a thousand feet). One could pick out the individual trees, buffalo, and termite hills, but rarely a village. Wearing my red
ear protectors as I peered out the open window, I came to better understand
the transformation that had taken place here a century earlier. If today, in
an era of explosive population growth, the savanna still remains so empty,
I thought, how much more desolate must it have been a hundred years ago,
after a pandemic of sleeping sickness?

Katanga was packed with ore, but there was no one to dig it up. In
the isolated villages a fruitless search was carried out for people willing to
work. From 1907 on, therefore, the companies began recruiting abroad:
each year, six or seven hundred Rhodesians came to work the Katangan
copper mines.[39] By 1920 their numbers had risen to many thousands;
they accounted for one-half of all the African laborers. The workers were
employed for stretches of no longer than six months, they lived in compounds, as at the South African mines, and were not allowed to bring their
families along.

There are almost no firsthand accounts from those early mineworkers,
with a few exceptions. "I came to Katanga on May 4, 1900. I was hired by a
Mr. Kantshingo," an old man recalled. He had to undergo a medical exam
and was given a worker's pass with his thumbprint on it.

There were no houses of stone or brick. The blacks slept in huts, the
whites in tents and in termite mounds [sic]. Many of the whites were
Italians. The crew bosses came from Nyasaland [Malawi]. The language we used was Kikabanga. A pick was called a *mutalimbi*. A shovel
was a *chibassu*, a wheelbarrow a *pusi-pusi*, a hammer a *hamalu*. At four
in the morning we left for work. We started at six and stopped at five,
six, seven o'clock at night. The workers were beaten very often. . . .

We used Rhodesian money. The beer we drank was called *kataka* and *kibuku*, it was made from corn or sorghum.[40]

In 1910 Katanga was linked to the rail network that the British had built in their southern colonies. From then on there was a direct connection between Katanga and Cape Town. Around the little village of Lubumbashi, close to the mine that the prospectors called Star of the Congo, a city quickly arose: Elisabethville. In 1910 there were three hundred Europeans and a thousand Africans living there: one year later, there were a thousand Europeans and five thousand Africans.[41] From the very start, the city was more South African than Congolese. The straight roads lined with trees reminded one of Pretoria; the cozy white house fronts were more like Cape Town. The Rhodesian workers and British industrials saw to it that English became the dominant language and the pound sterling the prevailing currency.

There is an extraordinary document that helps us to understand that earliest phase of the Katangan mining industry from an African perspective. In the 1960s André Yav, an old man who had worked all his life as a boy in Elisabethville, wrote down his recollections:

When bwana Union Minière began, the first people who came to work there were from the nearby villages. Those were Balamba, Baseba, Balemba, Baanga, Bayeke and Bene Mitumba people. There were not very many of them, and they didn't really want to leave their villages for too long. They would work for two or three months and then go home. After a time, the places where there was work to be had became big. Then they started calling in people from Luapula and Northern Rhodesia [present-day Zimbabwe and Zambia], and others came as well: Balunda, Babemba, Barotse and also boys from Nyasaland. They were strong enough to do the work, but couldn't leave their villages for a long time either. After six or ten months, they would go home again.[42]

Things did not stop there, however. Recruiters moved farther and farther into Katanga in search of young, able-bodied men. Some recruiters worked for official organizations, but there were also very many private contractors, white adventurers who did their best to lure as many young

people as possible to the mines. Some of them even went as far as Kasai or
Maniema, journeys of eight hundred kilometers (about five hundred miles).
Their recruitment methods were often dubious: they would bribe village
chieftains with European luxury goods such as blankets and bicycles and a
bonus for each worker supplied them. Concerning the working conditions
in the mines, they remained prudently silent. They bought up workers in
order to sell them on again. Force was often used as well. In fact, their working methods differed little from the recruitment tactics of the Force Publique around 1890, or the Afro-Arab slave traders in 1850. In his memoirs,
our retired boy was perfectly clear about that:

> In that way, *bwana* Changa-Changa [the African nickname for Union
> Minière] and the other whites were able to set up their mining com-
> panies. . . . The misery we suffered was unimaginable; we slept on the
> ground, were bitten by snakes, by mosquitoes, by all kinds of insects.
> That's the way it was to work for the white people, and all that just
> to find ore in Katanga, and things were even worse with the whites
> of the Comité Special [du Katanga, active until 1910]. They made us
> walk around, go prospecting, look around in the bushes and on the
> hillsides for all kinds of stones. And what's more, we, the boys, had
> to go with the white people along all the rivers of Katanga, of Congo,
> everywhere.[43]

The housing provided for the first generation of mineworkers was
often abominable. The miners were placed in work camps, far from where
the whites lived in the city center. This spatial segregation was established by law from 1913.[44] Their neighborhoods looked more like military encampments than urban districts: rectangular and almost without
shade. Traditional huts were arranged in serried ranks. Four workers were
assigned to each hut, with four square meters (about forty-two square feet)
of living space each. Latrines were provided, at least in theory. In reality, the exhausted miners were forced to live under harsh and unsanitary
conditions. At the Kambove mine, the camp inhabitants sometimes literally had to wade through the dreck. Drinking water was scarce. With
its steam engines and drilling installations, the mine itself used up most
of the water. During the dry season, workers drank from stagnant ponds
or muddy streams.[45] And the diseases arrived. Dysentery, enteritis, and

typhoid fever took their toll, and local influenza epidemics broke out at
Elisabethville, at the Star, and in Kambove. At those three places in 1916,
322 workers out of a total of 5,000 died. Hard labor in the dusty mines also
caused many workers to contract pneumonia and tuberculosis. One quarter to one third of them fell ill, but health care remained minimal.[46] In 1920
there were some seventy physicians and one dentist for all of Congo: they
were there largely to serve the white population.[47] The miners worked
long hours and were paid a pittance. Many of them became apathetic and
depressed and longed for home. They organized themselves only in ad hoc
fashion and often along ethnic lines, to care for their sick, bury their dead,
to drink, and to sing. Some of them deserted, others did not dare. Until
1922 corporal punishment was allowed.

It was, all things considered, a grim situation. Southern Katanga had
never been bothered much by the Free State's rubber policies, but now the
region was dragged along by a relentless wave of industrial capitalism.
This caused André Yav, the retired boy, to draw an extremely remarkable but also very telling conclusion: he decided that King Albert I was
far worse than Leopold II, who had at least "honored the laws of Africa
and Congo"! That called for a bit of explanation: "In the days of King
Leopold II, the 'boys' ate with the white people at the same table. The
white people saw them as employees. They were not like the whites who
came after Leopold II. When he died, he was succeeded by King Albert I.
Those whites made hard decisions, and those decisions were really very
bad. They were the ones who brought a bad kind of slavery for us, the
Congolese."[48]

No less trying were the conditions at the Kilo-Moto gold mines in Orientale province. Only one out of every eight workers was there voluntarily,
the rest had been press-ganged in local villages: human trafficking, in other
words, and forced labor. Recruiters would pay a local chieftain ten francs
for each laborer and take the young men away in chains. They were bound
together at the neck by a wooden yoke or a noosed rope. In 1908 there were
eight hundred workers, by 1920 more than nine thousand.[49] In 1923, in
diamond-rich Kasai, some twenty thousand Africans were working in the
service of two hundred whites.[50]

Between 1908 and 1921, in other words, Congo experienced its first
wave of industrialization, thereby prompting the proletarianization of its
inhabitants. Men who had once been fishermen, blacksmiths, or hunters

became wage laborers for a company. Even in this earliest phase, their numbers were large. In Katanga, where 60 percent of the laborers worked for Union Minière, the body of mineworkers grew from 8,000 in 1914 to 42,000 by 1921, and the number of railroad workers from 10,000 to 40,700. Together, Kasai and Orientale province were good for 30,000 workers, while another 30,000 migrant workers lived in Kinshasa and Léopoldville. The reason for this massive recruitment of African workers was simple enough: sweat costs less than gasoline.[51]

This proletarianization was not limited to industry alone. Agriculture too had need of manual laborers, especially now that white farmers had started coffee, cacao, and tobacco plantations. The most extensive agrarian employment of all, however, was in the palm oil sector. In Liverpool in 1884, a man named William Lever had started making soap on an industrial scale. The bars rolled from the production line, and he christened his product Sunlight. This company's rise to become the Unilever multinational was due in part to the contribution of Congo. At first, the soap was made from palm oil that Lever purchased in West Africa. But when the British colonial administration there stopped providing favorable conditions, the Belgian state granted Lever an extremely sizeable concession in Congo in 1911. At his own discretion, he was allowed to stake out five circles with a sixty-kilometer (about a thirty-five-mile) radius in those areas where wild palms flourished, for a total holding of some 7.5 million hectares (about 29,000 square miles), two and a half times the size of Belgium. This was the start of the Huileries du Congo Belge (HCB), an enterprise that was particularly active in the south of Bandundu and grew to become a massive concern. Close to Kikwit, the town of Leverville arose. For the harvesting of palm nuts the company made use of thousands of Congolese, who climbed the trees in the traditional manner to cut down the clusters. Lever had a reputation as a great philanthropist, but very little that his company did bore witness to that in Congo. His employees earned a miserable twenty-five centimes a day and lived under primitive conditions. Press-ganging and the bribing of village chieftains took place. Dozens of villages had to pack up and move for the industry's purposes. People today in Kikwit think back on that period with bitterness: this was worse than what the region had experienced under the rubber regime.[52] It was something King Albert could hardly have suspected in 1912, when William Lever presented him with an ivory box containing the first bar of Sunlight soap manufactured with Congolese palm oil.

"I EARNED THREE FRANCS A MONTH," Nkasi had told me. It was the first time in his life that he earned a wage, which was why he remembered it so clearly. The budding industrialization of Congo led not only to an initial form of urbanization and proletarianization, but also to a far-reaching process of monetization. For the first time, on a major scale, the population became involved with a concept as abstract as money. Formal currencies were nothing new: in Bas-Congo people had been paying with white seashells since time immemorial; in Katanga the currency took the form of crafted copper crosses; in other parts of the country they paid with *mitakos*, the copper bars introduced by the first colonizers. But such currencies were brought to bear only for very special transactions. There was as yet nothing like a widespread monetary economy. But that changed soon enough. In the year 1900 no more than few hundred workers in Bas-Congo, most of whom worked on the railroad, were on a payroll; by 1920, when Nkasi moved to Kinshasa, there were already 123,000 such employees scattered across the country. And that was before the real employment boom began: in 1929 there were some 450,000 paid workers. Congo became a monetary economy.[53]

That monetization had a major impact. Once again, the state was overtly intruding into everyday life. One could no longer buy a chicken from the neighbors without symbolic government involvement. The centuries-old barter system, a transparent system of exchange between individuals, was pushed out by an abstract system imposed by the state. People had no choice but to assume that these strange pieces of paper showing a white woman in a white tunic had any effective value at all. "Banque du Congo-Belge" one read, printed in stately letters on that first Congolese banknote, "un franc"—at least, for those who could read. The woman, who had rather Hellenic look to her, wore a tiara. Her left arm rested on a big wheel, and with her right she embraced a sheaf of wheat.[54] This was obviously intended as an allegory for agriculture and industry, but the average Congolese was not very familiar with neoclassical graphics and kitsch. In the early 1920s, however, the local coins came one step closer to local reality: they bore the imprint of an oil palm, *m'bila* in a number of the native languages.[55] The people recognized it as a literal link between state and industry: Lever's concern soon became known as *Compagnie m'bila*. Money, that was barter with the factory. You gave them your body, they gave you your wages.

An advantage of all this, however, was that it now became much easier to collect taxes. One no longer needed to pay in kind or with labor for his mandatory membership in the state. Gone were the days of toting burdens, paddling upriver or collecting rubber to serve the white man; gone was the rule that one had to serve the state for forty hours each month. When Belgium assumed control of Congo, it first introduced a system whereby goods other than rubber could serve as tax—the colonial revenue department was equally pleased to received bars of manioc, copal, palm oil, or chickens—but went on after a time to express its preference for taxation in the form of hard cash. When a missionary asked him in 1953 to describe the course of his long life, Joseph Njoli, a man from Équateur, remembered that quite clearly:

> After rubber, they imposed on us a tax in fish and manioc. After the fish came the palm oil and wood that we had to bring to the regional administrator at Ikenge. His name was Molo, the white man who lived at Ikenge with the river people. The chores we were given to do took many forms. . . . Then there came another white man, Lokoka. He stopped the work we had been doing and brought us money. He said: "You people may now pay taxes with money. Everyone has to pay four-and-a-half francs." That was how money was introduced to the black people. And now we still live in slavery to the Belgians.[56]

Four and a half francs a year: that was not particularly exorbitant. The tax burden was kept light on purpose. In 1920 that sum was the equivalent of six kilos (thirteen pounds) of rubber, or forty-five kilos (a hundred pounds) of palm nuts, forty-five kilos of palm oil, thirty-five kilos (seventy-seven pounds) of copal resin, nine chickens, half a goat, or a few dozen loaves of manioc.[57]

On paper, the Belgian Congo wished to put an end to the noxious practices of the Free State, but in actual practice things went quite differently. In those zones where international big business had settled, new forms of exploitation and bond service arose. Migratory waves were set in motion that did more to disorganize the country than to help it recover. Young men ended up in shabby workers' camps, while only women and old people remained behind in their villages. Much of the misery in the period from 1918 to 1921 could be blamed on the four long years of World War I, but a great deal of misery had already gone before. It would be a mistake to pass the blame off on that accursed conflict. The Great War was not the cause, but it did make the situation worse.

ON NOVEMBER 11, 2008, the rain was pounding down over Kinshasa. This was, even by equatorial standards, an unusually heavy downpour. What fell were not drops, but rivulets of glass, liquid test tubes. Traffic in the city ground to a halt, horns honked incessantly as though ordering the puddles to dry up, and the courtyard of the Maison des Anciens Combattants was reduced to a swimming pool. During the 1950s this building had been an open-air movie theater; now it served as a club for the veterans of the many wars Congo has seen, a daily gathering place for its members. "It's unbelievable," a Belgian soldier in uniform remarked, "nothing in this country is watertight, the rain gets in everywhere, but here it just collects, no problem." He looked at the tiled courtyard. Seemingly without success, a dozen boys with buckets were trying to bail it out. The water must have been a foot deep. "You could raise goddamned *koi* carp here."

Meanwhile, the crowd flowed in. Women wrapped in beautiful robes: the heels of their dress shoes made little indentations in the ground. Men with glistening brass instruments. Gentlemen in three-piece suits. Former military men in green uniforms. This of course was their big day. There weren't many of them left anymore. They stood beneath the shelter of roofed-in gallery, assaying each other's medals, seizing them from each other. "Saio? You weren't even there. Give here." Amid petulant grumbling, medals changed jackets. That went on for a long time, until everyone who wanted to wear something shiny actually had something to pin on. André Kitadi told me: "None of these people were there. There are only four veterans of '40–'45 still alive in Kinshasa." He was one of them, I had interviewed him already. He didn't give a hang about medals.

Today was the ninetieth anniversary of the armistice that ended World War I.

The invitees waited beneath awnings until the courtyard was dry again. The ceremony was supposed to have begun at eleven, but it was already twelve-thirty. Finally someone showed up with a pump. Half an hour later they found the diesel fuel as well and fifteen minutes later the pump actually kicked over. After five minutes of noisy slurping, the courtyard was dry and the garden behind the Maison des Anciens Combattants had become a mud puddle. The ceremony could begin.

In 1914 Congo was as neutral as Belgium. They had to be: both countries had been conceived to serve as a buffer state between rival powers. For Congo, that neutrality proceeded from the final act of the Berlin Conference. But that neutral stance came to an end on August 15, 1914, eleven days after the German attack on Belgium. Across from the village of Mokolubu, on the Congolese side of Lake Tanganyika, a steamboat suddenly appeared. It was coming from the far shore, the German shore. The ship opened fire on a local café and sank fifteen canoes. A detachment of German soldiers landed and cut the telephone lines at fourteen places.[58] One week later, the port of Lukuga was attacked as well. That was how World War I began in Congo. The country's territorial integrity was threatened, the imperative of neutrality no longer applied.

Colonialism made it possible for an armed European conflict to become a world war. Large parts of Africa therefore became caught up in the international conflagration. The German colonies in East Africa (later Rwanda, Burundi, and Tanzania) and West Africa (later Togo, Cameroon, and Namibia) were bordered on all sides by French, British, Portuguese, and Belgian overseas property. In the northwest, Congo shared a few dozen kilometers of border with Cameroon, to the east more than seven hundred kilometers (430 miles) with German East Africa. Little wonder then that, for quite some time already, Berlin had been showing interest in the Belgian Congo. It wished to establish a bridgehead between its eastern and western colonies, in part at least to squelch the British axis "from Cape to Cairo." After all, wasn't colonization a task that should be left up to the major powers? Could one really leave such responsibility in the hands of piddling dwarf states like Belgium?[59] As early as 1914 Germany had approached Britain with the formal proposal to divide the Belgian Congo between them. The English, however, were not interested: they knew all too well that the French, with their historical *droit de préemption* (right of first refusal) to Congo, would never allow that.[60] But there were those, even in Belgium, who wondered whether Belgium might not appease its ravenous eastern neighbor by giving it half of Congo as a present. An area of 680,000 square kilometers (265,200 square miles) of jungle: wouldn't that allay the Teutonic hunger just a bit?[61]

But war was what came, in Africa as well. Not a single native knew who Archduke Franz Ferdinand von Habsburg was or why a lucky shot in Sarajevo had to lead to bloodshed on the savanna, but they saw the whites acting very serious about the whole thing. The military operations in Africa, however, in no way resembled the immovable war of positions to which Europe was to be submitted. There was no clear and continuous front, not like the line extending from Switzerland to the North Sea. There were no trenches, no mustard gas attacks, no fortifications undermined with dynamite, no Christmas truces with soccer matches in no-man's land. The sheer scale of the African continent, the lack of roads, the shortage of troops, and the often extremely trying topography led to a very different kind of warfare. It was not regions that were conquered, but strategic spots. One did not break through a solid front line, but defeated a local regiment. Zones were not seized but roads were controlled. The intensity of the conflict was much lower. In German East Africa, General Paul von Lettow-Vorbeck made a four-year stand with three thousand German troops and eleven thousand Africans, numbers that were run through at Verdun in the course of a morning.

From Brussels, the governor general heard that he was allowed to make use of the Force Publique to defend the colony. Later, when the Belgian government in exile was staying at Le Havre, France, intensive communication took place with the colonial administration at Boma. But the one-way flow of governmental directives from Europe was interrupted: while Belgium was more or less completely overrun by German troops, the territory of the colony itself remained virtually intact throughout the war. The balance had suddenly shifted.

The Congolese troops fought on three fronts: Cameroon, Rhodesia, and German East Africa. The first two called for relatively small-scale efforts. In 1914 six hundred soldiers and a handful of white officers came to the aid of Allied troops in the battle for Cameroon. One year later, when Germany threatened Rhodesia, 283 Congolese and seven Belgian soldiers joined forces with the British colonial troops. But the most intense show of military strength by far took place in the east of the colony. In the region

MAP 5: BELGIAN CONGO DURING WORLD WAR I

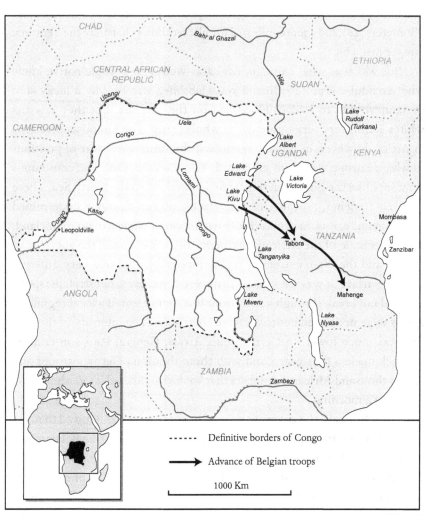

of Kivu the border between Belgian and German territory had been established only in 1910. As from 1915, however, German troops made repeated
attempts to invade Kivu, from where they could move to the Kilo-Moto
gold mines in the Ituri rain forest. Those attempts failed. They did, however, succeed in gaining control over two of the Great Lakes: Lake Tanganyika and the much smaller Lake Kivu. Their gunboats—the *Kingani,*
the *Von Wissman,* and most notably the (one-hundred ton) *Von Goetzen*—
patrolled the Congolese shores. On Lake Kivu they took control of the

island of Idjwi, the only part of the Belgian Congo to fall under German occupation.

The struggle for Lake Tanganyika was one of the most epic of World War I. From South Africa, British troops smuggled the parts for two fast and maneuverable gunboats of their own to the shores of the lake. Carrying ships in overland: it was like a repeat of Stanley's day. Under their code-names *Mimi* and *Toutou*, the ships played a decisive role in weakening the Germans' naval power. Even more audacious, if possible, was the initiative to reinforce Belgian troops on Lake Tanganyika with four aquaplanes. Air travel was still in its infancy, especially in the colonies. No one knew how these lightweight planes would react in the warm tropical skies. No one knew a thing about wartime flying, let alone about flying fragile biplanes that took off from water. The four planes arrived in pieces at Matadi. The train then took them to Kinshasa, whence they were loaded onto a freighter bound for Kinsangani. One month later they reached Kalemie. The shipment comprised five hundred metric tons (five hundred fifty U.S. tons) of hardware, fifty-three thousand liters (14,000 gallons) of fuel and oil, four machine guns, and thirty thousand rounds of ammunition. Lake Tanganyika was too turbulent for takeoffs and landings, so the planes were brought to a landlocked lagoon thirty kilometers (about nineteen miles) farther away. The lagoon was completely out of the enemy's sights and the water was placid. In 1916 the planes flew out a number of missions over Lake Tanganyika, primarily with the intention of bombarding the *Von Goetzen*. On July 10 of that year, the bombs found their target. (The *Von Goetzen*, however, did not sink; in 2010 it was still in service, as a ferry-boat on the same lake where it had met its inglorious end as gunboat.) The defense of the German coastline, and particularly of the town of Kigoma, had been broken.

Meanwhile, the infantry came into action as well. The commander of the Force Publique, General Charles Tombeur, assembled a large contingent on Congo's eastern border. He brought together fifteen thousand men, all equipped with rifles and ammunition. Moving all that matériel into place must have been a logistical nightmare. The transport was seen to by thousands and thousands of local bearers; for every soldier who went into battle, some seven porters were needed. During the four war years, no less than 260,000 native bearers—out of a population of less than ten million inhabitants—took part in the effort. Many of them suffered from malnourishment. Drinking water was scarce. People drank from ponds; people drank their own urine. There was a dire shortage of victuals, tents, and blankets, even as the troops moved through the Kivu highlands with their chilly nights. An estimated twenty-five thousand porters died along the way. Another two thousand soldiers succumbed; at the height of the struggle, the Congolese troops numbered twenty-five thousand men. But unlike the campaign to Sudan in 1896, there was almost no desertion or mutiny, partly because the white officers behaved more mildly toward their African aux-iliaries, partly because this was a victorious campaign that boosted the sol-diers' morale.

In March 1916 Tombeur felt the time was ripe for an offensive. His troops crossed the border into German East Africa and the advance on Kigali, later the capital of Rwanda, could begin. The city fell on May 6. From there the campaign moved on to Tabora, administrative nexus of the Germany colony. That city lay six hundred kilometers (370 miles) farther as the crow flies, and the soldiers advanced on foot, once again with tens of thousands of bearers. A second Congolese column departed from the shores of Lake Tanganyika. With a number of big hotels, mercantile houses, and industry, Tabora was a substantial city on a dry, open plain at twelve hun-dred meters (3,900 feet) above sea level. The battle formed the climax of the Belgian colonial efforts during World War I. On September 19, after ten days and nights of heavy fighting, the city fell into Belgian-Congolese hands. The German troops fled; the Belgian tricolor was raised above their fort. One year later, in 1917, the city would serve as base for another success-ful campaign against Mahenge, five hundred kilometers (310 miles) in the direction of Mozambique. The Force Publique now controlled one-third of German East Africa. A few scattered troops actually pushed on to the Indian Ocean, but Tabora was the name everyone would come to know. General Tombeur was raised to the peerage—most appropriately, his new name became Tombeur de Tabora—and a stylized monument was raised to him in Sint-Gillis, close to Brussels. In Congo, the name Tabora took on the connotation of a mythical conquest, heard of by generations of schoolchil-dren to come. "[King] Albert watches the enemy," the pupils of the Marist Brothers in Kisangani sang, "Unflaggingly / In Europe, in Tabora town / He keeps his eye on them."[62]

Martin Kabuya, the ninety-two-year-old veteran whose grandfather was buried alive during the Sudanese campaign, was two years old when the war ended. His other grandfather, from his mother's side, saw the fighting from close by. One suffocatingly hot day, as I sat in his garden, he told me the following: "My grandfather's name was Matthias Dinda and he was born in 1898. He was a Zanda, from the north of Congo. Our tribe originally comes from Sudan, in fact we are all Sudanese. He was a very strong man, he hunted leopards. He joined the Force Publique and was promoted to *soldat de première classe* [soldier first class], the highest rank for a black man. From Goma he went to Rwanda, and Burundi and Tanzania, all those German territories. He was there when Tabora fell." Kabuya was quiet for a moment. An orange-headed lizard flashed across the wall. "My grandfather was a friend of the men who planted the flag there. He even provided them with cover. He was a very great soldier."[63]

I saw Kabuya again at the Armistice Day commemoration in the Maison des Anciens Combattants. The few dozen invitees took their places in the now-dry courtyard. He sat up in front, with the veterans. Plastic lawn chairs had been set up for them. The podium, packed with nicer chairs, was soon filled with military and civilian dignitaries. As the brass band launched into the Belgian and Congolese national anthems, everyone stood and remained saluting the soldiers and officers. It was gripping to celebrate a truce in Kinshasa while, in the east of the country, Nkanda's rebels were in the midst of their most concerted offensive. During his speech, one of the World War II veterans said: "This riles us and fills us with horror. If we were as young as we were in 1940, we would take up our weapons and go and disarm these troublemakers."[64]

After the speeches it was time for the annual *remise des cadeaux* (distribution of gifts). A deputy cabinet minister presented the chairman of the veterans' club with a refrigerator, another decorated veteran received ten kilos (twenty-two pounds) of manioc flour from the Belgian military attaché. But the most important gift—a boom box, imported from China—went to a fragile old woman referred to by all as "la veuve" (the widow). Her name was Hélène Nzimbu Diluzeyi; she was ninety-four and the last remaining widow of a veteran of World War I.

Afterward, for a full thirty minutes, a band played "Ancien combatant" by Zao, a singer from Congo-Brazzaville, perhaps the loveliest song in Congolese pop music. "La guerre, ce n'est pas bon, ce n'est pas bon"

(War, it's not good, not good) it went. Elderly soldiers began dancing in the courtyard, while beer and soft drinks and snacks were brought out. Some of them shuffled cautiously to the beat, others played war: one man held an umbrella and pretended to be shooting it, another fell in slow motion to the ground, shook his arms and legs in time to the music, then pretended to be dead. *"La veuve"* watched in amusement, clapped her hands, and couldn't help laughing out loud now and then at this brilliant pantomime.

When the party was over, I walked her home. She lived in the Kasavubu district. We crisscrossed the muddy streets of the *cité*, avoiding the larger puddles. She held tightly to my left arm; under my right I held the monstrously huge box containing the boom box. It was the first time I had ever walked arm-in-arm with a veteran's wife. When we reached her yard, we sat down together beneath the line hung with laundry. Children and grandchildren came and gathered round us. Her son came to interpret. "My husband's name was Thomas Masamba Lumoso," she began. "He was born in 1896. He came to Kin when he was ten. The Protestant missionaries taught him English, then they handed him over to the army. That's where he got his uniform. In khaki."

"No, Mom, that was much later. Back then they still wore a blue uniform with a red fez."

"Really? *En tout cas,* he was eighteen when the war began. He operated the TSF, as a corporal."

TSF, I remembered, that was the *télégraphie sans fils,* the field radio.

"He went to wherever the war was. Everywhere. But he was never wounded. God protected him greatly."

"That's right," her son chimed in, "and he spoke a lot of languages. Swahili, Kimongo, Mbunza, Tshiluba, Kinzande, but also Flemish, French, English, and, because of the war, a little German too."

"German?"

"Yes, things like 'Guten Tag! Wie geht es? Danke schön!' I don't know what it means, but that's what he always said."[65]

It was the only time during my ten trips through Congo that I met someone who knew a few words of German.

That evening, at the house of the widow's other son, Colonel Yoka, I saw a photograph of the war veteran. In uniform, wearing his decorations and looking quite grave. In a report from 1921, his father was described as being "active and honest." But the most interesting document the colonel showed me was a letter from his Belgian superior: "The aforementioned Masamba from the village of Lugosi served as a supervisory noncom on the TSF from August 9, 1914, to October 5, 1918." Signed, on October 7, 1918, by someone named Vancleinghem, as far as I could read the handwriting. The dates said a great deal. Masamba's tour of duty coincided almost exactly with the duration of World War I. He entered the army five days after it started and was discharged one month before the Armistice.[66] The last of the veterans was also the one who had served the longest.

THE WORLD WAR effected more Congolese than the men of the Force Publique alone. In Katanga, the miners worked like mad. The excavation activities were running at full speed. The financial ties with Brussels had been broken, but the war caused the demand for copper to skyrocket. In the midst of the conflict, colonial copper exports rose from 52 million Belgian francs in 1914 to 164 million francs in 1917.[67] The British and American shells fired at Passendale, Ypres, Verdun, and along the Somme had brass casings made for 75 percent from Katangan copper. Parts of their cannonry were made of pure, tempered copper. Their bullet shells were made of nickel, which is 80 percent copper. Torpedoes and naval instruments were made from copper, bronze, and brass.[68]

Yet even outside the big industrial areas a great many Congolese experienced the fact that a war was on. In Orientale province, farmers were obliged to raise rice as victuals for the troops. In other parts of the country the government charged the population with the cultivation of cotton; that was not only good for exports, but also for the local textile mills. A whole system arose of *cultures obligatoires* (mandatory state crops). This evoked many unpleasant memories. In their village in Bas-Congo, Nkasi and Lutunu may have noticed little of the war, but brought many people in the interior back under the colonial yoke. And, as has happened more often in Congolese history, the protests against that situation took on a religious form.[69]

In 1915 in the Ekonda region of Équateur, a woman by the name of Maria Nkoi had a mystical experience. She became convinced of her own powers of healing and her prophetic duties. From then on she was known as Marie aux Léopards (Marie of the Leopards).[70] She began treating the ill and preaching. In addition, she called for a revolt against the colonizer and predicted that Congo would soon be liberated by the "djermani," the Germans.[71] These inflammatory words caused a run-in with the local administrators. She was jailed. Her story is reminiscent of the woman who, in 1704, amid the ruins of the cathedral at Mbanza-Kongo, had come up with an alternative form of Christianity and was prosecuted for that. Then too, European authority had been experiencing a crisis, and then too people feared the consequences of such a religious revival.

Liberated by the Germans? Albert Kudjabo and Paul Panda Farnana would have had something to say about that! After all, they had been taken prisoner by the Hun! Kudjabo and Panda were among the very few Congolese to fight in World War I in Belgium itself. As early as 1912, a man named J. Droeven had joined the Belgian army; he was the son of a Belgian arms manufacturer who was murdered in Congo in 1910 and a native woman. Droeven was the first man of color in the Belgian army, but less than three months after the war started he deserted and went off to live a life of debauchery in the cafés of Paris.[72] Kudjabo, on the other hand, was part of a Congolese Volunteer Corps that had offered to help the beleaguered Belgian forces in 1914.[73] Most of the corps consisted of former European colonials; its leader was Colonel Louis Chaltin. These were the only Belgians with previous combat experience, gained during the Arab campaigns and the Sudanese expeditions. But even that didn't matter. They helped to defend the city of Namur from the advancing German army, but not very successfully. *Das Heer* steamrolled Belgium, and twenty-one-year-old Albert Kudjabo, along with Paul Panda, were captured. Sent to Berlin as a prisoner of war, he suddenly found himself amid soldiers from all over the world. A handful of anthropologists and philologists became fascinated by this impromptu ethnographic assemblage; they set up the Royal Prussian Phonographic Committee and made almost two thousand recordings of this band of exotics. Albert Kudjabo was asked to sing a song. He drummed, whistled, and sang in his native language.[74] Those recordings have been preserved. It is a moving experience to hear them: the only Belgian soldier from World War I whose voice we know is a Congolese.[75]

WORLD WAR I had far-reaching consequences for the Belgian Congo. Territorial ones, first of all. The 1919 Paris Peace Conference that produced the Treaty of Versailles decided to divvy up the German colonies among the victors. Cameroon became French and British, Togo became French and British, German East Africa turned British, and Namibia was mandated to the British dominion of South Africa. Belgium received guardianship

over two minuscule countries on its eastern border, the historical kingdoms of Rwanda and Burundi (still Ruanda and Urundi at the time). In 1923
the League of Nations ratified these territorial mandates. A trust territory
was, on paper, not a colony, but in actual practice there was little difference.
Here too the rigid and only recently developed tenets of anthropology were
applied. In the protectorates, too, people reasoned, one had "races." Those
were absolute: you were either Tutsi or Hutu or Twas (Pygmy). From the
1930s on, this was also printed in one's passport. For centuries, the borders
between these tribal groups had been diffuse, but people forgot about that.
The consequences of this neglect, during the second half of the twentieth
century, were catastrophic.

In Congo, the war comprised a sort of pause button in the country's
social history.

The tentative attempts to improve the living conditions of the natives,
by means of better housing at the mines or large-scale campaigns to combat sleeping sickness, were put on the back burner. After four exhausting
years, the public health situation was once again extremely precarious.
In 1918–19 the Spanish flu claimed fifty to a hundred million victims
worldwide; half a million died in "the Spanish fever," ninety-two-year-old
Kabuya told me, "now *that* killed a lot of people." The decimation of 1905
seemed to have returned. The pause button was a rewind button as well.

But, in the eyes of the Belgians, something really had changed. For the
first time they began viewing the fate of the Congolese with compassion.
The Belgians realized that these people had suffered greatly under a war
that was not their own. In addition, the experience of war had resulted in a
feeling of soldierly camaraderie. That caused a Belgian officer in the Force
Publique to wax lyrical: "No, these men, they have fought, suffered, hoped,
desired, forged ahead and triumphed along with us, like us, these are no . . .
these are no longer wild men or barbarians. If they could be our equals in
suffering and making the greatest sacrifice of all, then they must, then they
shall be that too when it comes to being civilized."[76] The soldiers of the Force
Publique had shown great courage and loyalty, even under the worst conditions. That called for greater mildness and, yes, greater involvement in the
natives' fate.

For the Congolese themselves, however, it was an ambivalent experience. Many soldiers entered fully into the undeniable Belgian military successes. The euphoria of victory was sweet and forged new bonds that were

certainly sincere and warm. The Belgians could fly through the air and land on water! But for many normal Congolese, the war effort had been grueling. In addition, and this was the most sobering of all, they had seen how the whites—who had taught them not to kill anymore and to stop waging tribal war—had applied an awesome arsenal for four whole years to combat each other for reasons unclear, in a conflict that claimed more lives than all the tribal wars they could ever recall. Yes, that did something to the respect they felt for these Europeans. It began to crumble.

CHAPTER 4

IN THE STRANGLEHOLD OF FEAR

Growing Unrest and Mutual Suspicion in Peacetime

1921–1940

D URING THE INTERWAR YEARS, THE MAJOR SOCIAL UPHEAV-
als that began during the first decade of the Belgian Congo continued unabated. Industry picked up its pace. More and more people left
their villages and went to work for an employer. The first cities arose. There,
tribes became mixed and new lifestyles gained popularity. On Sunday afternoons, people danced to the music of Tino Rossi; the generation before them
had still done so to the rhythm of the tom-tom. But in the countryside, time
had not stood still. The system of mandatory crops introduced during World
War I was now applied everywhere. The mission posts expanded their hold
over the people's souls. Schools and hospitals were built even in remote
areas. The teams combating sleeping sickness moved into even the smallest
villages.

In that light, everything was tending toward expansion, a process that
served both colonizer and native. Or, at least, that is how people preferred
to view it. "Since the world war of 1914–18, the calm in Congo has never
been seriously disturbed," wrote a Catholic school headmaster in a Flemish
backwater.

A few minor disturbances, provoked not seldom by secret sects and
sorcerers, sometimes served to make a certain area unsafe. . . . The
Bula-Matari, as the natives call the Belgian administration in Congo, is
generally able to rely on the Negroes' submissiveness and deference to
authority, at least in so far as the persons in charge themselves attend

to the *requirements for a good colonial official,* and excel in an *orderly and virtuous life,* by means of *sincere charity and redoubtable willpower.*"[1]

That was a gross exaggeration. The colonial officials could apply all the sincere charity and redoubtable willpower they pleased, they were still unable to reverse the tide of growing irritation amid the native population. This was not about "a few minor disturbances" in a "certain area," but about significant popular movements that were able—despite heavy-handed repression by the colonial government—to expand across large parts of the colony. The fever of independence that manifested itself beginning in 1955 was not new at all, but had a very long incubation period. But to understand that, we must first pay a visit to Nkasi's younger brother. And to the Holy Ghost.

GOD'S WAYS MAY BE MYSTERIOUS, but the roads leading to the Holy Ghost are pretty hopeless, especially now that he has moved to Nkamba. From Kinshasa to Mbanza-Ngungu, formerly Thysville, the roads are excellent. A few years ago the Europeans and the Chinese joined forces to provide Congo with at least *one* decent road, leading from Kinshasa to the port of Matadi. But as soon as we leave that highway, the road becomes a sandy track, the sandy track becomes a mud puddle and our progress becomes snail-like. The distance from Mbanza-Ngungu to Nkamba is eighty kilometers (fifty miles) and we finally cover that in three hours. A new speed record, people tell us later. Yet the road to Nkamba is definitely no dirt track used only once in a great while. Each year, thousands and thousands of pilgrims go up it in search of spiritual renewal. They refer to it not as Nkamba, but as the Holy City or *la nouvelle Jérusalem.*

Simon Kimbangu first saw the light of day on September 24, 1899, only a few years after Nkasi was born. His childhood and adolescence were not so very different from those of his contemporaries, but he would go down in history as a major prophet. Few are those who have a religion named after them, but Simon Kimbangu was to join the ranks of Christ and Buddha: today, Kimbanguism is still a living religion in Congo, accounting for 10 percent of the country's believers.

Nkasi had said so himself: "Kimbangu, that was no magic. He was sent by God. A sixteen-year-old girl who had already been dead for four days, he brought her back to life."

Congolese and colonizers first heard about this remarkable man in 1921,

the year of the alleged resurrection, but Nkasi had known about him long
before that. They came from the same area. Nkamba and Ntimansi, their
native villages, were within walking distance of each other. "Oh . . . so when
did I see him for the first time? *Bon* . . . I knew Simon Kimbangu back in the
1800s already. If he said: 'Now we're off to Brussels,' then one second later he
really was in Brussels. After all, he even healed my younger brother!"

The road is rough, but it is a relief to arrive in the Holy City. The area
around the town is hilly. Eucalyptus trees rustle in the valleys and the shade
is soothing. Nkamba itself is on a hilltop overlooking huge stretches of Bas-
Congo. A lovely breeze is blowing. Still, one does not enter town just like
that. References and traveling papers from Kinshasa and a young adept from
Mbanza-Ngungu are required to pass the three roadblocks manned by Kimbanguist security personnel. There is something peculiar about them: they
are all dressed neatly in uniform, with green berets and facings, but they are
also barefooted. No boots, no clodhoppers, no sandals, nothing. Kimbanguists don't believe in shoes. Once inside, the visitor is immediately struck
by the peace and serenity of the place. Kimbanguism is the most Congolese
of all religions and at the same time you feel like you're in a different country.
Everyone walks around barefooted, dressed soberly, radios and boom boxes
are forbidden. No one shouts. Alcohol is taboo. What a contrast with Kinshasa
with its extravagant dress, its everlasting shouting and swearing, its pushing
and shoving in line for the taxi buses, its honking, and its busted loudspeakers!

The most striking building in town is the temple, an enormous rectangular thing built in eclectic style built by the believers themselves between
1986 and 1991. Putting together a building like this in five years' time is no
mean achievement. In front of it is the mausoleum of Simon Kimbangu and
his three sons. First venerated as a prophet, the founder today enjoys divine
status. That same status also applies to his three sons, who are nothing less
than the embodiment of the Holy Trinity. A young female Kimbanguist once
explained it to me at poolside in Kinshasa. I still have the scrap of paper that
she scribbled all over then. "Kisolokele, born in 1914 = God the Father; Dial-
ungana, born in 1916 = Jesus Christ; Diangienda, born in 1918 = the Holy
Spirit." The Kimbanguists no longer celebrate Christmas on December 25,
but on March 25, the birthday of the second son. When the founder died in
1951, Diangienda Kintuma, the youngest of the three, assumed spiritual leadership of the movement. He kept going for a very long time: from 1954 until
1992. Now that position is occupied by a grandson, Papa Simon Kimbangu

Kiangani, but the succession was not a perfectly smooth one. His cousin Armand Dingienda Wabasolele, another of the prophet's grandsons, felt he was entitled to be the spiritual leader of Kimbanguist Church and, in addition to a schism, this contention has led to a great deal of musical rivalry. The Kimbanguists attach a lot of importance to music: in addition to beautiful choruses, their liturgy is characterized by the generous use of brass bands. In Kinshasa, the former pretender to the throne, Wabasolele, is the leader of a two-hundred-man symphony orchestra; in Nkamba, his cousin, the current spiritual leader, Kiangani, prides himself on his philharmonic orchestra. I once attended an open-air performance by the symphony orchestra in Kinshasa; I have no idea where they obtained their glistening instruments in that shattered city, but their performance of Carl Orff's *Carmina Burana* was a steamroller that easily outvoiced the honking of the evening rush hour. Wherever the truth lies, today it is Simon Kimbangu Kiangani who is venerated as the Holy Ghost.

That veneration is something to be taken quite literally. Darkness is already falling when I settle down on the square before the cathedral for evening prayers. I sit with my back to the spiritual leader's official residence. To the right I see the monumental entranceway. Its pillars are hung with colorful textiles, a throne stands on carpets that have been rolled out over the concrete. A brass band plays uplifting martial music; the musicians are wearing white and green uniforms and marching in place. Kimbanguism is an extremely peace-loving religion, yet brimming with military allusions. Those symbols were not originally part of the religion, but were copied in the 1930s from the Salvation Army, a Christian denomination that, unlike theirs, was not banned at that time. The faithful believed that the *S* on the Christian soldiers' uniform stood not for "Salvation" but for "Simon," and became enamored of the army's military liturgy. Today, green is still the color of Kimbanguism, and the hours of prayer are brightened up several times a day by military brass bands.

Those bands, by the way, are truly impressive. It is a quiet Monday evening when I find myself on the square. While the martial music rolls on and on, played first by the brass section, then by flutes, the faithful shuffle forward to be blessed by the spiritual leader. In groups of four or five, they kneel before the throne. The spiritual leader himself is standing. He wears a gray, short-sleeved suit and gray socks. He is not wearing shoes. In his hand he holds a plastic bottle filled with holy water from the "Jordan," a local stream. The believers kneel and let themselves be anointed by the Holy Spirit. Children open their mouths to catch a spurt of holy water. A young deaf man asks for water to be splashed on his ears. An old woman who can hardly see has her eyes sprinkled. The crippled display their aching ankles. Fathers come by with pieces of clothing belonging to their sick children. Mothers show pictures of their family, so the leader can brush them with his fingers. The line goes on and on. Nkamba has an average population of two to three thousand, plus a great many pilgrims and believers on retreat. People come from Kinshasa and Brazzaville, as well as from Brussels or London.

Thousands of people come pouring in, each evening anew. For an outsider this may seem like a bizarre ceremony, but in essence it is no different from the long procession of believers who have been filing past a cave at Lourdes in the French Pyrenees for more than a century. There too, people come from far and near to a spot where tradition says unique events took place, there too people long for healing and for miracles, there too people place all their hope in a bottle of spring water. This is about mass devotion and that usually says more about the despair of the masses than about the mercy of the divine.

After the ceremony, during a simple meal, I talk to an extremely dignified woman who once fled Congo as a refugee and has been working for years as a psychiatric nurse in Sweden. She loves Sweden, but she also loves her faith. If at all possible, she comes to Nkamba each year on retreat, especially now that she is having problems with her adolescent son. She has brought him along. "I always return to Sweden feeling renewed," she says.

THE NEXT DAY I finally meet Papa Wanzungasa, Nkasi's younger brother, the one I came to Nkamba for. He is only one hundred years old, but still active. What a family! His 60-year-old cousin looks like he's 45, his 126-year-old brother is one of the oldest people ever, and at the end of his first century he is still a member of the upper ministry at Nkamba and first deputy when it comes to evangelization, finances, construction, and materials supply. He has been registered with the Kimbanguist Church since 1962 as Pasteur No. 1. In 1921, when Simon Kimbangu's public life began, he was a boy of thirteen. Kimbangu was thirty-one.

No other area in Congo was so impacted by the arrival of Europeans as Bas-Congo. Slavery was abolished, the demand for porters and laborers on the railroad severed the traditional pattern of life, farmers had to raise manioc and peanuts for the colonizer, and money and taxes were introduced.

Europeans repeated time and again that they wanted to open up Congo and civilize it, but for the Africans the immediate results were disastrous. Sleeping sickness and the Spanish flu had killed an estimated two-thirds of the population, and European medicine proved powerless. That produced a deep-seated suspicion among the local population: these white people brought more sickness than they did healing.

It was at the mission post of Gombe-Lutete, twelve kilometers (about 7.5 miles) from his native village, that Simon Kimbangu was baptized by British Baptists and became a catechist himself. In 1919 he went looking for work in Kinshasa, just as Nkasi had. He applied for work at William Lever's Huileries du Congo Belge, without success. But he found himself in a world of Africans who had traveled and could write and do arithmetic. Thousands of black employees there were working for some twenty companies. By that time he was already hearing voices in his head, and receiving visions that summoned him to great deeds. For the time being, though, he paid them no heed. But a year later, when he returned to his village and found that the British Baptists had appointed someone else as their official catechist, something snapped.

On April 6 he heard talk of Kintondo, a woman who was seriously ill. He went to her, wearing a hat on his head and clenching a pipe in his teeth . . . almost, one would say, like a missionary. When he arrived he laid his hands on her and commanded the deathly ill woman to rise up, which tradition says she did, the very next day. Rumors of the miraculous healing spread like fire. The stories grew wilder all the time. In the weeks that followed, Kimbangu was said to have healed a deaf man and a blind man. That's right, and they said he had even caused a young girl, who had died a few days earlier, to rise from the dead! Here at last was someone with much more power than those white people with their injections against sleeping sickness that actually made you sicker than you were before. Redemption was nearing. From all over the region, people abandoned their fields and hurried to Nkamba.

As did the parents of Nkasi and Wanzungasa. Nkasi was in Kinshasa by then, shoveling dirt, but his brother saw it all at firsthand.

We settle down in the green leather easy chairs in Nkamba's state chamber, to talk about that distant past. As behooves a Kimbanguist, Wanzungasa speaks in a quiet, friendly voice.

Our parents were both Protestants, they were farmers. As a child, I had a hunchback. My mother heard about a man in Nkamba who

healed all kinds of disorders, the blind and the deaf, and who had even
brought the dead back to life. She took me along, and we arrived here.
Nkamba was full of people. They were called to the front in the order
in which they came. When it was my turn, I was called up along with
my mother. We knelt in front of Simon Kimbangu. He placed a hand on
my head and said: "In the name of Jesus, stand up, straighten your back,
and walk." I did that, and saw that my hunchback had disappeared. It
didn't even hurt.

He tells his story calmly and factually and makes no attempt to prosely-
tize his listener. The facts are there, for those willing to believe.

My mother was overjoyed. Simon Kimbangu said that we should go
and wash ourselves in the holy water. We stayed for three more days,
to be sure I was completely healed. Today the doctors say that I had
tuberculosis, but that's not true. I walked completely bent over. I was
healed by my faith. That's how it goes in my family, otherwise my
brother could never reach the age of 126, could he? There were many
more sick people in our village. The news of my healing traveled fast.
Then everyone went to Nkamba and became Kimbanguists.[2]

The colonial government was worried by this sudden abandonment of
the countryside. The Cataractes district of Bas-Congo was a major bread-
basket for Kinshasa, but suddenly the markets were empty. Rumors even
reached the big city. Some people put down their work and returned to
their native regions. There, the first to become concerned were the Protestant missionaries: many of Kimbangu's initial followers, after all, came
from their mission posts. And even though the Protestants advanced a much
more individual form of worship, they wondered whether Kimbangu wasn't
taking things too far. Kimbangu had ignited a fire that flashed across the
countryside. All over Bas-Congo, new prophets were shooting up like mushrooms. They were called *bangunza*, or *ngunza* in the singular. Their rallies
led to frenzied scenes. One Swedish missionary, who had lived in Congo for
years, noted in his diary:

Today I attended the *Ngunza* gatherings. It is extraordinary. You should
see them shudder, stretch their arms, point them in the air, look at

the sky, straight into the sun. You should hear them shout, pray, beg,
softly whisper "Jesus, Jesus." You should see Yambula [one of his best
evangelists] leap, run, spin on his axis. You should see how the crowd
comes together, strides along, kneels beneath the shaky hands that hold
the *bangunza* above their heads—Listen to what is happening here! Go
away, cast off these graven images.[3]

Two aspects cannot be emphasized enough. First, the followers of the
new faith did not turn against Protestantism, but in fact used it to their own
ends. This was no rupture with Christianity, but a specific coloring-in, yes,
an intensification of it. This was no return to precolonial religious practices;
the new believers, in fact, explicitly renounced the ancestral belief in witchcraft. But at the same time—and this is the fascinating thing—the followers
made use of religious symbols and gestures that hearkened back to traditional healing (trance, charms, incantations). They were against cult objects
images, but behaved like native healers. What they found, in other words,
was an African form for an imported faith. Second, even though this sudden
religious revival was not without a link to social conditions, it was foremost
an exclusively spiritual phenomenon. Kimbangu was no political rebel: he
made no anticolonial tirades and his doctrines were not directed against the
Europeans. But the colonial authorities had a very hard time believing that.

Less than three weeks after Kimbangu's first public appearance, district
commissioner Léon Morel sounded the alarm. That was altogether understandable: for a colonial administration that was trying to introduce a standard monetary economy and a classic work ethic in Congo, these day-long
gatherings of the willfully unemployed were absolutely disquieting. Ever
since 1910 the colonizers had been dividing the population into safe little
chefferies; now they were suddenly converging by the thousand to take part
in bizarre rituals. A meeting was organized in Thysville to which Catholic
and Protestant missionaries were invited. The Catholics, most of them Belgian, agreed with the colonial rulers and accused the Protestants of laxity
in their dealings with the natives. They called for a vigorous and drastic
government intervention. The Protestants, on the other hand, favored a soft
approach. This was, after all, a form of Christian popular devotion, they
felt, and it couldn't be all bad, could it? A number of their most cherished
converts were involved, people they had known for years and for whom they
felt friendship. Heavy-handed tactics would completely alienate them from

the mission post. And besides, wouldn't such repression simply serve to fan the fires?

As was often the case, the standpoints and behavior of the Protestant missionaries were a great deal more subtle and humane than those of the Catholics, but a head-on collision with the mammoth alliance of Catholic Belgian missionaries and Belgian colonial administrators was useless. On June 6 a detachment from the Force Publique, along with Léon Morel, moved on Nkamba to arrest Kimbangu, which resulted in skirmishes and looting. The soldiers stole the mats, the clothing, the chickens, the Bibles, the hymnals, and the little bit of money the faithful had with them. They fired with live ammunition. People were wounded and one person was killed. Afterward the army carted off the movement's leaders to Thysville, but Simon Kimbangu himself was able to escape. For his followers, this was just one more proof of his supernatural gifts.

He remained in hiding for three months and continued to spread his faith in the villages where colonial officials rarely came and where no one would betray him. That says something about his popularity and about the generally mounting resentment toward the white rulers. In September 1921 he turned himself in—just like Jesus at Gethsemane, his followers felt. The ensuing trial they compared to Christ's prosecution by Pontius Pilate. And not without reason. It was, after all, a mockery as well. That Kimbangu would be found guilty was a foregone conclusion. A watered-down version of a state of siege had even been imposed, to make sure he would appear before a military tribunal rather than a regular (and milder) civil court. This meant he had no legal representative nor any right of appeal. His fate was decided within three days. Reading the case files today, one is amazed by the tendentious nature of the magistrate's questioning. The sole objective was to prove that Kimbangu was guilty of undermining public security and disturbing the peace; that was the only crime with which he could be charged and which bore the death penalty.

Commander Amadeo De Rossi chaired the court-martial: "Kimbangu, do you admit that you have organized a revolt against the colonial government and that you have characterized the whites, your benefactors, as being terrible enemies?"

Kimbangu replied: "I had not created any revolt, neither against the Belgians, nor against the Belgian colonial government. I have only tried to preach the gospel of Jesus Christ."

But the presiding judge was not to be swayed: "Why did you call on the people to lay down their work and stop paying taxes?"

Kimbangu: "That is untrue. The people who came to Nkamba did so of their own free will, in order to listen to God's word, to find healing or to receive a blessing. Never once have I asked the people to stop paying taxes."

The judge tried a different tack, and suddenly became overly familiar. The tone grew sarcastic: "Are you the *mvuluzi*, the redeemer?"

"No, that is Jesus Christ, our redeemer. He has given me the mission of spreading the news of eternal salvation to my people."

"Have you brought the dead back to life?"

"Yes."

"How did you do that?"

"By applying the divine power given me by Jesus."[14]

Those were precisely the answers the court wanted to hear. They confirmed the suspicion that Kimbangu was a subversive fantasist. Because the hymns sung at Nkamba spoke of arms, the court tried to pin on him the charge of summoning people to violence. Kimbangu replied that the Protestant missionaries were not arrested, even though their hymns spoke of "Christian soldiers." The court tried to trip him up by citing him as having said: "The whites shall be black and the black shall be whites." Kimbangu said that did not literally mean that the Belgians were to pack up and leave. And besides, since when was egalitarian discourse a racist position? It was suspected that during his stay in Kinshasa he had come in contact with black Americans who were followers of Marcus Garvey, the radical Jamaican activist who believed that Africa was exclusively for the Africans. Kimbangu denied that charge: "Cela est faux" (That's false).

But it was to no avail. It didn't help either when, halfway through the proceedings, Kimbangu went into a trance and began raving and shaking all over. Epilepsy, we would assume today, but the court-appointed physician prescribed a cold shower and twelve lashes. The final verdict was what one would have expected after all this: on October 3, 1921, Kimbangu was sentenced to death, his closest associates to life imprisonment and hard labor. The court order made no bones about the real motives: "It is true that the animosity towards the powers that be has been limited till now to inflammatory songs, insults, forms of defamation, and a few, unrelated cases of insurrection, but it is also true that the course of events could have led, with fatal consequences, to a major uprising."[5]

Kimbangu was to be made an example, that much was clear. His prosecutors would have liked to see him executed as quickly as possible but, to the amazement of all, Kimbangu received a pardon from King Albert in Brussels. His sentence was commuted to life imprisonment. Kimbangu was taken to the other side of the country, to the prison at Elisabethville in Katanga. There he remained behind bars for the next thirty years, until his death in 1951. Unusual punishment for someone who, for a period of less than six months, had brought a little hope and comfort to a few stricken villages. His term of imprisonment was one of the longest in all of colonial Africa, even longer than that of Nelson Mandela. He spent most of that time in solitary confinement. He had never committed an act of violence.

A PEACEFUL INTERLUDE, this interwar period? Only a few minor disturbances? The excessive sanction imposed on Simon Kimbangu showed that, behind the manly, apparently unruffled facade of the colonial administration, there was a great deal of skittishness. The colonizers were terrified of disturbances. That manifested itself in the way Kimbangu's followers were persecuted.

From 1921 on, the government began banishing key figures in Kimbanguism to other provinces, with the intention of breaking apart the movement. Old Wanzungasa knew all about it. His uncle was picked up and forced to serve for seven years in the Force Publique. His youngest brother, still only a child at the time, was forced to undergo a Catholic mission education and was baptized against his will, making him the only Catholic in an otherwise Protestant nest. But his in-laws-to-be suffered most of all. "They were banished to Lisala, all the way in the eastern part of Équateur." Why? Because the mother of his future fiancée was related to Marie Mwilu, Simon Kimbangu's wife. "Her father died there in exile. My wife was only a girl at the time, she stayed here."

Initially, the measures directly affected a few hundred families, but in the course of the colonial era their number rose to 3,200. Today the Kimbanguists claim that 37,000 heads of households were forced to move, a total of 150,000 individuals, but the administration's records speak of only one-tenth of that. Internal exile, by the way, was one of the government's standard punitive measures: during the entire colonial period, some 14,000 individuals were banished to other parts of the country, most of them for political-religious reasons. The official explanation was that this was for the purposes

of reeducation, but in actual practice the deportations were often permanent. The details sometimes remind one of Europe in the 1940s. The Kimbanguists were taken away in closed cattle cars. Hunger, heat, and disease
took their toll along the way. Many of them died as a result of the hardships
during the journey itself. One man lost his three children before they could
even arrive at their final destination; they were buried in a grave beside the
river.[6] The Kimbanguists were banned to the rain forest of Équateur, to
Kasai, to Katanga, even to Oriental province. There they lived in isolated
villages where their faith was outlawed. Beginning in 1940, the highest-risk
exiles were sent to agricultural colonies, work camps surrounded by barbed
wire where men and their families were put to forced labor and watched
over by soldiers with guard dogs. The mortality rate there was sometimes
as high as 20 percent.

   None of this, however, had the desired effect. Kimbanguism was not
crushed by these drastic measures, on the contrary. Banishment made the
people even firmer in their beliefs. Each obstacle thrown up only bolstered
their conviction that Simon Kimbangu was the true redeemer. Under such
difficult conditions their faith provided them with comfort and something
to hold on to, to such an extent in fact that it proved infectious for their surroundings. The local inhabitants were impressed by this new faith. In this
way, Kimbanguism spread throughout the interior. Exile did not undermine
the movement, but caused it to multiply. There were tens of thousands of
followers.

   Meanwhile, close to Nkamba, the religion went underground. Meetings were held by night in the forest, where Marie Mwilu, Kimbangu's wife,
talked about Papa Simon and taught the new believers to sing and pray.
People even came downriver from Équateur for these gatherings. Coded
messages were used to communicate with exiles in other parts of the country. This clandestinity may have been an obstacle, but it was also a fantastic
learning experience that served to stimulate and consolidate the movement.
The energy and fervency of those underground years is sometimes reminiscent of the experiences of the early Christians under Roman rule. As a teen-
ager, Wanzungasa experienced it firsthand: "We could only pray at night in
the jungle, amid the 'spiders.' Those were other Congolese who spied for
the whites. During the day we took other routes through the forest, but we
exchanged secret signs. At night we came together to sing. Sometimes the
Belgians surrounded us during prayers. They had heard our singing, but

couldn't see us. We could see them, though, but we were invisible to them." The early Christians in Rome during their persecution also kept up their spirits with magical stories. If the earthly powers don't give you the respect you deserve, you look for it at a higher level.

This tough approach to Kimbanguism was one of the most serious mistakes made by the administration; the colonial authorities misjudged the situation completely. They combated symptoms, but not the cause. The real problems that gave rise to such a massive religious revival they ignored completely. Hard repression of the form took precedence over any empathetic concern about the contents. And that backfired on them. A radical version of Kimbanguism arose in Bas-Congo in 1934, Ngunzism, and that *was* openly anticolonial. Its adherents called for an end to taxation and for the Belgians' withdrawal. Shortly afterward came Mpadism or Khakism, initiated by Simon-Pierre Mpadi, who added the soldier's khakis to Kimbanguism, to say nothing of a much more radical train of thought. He turned against the colonizer, advocated polygamy, and organized gatherings where the crowd engaged in ecstatic dancing. At the start of World War II, he hoped that Congo would be liberated by the Germans. Matswanism was another phenomenon, one that blew in from Congo-Brazzaville. André Matswa (or Matsoua) was a World War I veteran who had served in France with the legendary *tirailleurs sénégalais*, the French colonial troops. While still in France he had set up a fraternal society and emergency fund for Africans. When he returned to Brazzaville he was venerated as a messiah and that movement made its way across the river. Matswa was ultimately deported to Chad, where he died in 1942. But, despite all the repressive measures, new messianic religions kept popping up. That stubborn resilience is telling indeed. It comprised, after all, the first structured form of popular protest and showed how many people were longing to be set free.

And it was not limited to Bas-Congo. New religious movements sprang up all over the country. In the mines of Katanga there arose the Kitawala, a corruption of "the Watch Tower," the name originally used by the Jehovah's Witnesses. That faith, which started in the United States in 1872, had migrated to South Africa and from there, beginning in 1920, to the Katangan copper belt.[7] In Congo it took on an explicitly political character. Spreading bit by bit across the colony, it thrived mostly underground. Still, it was to become the largest religious movement in Congo after Kimbanguism. In other parts of the country, smaller, secret sectarian societies arose. In the district of Kwango there was the Lukusu movement, also known as the snake sect. In Équateur there arose the Likili cult, whose members rejected Western beds, mattresses, sheets, and mosquito netting—all items held accountable for Congo's falling birthrate.[8] Along the upper reaches of the Ariwimi in Orientale province originated the sinister Anioto society, whose members were known as the leopard people. That movement spread across the northeast of the country. The leopard people performed random acts of terror and murdered dozens of natives. Their motives were not always clear, but the tenor of the movement was clearly anti-European.[9] During the 1920s and 1930s, therefore, there arose some fifty religious movements. Their methods varied from pacifistic to terroristic, but the underlying rancor was more or less the same.[10] In Congo, religion was the *pilipili* (hot chili pepper) of the people.

"We are God's people," Wanzungasa said to me at the end of our conversation in the green leather armchairs of the Holy City's state chamber. "We are not allowed to do evil, not even toward those who have done evil to us in the past. We do not demand an eye for an eye. We wield brass bands, not machetes." He paused for a moment. I looked up from my notebook and saw his serene, lined face. He had been born in 1908, the year the Belgian Congo was established. His religion was officially recognized by Belgium only on December 24, 1959, some six months before independence. He was probably thinking back on the first half of his life, his first half-century. In a quiet voice, he concluded: "There was no freedom then. During the colonial period, people were bought and sold. We were like slaves. Truly, the only color colonialism ever had was that of slavery."

IN KINSHASA I had the opportunity to speak at length with Nkasi about the 1920s and 1930s and the gathering resistance. He who had looked up so often to the white people later in life had to admit that those had been troubled times. "The old people were very tough. The white man, that wasn't your comrade back then!"After his period as a manual laborer in Kinshasa, he returned to his native region. In those days very few people remained in the city permanently; wage labor was seasonal labor. After his little brother had been miraculously cured by Kimbangu, the obvious thing was for him to become a Kimbanguist as well, despite the inherent dangers. "In Nkamba, Monsieur d'Alphonse was appointed *chef de poste*," he said with a singular lack of enthusiasm. That colonial official had been charged with pacifying the area after the Kimbanguist upheaval. To that end he appointed Lutunu, the freeman-boy-cyclist-drunkard-and-assistant-regent of old, as his native *chef.* Lutunu, after all, got along well with the whites.[11] Monsieur d'Alphonse shuttled back and forth between the administrative center at Thysville and his post in Nkamba. Nkasi remembered that all too well: "I had to help carry him. On my shoulders, that's right! There were two of us to carry his litter, and he shook back and forth terribly." Nkasi could nevertheless laugh heartily about it now. Sitting on the edge of his bed, he imitated the white colonial's shaking in the *tipoy.* He flapped his arms at his sides, sloppily and uncontrolled, as though he himself were seated in the sedan chair. Humor must have come to their assistance back then too. The journey covered more than eighty kilometers (fifty miles) and Monsieur d'Alphonse proved a harsh man. "My uncle was one of the local worthies, but Monsieur d'Alphonse gave him two hundred lashes. That was in 1924, I think. My uncle had said: 'Mundele kekituka ndonbe, ndonbe kekituka mundele.' The whites shall be black and the blacks white." Lashes, most probably fewer than two hundred, for a phrase that happened to be the slogan of the Kimbanguists. "The Force Publique soldiers lashed him across his bare buttocks. My uncle had two wives, but directly after those two hundred lashes he became a good Christian, a Kimbanguist. That's why he ended up with no lashmarks, wounds, or bruises on his buttocks, nothing at all."

It was during that period that the Matadi-Kinshasa railroad was broadened and made ready for electrification. The slow train puffing along its narrow gauge tracks was no longer sufficient, now that Congo was industrializing at a rapid pace. And air travel, of course, was still in its infancy: the first plane from Brussels landed at Léopoldville only in 1925; it was a biplane and it had taken twenty-five days to complete the journey, more than twice as long as the packet boat.[12] The work on the rails lasted from 1922 to 1931, with workdays of up to eleven hours. The route was adapted here and there, three new tunnels were dug, old bridges were replaced. The entire journey was reduced from nineteen to twelve hours.[13] Nkasi, who had seen his father work on the first railroad, was now part of it again. After all, hadn't he gained experience with a shovel in Kinshasa? "Now I had to work with a pick." With the *piccone* was what he said—in Italian, because many Italians were involved in the revamping. His foreman, Monsieur Pasquale, was one of them. "I got ten francs a month and a bag of rice. But one day Monsieur Pasquale said to me: "*Tu dormi, toi?* So you're napping, are you?'" Nkasi could still imitate the

Italian's broken French. "I told him: *Je travaille!* I'm working!' He took me to his home and I became his boy. He showed me how to make the bed and set the table. And for that work I got twenty francs a month!" He still beamed when he thought of it. Never in his working life had he had such a stroke of luck! "Those Italians were used to the sun. They were all single, they never had their wives with them. And they didn't take a black woman, oh no!"

Of the sixty thousand Congolese workers on the new rail project, no fewer than seven thousand died. Nkasi's new position, however, placed him in a financial position that allowed him to start thinking about marriage. Since the introduction of currency, the price of dowries had shot up. Marriage was reserved only for the wealthy. The rich were sometimes able to take several wives, while young men couldn't afford to marry at all.[14] Nkasi was almost forty by then. In his native village of Ntimansi he met Suzanne Mbila, a Kimbanguist like himself. Their first son was born in 1924, and they married in 1926. Their family grew steadily and he found himself once again living among his own people; that situation showed no signs of changing.

Unless, of course, one took the American stock exchange into account.

The Wall Street crash of October 1929 was felt all the way to the forests of Bas-Congo. The world economy had become so intertwined that the doubts and panic of investors in New York determined the further life of a man and his family in a piddling village in Congo. The effect was, of course, not immediate. The causal chain went as follows: the stock market crisis put a drag on the economy and caused a fall in the demand for raw materials worldwide; the Congolese mining industry, motor of the colonial economy, broke down; exports from the colony fell by more than 60 percent;[15] in 1929, this resulted in a gigantic budget deficit; the Belgian government, realizing that the colonial budget was too dependent on income from the mines, decided in favor of diversification; agriculture provided an alternative, particularly agriculture aimed at export; the large-scale cultivation of tobacco, cotton, and coffee, however, required time and investments; the goal of protecting companies during this time of crisis meant that an easier way to generate revenues was to raise taxes—those paid by the natives themselves, that is; a higher taxation of natural persons had an added advantage: it would cause the demand for money to rise, the Congolese would be forced to accept employment and that could only have a civilizing effect. More revenues for the state and at the same time a better grip on a population that was starting to show signs of dissatisfaction—wasn't that what they called killing two birds with one stone?

And so it came about. In 1920 the colony yielded only 15.5 million Belgian francs in tax revenues. By 1926 that had already grown to 45 million. And in 1930, in the midst of the crisis, that sum had increased to 269 million. Within four years, tax revenues had increased sixfold. By 1930, direct taxation accounted for 39 percent of the colonial budget, while taxation on the profits of the big concerns, which had still booked enormous profits in the previous years, accounted for only 4 percent.[16] What's more, many ailing private firms were now *receiving* money from the colonial government, because they had originally been lured to Congo with financial guarantees: in the event of a downturn, they were to receive a fixed dividend of 4 percent from the colonial treasury.[17] The hole generated by the crisis had to be filled, in other words, with money from the Congolese common man, in addition to a capital injection from the Belgian state treasury and revenues from the colonial lottery. This did not mean that every Congolese worker was suddenly required to pay six times as much (the tax pressure in the cities had already been increased gradually, but sorely), but that the tax department was now extending its tentacles farther into the interior. The bludgeon of personal taxation in this way drove thousands of people into the mines, onto the plantations, or into government service. In 1920 123,000 Congolese were on payrolls; by 1939 that had risen to 493,000.[18] Anyone refusing regular employment and wishing to keep farming for themselves was forced to raise certain crops and sell them to private colonial companies. By 1935 an estimated 900,000 people were involved in the cultivation of cotton.[19]

Nkasi, too, felt compelled to undertake something. "Well, and then the crisis came. . . . And the lack of money . . . [W]hen the administrator of Mbanza-Ngungu, Musepenje, came by Ntimansi I applied for a job with the provincial government."

One can hardly overestimate the portent of this. The Kimbanguists had gradually come to despise anything that reeked of the colonial government. They hid away in the forests and warmed themselves secretly at the glow of their religion. They wanted nothing to do with the whites. But now they had to go to work for them. Operation Tax Hike was a complete success.

In no time, however, Nkasi would become enamored of European culture.

He had been lucky to meet this fellow Musepenje. That, at least, was how I had scribbled down the name phonetically in my notebook. *Musepenje. Muzepenjet?* Whenever I heard a word that I didn't understand during an interview, I tried to jot down the sound of it as faithfully as possible. And

Nkasi was often hard to understand. "Monsieur Peignet?" I wrote next to it. At home, it took me days to figure it out. But in the colonial yearbooks for the 1930s I found a certain Firmin Peigneux, provincial administrator in Nkasi's region, the colonial official in closest contact with the population. He traveled from one *chefferie* to the next, talked to village chieftains and ruled on conflicts concerning property rights. Monsieur Peigneux, in other words. Most Bantu speakers pronounce the *eu* as an *è*. I should have known. At the Africa archives of the Ministry of Foreign Affairs in Brussels I was able to view his personal files.[20] This man, I quickly realized, was cut from a very different cloth than a brute like Monsieur d'Alphonse.

Peigneux, who hailed from the province of Liège, had gone to Congo alone in 1925 at the age of twenty-one. He quickly gained a reputation for his empathy. After his first year of service, his superior evaluated him as follows:

This official truly possesses the qualities needed to become an elite administrator in the short term. . . . In his contacts with the natives, Monsieur Peigneux exercises a level-headed policy that has won him the confidence of the chiefs and notables. He is interested in the study of social affairs and already possesses to a high degree the art of dealing in a careful and well-considered fashion with the primitives who surround us, without snubbing their secular opinions and customs. . . . The government may rightfully entertain the highest expectations for this official's future effectiveness."

That proved no exaggeration. Peigneux went on to build a brilliant colonial career and in 1948 was appointed to the post of provincial governor, the second-highest position in the official hierarchy after that of governor general. It reflects on his unceasing social involvement to note that, after he was sent back to Belgium for health reasons in the 1950s, he became a member of the board for the Fund for Native Welfare.

Nkasi still spoke of Monsieur Peigneux with great affection. "*Musepenje, c'était mon oncle.* Musepenje was like an uncle to me. He even drank palm wine with us! He and Monsieur Ryckmans, those were the only friendly whites." André Ryckmans was the son of Pierre Ryckmans, the best governor general the Belgian Congo ever had. He served from 1934 to 1946, and stood out by reason of his great intelligence and moral integrity. In terms of appearance he bore a great resemblance to Albert Camus; in terms of humanity he did in many ways as well. His son André was a provincial administrator who also got along very well with the local population. He learned their sayings and their dances and spoke fluent Kikongo and Kiyaka. Shortly after independence, he was murdered under tragic circumstances.

And so Nkasi went to work for Monsieur Peigneux. He learned carpentry and became a cabinetmaker. A few years later, when Peigneux was transferred to the district of Kwango as assistant district commissioner, Nkasi went with him. He and his family moved to Kikwit, where they would stay for more than twenty years. His eldest son, Pierre Diakuana, himself eighty-four years of age now, confirmed that. I found him in one of Kinshasa's back neighborhoods: "I was born in Ntimansi, but I was still young when we moved to Kikwit. The lower part of the city there, my father built that. We lived in the neighborhood for the blacks, at rue du Kasai, *numéro 10*. We had a big house made of unbaked brick. Papa became an *évolué* [Westernized African] then. I had Belgian friends."[21]

Nkasi himself thinks back on the period with great pleasure. "I worked for the state. I was the chief carpenter. It was my job to build *le nouveau pays des mindeles*, the new country for the whites." That was accurate enough. Kikwit had only recently been promoted to capital of Kwango district. Before that the capital had been Banningville (today's Bandundu), in the far north of Kwango. Social upheavals, however, prompted the administration to move to the district's center. In personal terms as well, it was a remarkable time for Nkasi. "In Kikwit I had four children, one of whom died. In 1938 my father died, on Happy New Year's Day. He was old, very old." During his long stay at Kikwit he got to know European culture from nearby. "I was *tout à fait mindele* back then, completely white. I had one wife. I had a suit with a tie and white shoes, I ate at Monsieur Peigneux's house. I interpreted for him, from Kikongo into French. Monsieur Peigneux even went to pick up my wife at the station. I was hired as an agent of the state, as a manager, just like a European manager. That's why they gave me the *carte civique*." From 1948 on, the *carte de mérite civique* (certificate of civil merit), was given to those Congolese whose lifestyle was considered sufficiently advanced. Thanks to the tax pressure of the 1930s, the follower of a subversive religion from the 1920s emerged in the 1940s and 1950s as someone who could speak proudly of his quasi-European status. And he still did today, even though very little is left of that prosperity.

But Nkasi's memories of Kikwit are especially interesting in another way. "In Kikwit I also built the prison," he told me. "The warden at the time was Monsieur Framand, a fat man." In the last few years I had visited the prison at Kikwit on a number of occasions. An altogether miserable place, it is still in use today. The prisoners wear rags, sleep on the floor, and can only eat because their chaplain, an elderly Flemish missionary, has set up a food-sharing system with a number of surrounding parishes. There are no toilets: the inmates squat on an empty stretch of concrete in an empty cell. Human feces lie to the left and right of them. The prisoners are all young men, with the exception of one young female, a beautiful, taciturn woman with a two-year-old child. No idea whether it was conceived before or after her internment. Chiseled in blue stone above the prison entrance is a date: 1930. Almost all the prisons in Congo were built between 1930 and 1935, a period in which the judicial system was buttressed to deal with the growing number of uprisings. More courts came, more judges, more legal actions, and more prisons.

"I built a gallows once in that prison," Nkasi said. "It was for the hanging of two young boys. They had stolen clothing from a shop and murdered the owner, who was sleeping in the back. That was in 1935, I think."[22] In the Belgian Congo the death sentence was pronounced repeatedly and during the interwar years it was often carried out as well. In 1921, the same year in which Kimbangu was sentenced to death, a group of some ten leopard men from the Anioto sect were hanged at Bomili, in Orientale province. In 1922, in Elisabethville, a man named François Musafiri was strung up after having stabbed to death a white man who was reportedly his wife's lover. The execution was attended by hordes of people. Some four thousand spectators arrived to see it, approximately half the city's population: three thousand Africans, including children, and a thousand whites, almost a tenth of the entire European population of Congo.[23] Public executions, it was felt, had an edifying effect. They caused the black man to toe the line and instilled in him respect for the colonial administration. One wonders whether it always worked out that way. In 1939, at the hanging of Ambroise Kitenge, things went wrong from the start. When the trapdoor fell open, the rope—borrowed from the local fire department—broke. Such bungling hardly jibed with the stern image the colonizer wished to project. How often was the death penalty carried out? The figures are incomplete, but we know that during the period 1931–53 some 261 individuals were sentenced to death and that the sentence was carried out in 127 of those cases.[24] That comes down to an average of one execution every two months, but between the wars the frequency must have been higher. Important to note: not one Belgian was ever sentenced to hang.

NKASI NEVER MENTIONED IT ONCE, but the reason Kikwit suddenly became the district capital had everything to do with an extremely serious popular uprising in the surrounding area, so serious that the government fearfully kept it hushed up. In 1931 a revolution was sparked off among the Pende, leading to the worst disturbances of the colonial period before the struggle for independence. The Pende were an ethnic group, many of whom were employed by Huileries du Congo Belge, the Unilever subsidiary. That company worked a region with an extreme abundance of palm trees but also an extreme paucity of workers. In the area where the Pende lived, however, the situation was reversed. The Pende were forced, often at gunpoint, to enter service as bearers or harvesters and transferred to another part of the country. The work was extremely strenuous. The men were expected to deliver thirty-six clusters of palm nuts each week; if they did so—on top of their measly wage of 20 centimes per kilo (2.2. pounds)—they were then given a 2.10 franc bonus and three kilos (6.6 pounds) of rice. This meant that they had to find five to eight ripe clusters each day; to harvest them they had to scale the branchless trunks of the palms to heights of sometimes more than thirty meters (nearly one hundred feet), and cut down the clusters with a machete. The Unilever operators assumed that all blacks had the ability to perform such acrobatics effortlessly, while in fact it required a highly specific set of skills not given to all. Fatal accidents occurred. And once the clusters were on the ground, the work was not over yet. The nuts had to be carried to the collection point; in actual practice, this meant that Pende women often had to cover up to thirty kilometers (about nineteen miles) on foot along forest trails, balancing a cluster of twenty to thirty kilos (forty-four to sixty-six pounds) on their head.

When the economic crisis broke out, Unilever took a beating as well. A kilo of palm oil went for 5.9 francs in 1929; by 1934 that was down to 1.3 francs.[25] The company felt obliged to recoup a portion of the losses from its workers. By the mid-1930s, they were paying only three centimes per kilo of palm nuts, rather than twenty.[26] That led to a great dissatisfaction. The state boosted taxes while the company scuttled its compensation. Things could not go on that way for long.

Here too, socioeconomic unrest manifested itself in the form of popular religion. After a woman named Kavundiji began receiving visions, the

Tupelepele (literally: floaters) sect arose. The actual leader of the move-ment was Matemu a Kelenge, who went by the nickname Mundele Funji (White Storm). Its followers hoped for the return of the ancestors, who would restore balance and ring in a new era of prosperity. In anticipation of that, the followers were to throw off all things European. Identification papers, tax receipts, banknotes, and employment contracts were tossed into the river. On the banks the people were to build a shed in which the ances-tors would leave behind goods for them, miraculous goods, such as peanuts so fertile that one needed to sow only one to cultivate an entire field. The hope of redemption could hardly have been expressed more poignantly. One inhabitant of the region at the time summarized the situation lucidly:

> The whites have turned us into slaves; to get palm nuts from us, they
> have not hesitated to whip or beat us. They entertained themselves
> with the women and girls from our villages. Our lives were no longer
> those of men, but of beasts. Our whole existence stood in the service
> of working for the white people: we slept for the whites, we ate for the
> whites, we got up for the whites and for the white man's work. We
> were tired of always having to work for the white man, who subjected
> us to inhuman conditions. That is why we heeded and accepted the
> messages of Matemu a Kelenge, later known as Mundele Funji, when
> he asked us to stop paying taxes, to stop working for the white man and
> to chase him away from us.[27]

Just as they had in the case of Simon Kimbangu, the colonial authorities sent in troops. The situation seemed under control until June 6, 1931, when Maximilien Balot, a young Belgian official, went to the region by car to col-lect taxes along with a few African assistants. In the village of Kilamba he drove onto the road leading to the shed where the ancestors were expected to return. There he happened upon Matemu a Kelenge, the sect's leader. Kelenge shouted that there was no more money, and that he would kill the white man and his henchmen. At that point, Balot fired a shot in the air. Many of the people ran away, including most of his own assistants. A sec-ond shot wounded a villager. "You see? The whites want to kill us!" Matemu shouted. "So kill me!" Balot fired and missed. Matemu scrambled to his feet and slashed the white man across the face with a large knife. Balot then pounded him senseless with the butt of his rifle and walked away. But an arrow shot by one of the villagers caught him in the neck. Matemu ran after him and hacked with his machete at the white man's shoulder. Balot's arm was now dangling at his side. Three villagers, including a chieftain, took aim with their bows and shot at him. When Balot fell to the ground, the chieftain saw that he was still alive. He cut off his head and took it as a trophy. The next day Balot's body was cut into pieces and distributed among the dignitaries of eight different villages. His bags were plundered.

Never before had an official of the Belgian Congo been slaughtered so brutally in the execution of his duties. The colonial administration's reaction was grim. It set out to grind the uprising beneath its heel. A punitive expedition, unlike anything seen since the worst years of the Free State, headed for the Kwango. Three officers, five noncoms, 260 soldiers, and seven hundred bearers occupied the region for months. Heavy fighting ensued. Rebels were captured and brutally tortured; even women were taken prisoner and raped. An investigative committee of the Belgian government later confirmed the gruesome tally. At least four hundred Pende had been murdered, perhaps many times that number. The Pende uprising had been broken, but the people's frustration was none the less for it.

When she arrived back in Brussels, Balot's widow, with an almost preternatural mildness and grandiosity, said: "The agents of the private companies treat the blacks badly and exploit them. People need to know that. What happens there must stop, otherwise there will be uprisings everywhere. Private companies have granted themselves rights that should be reserved only for the government. What's more, many district officials have not behaved as they should. My husband has paid for those others."[28]

IT MAY COME AS SOMETHING OF A SURPRISE that the first forms of popular protest took place in the countryside, among the farmers of Bas-Congo and the palm-nut harvesters of the Kwango. A thoughtful observer traveling around the country in 1920 would probably have predicted that the fires of unrest would first ignite in the cities, with their rudimentary workers' camps and their hard and unhealthy labor. But that was not the case. Why not?

There are, roughly speaking, two answers to that: the quality of life in the cities was improving, so that many Africans had begun feeling at home there, *and* the European population did everything to keep the masses calm. For as long as it lasted . . .

During the interwar period, the proto-urban agglomerations grew into real cities. Their populations showed a spectacular increase. Between 1920 and 1940, the population of Kinshasa doubled to fifty thousand inhabitants.[29] The population of Elisabethville grew from sixteen thousand in 1923 to thirty-three thousand by 1929, a doubling in the space of only six years.[30] More and more Congolese were moving to the cities. The forced recruitment of laborers was coming to an end, but many migrated of their own free will. In Kasai, Maniema, the Kivu, and, yes, even in Rwanda and Burundi, thousands of villagers let themselves be talked into going to the Union Minière mines in Katanga. In 1919 that company employed some eighty-five hundred local workers; by 1928 their numbers had grown to seventeen thousand.[31] From Bas-Congo and Équateur, people went to Léopoldville; Stanleyville owed its growth largely to the arrival of workers from Orientale province.

Most of those who packed their bags and went to work for a boss were young people. What made working in a mine, on a plantation, or in a factory so attractive to them? Often they were anxious to get out of the village with its poverty, its corrupt chieftain and powerful elders who married all the young women. Away from a miserable farming existence and the raising of state-ordained crops. Away from mandatory road building and primitive village life. Away from that world of deprivation, with no future in store for them.[32]

What's more, the cities and mines were no longer the horrors they had been until recently. The mortality rate at Union Minière in Katanga fell dramatically. In 1918 20.2 percent of the workers had died of the Spanish flu; one year later the mortality rate had fallen to 5.1 percent and by 1930 to only 1.6 percent.[33] Mineworkers also contracted fewer illnesses.[34] They were inoculated against smallpox, typhoid fever, and meningitis. Hospitals and medical centers were built. Housing, clothing, and nutrition improved considerably. The same went for the diamond pits in Kasai. A worker in the gold mines of Kilo-Moto in those days received a daily ration of 179 grams (about 6.25 ounces) of fresh meat or fresh fish, 357 grams (about 12.5 ounces) of rice, 286 grams (about 10 ounces) of beans, and one and a half kilo 3.3 pounds) of bananas, in addition to salt and palm oil.[35] In his village, he could only dream of such a rich and varied diet.

In addition to health standards, the pleasure quotient also improved. Life in the workers' camps of Katanga took a major turn for the better from 1923, when Union Minière began allowing workers to bring along their wives and children. In 1925 18 percent of the mineworkers were married; by 1932 that proportion was 60 percent.[36] The feeling of being uprooted, which had characterized an earlier generation, was dwindling rapidly. Many chose to prolong their contracts voluntarily. Beginning in 1927, mineworkers were allowed to sign three-year contracts, as opposed to the six-month maximum only a few years before. Many workers took advantage of that: by 1928, 45 percent already had a longer-term contract; in 1931 it was 98 percent.[37] Working in the mines was no longer an ordeal. When the economic malaise of 1929–33 forced the company to lay off three-quarters of its personnel, the protests were aimed not so much against the sudden unemployment, but against the prospect of having to return to the villages. The laid-off workers had to leave their company houses, but rather than return home they chose to settle in the immediate vicinity of Elisabethville, where they cleared fields and turned to farming until the economy recovered.[38]

The Katangan mining industry was no longer peopled by overworked young men who lodged for a few months in gruesome workers' camps, but by young families who felt quite happy in their new surroundings. Wages rose; in the camps children were born who knew the village of their parents and ancestors only by word of mouth. Elsewhere in Elisabethville, the cité indigène swelled to become a lively, multiethnic universe with a dynamism and atmosphere all its own. Unlike the neat, increasingly comfortable workers' camps housing the employees of the big mining companies, the cité was inhabited by a ragtag population: carpenters, masons, woodworkers, smiths, and craftsmen, as well as nurses, clerks, and warehouse foremen. The operators of small- to-medium-sized businesses lived next door to government personnel.[39] The population density there was five times that in the white city center.[40] There arose, in other words, an extensive and permanent urban population of African origin. The colonial administration, at first, was less than enthusiastic. Wouldn't such a protracted gathering of proletarians lead to a subversive or, even worse, Bolshevist climate? Fear of the red menace was deeply rooted within the colonial government. Or, to put it more succinctly: "The fear of the black went in the guise of fear of the red."[41]

In 1931, however, the colonizer realized that communities had now been formed that were no longer traditional villages and would not become such. Their existence was recognized with a monstrous bit of officialese, a fit of jargon at which the colonial administration was, in fact, quite expert: the centre extra-coutumier, the extra-legal center, as it were. Those centers were given a structure similar to that of the classic *chefferie,* and a chief was appointed to act as intermediary between the masses and the powers that be.

The new lifestyle that arose in the cities was different from village culture, but it was also more than simply a copy of European urban culture, if only because the new African agglomerations in no way resembled their European counterparts. The colonial city was an entirely new experience, even for Belgians! There was more space and freedom, the distances were greater, the lanes broader, the lots roomier. From the very start, these cities were planned with the automobile in mind. It had something American about it, many whites felt. Léopoldville with its various urban nuclei but no clear city center looked more like Los Angeles than like the medieval towns of Belgium or the nineteenth-century middle-class neighborhoods of Brussels or Antwerp. The colonial city did not trot along in pursuit of the European model, but took the lead over it. When a Belgian journalist saw white women in Congo taking a plane to Léopoldville to have their babies, he crowed that in the colony "a new society, a new Belgium with new ideas is being born."[42] In the Congo of the 1920s, it seemed as though the 1950s had already begun.

For the Congolese as well, the colonial city constituted a new universe with a material culture of its own. An imaginary young family from Kasai who moved to Elisabethville, where the father went to work in the mines, moved into a brick house. The wife began preparing meals in enameled pots and pans rather than in terra-cotta, even though she retained her preference for cooking out of doors rather than in the dark kitchen at the back of the house. The family began using tables, chairs, and cutlery. New ideas arose concerning care of the body and personal hygiene: people wore European clothes, sometimes even shoes; they washed themselves with soap and used latrines. The parents slept beneath blankets that came from England; if their children fell ill they were given medicines from Belgium. If the woman became pregnant, she went to have the child in a maternity ward run by black nurses or white nuns. When the family on rare occasions visited their former village, they took their relatives such novelties as needles, thread, scissors, safety pins, matches, mirrors, and money. But during those visits it also became clear how much distance had grown up between them. As a salaried employee, the young father had acquired a new sense of autonomy. He was less impressed by what the village chieftain and elders told him. They listened to him now! He told them about the iron discipline in the mine, about the siren that called the workers in each morning, about working six days a week. His audience laughed at that, of course. Six days a week? He would have been better off staying in the village, then his wife would have worked the fields! They were only envious, he knew. Everyone viewed his clothes admiringly; he had noticed that already. On the way back to town he felt more energetic and motivated than ever. If he could only make his way up in the hierarchy of Union Minière, he may have thought, as a mechanic or machine operator for example, then after saving for a long time he could buy his family a bicycle, a sewing machine, or even, imagine that, a gramophone! On Sunday morning they could ride to church together, on the bicycle. He would sit on the saddle, his wife on the baggage carrier, the children on the crossbar and the handlebars. That was what was called prosperity, and it felt good.[43]

The moment in the week when the new lifestyle was truly celebrated was on Sunday afternoon. In Elisabethville, the miners went to watch the white man's teams play soccer.[44] In Boma, the dockworkers went out strolling, wearing starched collars and straw hats and carrying canes. Their wives wore flowery cotton textiles and hats that had long been out of fashion in Europe.[45] In peaceful Tshikapa, close to the diamond mines of Kasai, the tenor voice of Enrico Caruso could be heard coming from some of the huts.[46] Someone else played jazz records and Cuban tunes on his gramophone. At four o'clock in Léopoldville, the Apollo Palace quickly filled with dancers.[47] Western trousers were all the thing: the men gathered there in long trousers, short trousers, cycling shorts, jodhpurs, or soccer shorts, as long as they wore trousers. And the women wore dresses, long skirts and fancily draped *pagnes* (skirts), all of them dancing in heels, sometimes twelve centimeters (just over four and half inches) high. The occasional male wore a tuxedo with patent-leather shoes, but most of them went barefooted. The dancing proceeded carefully and in great earnest, fearful as they were for all those spiked heels. An orchestra played merengue or rumba music, complex, syncopated African rhythms beat out on bottles and drums. But one could also catch snippets of fandango, cha-cha, polka, and Scottish music, in addition to echoes of martial music and hymns.[48] The most important influence of all, however, was Cuban: 78 rpm records brought back music that felt somehow vaguely familiar to the Congolese. It was the music the slaves had carried with them across the ocean centuries before and that now, enriched with various Latin influences, had come home again. Singers in Léopoldville liked to sing in Spanish, or in something that passed for it. The clear vowels sounded like the phonetic patterns of Lingala: all you had to do was toss in a *corazón* or a *mi amor* now and then. The guitar became the most popular instrument, in addition to the banjo, the mandolin, and the accordion. Camille Feruzi, the greatest accordion virtuoso of Congolese music, composed peerless melancholy melodies. And aboard the boats bringing people from the interior to Léopoldville, young Wendo Kolosoyi tirelessly strummed his guitar: he would grow to become the founder of the Congolese rumba, the most influential musical style in the sub-Saharan Africa of the twentieth century. Léopoldville in those years was a kind of New Orleans where African, South American, and European popular music fused to form a new genre: the Congolese rumba, irresistible dance music that would wash over the rest of the continent, but that could be heard at that time only in the bars of the new capital. It was music that made people laugh and forget, that made them dance and flirt, that made them happy and horny. It was Saturday Night Fever, but on a Sunday afternoon. Why would you want to protest against such a dazzling, uproarious life?

BUT THE ADMINISTRATORS STAYED ON THEIR TOES. At a table in Elisabethville's Cercle Albert in the 1930s, one could often see three men engrossed in conversation.[49] Three white men. They spoke quietly and their expressions were serious. Their voices: basso continuo. Their conversation: completely inaudible. Above their heads floated clouds of cigar smoke, occasionally dispersed by a burst of good-humored laughter from their own midst. Officially, Africans were not forbidden to eat in European restaurants, but the extremely chic Cercle Albert was an exception. Still, it was here that decisions were made about the lives and futures of the black population. The three men were Amour Maron, provincial commissioner of Katanga, Aimé Marthoz, director of Union Minière, or one of his successors, and Félix de Hemptinne, bishop of Katanga. Hemptinne's stately white beard had the African population convinced that he was the son of Leopold II. Three Belgians. Each of them stood at the head of one of the pillars of colonial power: government, finance/industry, and church. The "blessed colonial trinity" it was sometimes called in jest. Who knows whether the bishop was able to laugh about that.

These three men had joined forces to ensure that life in the mining town of Elisabethville was run in an orderly fashion. Their respective agendas converged in many ways: industry wanted submissive, loyal employees; the government wanted no repeats of the Kimbangu affair or the Pende uprising; the church wanted to deliver pure souls in the hereafter—and that meant producing obedient citizens on this side of the divide. At other places in the colony, the three administrations became tightly intertwined as well. Although there were often tensions between the pillars of the colonial triad, there was one thing about which they were in full agreement: if the step from a tribal to an industrial lifestyle was not to end in bitter defeat, they had to keep a close eye on their black fellow man. Gradually, and above all circumspectly, the new urban Congolese citizens would be kneaded into willing workers, docile subjects, devout Catholics.

If no large-scale uprisings took place in the cities, then, that was due not only to the pleasurable prosperity experienced by the workers, but also and above all to the sophisticated arsenal of strategies employed by the colonial trinity to keep tabs on, to discipline, and if need be to punish the population. There may never have been anything like a grand master plan, but in actual practice church, state, and big money frequently toed one and the same line. Their philosophy—how do we keep them under control? how will they produce best? how must we instruct them?—manifested itself in a host of ways. In Léopoldville, the authorities were anxious about all that dancing and strongly advocated illuminating the streets of the *cité* at night, for how else could they "effectively supervise an agglomeration of twenty-thousand souls, with a handful of policemen lost in the dark?"[50] At Elisabethville they succeeded in imposing a lingua franca, Swahili, a language that was not indigenous and almost no one's native language, but which made it easier to exercise control over the ethnic melting pot.[51]

Schooling was still a privilege held exclusively by the missionaries, and it became a powerful instrument to mold the masses in the desired shape: the pupils were taught everything about the Belgian royal family, but nothing about the American civil rights movement. Even the French Revolution had to be handled with utmost care. European textbooks were too inflammatory: "There, the revolution is often not regarded with a properly critical eye. Some reforms, liberties, etc., condemned by the church are applauded too readily," wrote the influential missionary and school inspector Gustaaf Hulstaert. The pupils were in danger of becoming "liberal, then disinterested and atheistic."[52]

Meanwhile, African clerks also began reading French-language newspapers. Communist papers like the Belgian *Le Drapeau Rouge* were banned as from 1925, as were magazines with such evocative titles as *Paris Plaisirs*, *Séduction*, and *Paris Sex-Appeal*.[53] A similar urge to excise content became manifest following World War I, when the first movie theaters appeared. Film was seen as a dangerous medium, one that could cause foment among the lower, unlettered masses. In 1936, therefore, a separate film censorship board was set up especially for African audiences, which resulted in separate showings for Europeans and Congolese. Usually this meant that those films considered unsuited for white children were forbidden for black adults as well.[54] "Tous les coloniaux seront unanimes à déclarer que les noirs sont encore des enfants, intellectuellement et politiquement" (All those in our colony are unanimous in stating that the blacks are still children, both intellectually and politically), said the official papers that set out the new media policy.[55] In terms of civilization, as the recurring metaphor had it, the African was still a child: he could not be left to his own devices; his development had to be watched over carefully. Ultimately, the colonial trinity aspired to a form of emancipation, but only in the long term, in the very long term if need be. Things must not be allowed to become too exuberant. *Dominer pour server* was the motto of Pierre Ryckmans, governor general at the time: to rule in order to serve. Paternalistic? Far from it: that "to serve" sounded dangerously progressive to many. "To discipline" would have been better, or perhaps "to educate."

Growing up in the Léopoldville of the 1920s was an intelligent and sensitive boy who, after World War II, would develop into the first giant of Congolese literature: Paul Lomami Tshibamba. Shortly before his death in 1985, he looked back on the mood that dominated the interwar period:

> The colonizer did everything to convince us that we were big children, that we would remain that way, that we were under his guardianship and that we had to follow all instructions he gave, in order to educate us with an eye to our gradual integration into Western civilization, the most ideal civilization of all. And we, what else could we expect? My generation no longer knew our parents' traditions: we were born in this city founded by colonials, in this city where the life of a man was subordinate to the power of money. . . . Without money you ended up in prison. Money was used to pay taxes, to clothe yourself, and even to eat, which was unknown in the villages. It was the white colonizer who supplied that money, so you had to submit to whatever he said. That is the world into which I was born and in which I lived: you had to bow to what others asked of you.[56]

Yet monitoring the urban workers' environment alone was not enough; one had to intervene actively as well. In addition to the schools, club life and family policies were the instruments of choice. The decision to allow women and children into the work camps had a utilitarian motive: it was intended to boost the general willingness to work, to impede prostitution and alcohol consumption, to stimulate monogamy, and to promote the general tranquility of camp life. In addition, children in the work camps would grow up within the company culture. That, with the aid of the mission schools, would help groom them to become the next batch of disciplined employees.[57]

The church wielded an exceptional amount of political power. Around 1930 there were as many Catholic missionaries in the Belgian Congo as there were colonial officials.[58] Ecclesiastical authority and worldly power meshed seamlessly, as Lomami Tshibamba well knew:

> In the day-to-day life in which we grew up, the priest demanded our submission: the representatives of Bula Matari, in other words of the government or the territorial administration, all had authority and that authority came from God. As a result, total obedience was expected from us. That is what the priest advised! Being good, both to God and to the people of this new society, which was created by Bula Matari, required obedience, subjugation, and respect. We were reduced to servility—that was not the term they used, but that is what it came down to.[59]

Cultivating servility was also the motive behind the social policies of the big companies. Union Minière went furthest in that. Granted, the company built schools, hospitals and leisure clubs for the workers and their families. And yes, in the late 1930s a start was made with a system of retirement benefits. And indeed, the miner was surrounded from cradle to grave by the solicitousness of the company, more than with any other mining operation in Central Africa. But there can be no doubt about the fact that the company's paternalistic benevolence was prompted more by matters of efficiency than of philanthropy. The objective was to raise perfect workers: happy and tractable.

More than an employer, Union Minière was a state within a state, on occasion even a state with totalitarian features. Every facet of life in the workers' camp was supervised by the white camp boss. He kept a file on each worker and his family; he was responsible for the housing, the supplies, the salaries, and the schools; he mediated in conflicts and imposed disciplinary measures. The wife of a Union Minière worker who needed to go back to her native village first had to ask the camp boss's permission, even though she was not a company employee! From the age of ten, her children were obliged to follow classes in manual training, a matter of preparing them for their work later. If they were boys, the company helped them to save for a dowry. Union Minière was a total company, with the backing of church and state.[60]

Native organizations were regarded with great apprehension as potential breeding grounds for social unrest: "The club feeling is discouraged in as far as possible. The camp leadership keeps a close eye on all activities organized by the natives."[61] The Union Minière found sewing circles, choirs, and home economics courses preferable to the employees' own initiatives. The missions had churches in the working neighborhoods and were well-suited to the task. In Léopoldville this was organized largely by the Congregation of the Immaculate Heart of Mary, in Elisabethville by the Benedictines. The cathedral at Elisabethville was graced on Sundays by an excellent Gregorian-chant boys' choir, made up exclusively of African children.

In the cities, beginning in 1922, Belgian priests saw to the setting up of the first Boy Scout troops. The paramilitary character of Scouting, a movement originally secular in nature and, so, more in line with the state than with the church, was an exclusively Catholic phenomenon in the colony. It allowed the missionary to exercise control over his best pupils even after school hours. With activities such as trailblazing, tree climbing, knots, camping, and Morse code, the young people were taught both pride and discipline. The young Scout collected badges, said his pledge and cared for his uniform. The membership was never extremely large (around one thousand members in all of Congo), but it helped to cultivate a native elite with a sense of discipline and fidelity.[62]

A much larger group was reached through what was probably the most successful part of Belgian missionary work: soccer. Here too, Léopoldville and Elisabethville took the lead, starting around 1920. Missionaries in their cassocks explained the rules of the game, and in no time saw children and young people practicing with homemade balls and grapefruits in the dusty streets of the *cité*. The first teams were set up: Étoile and League in Léopoldville, Prince Charles and Prince Léopold in Elisabethville. In 1939, Léopoldville alone had fifty-three teams and six divisions. There were teams with shoes and barefoot teams—playing in bare feet entailed milder passes, but greater agility. The matches were held on Sunday afternoons. In addition to hundreds of players, this also attracted thousands of supporters. Friends, colleagues, wives, and children screamed themselves hoarse from the sidelines. Soccer was more than recreation. It also had a formative side. A Flemish Benedictine noted contentedly: "Rather than spending their Sunday afternoons squatting in their huts and drinking *pombo*, or going out drinking in bars amid women of dubious virtue, they participate freely and out of doors in the sports that interest them."[63] A Marist brother was equally enthusiastic: "It keeps them, at least for a few hours, away from dancing and lolling about and, after benediction, allows them to spend a pleasant Sunday afternoon."[64] Just as soccer was propagated at the Flemish academies and boarding schools as a pressure valve for the excess sexual energy of boys, in the colony it was introduced to quell possible social unrest. In addition to an exuberant game, soccer was also a form of discipline. One had to attend training sessions, develop skills, control one's reflexes, obey the rules and listen to the arbiter. Festive, yet restrained: an ideal colonial training ground. "Sport teaches the native . . . to comply with a discipline he takes upon himself,"[65] was the way it was put.

In the streets of Kikwit in 2007 I occasionally saw a timeworn, yellow scooter race past, driven by an old white man. In itself that was quite exceptional: the few Europeans one saw always went by car, particularly the elderly among them. The scooter rider in question turned out to be Henri de la Kéthulle de Ryhove, a Jesuit of noble origin, well into his eighties and still tirelessly in action—for the last few years in particular in the fight against sickle-cell anemia, a hereditary illness. Père Henri was also the nephew of Raphaël de la Kéthulle, perhaps the most famous missionary in all of the Belgian Congo. His uncle owed that fame not to heroic proselytization deep in the jungle and not to the evangelical brightening up of a miserable leper colony, no: Père Raphaël spent his life working in Kinshasa, teaching his people to play soccer. He was a Marist educator, part of the first batch of urban missionaries. The scion of an aristocratic Francophone family from Bruges, he himself had attended Sint-Lodewijks College. (It is a detail that makes me smile: I myself attended a former branch of that same college. At my school, too, three-quarters of a century and a Dutchifying shift later, soccer was still the major religion next to Christianity. Our paved schoolyard had the outlines of five or six soccer fields painted on it, there were five volleyball nets and two basketball hoops. We had four rather than two hours of mandatory gym each week. West Flemish Catholicism, despite the influence of Guido Gezelle—our own Gerard Manley Hopkins—still had more affinity with sports than with poetry.)

"My uncle was the founder of the Association Sportive Congolaise, Congo's first sporting club," Père Henri said once we were seated. His white hair was still blown back from his forehead after his scooter ride. A Kikwit coiffure, as it were. "He was the greatest promoter of soccer in Kinshasa." But that was not the whole story. "His club also promoted gymnastics, track and field, swimming, and even water polo." Raphaël de la Kéthulle must have been every bit as indefatigable as his nephew. Besides all manner of sporting initiatives, he also founded a number of schools. He was one of the original initiators of colonial Scouting and school drama programs, and founded a brass band and an alumni association. But above all he was the driving force behind the development of a decent sporting infrastructure in Léopoldville. Père Henri knew the story by heart. "He built three soccer stadiums, a huge sports park, tennis courts, and an Olympic swimming pool, which even had its own five-meter board. In that same pool, he also organized canoeing matches!" The absolute apogee of his urge to build was the Stade Roi Baudouin, later renamed the Stade du 20 Mai, a soccer stadium that seated 80,000 supporters and that, at its opening in 1952, was the biggest in all of Africa. It was here, in 1959, that the riots broke out that would ultimately lead to independence. It was here that Joseph-Désiré Mobutu addressed the people after his 1965 coup. It was here that Muhammad Ali fought George Foreman in their famous 1974 bout. Today, every Kinois still knows about Tata Raphaël, Daddy Raphaël, even if that is only because the huge stadium is named after him these days and because his image, which bears a striking resemblance to the logo of Kentucky Fried Chicken, sprawls across the front of the Collège Saint Raphaël. "Yes, he was quite energetic," Père Henri concluded, "though he did have something of a *bottine légère*." A light boot? "That's right, he was known to deal out a swift kick now and then, when necessary."

The club life facilitated by the Catholic missions offered the urban worker not only healthy recreation, but purposefully altered the social topography as well. Out of fear for ethnically tinted revolts like those among the Pende, tribal boundaries were broken down—the very same boundaries delineated so sharply before by mission-school education! Henri de la Kéthulle told me: "In the sports, my uncle mixed up the various peoples. His football competitions always consisted of mixed teams. He organized inter-Congolese matches, even the first international soccer match. A Congolese team played against a Belgian one. Beerschot it was, I believe."[66]

BLOOD, HOWEVER, WOULD NOT BE DENIED. Despite the well-intentioned sports initiatives and patronizing family policies, a certain hunger remained among parts of the Congolese urban population. The colonial administration may have shown a friendly face, but that lasted only as long as one toed the line. The masses were steered beneath the smiling countenance of the colonial trinity, but anyone stepping out of line was punished without pardon.

And so native organizations continued to exist.[67] The Kitawala religion spread among the mineworkers and infiltrated large parts of the countryside. From Katanga it reached Kivu, Orientale province, and Équateur. Operating clandestinely, it provided a mixture of mysticism and revolt. When adherents were arrested in Jadotville in 1936, they said of the Bible: "This book clearly states that all people are equal. God did not create the whites to rule over the blacks. . . . It is unjust that the black man, who does all the work, must continue to live in poverty and misery, while the wages of the whites are so much higher."[68] Many followers were banished, but exile served—just as it had with the Kimbaguists—only as a stimulus to the movement.

Ethnic organizations in Katanga, like those among the Lulua or the Baluba, featured a conviviality and hominess unrivaled by any Scout troop. They provided assistance to newcomers and helped young people to pay their dowries. There even arose certain forms of solidarity between people with the same first name. An old man from Lubumbashi explained: "If my name is Albert and your name is Albert, then you become my brother. . . . We take care of each other. We help each other to get food, we play sports together, we support each other in every way."[69] Starting in 1929 the financial crisis resulted in intensive forms of solidarity among the Congolese. André Yav, the former boy from Lubumbashi, said of that: "Everyone was very hungry back then. Unemployment rose incredibly. But this is what we did: if a man had work, then he was the father and mother of all his friends. They came to eat at his house and they came there to get clothes."[70] Such forms of spontaneous self-organization were ineradicable.

In the 1920s there were groups that called themselves Les Belges. With no lack of humor, the members decked themselves out with titles borrowed from the colonial administration ("district commissioner," "governor general," "king") and in their dances imitated white officials and missionaries. In addition to satire, they also engaged in finding housing for newcomers, distributing food, and organizing funerals.[71]

Following the financial crisis there also rose the first associations of Africans who had risen through the ranks. Organizations with names like Cercle de l'Amitié des Noirs Civilisés and the Association Franco-Belge brought together Congolese who had attended school, who enjoyed a decent income, and who spoke French together. They represented the start of a Congolese middle class, with all of the hopes and snobbism that went along with that. Their members often looked down on the life of the street, which they themselves had left behind only recently, and longed for a more European lifestyle, for cufflinks and respect. But if frustrated, that longing could backfire in the form of irritation and protest—which is precisely what happened in the 1950s. During the interbellum, however, their activities were not yet overtly political, although some wished to organize themselves independently of the church.

STARTING IN THE 1930S, at the border post with Rhodesia, a fascinating phenomenon could be observed several times a week.[72] Whenever a train arrived from the British dominions, it would stop with a loud hiss to allow the white engineer to step down. His colleague from the Belgian Congo would then climb aboard to continue the journey to Elisabethville. Those witnessing this for the first time must have rubbed their eyes in disbelief: was that new engineer really an African? Yes, he was. In the Belgian Congo, unlike in South Africa and Rhodesia, people prided themselves on the lack of a "color bar." Africans in the mines and factories were allowed to operate expensive and dangerous machines, albeit under the watchful eye of white foremen. Dedicated Union Minière employees could, to a certain extent, work their way up in the company. Hotels, restaurants, and cafés were, in theory, open to everyone. Only in the movie theaters was there a clear racial division. There was no formal ban on sexual relations between whites and blacks.

But the absence of a legal color bar did not mean that there was no invisible color bar.[73] And that latter phenomenon was perhaps the most stubborn of all. Africans could not build their careers in a way that admitted them to the top levels of a company. In administrative service, the position of clerk or typist was the highest achievable level. The cities consisted of strictly delineated white centers and black, outlying neighborhoods, supposedly to prevent the spread of malaria. But that was a specious argument. Graveyards, too, were racially segregated, and there one ran little risk of contracting malaria. There were also no Scouting troops of mixed race. And Congolese soccer teams were not allowed to compete against European ones, out of fear for riots after a defeat or humiliation after a victory. One of the most acute observers of the colonial period wrote of this: "Remarkably enough, the fact that there was no official color bar only aggravated the racial reflexes of the whites. Denied by law, racism confirmed itself with full force in the facts."[74] And that was true enough. Today, anyone reading the colonial papers published between the wars will notice how much the thinking was determined by an us/them logic, and how much fear went hidden behind their forceful rhetoric. After a white man was murdered by a Congolese, *L'Avenir Colonial Belge*, one of the colony's most popular papers, wrote:

Is our personal safety, the safety of the Whites, still ensured in Léopoldville?

One can reply in all sincerity: No! The acts of insubordination by the blacks are multiplying; their insolence is great and strikes fear into the hearts of even the bravest among us. Thefts are increasing in number and scope; the arrogance of the blacks towards the Whites is at times staggering; the fear we instill in them is null; the respect for the *mundele* is a thing of the past.

That is how things stand in the Year of Our Lord 1930.

But, we hear you say, is Stanley Pool a region once again in need of pacifying?

Well, why ever not?

What that repacification entailed, the paper felt, was clear enough: any African threatening the life of a white person, for whatever reason, should face the death penalty.[75] Valid self-defense, mitigating circumstances, involuntary manslaughter, irresistible compulsion; none of that mattered. The courts, fortunately, were more subtle in their thinking, but that a newspaper flogging such humbug could become one of the most influential in the colony shows how the majority of whites thought about the racial question. *Les noirs*, that was printed in lower case; *les Blancs*, that took a capital.

In essence, colonial society between the wars was ruled by mutual fear: the white rulers were terrified of losing their respectability in the eyes of the Congolese, while a great many of those same Congolese were afraid of the white man's authority and did all they could to earn his respect. It was a stranglehold of fear. How long could this be kept up?

ALBERT KUDJABO AND PAUL PANDA FARNANA spent four long years as prisoners of war in Germany, years that included much more than singing songs for ethnographers in Berlin. Years of sickness and forced labor. Years of mockery and humiliation. Kudjabo had been forced to work on a farm outside Stuttgart, where the farmer cheated him out of money. Panda ended up in Hannover, and was taken from there to Romania.

But now they were back in Belgium, the country for which they and a few other Congolese had risked their lives. And what did the veterans' magazine *Le Journal des Combattants* write about them? "Let us repatriate them and send them back to the shade of their banana trees, where they will certainly feel more at home. There they can relearn their Negro dances and tell of their war experiences to their families, who sit around them on chimpanzee skins."[76]

Is that what they had fought and suffered for? They weren't about to let it go at that. A reply came: "In the trenches the soldiers never tired of repeating that we were brothers and we were treated as the whites' equals. Nevertheless, now that the war is over and our services are no longer required, people would rather see us disappear. In that regard we are in complete agreement, but then under one condition: if you insist so vehemently on the repatriation of blacks, it would be only logical for us to demand that all whites now in Africa be repatriated as well."[77]

What nerve! No one in Congo dared to adopt such a self-assured tone. The reaction was written in a French more elegant than that of the original article. A new voice was truly being heard. A few weeks before the article in question appeared, on August 30, 1919, the Union Congolaise had been set up in Brussels, "an association for the assistance and the moral and intellectual development of the Congolese race." It resembled the organization André Matswa had set up in France. The association originally had thirty-three members, almost all of them veterans. The central figure was the former prisoner of war Paul Panda Farnana; his companion-in-arms Albert Kudjabo became its secretary. They set about helping the poor and sick, assisted in paying funeral costs, and arranged for free night-school training. But theirs was also an explicitly political line. As early as 1920 the Union Congolaise demanded that forced labor regimes be relaxed, that workers' wages be raised, and that schooling be expanded. Above all, they called for the Congolese to have more say in the colonial administration. Once again: this was 1920! In those days the authorities consulted at most with individual village chieftains they had appointed themselves. Much better, Paul Panda suggested, would be for the Congolese themselves to elect a council to advise the colonial government in Boma.

Panda's Union Congolaise grew steadily. Branches were set up in Liège, Charleroi, and Marchienne-au-Pont. The new members were often Congolese sailors who had jumped ship in Antwerp harbor. These young, single men, who had spent weeks in a deafening engine room as *lubricator,* fireman, or trimmer, rebelled against the fact that their white colleagues on arrival were paid twice as much for the same work. In Congo there were no white laborers, only supervisors, but aboard the big ocean steamers the huge contrast became visible for the first time. And while irritation on dry land led to religious ecstasy, disgruntlement on board led to more prosaic resistance: strikes. Tools were downed at the ports of both Antwerp and Matadi, especially after the African crew members were forbidden to supplement their measly wages with a private trade in bicycles and sewing machines. What's more, once on land they were not allowed to hang about in bars. The Belgian government greatly feared them ending up in the red-light district or, even worse, in red cafes. There were enough Communists in Antwerp as it was!

At first the Union Congolaise encountered a certain degree of sympathy. Paul Panda Farnana was an extremely eloquent intellectual who had mastered the rare art of presenting radical ideas as reasonable measures. In December 1920 he was allowed to address the first National Colonial Congress in Brussels, where his speech concerning the need for political participation by natives also met with a great deal of support from the Belgians present. Grant us power, that was his bottom line. And he received applause! As a gifted orator, he had not forgotten to pepper his speech with references to historic popes.

One year later, however, Panda Farnana took part in the second Pan-

African Congress, an African American initiative led by the radical American civil rights activist W. E. B. Du Bois. That gave Panda Farnana a bad reputation: the colonial press accused him of nationalism, Bolshevism, and Garveyism. Incorrectly. The Pan-Africanism of that day was out to liberate and emancipate the black race, all over the world. The congress, which lasted a week and was held in London, Brussels, and Paris, gave the lie to the accusations of Bolshevism. All the delegates wanted was to promote the equality of blacks and whites, in times of both war and peace. The delegates made a field trip to the colonial museum at Tervuren, where the African Americans were incensed by the collection, already huge by that time, of what they saw as the fruits of plunder. Panda Farnana had never thought about it that way before. The sessions in Brussels and Paris were chaired by Blaise Diagne, a Senegalese who had held a seat in the French parliament since 1914, the first African ever. That must have made a huge impression on Panda Farnana. While the French colonies were already allowed to send elected representatives to Paris, in the Belgian Congo one could attain no higher status than that of railroad engineer, choirboy, Boy Scout, or goalie. When it came to political participation, the status of *chef médaillé* meant nothing: that was no popular involvement, that was simply an excuse. A few years later he expressed his unvarnished, final opinion: "So far, the colonization of Congo has amounted only to 'civilization-vandalism,' in favor of the European element."[78]

In May 1929 Panda Farnana returned to the colony, where he settled in his native village of Nzemba, close to the ocean. There he helped to set up a school and a chapel. With his rare combination of life experience, acuity, and tact, he could have been a key figure in the negotiations for a more just colonial regime. But less than a year after his return he died, unmarried and childless. The Belgian Congo had lost its most brilliant voice of dissent. He was only forty-two years old.

CHAPTER 5

THE RED HOUR OF THE KICKOFF

The War and the Deceptive Calm That Followed

1940–1955

T HEY STOOD SWAYING IN A CIRCLE, SHIFTING THEIR WEIGHT from one foot to the other; something between cautious danc-ing and marching in place. The little group of veterans at the Maison des Anciens Combattants in Kinshasa were clearly enjoying themselves. Their brand-new uniforms were a gift from the Belgian army to Congo's present-day armed forces. The veterans wore them with pride, clapped their hands and sang in deep voices: "Saluti, saluti, pesa saluti, tokopesa saluti na bakonzi nyonso." A marching song. "Salute, salute, atten-shun, we salute all our leaders." Those leaders, as they explained to me afterward, had been Belgians. All their officers were Belgians back then. "Biso baCongo-lais, biso baCongolais," was how it went after that, "We Congolese, we Congolese, we have shown our strength. Today we have conquered Saio." A simple but catchy soldiers' song. Once you'd heard the melody, it stuck with you. A Congolese soldier had come up with the tune in 1941, shortly after the taking of the fortified garrison town of Saio in Abyssinia (present-day Ethiopia). It was sung in the backs of the trucks in which the Congo-lese soldiers drove back to Kisangani through the arid, open landscape of Sudan. Almost seventy years later the veterans still knew it by heart. It breathed a new sort of brotherhood. That's right, in those days the whites were still their superiors, but during the war something had changed. The Congolese soldier was extremely proud to present his white officers with the conquest of Saio.

That sense of pride, however, would not last for long. Much more even than World War I, World War II brought about rapprochement, followed by disillusionment. I talked about this to eighty-seven-year-old André Kitadi, one of the men who had sung along that day. He was deputy chairman of the veterans' association of 1940–45, a remarkable, quiet-voiced man of keen judgment. His office was empty, save for a metal desk, a Congolese flag, and a huge puddle of water. The previous night's rain had collected on the concrete floor. "We fought for Belgium, that much was clear. The Belgians used us to defend their interests. We took part because we had discipline. We had *la conscience de la guerre* [a sense of duty about the war]."[1]

After the German army rolled over Belgium during those eighteen days in the spring of 1940, the legal status of the Belgian Congo remained unclear for a few months. That was due to the general collapse in the fatherland itself. While the Belgian government fled first to France and later to England, where it aligned itself with the Allies, King Leopold III, great-nephew of Leopold II, bowed to the German victory. He was taken prisoner and remained in Nazi Germany until the end of the war. Which raised the question: who was the colonial administration to heed now? The king of a country that no longer existed as a sovereign state but still had a colony, or his minister of colonies in exile, who was effectively the administrator general of the Belgian Congo? In the colony itself, opinions differed. Conservative elements like Félix de Hemptinne, the influential bishop of Katanga, were royalists and resigned themselves to the German victory and a new fascist world order. Many industrialists also harbored ultra-right-wing sympathies. They hoped to continue supplying Germany with raw materials, which in fact they did during the war, by way of Portugal. Anti-Semitism reared its head here and there. In the El Dorado of Elisabethville, a small Jewish community had formed. The local rabbi, the only one in Congo, saw to his horror how the shop windows of Jewish merchants were daubed with swastikas and slogans like *sale juif* (dirty Jew).[2] But in the end, Governor General Pierre Ryckmans put his foot down: the Belgian Congo would unanimously take the side of the Allies and continue to fight against fascism. Officially his administration was answerable to the exiled minister of colonies, but in actual practice he enjoyed great autonomy. His personal courage was of more overriding importance than any directive from London.

The French colonies, too, hesitated about which side to take: most of them decided to support Philippe Pétain's collaborationist Vichy regime, while a few opted for Charles de Gaulle's Free France. In this way, the conflict between the Allies and the Axis powers was transferred to the African continent. Although Germany had lost its final overseas territory in 1918, large parts of Africa still moved within the National Socialist sphere of influence. What's more, Germany's new ally, Italy, still possessed African colonies. It had ruled ever since the late nineteenth century over Eritrea and Italian Somaliland in the Horn of Africa, areas along the Red Sea coast whose strategic importance had grown with the opening of the Suez Canal. In 1911 Italy had succeeded in taking Libya and in 1935 Mussolini invaded Haile Selassie's Ethiopia, the only state of any size in Africa that had never been colonized. Thanks in part to soldiers from the Belgian Congo, that instance of foreign rule would this time be short-lived as well.

When the Belgian government-in-exile sided with the Allies, Winston Churchill called for material and military support from the Belgian Congo. In Northern Africa, after all, Libya posed a threat to Egypt (which had gained independence in 1922, but was in many ways still dependent on England), while the Horn of Africa was a menace to British-held Kenya and Sudan. From those British colonies Churchill first sent his own troops to Abyssinia, but starting in February 1941 their ranks were reinforced by the eleventh battalion of the Force Publique. Some three thousand Congolese soldiers and two thousand bearers took part. There was one Belgian officer to every fifty Africans. By truck and boat they moved across the Sudan, where daytime temperatures often reached 45 degrees Celsius (113 degrees Fahrenheit) in the shade. From there they invaded the mountainous western portion of Abyssinia. The trucks received a new coat of paint; in order to better camouflage them, brown sand was mixed with the army green. But it was largely on foot that the soldiers traveled through the desolate region. During the day the troops almost collapsed in the heat; at night, at higher elevations, their teeth chattered from the cold. When the rainy season began a few weeks later, some of them had to bivouac in the mud. Towns such as Asosa and Gambela did not pose much of a problem. After brief but intense fighting, the Italian troops fled. Their officers did not even bother to take their sabers or tennis rackets with them. A much greater challenge was Saio, a major Italian garrison city close to the Sudanese border. After heavy shelling on June 8, 1941, the demoralized Italians demanded a truce, despite their clear superiority in numbers and arms.

The Belgian commanders agreed, on terms of a total surrender. No fewer than nine Italian generals were taken prisoner, including Pietro

MAP 6: BELGIAN CONGO DURING WORLD WAR II

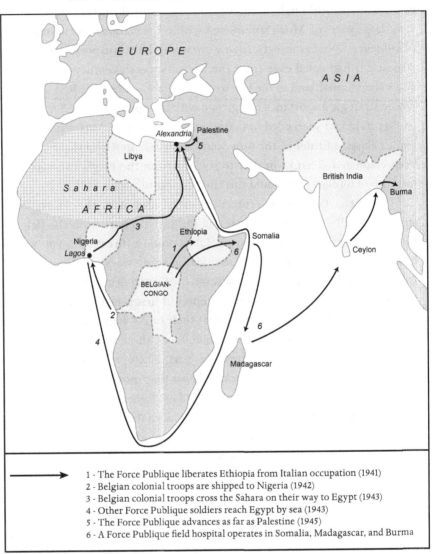

1 - The Force Publique liberates Ethiopia from Italian occupation (1941)
2 - Belgian colonial troops are shipped to Nigeria (1942)
3 - Belgian colonial troops cross the Sahara on their way to Egypt (1943)
4 - Other Force Publique soldiers reach Egypt by sea (1943)
5 - The Force Publique advances as far as Palestine (1945)
6 - A Force Publique field hospital operates in Somalia, Madagascar, and Burma

Gazzera, commander of the Italian troops in East Africa, and Count Arno-
covaldo Bonaccorsi, inspector general of the Fascist militias that had terrorized Mallorca during the Spanish Civil War. In addition, 370 Italian officers
(45 of them high-ranking) were taken prisoner, along with 2,574 noncoms,
and 1,533 native soldiers. Another 2,000 native irregulars were sent home.

The taking of Saio, however, was primarily of material and strategic importance. The Force Publique captured eighteen cannons with five thousand rounds, four mortars, two hundred machine guns, 330 pistols, 7,600 rifles, fifteen thousand grenades, and two million rounds of small-arms ammunition. In addition, Belgians and Congolese confiscated twenty metric tons (twenty-two U.S. tons) of radio equipment, including three complete transmitter stations, twenty motorcycles, twenty automobiles, two armored vehicles, 250 trucks, and—hardly unimportant in the highlands—five hundred mules. An army had gone out of business here, that much was clear. It was the most significant Belgian victory over Fascism and in fact the greatest Belgian military triumph ever, but the heaviest toll was paid by the Congolese. The Belgian casualties were 4 killed and 6 badly wounded, while 42 Africans were killed, 5 went missing, and 193 succumbed to illnesses or injuries. Among the bearers there were 274 fatalities; most of those died of exhaustion or dysentery.

The Force Publique's Abyssinian campaign was instrumental in Haile Selassie's return to the throne. Ethiopia had been a colony for only five years, from 1936 to 1941; now the centuries-old empire had been restored. The restoration gave new inspiration to the Jamaican Rastafarians, who had begun to venerate Emperor Haile Selassie as a deity in the 1930s. His divine status, however, was one that had been affirmed in ways more military than metaphysical. Congolese soldiers had freed Ethiopian towns like Asosa, Gambela, and above all Saio. Indirectly, therefore, Belgian colonialism contributed to the spiritual dimension of reggae. What Tabora had been to World War I, Saio was to World War II: a resounding victory that bolstered the troops' morale. It was no mean feat; here, for the first time in history, an African country had been decolonized by African soldiers. "The only people we saw were white," Louis Ngumbi, a veteran from eastern Congo told me, "we shot only at white people."[3] That was a bit of an exaggeration, but that the Force Publique took several thousand white soldiers as prisoners of war, including nine generals, made a huge impression. Saio was etched in the memory of an entire generation of military men. André Kitadi, deputy chairman of the veterans' association, knew the numbers by heart: "In Abyssinia we captured nine Italian generals, plus 370 Italian officers, twenty-five hundred Italian soldiers, and fifteen thousand natives."[4]

Kitadi volunteered in 1940. The war had already started, but that didn't bother him. In the army you could get a good education: he became a telegraph operator. During the Abyssinian campaign he remained on stand-by in

Orientale province, along the border with Sudan, ready to be deployed. But that never proved necessary. When the troops returned, singing, and were welcomed by cheering crowds, he was transferred to Boma. But not for long. Now that the Horn of Africa had fallen, the Allies shifted their focus to Western and, above all, Northern Africa. In fall 1942, after the Vichy French were rousted from Morocco and Algeria, Kitadi boarded a carrier that brought him and his comrades to Lagos in Nigeria. It was from that British colony that the battle for Dahomey (present-day Benin), a French colony still aligned with the Vichy regime, was to begin. "It took us four days to get to Nigeria. We arrived in Lagos and were taken to barracks three hundred kilometers (185 miles) away. We were trained there. For six whole months." The men of the Force Publique came in contact there with the British colonial troops. Although still under Belgian command, Kitadi was fitted out with a British uniform. In early March 1943 he received new marching orders. Following the Allied successes in Northern Africa, Dahomey had sided with De Gaulle. The last German-Italian stronghold in Africa was now Libya. From there General Erwin Rommel had struck at Egypt, to force a passage to the Suez Canal. The Allies wanted to stop him at all costs and began a troop build-up in Egypt. Kitadi now had to get from Nigeria to Egypt. That was, to say the least, no easy task as long as the Mediterranean was still controlled by Italy. Overland, then? Straight across Africa? In those days neighboring Chad, a French colony, was run by a black governor general, Félix Eboué. He supported De Gaulle and opened his territory for a crossing by Allied troops. The only problem was that this entailed a long journey through the desert. . . .

We left with ten, maybe fifteen columns. Each column consisted of a hundred and fifty trucks, with one Belgian officer and a Marconi operator. I was one of them. As *opérateur*, I was responsible for the wireless communication with the other columns. We went from Nigeria to Sudan, across the Nubian Desert. We could only navigate by compass. I'll never forget that crossing. We ran into sandstorms that blinded you sometimes for an hour or more. When the sand heated up, you saw things that weren't there. It took us more than a month. Sometimes we covered only twenty kilometers [about 12.5 miles] a day. There were ravines. Accidents happened there. . . . We lived on biscuits and cans of corned beef, and were given only half a liter of water a day. A lot of us fell ill. . . . Two hundred of the two thousand soldiers died along the

way. . . . We lived like animals, we couldn't wash ourselves. . . . The
whole journey from Lagos to Cairo cost us three months. We drove
thousands and thousands of kilometers.[5]

His voice faltered. He stopped. Never before had I heard about this
heroic crossing of the Sahara. I asked whether he had ever had his story written down. "No," he said, "this is the first time a white person has ever asked
about it."

There was, of course, another way to get to Egypt. Martin Kabuya, a
ninety-two-year-old whose grandfather had been in Tabora when it was
taken in 1916, took that other route. He too was garrisoned in Nigeria,
where he too was a wireless operator. His appearance was still imposing,
but his voice had grown brittle as an eggshell. He whispered his story to
me. "I was very, very good in Morse code. *Ti-ti-tiii-ti.* I never made a mistake, even when I worked by ear. If you're able to do that, the rest is easy.
On March 24, 1943, I received orders to board a Dutch merchant ship, the
Duchesse de Ritmond. We sailed south on the Atlantic to South Africa. There
we had to round the Cape of Good Hope, and then on to the Gulf of Aden
and the Red Sea to close to the Suez Canal. There were at least a hundred
ships. Around South Africa, a few of them were attacked by Japanese planes.
Twenty-seven men were killed on one of those ships. The soldiers slept all
bunched together in the hold. Conditions were bad."[6]

Both Kitadi and Kabuya took part in the fighting in Egypt. Kitadi spent,
by his own account, "a whole year" in the desert outside Alexandria, where
enemy positions and planes were fired on. The threat came from Libya and
Sicily. "During the day it was scorching hot, at night we had to wear gloves
to keep our hands warm. On Sundays we were allowed to go into town,
to Alexandria, but that had been bombed by the Germans. It was full of
flies." Kabuya was stationed at Camp Geneva, a huge depot close to the Suez
Canal; his job was to intercept and decode Morse messages from the enemy.
"I was in the *Section d'écoute,* the monitoring section; we eavesdropped on
messages about their troop movements."

The war brought them in contact with other peoples: British officers,
Nigerian soldiers, Arabs, and German and Italian prisoners of war. The her-
metic world of the barracks in the Belgian Congo lay far behind them. "There
were a lot of Italian prisoners of war in Alexandria," Kitadi said. "We kept
them out in the desert, behind barbed wire, but they dug tunnels. Our muni-

tions dump was a little further away. The Arabs tried to steal our munitions. They're real thieves," he said in amusement. Kabuya saw prisoners of war as well. "One time a German prisoner came after me, a big SS man, he must have been two meters [6.5 feet] tall. He had got hold of a revolver. I stabbed him in the stomach with my bayonet. Our bayonets were poisoned. They were very good weapons. That SS man was the only person I ever killed."

When the war was over both men went by truck to Palestine, but things were calmer there. The most strenuous task was the occasional stint of guard duty along the border at Haifa. The biggest danger Kitadi encountered there came from a case of food poisoning that put him in the hospital at Gaza: something to do with meat that been roasted after it was already spoiled.

The Force Publique's participation in the Allied campaigns is virtually unknown. Its contributions, in numerical terms, were less decisive than those during World War I. The many tens of thousands of bearers from the olden days had been largely replaced by trucks. That is why today, even in Congo, the memory of those events is quickly withering away. In Kinshasa, a city of eight million, only a handful of veterans is still alive. One of them is Libert Otenga, a man who can still sing "We're Going to Hang Out the Washing on the Siegfried Line" at the top of his lungs. I was keen to talk to him, because he was one of the very few who had served in the Belgian field hospital. In the course of the conflict, that mobile medical unit of Belgian doctors and Congolese medics made an incredible exodus past the most remote fields of battle. Their wanderings ended somewhere in the jungles of Burma (present-day Myanmar); the Belgian Congo, therefore, assisted the British not only militarily and with matériel, but also medically. The Belgian field hospital became known as the tenth BCCS, the tenth Belgian Congo Casualty Clearing Station. It had two operating tents and a radio tent. In the other tents there were beds for thirty patients and stretchers for two hundred more. During the war, the unit treated seven thousand wounded men and thirty thousand who had fallen ill. Even at the peak of its activities it consisted of only twenty-three Belgians, including seven doctors, and three hundred Congolese.[7] Libert Otenga was one of them. When I located him at last, he was still able to recount his odyssey effortlessly. His voice rang like an alarm bell, but he chose his words carefully.

I was a medical assistant. I joined the army in 1942. First we went to Somalia. I worked there with a Belgian surgeon. Thorax, abdomen, bones. We operated on everything. Then we left with British-Belgian troops for Madagascar. There were German prisoners of war there. The German is a special case, believe me! One of them badly needed a blood transfusion, and Dr. Valcke, one of the Belgian physicians, was willing to donate his own. But the German refused! Blood from an Allied sympathizer, he was having none of it. And from a black man, that was *entirely* out of the question. He wanted to preserve his honor, but we wanted to save his life. *Bon*, while he was asleep we gave him that blood anyway.

He still had to laugh heartily at the thought. I had never known that prisoners of war, even against their will, enjoyed the protection of the Third Geneva Convention regarding humane treatment. But Otenga went on marching straight through his recollections. "From Madagascar we went by ship to Ceylon. To Colombo. The hospital and the army were reorganized there. After that a ship brought us to India." That must have been to the Ganges Delta (present-day Bangladesh). "From there we took another boat, an inland boat, up the Brahmaputra River. When we disembarked, we had to go a long way on foot to the Burmese border." At the time, that area was the scene of fierce combat between Japanese and anti-Fascist forces, including the British. Japan had conquered Burma in 1942. "The border post was called Tamu. We moved into Burma and got to the Chindwin Valley. We followed that all the way to Kalewa. We set up the hospital there." The names of all the locations were still chiseled in Otenga's memory. He even spelled them out for me, in a military staccato. "Ka-le-wa, have you got that? We took care of people there. Soldiers and civilians. Many of them with bullet wounds. I remember a British soldier with shrapnel in his intestines. Things like that." The fact that Congolese paramedics cared for Burmese civilians and British soldiers in the Asian jungle is a completely unknown chapter in colonial history, and one that will soon vanish altogether. "Burma was where we stayed the longest. We carried out complex operations there. We even had an ambulance plane at our disposal. Finally, we were saved by the atomic bomb! Then Japan had no choice but to surrender Burma."[8] Then, to underscore that victory, he sang again the little song about the Siegfried Line.

IT COULD NEVER HAVE OCCURRED to Colonel Paul Tibbets at the moment he pushed the button. It was August 6, 1945. His plane was called the *Enola*

Gay. Within a few seconds the city below him would no longer be a city, but a name: Hiroshima. It would not have occurred to him that what he, as an American, was releasing over Japan in fact originated in Congo. The first American atomic bomb was made with Katangan uranium. When news of that terrible devastation finally reached the Burmese interior, Libert Otenga had no idea that he had been "saved" by an ore that came from under his own native soil. In Congo, too, the mineworkers at Shinkolobwe could never have imagined that the leaden, yellowish ore that was processed into "yellow cake" after they dug it up would lead to such destruction on the other side of the world. No one knew a thing. Operating in deepest secrecy, Edgar Sengier, then managing director of Union Minière, saw to it that Congo's uranium reserves did not fall into the wrong hands. Shinkolobwe had the world's largest confirmed deposit of uranium. When the Nazi threat intensified just before the war, he had had 1,250 metric tons (1,375 U.S. tons) of uranium shipped to New York, then flooded his mines. Only a tiny stock still present in Belgium ever fell into German hands. The potential military application of uranium was still unknown (it was used at the time mostly as a pigment in the ceramics industry), but in the late 1930s nuclear physicists announced that it could be used to unleash an unbridled chain reaction. Einstein considered informing Belgium's Queen Elisabeth of the situation—he knew her and they shared a love for music—but decided instead to turn to the Belgian ambassador in New York and finally to President Franklin D. Roosevelt himself. When the Manhattan Project started in 1942, the American scientists tinkering with the plans for an atomic bomb went in search of high-grade uranium. The Canadian ore they had used till then was quite feeble. To their amazement, it turned out that the Archer Daniels Midland Warehouses on the New York waterfront already contained a huge stockpile of the finest quality. The discovery led to spirited negotiations with the Belgian government in exile, which received $2.5 million in hard cash with which to finance Belgium's reconstruction after the war. In addition, Belgium was granted access to nuclear technology. A research center was set up in the Flemish town of Mol in 1952 and a small reactor in Kinshasa, the first of its kind in Africa.[9] The Americans also provided support for the construction of two large air bases in Congo; one on the coast in Kitona and the other at Kamina in Katanga. And once again: during World War II, almost no one in Congo knew about this. The strategic importance of uranium, however, was a prime reason for America's special interest in Congo, an interest that started during the war years, became decisive in the years surrounding independence and lasted until the end of the Cold War in 1990.

Yet it was not only about uranium. For the Allies, Congo was also one of the most important raw materials depots in their fight against Germany, Italy, and Japan. After destroying Pearl Harbor, the Japanese went on in early 1942 to conquer large parts of Southeast Asia, including Indonesia, Singapore, Malaysia, and Burma. Allied imports from those regions were cut off completely. Congo helped provide a solution. Ores and other raw materials once again came into great demand. Copper was needed for bullet shells and bomb casings. Wolfram was incorporated in antitank weapons. Tin and zinc served in the making of bronze and brass. Even products of vegetable origin such as rubber, copal, cotton, and quinine had strategic value. Palm oil was used in Sunlight soap, but was also applied in the steel industry.

Soldiers, therefore, were not the only Congolese to contribute to the Allied push. The colony's mineworkers, factory hands, and plantation workers also did their part. As it had during World War I, the Congolese economy was now rolling at top speed. There were some half a million payrolled workers in 1939, but their ranks swelled to eight hundred thousand, perhaps even a million, by 1945.[10] After South Africa, Congo became the second most industrialized country of sub-Saharan Africa. More textile plants, soap works, sugar refineries, cement factories, breweries, and tobacco manufacturing plants had been built during the interwar years. But an industrial sector in top gear did not bring immediate prosperity. Because of the war, less and less cargo was reaching the colony. There were no textiles, no machines, no medicines. The doctors had left, the little hospitals had no supplies, fewer boats plied the rivers. The more supplies shrank, of course, the higher prices rose. And because wages were fixed, the purchasing power of the average employee fell disastrously.[11] In remote Elisabethville, which was highly dependent on imports, the price of a length of textile from Léopoldville increased by more than 400 percent. Imported textiles from England or Brazil rose by as much as 700 percent.[12] In the mining town of Jadotville, the price of a blanket quadrupled.[13] Clearly a problem, when one realizes how chilly a Katangan night can be.

This dramatic inflation could only lead to social protest. At both the start and the end of the war, strikes and uprisings broke out. In November 1941 mineworkers in Manono, in northern Katanga, tried to take down the

Belgian flag and raise a black one in its place. The men wore crowns made
from palm leaves. Most of them were adherents of the Kitawala religion.
They had already slaughtered all their goats and dogs, convinced as they
were that a new age was about to dawn.[14]

One month later, large-scale protests were seen in the Katangan capital
of Elisabethville. White employees of Union Minière who had organized to
form a union protested against the unprecedentedly low purchasing power
and their dissatisfaction jumped the gap to the black workers' camps. There,
too, miners demanded substantial wage increases. Social protest in this case
did not assume the guise of a religious revival (as with Simon Kimbangu
in 1921) or ethnic revolt (as with the Pende in 1931), but was expressed in
the year 1941 in the form of a clear and very understandable wage claim.
Yet the colonial and industrial powers-that-be reacted in traditional fashion.
Trade unions for natives were still strictly forbidden. On the key day of the
strike, the workers gathered on the town's soccer pitch. It would be hard to
imagine a situation more laden with symbolism: the soccer pitch, the place
meant to teach the masses the virtues of discipline, now became the site
of popular protest and bloody repression. Amour Maron, provincial governor of Katanga, along with the personnel manager of Union Minière, tried
to hush up the strikers, but they were having none of it. Their leader was
Léonard Mpoyi, an educated clerk. One of the strikers present recounted:
"Maron said: 'Go back to work! We've already raised your wages.' We said
no. The people began to shout and rave. Maron asked Léonard Mpoyi again:
'So you people refuse to leave?' Léonard Mpoyi said: 'I refuse. We want you
to give us proof, a written document that shows that the company has raised
our pay.'" That document never came. But panic broke out. The soldiers of
the Force Publique came into action. "Maron ordered the soldiers to shoot
at the people. The soldiers obeyed, and fired without mercy."[15] At least sixty
people were killed and one hundred wounded. The first fatality was Mpoyi
himself.[16]

The violence with which that strike was crushed made a deep impression in Elisabethville. André Yav, the former boy from whom we heard earlier, wrote about it in his singular history: "It was a year deep in the war of
1940–45. Many, many people died. They died for a higher monthly wage.
That day there was great sorrow in Elisabethville, because of that *bwana*
governor."[17]

The big Elisabethville strike was a milestone in the social history of

Congo because it was the first, open expression of urban protest. Elisabeth-ville was the country's second city and the economic motor of the whole of Congo. Union Minière was the flagship of colonial industry, widely praised for its generous social provisions. But the paternalistic, finger-in-the-dike policy apparently did have its limits. The people would not put up with everything.

In Léopoldville's working neighborhoods during the war, several legends made the rounds that were, in all their inventiveness, extremely telling when it came to attitudes toward the white authorities. First there was the legend of Mundele-Mwinda, the white man with the lantern, an imaginary European who walked the streets of Kinshasa at night, holding aloft a lantern and looking for blacks. Anyone struck by his beam of light was immediately paralyzed. Then Mundele-Mwinda would take his victim to Mundele-Ngulu, another horrific creature. This white swineherd (*ngulu* means *pig* in Lingala) fattened up the victim until he turned into a pig. "And that pig was used to make sausages and hams, to feed the white people during the war."[18] That parents told their children such stories to keep them off the streets at night illustrates the antipathy directed toward whites by that time. It was a perfect inversion of the figure of Black Peter seen in Catholic Belgium to this day.

But adults too put stock in such legends. Under the influence of folktales about evil whites, they sought refuge in messianic religions; those stories bore witness to a deep distrust of the colonizer. In the barracks at Lulu-abourg in February 1944, the soldiers mutinied. The cause was bizarre: a vaccine. When military medics announced plans to inoculate the soldiers, a rumor quickly spread that this was a white man's trick, intended to anni-hilate them. A great many soldiers turned against their superiors, left the barracks, and spread out over a huge area. Mutineers and civilians began plundering. Tax offices, storerooms, and a number of houses belonging to white people were attacked. The repression of the mutiny was inexorably harsh. That an unfounded rumor could give rise to such large-scale protests shows how deeply the mistrust ran.[19]

At other places too, social unrest returned in full force at the end of the war. In spring 1944 in the area around Masisi in Kivu province, there was a socioreligious uprising by Kitawala followers. Many of the rebels worked in the local gold mines. Three whites and hundreds of blacks were killed, and the leader of the revolt was hanged. In November 1945 five to six thousand workers and boys in Léopoldville went on strike. Railway personnel spread the news to the port city of Matadi. The dockworkers joined in. They pulled rivets out of the rails and cut the telephone lines. Fifteen hundred strikers marched through the city, armed with iron staves, hammers, and clubs with nails in them. An unknown number of them, including women and children, were killed by soldiers. A military lockdown and a curfew were imposed. During the days that followed, the prison at Matadi became so packed that some rebels died of suffocation.[20] In Congo, the final days of the war did not feel like a liberation. When Brussels was freed, the Congolese danced in the streets of Léopoldville. They hoped that everything was going to be different. But the euphoria did not last for long.

IN THE CITIES the workers were asking for higher wages, but the war also made itself felt deep in the calm interior. In addition to the military mobilization, which plucked young men from their villages, there was also a far-reaching civilian mobilization. Each village had to contribute to the *effort de guerre* (war effort). The number of mandatory days in the service of the state rose from 60 to 120. This often placed great pressure on small-scale farming. In the equatorial forest in particular, the *effort de guerre* resulted in hardship. People were required to build roads through huge swamps and bridges across broad rivers. The villagers were required to gather palm nuts and copal fiber, and even to tap rubber once again. In 1939 Congo produced only 1,142 metric tons (just over 1,256 U.S. tons) of rubber, a fraction of the average harvest during the rubber boom, but by 1944 that had risen to no less than 11,337 metric tons (nearly 12,475 U.S. tons).[21] A tenfold increase within five years, right in the middle of the war.

An extremely vivid account of the war's impact on the countryside is provided by the wonderful diaries kept by Vladi Souchard, the pen name of Vladimir Drachoussoff, a young Belgian agricultural engineer of Russian extraction. His parents had fled to Belgium during the October Revolution of 1917; he himself was only a few months old at the time. At the age of twenty-two he left for the colony, in late May 1940, only a few weeks after the war broke out. Employed at first on a sugarcane plantation in Bas-Congo, he later joined the colonial civil service. As a young agronomist he traveled from village to village with the aim of boosting the war effort. His working territory lay in Équateur, close to Lake Leopold. Suddenly he, an immigrant's son from Brussels, was responsible for the farming activities in an area of tens of thousands of square kilometers, an area without roads or industry, some parts of which were characterized only by "a vague mixture of water, mud, and trees."[22] He traveled on foot, by bicycle, or canoe and visited villages where no colonials had been for years. His maps were outdated, villages had moved, and the government shelters were mostly in a state of sore neglect. During the war, there was no new crew of colonial officials waiting on the sidelines; he could expect no relief from his duties. Drachoussoff had to order communities to start cultivating rice and peanuts and to start harvesting rubber again. This latter directive caused the population to shudder. For it was in this region, after all, that "red rubber" had left the deepest scars. Youngsters had heard the stories from their parents or grandparents. Some of their accounts required no verbal confirmation. Drachoussoff saw it with his own eyes: "In the Lopori and close to Lake Leopold, I personally saw two old blacks who had lost their right hands and who had not forgotten those days."[23] Many villagers claimed that there were no rubber vines in their area, that they had never seen them, or that the vines had been exhausted. So began *la dure bataille du caoutchouc*[24] (the pitched battle for rubber), a struggle Drachoussoff nevertheless dared to call into question: "What right do we have to drag the Congolese into our war? None whatsoever. Yet necessity knows no law . . . and Hitler's victory here would install a racist tyranny that would make the abuses of colonialism look good."[25]

Those were ambivalent times, a fact of which Drachoussoff was well aware. He walked the line between necessity and impotence, between international politics and the jungle, between anti-Axis commitment and colonial reality. As an agronomist in a time of administrative scarcity, he juggled many tasks. And when evening came, he wrote down his experiences.

Wednesday, November 10, 1943. Mekiri.

At four A.M. I leave for Kundu on a borrowed bicycle. Two soldiers accompany me on foot. My companions travel on to Mekiri with our baggage.

I arrive in Kundu just before dawn, and bolt down a chunk of bread as I wait for it to grow light. A little before six I knock on the door of the capita [a Congolese go-between] and ask him to call all the men together to show me yesterday's harvest. The villagers are so surprised that they all show up, both those who have latex and all the rest. I hand

out a few encouraging words and three fines, and around the necks of
the four worst cases I lay the rope [that expression is symbolic: what
one does in fact is to tie a twenty-centimeter (about eight-inch) length of
kekele—a very sturdy cord made from tree bark that causes no pain but
serves to symbolize the arrest—around the person's neck]. Then I leave
triumphantly with my "convicts" to catch up with the caravan.

There was no prison in the wide surroundings. Incarceration meant
that you had to travel along with the colonial official for a few days. A hike
as disciplinary measure, the freedom of nature as detention.

On the road to Ngongo I find the soldiers and hand the prisoners over
to them. Justice has been done, Kundu shall make its contribution to the
war effort.

A ways past Ngongo I catch up with the rear guard of our caravan.
This leg of the journey covers twenty kilometers [about 12.5 miles],
straight across huge sandy plains where only a few Borassus palms will
grow, punctuated by wispy forest along the banks of the river. We monitor
the rubber production in the villages we pass: it is none too impressive,
and I write out a number of tickets.

In the village of Mekiri, the men to whom my arrival was announced
last night are waiting for us with latex, so that I can demonstrate how they
are to make it coagulate. As I give my little presentation, I send Faigne and
Pionso out to check and measure the fields. That evening, as a downpour
thrashes the place where we are to spend the night, turning the roof into a
sieve and drenching beds, clothing, and food, I deliver verdicts and rapidly
hand out a number of convictions and acquittals.

The proceedings require a sea of paperwork. I have been appointed a
magistrate with limited jurisdiction (to wit, I am allowed to rule only on
economic offenses) and ambulant prison guard (to wit, I am allowed to
let those I convict accompany me as I go). The maximum penalty is seven
days for not carrying out works of an educational nature, for the chopping
down of protected trees, and for hunting violations, and thirty days for
failing to contribute to the war effort. I am, of course, also an officer of the
legal constabulary, with limited jurisdiction on the basis of my position as
district agronomist.

Procedure demands that I, as officer of the legal constabulary, first report an offence and then address that report myself as magistrate. While I am changing costumes, I pass the verdict, after an interrogation that is often surrealistic in the extreme.

A man appears because he has failed to plant ten ares of peanuts. He either has a verifiable and valid excuse and I send him home (some substitute clerks even demand that we then draw up a verdict of acquittal), or he comes up with something. That results in the following dialogue, which is scrupulously included in the minutes of the court.

> Why didn't you plant any peanuts?
> Because I was ill.
> For how long?
> Two days.
> You had three months to prepare your field. It can't be those two days that prevented you from doing what you had to do.
> That is true, *mundele*. But there was something else. . . .
> What?
> My father's second wife had a baby.

Good Lord, it's impossible to be familiar with all the customs of the thirty or forty peoples who live around the lake, but birth celebrations certainly don't last for weeks. So then:

> *Bon*, that will be five days in jail for you.
> Yes, *mundele*.

Some of them argufy. Others come right to the point.

> *Mpua na nini asalaki bilanga te?* (Why didn't you cultivate your fields?)
> *Mpua na koï-koï* (Out of laziness). . . .

I would dearly love to acquit him then, but if I do they will all give me the same answer tomorrow.[26]

Drachoussoff was part of the colonial administration but, unlike most of his contemporaries, he also felt empathy for the local point of view. These people are satisfied with the forest and the rivers, he noted; money interested them only marginally. "Because there is little monitoring in this area, most of the farmers prefer to trade eight days of mild imprisonment for 357 days of peaceful living. Can I really blame them?"[27]

As it had in the nineteenth century, the demand for rubber drove people farther into the jungle, despite the predators and tsetse flies. Sleeping sickness, vanquished in its earlier epidemic form, began claiming more victims. Perhaps as much as 20 percent of the population of the equatorial forest became infected. Many of them also suffered from intestinal parasites because, far from home, their only drinking water came from swamps.[28]

Drachoussoff's diary is fascinating, because it allows us to hear the voice of a colonial whose worldview is being shaken. While most whites simply waited for the war to be over and then resumed their old lives, he realized that "Europe's enfeeblement can only serve to elicit centrifugal forces."[29] Things would never be the same. Despair began creeping in. This child of Russian émigrés was much more sensitive than the average Belgian to sudden historical turnabouts. The most brilliant passage in his diary was nothing short of prophetic:

> What have we come here to do, anyway?
>
> To "civilize" in the name of a civilization that is falling apart and no longer believes in itself? To Christianize? . . . But then why are we here?
>
> We bring peace and guard it, we fill the landscape with roads, plantations, factories, we build schools, we care for the people's health. In exchange for that we use the riches of their soil and substrata and we let them work for pay . . . modest as it is. Service and a service in return: that is the colonial pact in its entirety.
>
> And tomorrow? What will the black baby be like then, the baby bound tightly to its mother as she passes my barza, this young offshoot of a colonized Africa? Will he be willing to accept the power from our hands, or will he yank it away? How far away that seems today, deep in this jungle . . . but still, there are those moments when history accelerates: when my father was a child, he also believed in the eternity of the patriarchal world that surrounded him—and that was twenty-five years before 1917! Sooner or later—and I hope for Congo that it will not be too soon—a man will rise up. Will it be a chef coutumier who can deal with the modern techniques of exercising power, without falling back on

the traditional ones? Will it be one of those boys who sing "Vers l'avenir" [Toward the future] at the graduation ceremony? Many colonials today don't even think about that, even though our colonization will finally be judged less by what it has created than by what will remain of it once it has disappeared.

And he continued his lucid musings:

Let us suppose—a supposition that is consciously absurd—that Congo will be independent in 1970. But what problems would that create! In Europe, we have never known an insurmountable conflict between our social organization and our technical surroundings: both developed more or less hand in hand. But in Africa an archaic social organization collides with the supremacy of a technical civilization that causes the former to fall apart without replacing it.

Of course, Congo is moving bit by bit into the modern age. . . . But doesn't that come at the cost of a traditional world that is obsolete but still needed and—for the moment at least—irreplaceable? In the name of what? In the name of that lovely civilization, the fruits of which we reap right now in Europe? . . . That is why it is so hard to keep a clear conscience. Simply by being ourselves, we destroy traditions that were sometimes hard but venerable, and we offer as a replacement only white trousers and dark glasses, in addition to a little knowledge and a vast longing.

But wasn't education a form of emancipation? Didn't colonization actually lead to a gradual growing-up, the way the colonial trinity liked to claim?

Do we have the right, even the most open-minded among us, to punish and to educate, when education is all-too-often synonymous with debasement?[30]

Drachoussoff's diary is an unsung masterpiece of colonial literature: stylistically beautiful, subtle in tone, literary despite itself. For him, his war years in Congo were a lesson in humility. "Africa is a training ground for the character, but also a graveyard for illusions," he wrote at the end of the war.[31]

WHEN THE THIRD REICH FELL IN 1945, André Kitadi, the wireless operator who had driven through the Sahel, was still in Palestine. What to do, so far from home? A chaplain took him and his comrades along on a tour of all the holy places. "We went to Jerusalem, Bethlehem, and Nazareth. . . . [S]ome of us even had ourselves rebaptized in the River Jordan." Libert Otenga, medic at the military hospital in Burma, also took the opportunity to see a bit of the world, although he stuck to a worldlier brand of sightseeing. "After Burma we went back to India. To eat, to drink, and to dance. And to pick up girls." He roared with laughter. "They were good."

After every war, veterans are always a troublesome category. Anyone risking his life for a country hopes for something in return afterward. Recognition, honor, money. The toll the struggle has taken, it seems, is often felt only after it is over. Back in civilian life, veterans realize what they have endured. Injuries, mental as well as physical, have not nearly healed—if they will ever do at all. Young men have lost limbs and dreams. Memories come back, traumas smolder on in silence. They see how placidly those who stayed at home have gone on living. It's for them that the veterans have suffered, the people who can never feel what it is they have been through. Veterans are always a fractious factor, but in a colonial army they are absolutely explosive. There they fight less for their own people than for a foreign ruler. Congo was no different. "We fought the war as a Belgian colony," Otenga blared. And that called for generous compensation: "They should have made us Belgians afterward! That would have been fair."[32] Another person I talked to felt that they had been sent home after their glowing victories "like a mean dog the hunter sends back with nothing to chew on."[33]

The veterans returned with a host of new impressions. They had become more worldly-wise and less prone to be impressed by the colonial regime in the Belgian Congo. They had taken white generals prisoner in Abyssinia! In Nigeria they had seen a different form of colonialism! André Kitadi expressed it with his characteristic forcefulness: "The British treated us very well. We were well-dressed and well-fed. In Lagos they cooked for us, for the soldiers. Tea, bread, milk, jam . . . While back in Congo we had to scavenge for our own food in the *brousse*! We also saw that the British already had African officers, even majors and colonels. They sent good pupils to secondary school in England. There was none of that in the Belgian Congo. Such discrimination! They kept us under their thumbs! That produced a lot of irritation and suspicion, yes, even a certain rebelliousness. After the war we said: 'We want that too!' We wanted a transformation, but we weren't even allowed into their shops. We didn't like that. We had learned English. We put on English suits. We pretended to be Americans and walked into the Portuguese restaurants, talking loudly. 'So, what do you drink?' we asked each other. 'You want to eat?' "[34]

The whites' authority was being challenged, albeit subtly. Something had changed in the balance of power. Many Congolese were very well aware that the colony had proved stronger than the metropolis. Belgium had been crushed; Congo had remained on its feet and achieved military triumphs. Just like in 1914–18, the Force Publique had been more successful than the Belgian national army. Occupied Belgium, via its government in London, had only survived thanks to its colony. When it came to postwar reconstruction, the shattered mother country would lean heavily on its colony as well. The Belgians, in other words, needed Congo more than Congo needed the Belgians.

In fact, the new postwar world order did nothing to prove the Congolese wrong. In Yalta, the victors chalked out the boundaries of a new world. America, as a former colony, had little sympathy for Europe's colonial adventures. On the basis of a proletarian ideal the Soviet Union was against all forms of subjection. Colonies, once an inexhaustible source of noble daydreams and bombastic ideals, suddenly seemed to belong to another age, to be outdated. To say nothing of suspect. When fifty countries from all over the world gathered in San Francisco in 1945 to draft the United Nations Charter, the term *colony* was relegated to the wings of history. People spoke of *nonautonomous territories*. That term had something accusatory—for the countries with colonies—but also something hopeful—for the colonies themselves. Their subjugation would not go on forever. Article 73 left no doubt about that:

> Members of the United Nations which have or assume responsibilities for the administration of territories whose peoples have not yet attained a full measure of self-government recognize the principle that the interests of the inhabitants of these territories are paramount, and accept as a sacred trust the obligation to promote to the utmost, within the system of international peace and security established by the present Charter, the well-being of the inhabitants of these territories, and, to this end . . . to develop self-government, to take due account of the political aspirations of the peoples, and to assist them

in the progressive development of their free political institutions, according to the particular circumstances of each territory and its peoples and their varying stages of advancement.

And then?

Was this the great turnaround? Against the background of such an international climate, one might expect things to suddenly take off. For the veterans to begin shaking the tree of power, for clerks to feel empowered by such international backing, for workers to raise their voices, and for farmers to wave their pitchforks, or perhaps more aptly, their machetes.

But none of that happened.

After the tumultuous strike in Matadi, everything suddenly fell silent. A remarkable stillness descended over the colony. For ten years, from 1946 to 1956, calm reigned in Congo. There were no religious revivals like in the 1920s, no farmer's uprisings like those in the 1930s, no mutinies like the ones in the 1940s. There were no strikes.

How could that be? Had Belgian colonialism changed its standing overnight? Somehow, yes—at least in people's minds. In his farewell speech in 1946, governor general Ryckmans said: "The days of colonialism are over," by which he meant primarily the old system of overt exploitation. "Like integrity in diplomacy, altruism is the best policy in colonization."[35] The colony should at least reap the benefits of its own riches. This was not yet about the road to independence, but about "developmental colonialism."[36]

This new spirit was also reflected in the overblown slogans that became popular. After the war, the colonizer could hardly stop talking about Congo as "Belgium's tenth province." It was an attempt to replace the condescension of old with a more egalitarian way of dealing with Congo. The colony was no longer some remote outpost, but had become an integral part of the mother country. But it was also a ridiculous notion: how could a gigantic country that a twist of fate had made the colony of a dwarf state become one of its *provinces*? Congo was a thousand times the size of the provinces of Limburg, Brabant, or Hainaut!

Another attempt at rapprochement was seen in the concept of a "Belgian-Congolese Community." The idea came from Léon Pétillon, governor general from 1952, and was intended to obliviate the old *dominer pour server*, which by then sounded all-too paternalistic. Hand in hand, the Belgians and the Congolese would work together to build a new, modern world.

Just as the British had transformed their empire into the Commonwealth, and the French had redefined their overseas territories as the Union Française, Belgium too would from now on strive for equality with this Belgian-Congolese Community.

Some politicians paid explicit lip service to the newfangled discourse about "native welfare." The Permanent Commission for the Protection of Natives went further than others: "The future of the race and the happiness of our Congolese population groups are of an importance paramount to any other objective," it said.[37] Belgian opinion leaders from across the political spectrum chimed in in agreement. "Colonization must first of all entail a project of civilization on behalf of the peoples," one Catholic said.[38] "Whether we like it or not, our fate in Congo depends on that of the blacks," a socialist had already realized.[39] "Everything for, everything by the natives," was the way one European Liberal politico summed it up.[40] This unanimity may seem surprising, in light of the far-reaching socioeconomic compartmentalization in postwar Belgium. But many people had realized that the Congolese population had suffered greatly.

The Belgians set about assertively drawing up a new chapter in their colonization, optimistically and with greater pride than before. They would pilot the colony into the modern age, edify the population and, at the same time surpass themselves. Beginning in 1949 an ambitious ten-year-plan was to provide the colony with a modern infrastructure in all areas.

It was the age of highways, nylon stockings, and potted plants. The new world order prompted a certain belief in progress, yea, perhaps even a certain good cheer. Walloons and Flemings left for "the Congo" in great numbers. This was the *relève*, the fresh blood for which men like Drachoussoff had waited so anxiously during the long years of the war. By the end of that war there were only 36,080 white people in Congo; by 1952 their numbers had risen to 69,204, more than ever before.[41] Colonial officials and highly trained industrial workers, all of them men, began bringing their wives to Congo in increasing numbers. To the great relief of the church the era of the *menagère* was drawing to a close, although this left behind a few thousand children of mixed parentage, who often had no place in either world. The mother was almost always Congolese, the European father usually a Belgian in government service, but Greek and Portuguese men sired children by native women as well. Those Greeks and Portuguese were usually self-employed shopkeepers or restaurateurs. If the father acknowledged his natural son or daughter, the child would receive a European upbringing and passport. If not (and that was in nine cases out of ten), the child would remain with its mother in the neighborhood or village, where it was usually regarded as an outsider: too white to be black, too black to be white.[42] After the war, however, the number of Eurafrican births fell sharply. The newcomers from Belgium brought their families with them or had children in the colonies: blonde, fair-skinned, and freckled children in short pants, who chased lizards on the lawn before their villa and were more familiar with mangoes than with apples.

But for the Congolese population the changes were quite few. Essential reforms aimed at more rights (with regard to political participation and socioeconomic position) were very slow in coming.[43] In daily life there were no indications of any new pact between blacks and whites. The colonial trinity still championed the gradual education of the broad masses. Technically speaking, an elite could very well have been cultivated within a short period, but the authorities feared that such an elite would become alienated from the rank and file. All the people, the colonizer felt, should first ascend to an initial level of "civilization" before the next stage began. Teaching the masses to read and write seemed more prudent than cultivating a thin top layer that would then receive political rights.[44] Besides, had the bulk of the Congolese themselves ever asked to take part in government? Well, there you had it!

The fact that they did not ask for political power, however, did not mean that they were happy as ever. The native's political apathy was more an indicator of a lack of education than of any surplus of satisfaction.

In addition, daily life bore no signs whatsoever of rapprochement between Belgians and Congolese. Instead, the gap was widening. The fresh batch of colonial arrivals snuggled down in new and comfortable villas and lived in greater luxury than ever before. Their residential neighborhoods reminded one more of Knokke or Spa back in Belgium than of Central Africa. At the end of the working day they spent their time with their families; on the weekend their friends came by to barbecue or play bridge. Beer was kept in a refrigerator. (Electric fridges, no less: the age of pioneers was truly over!) An increasing number of them had cars, which they washed on Sunday mornings with the garden hose. The Europeans' Congo began resembling the middle-class, suburban California of the 1950s. Convivial enough, without a doubt, but an expatriate community that talked more *about* Africans than *with* them. Interest in the local culture waned and the working knowledge of one or more native languages disappeared. Vladimir Drachoussoff viewed this with regret:

> Officials who, outside their professional duties, show any interest in the native are few and far between. Family life, more comfortable furnishings, the possibility of (and consequently the desire for) a life almost as one would live it in Europe have edged out the old *broussard,* with all his weaknesses and faults, who went from post to post, talked to the village elders and finally understood them and let himself be understood.[45]

The Belgian-Congolese community became a fantasy, gradually overtaken and outstripped by an increasingly closed Belgian colonial community. The pith helmet was discarded, the tall tales by a glass of whisky beneath the Coleman lantern disappeared. The Congo became petit-bourgeois. Many of the women never went to the *cité,* the only blacks they knew were the boy and the chauffeur. White children often grew up in an atmosphere of latent racism. By 1951 things had reached such a point that the Permanent Commission for the Protection of Natives drew up a desideratum, calling for "schooling and games that will teach white children respect for individual humans, as that concerns the native family and black children."[46] That a venerable institution like the commission had to turn its attention to matters such as games of tag and hide-and-seek said a great deal.

Rare were the Europeans who succeeded in summoning up deep empathy for the Congolese perspective. And no one took that empathy as far as the Flemish Franciscan Placide Tempels. He was active in Katanga, in ways that included an attempt to fathom the profound disgruntlement of the mineworkers there. As early as 1944 he turned his attention to the uprisings in the colony, and wrote a courageous, but much-maligned essay entitled "La philosphie de la rébellion":

> This is the apogee of native disillusionment. He [the native] has allied himself with us in order to become one of us; but instead of being regarded as a son of the family, he is seen as nothing but a wage slave. Now he knows himself to be rejected for good, turned away as a son, classified as non-incorporable.[47]

No one had looked at it that way before. His standard work, *Bantu Philosophy*, appeared in 1945. The English and French translations made him world famous; Jean-Paul Sartre read his book with interest. His attempt to understand African cultures from the inside out introduced the concept of "force" as central principle. His insights called for a totally different brand of colonialism: "We thought we had to educate big children, which would have been relatively easy. But suddenly it becomes clear to us that we are dealing with fully developed human beings, with self-aware sages pervaded with an all-inclusive philosophy of their own."[48] His razor-sharp analysis caused Tempels to run into trouble with the ecclesiastical authorities. He was recalled to Flanders from 1946 to 1949. It was a sort of *relegation* (forced exile), this time not of a Kimbanguist to a village in the jungle, but of a visionary Catholic to a monastery in Sint-Truiden.

Things did indeed remain calm in Congo between 1946 and 1956, but it was a ghastly calm, a relative repose that spoke more of old fears than of new hope. Above the gardens of the colonial villas where on Sunday afternoons the sound of tinkling glassware arose, dark clouds were gathering. But no one saw it, not even the freckled son on the lawn, holding a lizard prisoner beneath a glass jar. It was the quiet before the storm.

WHERE WOULD THE TEMPEST OF RESENTMENT FIRST BURST LOOSE? The countryside had reason enough for protest. The rural population still lived under miserable conditions. The fields lay neglected. The prodigious war effort had posed an obstacle to subsistence farming. Malnutrition was widespread. Hunting had ground to a halt. Colonial officials had to encourage the people to resume the gathering of caterpillars, termites, and grubs, a traditional source of proteins.[49] At those spots where cattle were raised, after all, the beef was systematically reserved for the mines. The ten-year plan included an extensive program for getting native agriculture back on track. The goal was to help local communities with the use of modern agricultural techniques and means of production (the so-called *paysannats indigènes*), but those efforts met with little success. The countryside was and remained dirt poor. Rural impoverishment in Congo appeared not after independence, but during the colonial period itself. Birthrates were extremely low. Although rampant population growth is a problem in Africa today, dwindling natality was a cause for permanent concern in the Belgian Congo during the first half of the twentieth century.

So much misery might have led to social protest, but it did not. Or rather: that protest assumed a different form. People did not rise up, they ran away. The postwar years in Congo were characterized by the massive abandonment of the countryside. On an unparalleled scale, people began moving to the urban agglomerations. Kinshasa, with its 50,000 inhabitants in 1940, mushroomed into a city of 300,000 by 1955.[50] Young people had already begun migrating voluntarily to the cities in the period between the wars, but now they left en masse. After the war, 70 percent of the countryside had a population density of fewer than four inhabitants per square kilometer (about 6.5 inhabitants per square mile).[51]

Who was supposed to take the initiative for protest? Those with dreams went and pursued them elsewhere. Those who remained behind were often exhausted and apathetic. The rural population was aging rapidly; by 1947 an estimated 40 percent was over fifty.[52] An enormous percentage, in light of the relatively low life expectancy. Those elderly people were uneducated and bowed passively to colonial authority. There were no agricultural cooperatives or unions, and no social structures that could watch over the interests of people in the countryside. The only form of social organization they knew was tribal, but that was in a state of advanced decay almost everywhere. The chief no longer had any moral authority, but was now an arriviste who did the colonizer's bidding.

So what about the cities? Was sedition running hot there? Did the confluence of dreams result in a fist clenched in defiance? Not right away. For many, the move to the city truly did provide new opportunities. Not that the cities flowed with milk and honey, but in any case they were better than where they came from. And some of those new urbanites had the devil's own good fortune.

LONGIN NGWADI WAS EIGHTY when I found him in Kikwit. I had been searching for him for months, hoping he was still alive. When I finally met him he was washing himself in the brown water of the Kwilu River. His torso was skinny and sunken, his washcloth a completely ragged piece of green cloth. It was not simply threadbare, it was nothing *but* thread. Was this the man I'd been looking for? His face seemed longer than what I remembered from the historic photo. Only when he walked could you see that he had once been a fanatical soccer player. He had the soccer player's typical bowed legs and waddling walk.

He lived in a clay house. A huge eucalyptus tree grew beside the path to his door. Chickens pecked at the red earth, a goat kid wandered about, bleating at nothing in particular. Laundry was hanging in the sun to dry. As the wind picked up, the colorful fabric began to billow. Trouser legs snapped. Sleeves flapped. It looked like a crowd cheering along the sidelines, or along a boulevard where royalty or a celebrity was passing by. I looked at the sky. It just might rain. Longin invited me into his house, had me sit down in a plastic chair. It was very dark inside. I moved closer to the door, so that I would have enough light to write by. A few of his grandchildren stood in the doorway, staring at me with big eyes. When he chased them away, they scattered in every direction, reeling with laughter. The first drops began to fall.

"Rain! For the first time in two weeks!" He beamed. "This is a blessing. The good Lord blesses this conversation."

He was born in Luzuna, a village along the Kwilu, in 1928, and was baptized by the Jesuits at the Catholic mission post at Djuma. His father was a carpenter. "Just like Joseph!" He built chairs, doors, and school desks for the Belgian missionaries. His mother tilled the soil and raised manioc. They still ate well in those days. Rice, manioc, and fish, but also crayfish, grubs, mushrooms, and zucchini. What a difference with today! "Now we only eat once a day. It's always rice and beans. Or manioc and beans. We have meat only rarely. And we never eat fish anymore."

The sky clouded over. In the distance we could hear the rumble of thunder. It became so dark that I could barely read my own notes.

Longin went on talking imperturbably. His parents were already Catholics by the time he was born, he said. He was the second of three children. It was in Djuma that he first saw a car, a pickup that belonged to the nuns. "The white man is intelligent, I told myself. I congratulated the priest." That was also where he went to school. The missionaries ran the primary schools throughout the colony, greatly assisted by local teachers. Secondary education was limited either to vocational training or—for an infinitely smaller group—to seminaries. The classic form of secondary school, aimed at providing a broader education, did not exist yet. The first such schools were set up only in 1938. But for a long time, in other parts of Congo, one simply became either a cabinetmaker or a seminarian. Longin followed a technical curriculum. "I was supposed to become a mechanic, to work on the Lever concessions, but I didn't feel like being dirty all the time." At the age of sixteen he left for Kikwit. He badly wanted to become a priest. "But the padre said: You're already too old for that. So I quit school and went back to my village."

It's hard for us to imagine just how frustrating that rejection must have been. Going to seminary was not only the sole possibility for continuing one's studies, but priesthood was also the highest social position a Congolese could occupy. Then you were *monsieur l'Abbé*.

Longin showed me an old color photograph of himself. In it he was wearing a purple bishop's robe and sitting on a throne, looking earnestly at the camera. "That cassock is worn out, but I used to wear it every Sunday around town. Whenever I had a vision, I told people about it. Back then everyone in Kikwit called me *Monseigneur*." He has always had something with religion. Christianity, of course, *his* Christianity.

Just as Simon Kimbangu had begun to preach once the Protestants no longer wanted him as catechist, so Longin Ngwadi adopted the cassock after the Catholics refused to consider him for the priesthood.

The first drops began to fall, fat, heavy raindrops that made dents in the earth the size of marbles. Then the storm broke loose. The rain gushed over Kikwit and whipped at the thin roofs of huts and houses. Thunder and lightning crashed down in tandem. The sky burst open. In every tropical storm there comes a moment when the thunder no longer growls but shrieks. That moment had now arrived.

Longin threw his hands in the air and prayed to the Almighty, as a thin tendril of saliva rolled down his chin: "*Seigneur,* you are the one who has sent us Papa David. We ask of you: please, could you make a little less noise, so that we can continue our conversation? *Merci et amen!*"

Then, as though nothing had happened, he went on: "In 1945 I went to Kinshasa. I was seventeen. My father gave me money for the boat, my mother gave me food to take along. From Luzuna I walked to Djuma. There I took the packet boat. The boat trip took four or five days. First over the Kwilu, then the Kasai, and finally over the *fleuve* [Congo River] itself."

Longin was one of many tens of thousands of young men who left for the capital. Most of them moved in with family or friends who were already living there, but he had no contacts. "I didn't know anyone when I got to Kinshasa, no one at all. But a night watchman called to me to come onto the patch of ground he was guarding. It was someone from my own region. I was allowed to sleep on the ground, out in the open."

It didn't seem like a particularly propitious start to his life in the city.

"Soon after that I got my first job, with Papa Dimitrios. He was a Greek Jew and he owned a department store. He had me do some arithmetic to test me, then said I could stay. My job was to sell trousers and shirts, women's textiles, soap, sugar, all kinds of things. He found a room for me, close to the Jardin Botanique. After three months I already had a mattress, sheets, blankets, two chairs, pots and pans, and cutlery. Dimitrios gave me a lot of presents. I worked for him for three years. After that I started working at the Économat du Peuple, a big shop with seven floorwalkers. I only stayed there for a year. They threw me out because I sold some sausage that was already spoiled."

It was nothing compared to the office of the priesthood, but he was pleased with his new life in Léopoldville. His dubious success as sausage salesman was more than compensated for by a very different talent. "I played for Daring for four years. Under Tata Raphaël." Daring was one of the city's most successful soccer clubs, set up by Father Raphaël de la Kéthulle—a name familiar to us by now—in 1936. The club still exists today under the name Daring Club Motema Pembe, and is the premier soccer club in Congo. "I played on the same team for a long time with Bonga Bonga, the first Congolese to play in the Belgian soccer competition. He played for Charleroi, for Standard. He was our Pele! Our matches in Kinshasa were always held at the Kintambo velodrome. I was number 9, I was a striker. Tata Raphaël would stand on the sidelines and watch me play, smoking his pipe and shaking his head. He couldn't believe his eyes. I was like a snake!"

To underscore his words, he hopped up and began—on his octogenarian legs—to dribble around his darkened living room. Beneath the low ceiling he performed a whole series of fakes. He still had it. Left, right, a backheel kick, a spin. He illustrated it all in slow motion, while outside the thunder roared on incessantly. Meanwhile, I could see the rainwater running down his living-room walls. It was not trickling, it was running. Longin paid no attention to it. "My nickname was Élastique, the rubber band. That's what everyone called me back then. Élastique the forward, number 9 for Daring."

But that was not the end of his remarkable life story. In the early 1950s the city had another surprise in store for him. "Pétillon was appointed governor general." That was in 1952. "He asked five people to come to the Maison des Blancs. That was where all the secrets of the Congo were kept. The

white people gathered there to govern Congo. It was right beside the Memling Hotel. Only calm, intelligent and serious people came there. It was the
cercle des européens. It was my job to wait on them. 'S'il vous plait.' 'Merci.'
'S'il y a quelque chose, vous me le dites.' [Please. Thank you. If there's anything you want, let me know.] The hours were long, but I got fifty Congolese
francs when I was finished. Of the five Congolese who were called in, I was
numero uno. I was the most polite, the most well-mannered. So Pétillon said
that I could become his *boy maison* [house servant]. I went with him to the
governor's mansion."

The carpenter's son who was not allowed to become a priest, the salesman of household textiles and moldy liverwurst, the lightning forward for
Daring, now became manservant to the next-to-last governor general of the
Belgian Congo. "I worked for him for four years. He called me *mon fils*, my
son." Léopoldville was truly a city of opportunity.[53]

LONGIN NGWADI'S STORY WAS EXCEPTIONAL, of course, but many newcomers experienced a new sense of freedom in the city. That certainly
applied to many women. After her husband died, Thérèse from Kasai
moved to Léopoldville. An uncle took her in and helped her to set up a little
business. On the street market in Kinshasa she sold manioc beer, and later
fruit juice she made herself from ripe bananas. After one year she had her
children come to the city, a few years later she remarried, this time to a
worker she had come to know, someone of her own tribe who had ended up
in the city as well.[54] In the *cité*, a "free woman" was no longer a prostitute,
the category once referred to in official documents as "the healthy women
of native extraction who theoretically live alone," but simply a person trying to get by on her own.

Or Sister Apolline. She was Longin's age. I met her at the Franciscan
convent in Kinshasa. She came from a mixed family in the interior—her
father was Congolese, her mother Tanzanian; they had met during World
War I while her father was with the Force Publique, fighting in German
East Africa. When she turned twelve, her parents found a suitable partner
for her to marry. But she had different plans. She wanted to enter the convent, she felt freer there. The life of a nun took her to the big city. "I worked
in Lubumbashi for twenty-nine years. I was the headmistress of a primary
school there. And many years later I became the first black member of the
religious provincial council. I've always lived in the city."[55]

Or Victorine Ndjoli. She was the first Congolese woman to get her driver's license. "I went to home economics school at the Franciscan sisters. Sewing on buttons, needlework. Later, at the *foyer social* [community center], I
learned to make baby clothes and hats. Back then the white people were
looking for pretty girls for their advertisements. I was a photo model for a
brand of bicycle, for sherry, for milk. I liked that, but I wanted something
more. I ran away in order to take driving lessons. My father didn't want me
to at first, but afterward he was proud of me. I had my license within a week.
It was 1955, I was twenty. I took lessons in a Dodge, but I've never had my
own car. The men in the family were against it."[56]

Victorine also took part in the first beauty contests in Léopoldville, organized by the dance-school owner Maître Taureau (Master Steer). It would be
hard to imagine a name more macho than that. I asked him about it as we
sat in front of his house in Yolo, a working-class neighborhood where every
passerby knew him. "No, my real name is François Ngombe. *Ngombe* is Lingala for bull. And *maître* because I am the master of Life without a Master!"
He roared with laughter. "At my dance school I taught the cha-cha, bolero,
rumba, and charanga, but also swing and rock 'n' roll. As a sideline, I organized *Miss Charm* contests in the neighborhoods. The Greek and Portuguese
merchants gave away free textiles. The girls wore them as *pagnes* [skirts],
which worked as a kind of advertizing. There was a contest, and one of the
girls was chosen."[57]

Kinshasa became a city of fashion, elegance, and coquetry. Young
women wore long, colorful *pagnes,* a custom introduced by the nuns at the
missions. By way of Europe, batik textiles arrived in Central Africa. The
girls wore their hair short, but from around the age of ten they let it grow.
A dozen African hairdos arose at this time, some of them taking up to three
hours to create.[58] Women played a key role in the creation of a new urban
culture. They controlled small trade, they determined which clothing,
music and dances were fashionable and they gave form to a new, modern
African lifestyle.[59]

A number of women were able to break through to prestigious positions. In 1949 Pauline Lisanga was hired as announcer for Radio Congo
Belge. The station had begun with broadcasts for the African population.
Lisanga became Africa's first black female radio announcer.[60] Few Congolese owned a radio, but passersby and neighborhood residents would gather
around loudspeakers set up at many spots in the city. There they heard

Lisanga's voice. They listened to news programs, edifying sketches and religious programs, but also to traditional Congolese music and light Western music. There were even slots for contemporary Congolese music.

Léopoldville in those days was teeming with bands that provided entertainment for weddings, funerals, and parties. Their lively rhythms, virtuoso guitar arrangements, falsetto vocals, ingenious song lines, and light-hearted lyrics made for irresistible dance music. This was the rock 'n' roll of Central Africa. In Congo, the major dance venues were in the hands of Greek immigrants. In Kinshasa one had (and still has) the Akropolis; in Kisangani there was the Bar Olympia. A number of Greek entrepreneurs also began opening recording studios. There, the wondrous dance music of a number of Congolese orchestras was preserved for posterity. Radio resulted in the rise of new popular heroes. Kabaesele's African Jazz and Franco's OK Jazz became the most popular bands of the 1950s.

Yet urban life had more to offer than beauty contests, manioc beer and dance recordings. At the shipyards of Léopoldville, in the chemical and metallurgical plants of Katanga, and at trading firms in the urban centers, a new generation of Congolese men like Longin Ngwadi were finding their first jobs. There they made acquaintance with a demanding modern economy. There were no strikes, but here too reigned the deceptive silence before the storm. Only a few years later, when the fever of independence broke out in full force, many people dreamed of never having to work again after power had changed hands. But for the time being things remained calm, ominously calm. After all, how could any rancor have risen to the surface? Trade unions provided no solution; until 1946, in fact, they were forbidden for black workers. White civil servants had set up their first associations as early as 1920, but they admitted no Congolese members. A trade federation exclusively for trained personnel, the STICs, the Syndicats des Travailleurs Indigènes Spécialisés, was established after the war, but effectively excluded 90 percent of all Congolese workers. Later came the APIC, the Association du Personnel Indigène de la Colonie, which was a much more militant organization. But with the stipulation that such organizations be supervised by white advisers, the colonial administration was able to keep almost every trade union organization on a short leash.[61] Always having a civil official or padre looking over one's potentially rebellious shoulder effectively quashed all autonomy. Trade union activities were expected to be constructive and calm. At best, the colonizer saw the associations as a useful *éducation sociale* for the worker.[62] A sort of soccer, in other words, but then indoors: you learned to hold meetings, to draw up an agenda and take minutes, to discuss a budget. . . . The trade union was considered a form of training, not a legitimate forum for opposition and protest. When Belgian trade union organizations—both Catholic and socialist—tried to gain a foothold in the colony, their attempt was doomed to failure. The Congolese worker felt no affinity with them. It felt like something was being imposed on them from above, something white. Of the almost 1.2 million Congolese on payrolls in 1955, only 6,160 belonged to a union, less than one-half of one percent.[63]

The government did, however, stimulate the larger companies to establish works councils in which Congolese could have their say. These were easier to monitor than autonomous trade unions. The provincial councils also took on their first black members, and from 1951 the colonial administrative council, an informal advisory body without real powers, numbered eight Africans—most of whom came from the countryside and did not belong to the new urban middle class. These were tentative attempts to hear the grievances and complaints of colonial subjects, but they also attested to the opinion that there was still a world of time in which to arrive at more substantial measures.[64] Everything was still going swimmingly. Or so they thought.

HOW COULD ANYONE HAVE SUSPECTED that a revolution was brewing? The rural population remained docile, the city dwellers seemed satisfied enough. In fact, there was even a real caste of *évolués* on the rise who wanted to live in the most European fashion possible, who were wild about anything Belgian, and who loudly voiced their praise of the merits of colonialism. Today the term applied to these Westernized Africans seems rather problematic, but it was very much a title they chose for themselves.[65] And these *évolués*, the Belgian colonial was sure, posed no threat whatsoever. Given, there was at times something ludicrous about it, about the whole business of tidy suits and mannered French. But these were the true social climbers, the ones reaping the bulk of the fruits of that noble task of spreading civilization. There could hardly be more loyal subjects.

But it was precisely from within the circles of the *évolués* that the bomb would finally go off. Most of them had been born in the cities between the wars. They had only secondhand knowledge of village life. They attended the mission schools, went to work for European businesses, respected the colonial government, and therefore looked up to their white rulers as the only social role model they had ever known. Many of them went to great lengths to be taken seriously. They studied in the public libraries, read the newspapers, listened to the radio, went to the movies and to the theater, and read books; it was the white man's intelligence, even more than his prosperity, that they envied. The latter was nothing but an expression of the former.

A lively culture of clubs and associations arose. Still under colonial supervision, these organizations were nonetheless of great historical importance: in the alumni associations, academic clubs, and tribal organizations, after all, lay the seed of the political awakening to come.[66] The former pupils of Tata Raphaël's school came together in the Adapes (Association des Anciens Eléves des Pères de Scheut), later an important breeding ground for the first generation of Congolese politicians. In the *cercles des évolueés* (*évoluées'* clubs) they gathered to discuss books and organize debates; these were a sort of informal night school, and they shot up like mushrooms. In 1950 there were three hundred of them all over Congo. The tribal associations in the cities were now more than simply emergency coffers; they became cultural organizations that would soon develop political ambitions as well. In Elisabethville, tensions grew between the Baluba from Katanga and the Baluba from Kasai: the latter group, to the locals' great irritation, had come down to work in the mines in huge numbers. New clubs were set up as a result. In Léopoldville, the Bakongo felt threatened by the growing influx of Bangala, tribespeople from Équateur who were active in the military and in commerce. Lingala was replacing Kikongo, the original language of the area around the capital, and so the Abako, the Alliance des Bakongo, was set up; a purely cultural association that promoted the language of the Kongo people. Its founder, once again, was a young man rejected for the priesthood.

An *évolué* was a man (never a woman, except as partner) who had enjoyed a certain level of education, had a fixed income, displayed great seriousness about his profession, was monogamous, and lived in European fashion. As the children of two of them explained to me once, the *évolué* also owned a Raleigh bicycle, preferably with gears. "That was the black man's Mercedes in those days." In his home he had a Coleman lantern. He had a record player, which he used to listen to Edith Piaf. Wendo Kolosoyi records were all right as well, because that was calm music. "But definitely not any music that might give rise to lewd dancing. My parents went dancing on Sundays, my father always wore a derby." The *évolué* sent his wife to prenatal care at the health center. Their baby was weighed. At home they abided by the nutritional advice given by the white nuns. They rejected traditional medicine and ancestor worship, but the gap between male and female was sizeable. The former was educated and worked for an employer, the latter uneducated and jobless. Only two or three women in all of Stanleyville around that time were able to carry on a conversation in rudimentary French.[67] One of the *évolué* children told me: "I often heard my father tell my mother: 'You, you're a real Negress, you know! The white people don't live like that!'"[68]

The number of *évolués* was never very large (fewer than six thousand in 1946, and a little under twelve thousand by 1954), but their articulateness tipped the scales in their favor. Tragically enough, what they desired was closer contact with the Europeans, at the very moment when the Europeans were withdrawing more and more to their villas, swimming pools, and tennis tournaments. Yes, in the Belgian Congo there were black truck drivers and telegraph operators, but in cafes and restaurants the color bar was more pronounced than ever. If a white journalist in Léopoldville dared to take a black colleague along to a European bar, conversation would stop. Trains and riverboats may have been run by black engineers and captains, but the passenger compartments were strictly divided into black and white. If a black man jumped into a swimming pool, the whites would get out. Corporal punishment with the *chicotte* was still applied to all Africans, even those who could distinguish the Latin dative case from the genitive and read De Gaulle's speeches. The writer Paul Lomami Tshibamba worked for *La Voix du Congolais*, a government-monitored magazine for *évolués*. For the second issue, published in 1945, he wrote a controversial but by all means moderate piece entitled "Quel sera notre place dans le monde de demain?" (What Will Be Our Place in Tomorrow's World?). By his own account, its publication resulted for him in "countless legal sittings, accompanied by endless lashes."[69] The *chicotte* cracked while, elsewhere in the city, synchronous but far more lazy, the tennis balls thunked against the backboards. Meanwhile, white colonials went to the horse races and organized bicycle races. Festive kermis competitions, with amateur cyclists riding cheerfully under banners advertizing Martini and Rossi vermouth.

The painful yearning felt by the *évolué* was never clearer to me than during those few seconds of historical footage in *Heimweh nach den Tropen,* a gripping documentary by Luc Leysen. It is 1951 and the whites are lined up to judge a contest in Léopoldville. Yet these are not poodles or poultry being judged, but families. Before an exclusively white audience, Congolese families are parading past the jury. The father in short pants, his wife beside him, then the children neatly lined up according to age. The youngest child carries a sign with the contestants' number. The audience applauds politely. Then they walk offstage gravely. . . . So much despair in so few seconds.[70]

The *évolués* desired a special legal status that would do justice to their unique place in society. That was understandable. They had, after all, become "social mulattoes," people who dangled between two cultures.[71] The *évolués* of a small town like Luluabourg expressed it most grippingly:

> We ask the Government to kindly recognize that native society has evolved powerfully in the last fifteen years. Beside the native masses who are rated less important or who are uneducated, a new social class has been formed which constitutes a sort of native middle class.
>
> The members of this native intellectual elite do everything possible to advance themselves and to live in a respectable fashion, as respectable Europeans do. These *évolués* have realized that they have responsibilities and duties. But they are convinced that they deserve, if not a special legal status, then in any case special protection from the Government against measures or treatments applied to the ignorant and backward masses. . . . It is painful to be received as a savage, when one is full of good will.[72]

It is also painful to think that anyone who writes so eloquently could still be subject to flogging with a strip of hippopotamus hide. The subservient, almost servile tone bespeaks a great longing. The *évolué* did not wish to tear down the wall between black and white, but to be lifted over it. He did not fight against the color bar. He did not demand rights for "the Congolese people," or for his tribe, but only for the circles to which he, after great effort, had gained access. Was that egotistical? Definitely. Was there something denigrating about it? Yes. But in the final analysis, in their desire for assimilation, they had even adopted the perspective from which most Europeans regarded the natives.

The Belgian colonial authorities hesitated for a long time. After all, they had never set out to cultivate an uprooted elite, had they? Everything in good time, that was the motto. It was not until 1938 that a hesitant start was made with general secondary schools, and not until 1954 (only six years before independence, but no one knew that yet) that the first university, Lovanium, was set up, an auxiliary branch of the Catholic University of Louvain. During its first year, the new university had thirty-three students and seven professors. You could study natural sciences, social and administrative sciences, education, and agronomy. A law school was started only in 1958.[73] No big hurry, in other words. Was it then really necessary to recognize a privileged caste?

In 1948 the Belgian administration found a provisional solution: the *évolué* could apply for a "certificate of civil merit." Anyone without a criminal record and who had never been deported, who had sworn off polygamy and sorcery, and who could read, write, and do arithmetic was eligible. Those who held such a certificate could no longer be administered corporal punishment and would, in the case of a trial, be tried before a European judge. They had access to separate wards in hospitals and were allowed to walk through the white neighborhoods after 6 P.M.[74] This made a great impression on the average Congolese. In Boma, Camille Mananga, a man who was thirteen when the certificate of merit was introduced, told me: "That was reserved for the truly prominent. They were allowed to go shopping and drink along with the whites. That was a very great distinction. I was still much too young. The sky was more within my reach than a certificate like that!"[75] But for people who had been working their way up the ladder for years, it represented fairly minimal privileges that stood in no proportion to their efforts. Structural wage inequality still existed. As a former *évolué*, Victor Masunda, another inhabitant of Boma, could still get wound up about that: "Of course I didn't apply for that card. It really didn't mean any higher wages. A lot of people groveled, but I refused to lower myself. Applying for the certificate of merit was degrading. Was I supposed to become their little brother? No. I could get hold of my red wine and whisky on my own."[76]

It was for this reason that, in 1952, the *carte d'immatriculation* (registration card) was introduced. This new document gave the *évolué* the same rights in public life and in the eyes of the law as the European population. The most important advantage was that the *évolué* could now send his children to European schools, an exceptional social promotion that also guaranteed a decent education. But the skepticism among large parts of the colonial elite was so great that extremely stringent requirements were posed for obtaining such a card. Those requirements were often humiliating as well. During the period of application, an inspector was allowed to pay surprise visits to the family home, to see whether the candidate and his family lived in a truly civilized fashion. The inspector would look to see that each child had a bed of its own, that the family ate with knives and forks, that the plates were uniform in size and type, and that the toilet was clean. Did the family eat together at the table, or did the mother sit in the kitchen with her off-spring while the man dined with his visitor, in the old style? Only very few applicants lived up to the these criteria. The result, therefore, was that years of palaver were invested in drafting a legal status from which almost no one profited. In 1958, within a population of almost fourteen million, only 1,557 "civil merits" were handed out and only 217 "registration cards."[77] That led to frustration. For sooner or later yearning turns to distaste, yes, even to hostility.

TYPE IN A SEARCH for "Jamais Kolonga" on YouTube and within seconds you will hear one of the great classics of Congolese rumba. It could have come from the Buena Vista Social Club, but it was composed by African Jazz, the most popular band in Congo in the 1950s. That legendary orches-tra was led by Joseph Kabasele, nicknamed "le Grand Kalle." The song itself was written by his gifted guitarist, Tino Baroza. It became one of African Jazz's biggest hits. "Oyé, oyé, oyé," the refrain went, "hold me tight. Jamais Kolonga, hold me tight. Let me go, and I might fall." The part about holding tight was open to multiple interpretations.

I climb out of the car in a narrow, dusty alleyway in Lingwala. Could this be it? In colonial times, Lingwala was the neighborhood of the évolués. All the old people I spoke to knew Jamais Kolonga. Of course! But hadn't he passed away? Hadn't the local press run an alarming article? "Le vieux Jamais Kolonga laminé par la maladie!" (Old Jamais Kolonga flattened by illness!) was the headline. They had read that the man "who as bon vivant, with his wisecracks and pranks, served as the embodiment of the vitality of Kinshasa in the 1960s" was now critically ill.

But, after a series of dead ends and a fortune spent on call minutes, I had finally come up with a street and a number. The yard I walked into was surrounded by a crumbling wall and contained a patch of corn, with-ered and dry as dust. From a cinderblock house a man appeared, wearing short pants, walking on crutches.

"Are you Jamais Kolonga?"

"The one and only!"

One had informants who had seen a lot but had little to say, and one had informants who had little to say but talked a lot anyway. Kolonga belonged to neither category. He had seen everything and he was a fantastic story-teller. He didn't think so himself: "I've just had a hip operation. It's not going too well. It hurts a lot, even with all the medicines I have to take." He pulled up his pants leg to show me an impressive scar.

"Is there something I can do? Do you need anything?"

"Wine! If you've got some money, I can send one of my grandchildren out for wine."

"Wine? In your condition? Are you sure?"

I spent three whole afternoons talking to that little, sharp-witted man, sometimes in his living room, at other times in the shadow of his house. He was excellent company, with a remarkable sense of humor, an unsinkable joie de vivre and a spectacular memory. One time I went to visit him in a little hospital where he had been admitted to convalesce for a few days, and where he flirted with the nurses nonstop. His hip was getting better every day. But now, I asked him, what was the story with that white woman?

"That was in 1954. I was eighteen and had just started working for the Otraco."

"The Office des Transports au Congo?

"Exactly. My father worked there too. First they put me on the docks here in Kinshasa, but until I turned twenty-one, the wages were paid to my father's account. That was not exactly ideal. I couldn't even buy my own liquor. That's why I asked to be transferred to the interior." While everyone was moving to the city, he ran away from it. "I had to go to Port Francqui, which is what they call Ilebo now. It's close to Kasai. When you travel from Kinshasa to Lubumbashi, that's where you transfer from the boat to the train. In those days I even put up Simon Kimbangu's children; they were on their way to visit their father in the prison! *Bon,* so I worked there as clerk. And because of my father, I was the only black man allowed in the white people's shops. I drank Portuguese wine and whisky. That's right, even back then."

While he was talking, one of his granddaughters had gone to the local shop and come back with a cheap carton of wine, which she set down in front of us. Don Pedro. I stuck to my cola.

One day Kabasele was passing through with his orchestra. But his train ran off the tracks, and they missed the boat. They were stuck in Port Francqui for fifteen days! I knew that my Flemish boss's daughter was getting married soon, and arranged for Kabasele to play at the wedding. No sooner said than done. The party was gearing up. That evening I wore a navy-blue suit with a red tie. There were only three *évolués* there. I had to arrange special permits for the musicians, otherwise they wouldn't have been allowed into the white neighborhood after dark. I stood at the bar and saw a Portuguese lady. She danced very well. You have to realize that in 1954 a black man wasn't allowed to touch a white woman. We couldn't even talk to them! The only white women we saw were the Catholic nuns. The boys were the only ones who came in contact with married European women. But okay, I'd taken a good look at her while she danced and I asked her husband if I could cut in. Just like that! It was an impulse, an obsession. But her husband nodded. So I walked up to her and I asked her to dance. Then I danced with her, for a whole song. Afterward the whites all clapped, even the provincial governor! Later, Kabasele wrote a song about it: "Jamais Kolonga."

He poured himself a little more wine. Once an *évolué*, always an *évolué*. "So tell me about your father."

"He was born on January 1, 1900, in Bas-Congo."

"Oh really? Is that an arbitrary date, something the missionaries came up with?"

"No, that was really his date of birth. That same day someone was mauled by a lion, a black man. When my father was baptized, the white people still remembered that. There were a lot of lions and buffalo, even elephants, back then."

Now there are no big animals anymore. In terms of wildlife, Bas-Congo is empty. But what a rapid evolution! Only half a century before Jamais Kolonga danced at a European wedding, there were still lions mauling people in Bas-Congo. And missionaries, out after living prey of their own.

When [my father] was twelve or thirteen, Reverend Father Cuvelier came to the village. He said to my father: "I want you to shine my shoes. Where is your father?" And to my grandfather he said: "Can you give me your son?" "All right," my grandfather said, "I'll let him go with you, as long as he comes back to see me sometimes." My grandfather himself was a Catholic, you see? When he got married in the church, he sent away two of his three wives. He kept the children himself, of course. Anyway, my father went along to the mission post and was baptized on December 13, 1913. After that they registered him with the Redemptorist school in Matadi, and six years later he went to the new secondary school in Boma. So he was, ipso facto, one of the first students to graduate there.

It was the first time in all my journeys that I had heard a Congolese use the term *ipso facto*.

Around 1927 or 1928 [my father] was picked out by an official from Otraco. They needed intelligent people. Until he reached retirement age in 1958, my father worked for Otraco, always as an office clerk. When the company moved its headquarters from Thysville to Léo-poldville, he moved here. My father became an *evolué*. He managed *la cité Otraco*, the housing district for the native personnel. He was in charge of masons, carpenters, the men who worked with reinforced concrete. He visited the homes of the Otraco workers and every Saturday he gave a prize for the neatest, prettiest house. My father drank wine, he was one of the first Congolese who was allowed to do that. On holidays he gave speeches for the governor general, for Ryckmans, Pétillon, or Cornelis, he knew them all. In 1928 he even gave a speech for King Albert, when he came here! So of course they gave him a certificate of civil merit and later a registration card. Back then there were only forty-seven *immatriculés* in all of Congo!

That had made a great impression. Even old Nkasi remembered him. "Joseph Lema, he was completely *mundele*." Kolonga's father was appointed to the Otraco works council and later to the provincial council. He belonged to the first group of Congolese with even a slight say in administrative matters. Kolonga rummaged around in a grubby brown envelope and pulled out a black-and-white photo that had been eaten away by moisture and termites. The picture was crumbling in his hands.

"Look, this is my father. And this is my godfather, Papa Antoine." A man in uniform, heavily decorated. "He was a World War I veteran, and a good friend of my father's." On the back of the photograph I saw his father's handwriting. Extremely graceful and regular, brimming with self-confidence.

I was born in 1935, in Kinshasa. I spoke French with my father, Kikongo with my mother, and everywhere else it was Lingala. My parents came from the same village. My mother was married to an *évolué*, but she went back to the village each year for six weeks. It must have been there that she was bitten. She died of sleeping sickness in 1948. By then I was attending school at Saint-Pierre's, the primary school run by the Reverend Father Raphaël de la Kéthulle. During the recess I was allowed to arrange the books in the school library. And when there was a big soccer match, I was allowed to fetch the ball from his office and lay it on the center spot. The band played martial music and I marched out to the center of the pitch, even though I was the littlest pupil. De la Kéthulle taught me to be brave.

He tried to demonstrate how he had done that, but his sore hip kept him from it.

"I wanted to become a priest. I studied Latin and Greek for two years at the preparatory seminary at Kibula, outside Kinshasa. That was with the Redemptorists. But then they kicked me out."

"Why?"

"Because I didn't like manioc bread. I really couldn't eat it. They thought I was putting on a show. Jacques Ceulemans was the name of the man who expelled me. I still remember his name. He showed no mercy. I really couldn't stand that stuff. It was the greatest disappointment of my early years, but after independence—by that time I'd become a spokesman—I turned around and threw *him* out. That was during the soldiers' mutiny."

Desire, frustration, revenge: a familiar psychological process. For Kolonga, too, the priesthood had been an ardent pipe dream, a dream from which he was rudely awakened.

"I finally finished school in Kinshasa, at Saint-Anne's, De la Kéthulle's secondary school. We were all there at the same time. Thomas Kanza, Cardoso, Boboliko, Adoula, Ileo. Bolikango too, but he was a bit older." Each and every one of the men he mentioned had occupied key positions after independence. Jean Bolikango went to Brussels to negotiate the terms of independence. Cyrille Adoula, Joseph Ileo, and André Boboliko all served as prime minister at some point, Kanza was the first ambassador to the United Nations, Mario Cardoso was minister of education. "Our school was run by the Scheut fathers. The other school in Kinshasa was a Jesuit *collège*. [Justin] Bomboko, [Cléophas] Kamitatu, Albert Ndele, they all went there, among others." More resounding names from the history of Congo. The first two were later to be ministers of foreign affairs, the latter became director of the Congolese Central Bank.

What a setting, what a portrait of an era . . . This was the jeunesse dorée of Congo. Their schools had served to prepare a young urban elite fairly bursting with ambition. No generation before or after them had ever received such a sterling education. A certain inferiority complex with regard to the whites still remained, but with them the fear felt by an earlier generation reversed itself in moments of daring, certainly for someone like Kolonga. He still purred with pleasure when he thought about Monsieur Maurice.

> I went to work for Otraco in '52. Monsieur Moritz was one of the bosses. There was an elevator for whites and stairs for the blacks, even for the white-collar workers. I always took the elevator anyway, because I had to go the fourth floor. One day I found myself in the elevator with the notorious Mr. Moritz. And I had wine on my breath, too. Because my father was an *évolué* . . . *Bon*. Moritz hit me and we got into a fight. It all ended up at the Otraco security office. I was really the company troublemaker, let me tell you.[78]

POSTWAR CONGO was making a complete turnabout, and the *évolués* were the clearest proof of that. The atmosphere was one of anticipation. The high point was, without a doubt, the famous tour King Baudouin made in May and June of 1955. For the first time, a Belgian ruler visited not only the colony's strongholds of power and its hunting preserves, but also took time to wave to the people. It was a roaring success, a euphoric experience without parallel. Young people climbed in trees to wave back to the king, women wore *pagnes* bearing the Baudouin's likeness, children loudly sang the Belgian national anthem.[79] The king and his retinue crisscrossed the country like a traveling circus and were welcomed everywhere with song and dance. In Stanleyville he was carried on a litter by Bakumu tribesmen. He was followed by the women of Elisabethville, who called out: "Our king is so young and so handsome! May God preserve him!"(*Our* king, they called him; it was the first time that had ever happened.) In Kinshasa someone came up with the idea of having him driven around by Victorine Ndjoli, the photo model with the driver's license, but the plan fell through. *Mwana kitoko* was what they called him, pretty boy, for he was quite young and still single. Everyone tried to catch a glimpse of him. To look him in the eye or touch him was believed to bring good luck. Children in the provinces who had never worn shoes now received their first pair, specially for that one day. "It made it hard to walk," one of them said, "but we certainly laughed a lot."[80] Today, in the homes of elderly *évolués*, one still sees their wedding picture hung beside a state portrait of Baudouin.

One of the places the king visited along the way was Lingwala, the district where the *évolués* lived. "He wanted to see them with his own eyes, the houses that had been built with state funding," Kolonga said. "And so he came to look at my father's house, which was here on this plot of land." He pointed out the window with his crutch at the spot where the corn now stood withering on the stalk. "The house is gone now, but back then Madame Detiège, Otraco's social assistant, came by to check the easy chairs and decorate the house. The walls got a new coat of paint and they put flowers on the tables. King Baudouin came here with the governor general. They talked to my father for at least ten or fifteen minutes."

It was hard to believe that only a few years later that same father would be visited daily in that same house by a man who would stoke the desire for independence like no other. That man was Joseph Kasavubu. A few years after that he would become the first president of an independent Congo.

A GREAT DEAL HAD CHANGED. After World War I there had been those who longed for a return to the time before the whites arrived. But after World War II, more and more people began longing to live like the whites themselves. There was as yet no fever of independence, but the world war had served as a catalyst of major proportions. The war had displayed the mother country's vulnerability and had resulted in a new world order in which colonialism was anything but self-evident. The latent tension this generated was never expressed more clearly than by Antoine-Roger Bolamba, journalist, poet, and *évolué*, in 1955. He was the greatest Congolese poet writing in French during the colonial period.

Before the meat of the struggle
I will wait
wait for the red hour of the kickoff
Above my head already whistles the arrow that carries further,
    much further,
the dizzying fire of victory[81]

CHAPTER 6

SOON TO BE OURS

A Belated Decolonization, a Sudden Independence

1955–1960

AND THEN, SUDDENLY, IT WENT LIKE LIGHTNING. IN 1955 not a single native organization dreamed of an independent Congo. Five years later that political autonomy was a fact. The speed at which it took place stunned almost everyone, not least of all the Congolese themselves. The Belgian colonialism to which they were subjected, after all, was permeated with the idea of gradual change. Step by step, Congo was to be withdrawn from its archaic origins in order to enter the modern age. As far as the Belgians were concerned, the finish was not nearly in sight. Yes, the country had been on the right path ever since World War II, but the "civilizing work" was not even halfway done. "Independence?" Sacred Heart missionary and future archbishop Petrus Wijnants sneered to his congregation in 1959. "Perhaps within seventy-five years, but certainly not within fifty!"[1]

But things would turn out differently. Gradual change made way for a stampede, doddering progress for chaos. Who were the ones responsible? No one in particular. Or rather everyone. This high-speed decolonization was not the work of any one specific person or movement, but of an extremely complex interaction between the various players. It was like a game of Ping-Pong that begins calmly, with a slow tapping of the ball back and forth, and then suddenly accelerates into a nervous rally full of focused volleys, lazy lob shots, grim smashes. and sly feints. Faster and faster the ball flies, so quickly that it becomes unclear for players and onlookers alike exactly what is happening where and when. No one can follow it anymore, no one has an overview, but everyone knows: it can't go on like this much longer. And that is how things went in Congo too. The only difference being that there were more than two players and actually more than one ball. During the process of decolonization, it was not simply a matter of the Congolese against the Belgians; the blocs were not that monolithic. On the Congolese side there were *évolués,* clerics, soldiers, workers, farmers. The ambitions of people from Bas-Congo were not identical to those from Kivu or Kasai. The dreams of those in their thirties were not those of people in their sixties. But sooner or later they all came to the Ping-Pong table. On the Belgian side one had not only the Belgians in the colony, but also the Belgians in the mother country. There were European liberals, Catholics, and socialists. The church and the royals had interests different from those of the industrialists or the trade unionists. In Congo itself, colonial officials desired different things than did plantation owners in the interior or missionaries in the jungle. All these special interest groups stood shoulder to shoulder, back to back, and face-to-face. And then there were the supporters: Russia, America, and the United Nations stood shouting loudly from the grandstands, surrounded by young states such as Ghana, India, and Egypt. The players didn't know whom to listen to first, but the Congolese players—as underdogs—clearly received the most encouragement.

And then there all those balls on the table: at least three, in fact. Did people really want independence? And if so, when did they want it? And what was this independent Congo supposed to look like? The latter question had to do not only with the country's internal organization (unitary or federal?) but also its external relations with Belgium (complete autonomy or still some form of federative ties?). The answers to these three questions led to widely divergent standpoints. On one side of the Ping-Pong table, for example, there might be the call for unconditional and immediate independence, whereby all ties with Belgium were to be severed and Congo would remain unified, while on the other side one had the advocates of gradual decolonization, with enduring ties with the mother country and great autonomy for the various provinces. And between them lay a whole gamut of standpoints.

It was as though an entire world championship of Ping-Pong was being played at the same time on one and the same table. The result was squabbling, irritation, tension, belligerence, euphoria, despair, and madness. And fast play, of course. The rules changed all the time. The only way to keep a cool head was to stay focused, to reduce awareness to one's own field of vision, to stick to one's own tactics, to pay attention only to one's own game.

All those involved did precisely that. But another expression for focus is tunnel vision, and it was precisely that which led to folly, on the part of each and every player. The tragic decolonization of Congo was a story with many blind spots and only a little lucidity from time to time. But then, hindsight is golden.

THE YEAR IS 1955, and we are still at the home of Jamais Kolonga. After King Baudouin's visit, Jamais Kolonga's father began receiving frequent visits from an impeccably dressed *évolué*. "Kasavubu came here every day, here, to this yard." He pointed to the old, crumbling cement floor. "He would come in the morning and in the evening to talk with my father. I served him wine. Kasavubu was a true gentleman." In photos from that day, the man's sophisticated allure is indeed plain to see. The neat suit, the fashionable glasses, eyes that seem more to smile than to laugh out loud. Whispered rumor had it that one of his ancestors was Chinese and had worked on the railroad between Matadi and Kinshasa in the 1890s. Hence those eyes, people thought.

Joseph Kasavubu was born forty years earlier in Bas-Congo, in a village a hundred kilometers (sixty-two miles) north of Boma, at the edge of the Mayombe forest. He learned to read and do arithmetic at the Scheutist mission school, and because he was good at it he was allowed to go to the minor seminary, with a view to possible priesthood. He studied Latin and French there and at eighteen was admitted to the seminary at Kabwe in Kasai. It was the first time he had been outside of Bas-Congo. After studying philosophy for three years, he came to the conclusion that his calling lay elsewhere. He left the seminary, became a teacher, then a clerk, and finally a civil servant, but that hint of sacerdotal unction never left him. He would never become the sort of inspired orator Patrice Lumumba was. His voice was brittle and high, his tone rather more uniform and boring. It was not easy for him to seize an audience's full attention. He was unmistakably intelligent, but that intelligence was more the result of hard work and plodding reason than of any inborn brilliance. By means of frequent conversations with others of like mind he molded his preferences into clear standpoints. Once those had been established, however, he possessed the skill to express them with great conviction.

Like so many young people, he moved to Léopoldville during the war. At the age of twenty-five he started work as administrative assistant for the colonial administration's finance department. With that position, he became a part of the new, black urban elite. After work he would talk with people like Kolonga's father about the status of the Bakongo language and culture in Léopoldville. That they were the original inhabitants of the area around the capital was something on which they fully agreed and they were upset by the fact that it was not their language but Lingala, the language of the Bangala who lived in the jungle upriver, that was becoming the city's lingua franca. The Bakongo had been the first ones here, hadn't they? And didn't they live here in greater numbers? So why should Lingala be the language used in the schools? Wasn't there something like a "right of occupancy"? That slogan was a golden find: a term taken from nineteenth-century colonial rhetoric, downloaded directly from the Berlin Conference, but Kasavubu applied it to the urban situation of the 1940s and 1950s.

They also mulled over social and racial issues. How could it be that the whites earned so much more than the *évolués*, even more than those who held a registration card? Here too Kasavubu kneaded his indignation into a bold slogan: "à travail égal, salaire égal": the same work, the same wages. Tough words for a man who spoke so timidly.

Kasavubu joined the capital-city chapter of Adapes, the Scheutist alumni association. After the war he became the association's general secretary, a function he would hold until 1956 and to which he owed a great many contacts with the city's young elite. At the time, the alumni club had somewhere between fifteen and eighteen thousand members.[2] In 1955 Kasavubu was also appointed chairman of the Abako, the tribal association that had for several years already been promoting the Bakongo language and culture in Kinshasa. His period of tenure saw a radical turnabout. Kasavubu would transform the Abako into an explicitly political club. With that, the cornerstone was laid for the politicization of the *évolués* and, in fact, for the start of decolonization.

There was also another event that made 1955 a pivotal year, although no *évolué* in the Belgian Congo would have suspected it at the time. It took place, after all, in Belgium and the Netherlands. In December of that year, the magazine of the Flemish Catholic Workers' Federation, *De Gids,* ran an article entitled "A Thirty-Year Plan for the Political Emancipation of Belgian Africa." The author was Jef Van Bilsen, a correspondent for the Belga press agency who had worked for a long time in Congo and taught at the colonial polytechnic. The article argued that the colony should finally set about the business of cultivating an intellectual upper crust. A generation of engineers, officers, physicians, politicians, and officials would have to be brought into readiness so that by the year 1985 Congo could more or less stand on its own two feet.[3]

Unlike what is often claimed, Van Bilsen's plan did not meet with full opposition from the word go. Sympathetic consideration was given to it in both Belgium and Congo, even outside more progressive circles. His notion of slow emancipation, after all, harmonized well with the policy of gradualism that the colonial trinity had been advocating for decades. His thirty-year plan would do for politics what the ten-year plan of 1949 had done for the infrastructure and economy: modernize the country, slowly but surely. He did not break away from the existing paradigm, but thought it through to its final consequences. That he set 1985 as a deadline suddenly made the whole thing quite concrete, but even then he was not thinking in terms of complete independence: after that date Belgium and Congo would still be bound together by the crown and would form a sort of federation of states, a commonwealth à deux, as it were.

The article appeared in a French translation in early 1956, and that set the ball rolling. Copies of the publication began circulating in the native districts of Léopoldville, the neighborhoods from which thousands of people went out each morning, often barefooted, to work in the department stores, soap factories, or breweries of the Europeans, the neighborhoods to which évolués came home each evening after their shift as typist or clerk for a white patron (boss), the neighborhoods where a few people talked late into the night about the state of the world, over a glass of Portuguese wine. Why did the boss always call you Victor or Antoine, and never Monsieur Victor or Monsieur Antoine? Why did every white person say tu to you and never vous, even when you wore cufflinks and a white collar? In those select circles, Van Bilsen's essays became a hit. A white person who thought aloud about the political emancipation of the blacks: was that really possible? A plan that spoke of higher education and new opportunities: was this some kind of dream? It was as though a ray of sunlight had broken through the massive cloud cover of their lives. Did this mean that this state of affairs was not going to last forever?

It was, in fact, nothing but a pamphlet, but it put Kasavubu in a very bad mood when he read it. Conscience Africaine was the title, the July–August 1956 issue. This low-circulation magazine of Catholic origin, which appeared sporadically and had existed for only a few years, was run by Joseph Ileo, a man from Équateur. The six editors included quite a few alumni of Tata Raphaël's school; one of them held a certificate of civil merit, another even had a registration card. The issue in question was taken up largely by a long and anonymous article with the bold title "Manifesto." The writers had clearly read Van Bilsen's plan, Kasavubu saw that right away. "The next thirty years will be decisive for our future," he read. "The Belgians must realize, above all, that their dominion over Congo will not last forever."[4] Entirely in line with Van Bilsen, the text spoke of political emancipation and gradual change; it made a plea for a joint Belgian-Congolese initiative and spoke of a fraternal atmosphere that would put an end to all forms of racial discrimination. After all, hadn't young King Baudouin himself set the good example during his visit? The text continued: "We ask the Europeans to abandon their attitude of disapproval and racial segregation, to avoid the ongoing aggrievement to which we are subjected. We also ask them to abandon their condescending attitude, which is an offense to our self-respect. We do not like to be treated as children all the time. Please understand that we are different from you and that, even as we assimilate the values of your civilization, we also wish to remain ourselves."[5] The évolué no longer wanted to simply hanker after the white lifestyle, as he had been doing for years, but now wished to rely on his own capabilities as well. And then, in block letters: "We want to be civilized Congolese, not 'Europeans with a black skin.' "[6]

Kasavubu felt sick. Not because he disagreed with these statements, far from it. It was having to read somewhere else what he had been thinking for years, that was what galled him. What's more, almost the entire editorial board of *Conscience Africaine* came from Équateur, while he, Kasavubu, had just become chairman of country's largest Bakongo association. Were these Lingala-speakers, these Bangala, now going to take the lead in the capital when it came to political ambition as well? Although it is not widely known, ethnic rivalry in the big cities played as great a role in decolonization as did the aversion to foreign rule, no matter how artificial many of these "tribes" really were. These "Bangala" who so annoyed Kasavubu were, as a homogeneous tribe, an invention of the Bureau International d'Ethnographie (they were, in fact, a crazy quilt of cultures in the equatorial jungle; there had never been any inclusive tribal bond). But, thanks to the mission schools, this ethnographic figment from the 1910s was very real in the Kinshasa of the 1950s.[7] The Bakongo had no desire to yield pride of place to the Bangala.

Within the next few weeks, Kasavubu summoned the members of the Abako to examine the *Conscience Africaine* manifesto and comment on it.

Their "countermanifesto" appeared in August 1956. It was intended to surpass the first text and preferably to pulverize it. The tone was much more radical and the content unequivocally revolutionary. With regard, for example, to the thirty-year plan advanced by Van Bilsen and *Conscience Africaine?* "We, for our part, we do not wish to participate in carrying out this plan, but only in doing away with it; its execution would lead to only more delays for Congo. In essence, it is nothing but that same old lullaby. Our patience is more than exhausted. The time is ripe, and therefore they must grant us that emancipation this very day rather than postpone it for another thirty years. History knows no belated emancipations, for when the hour has come the people will no longer wait."[8]

That part about "the people" was, of course, exaggerated. Kasavubu did not have the Congolese people behind him, and even large sections of "his" Bas-Congo had never heard of him. He spoke, at best, on behalf of the Kikongo-speaking *évolués* of the capital. In colonial circles, however, this text exploded like a bomb. It was the first time that a group of Congolese had so openly called for more rapid emancipation. A federation of states obviously did not appeal to them at all. And the colony's unity did not seem particularly sacred to them either: they seemed only to be standing up for Bas-Congo. Many colonials went into a tizzy. They spoke of "madness," of "a race toward suicide" and a "racism worse than that which they claim to be combating."[9] Jef Van Bilsen became the whipping boy. It was he who had opened Pandora's box, they felt.

For the colonials this call for independence came like a shot out of the blue, which says a great deal about the closed world they inhabited. Following World War II, after all, a first wave of decolonization had already swept Asia. Within the space of only three years, between 1946 and 1949, the Philippines, India, Pakistan, Burma, Ceylon, and Indonesia had become independent. That same spark jumped the gap to North Africa, where Egypt threw off the British yoke, and Morocco, Tunisia, and Algeria began agitating for greater political autonomy. Figures such as Nehru, Sukarno, and Nasser maintained close contact. In 1955 that relationship had culminated in the seminal Bandung Conference on Java, an Afro-Asian summit where new countries and countries longing for independence unanimously relegated colonialism to the scrapheap of history. "Colonialism in all its forms is an evil which should speedily be brought to an end," the closing statement read.[10] No Congolese delegation was present at Bandung, but there was one from neighboring Sudan, which became independent a few months later. In addition, after the conference, radio stations on Egyptian and Indian soil began broadcasting the message of anti-imperialism. On the shortwave frequencies, people in Congo could listen to La Voix de l'Afrique from Egypt and All India Radio, which even featured broadcasts in Swahili.[11] The message was spread by means of a technical innovation: the transistor radio. The introduction of this tiny, affordable piece of equipment had major consequences. From now on, people no longer needed to stand on market squares and street corners to listen to the official bulletins from Radio Congo Belge, but could remain in their living rooms, secretly enjoying banned foreign broadcasts that kept repeating that Africa was for the Africans.

TO DEAL WITH THE GROWING UNREST, Brussels decided at last to introduce a nascent form of participation. For ten years politicians had been squabbling over possible forms of native involvement in the cities, but in 1957 a law to that effect was finally passed. The native boroughs of a few large cities were to have their own mayors and city councils. On the lowest administrative rung, therefore, actual power was being granted to the Congolese for the first time. From experience the administrators had already seen that informal neighborhood councils could be effective in solving local problems, particularly when their members were chosen by the community.[12] From now on those members would be chosen in formal elections, although the borough mayors continued to be answerable to a Belgian "first mayor." The first elections in the history of the Belgian Congo were held in late 1957, but were limited to Léopoldville, Elisabethville, and Jadotville. Only adult males were allowed to vote.

Congo, at that point, was one of the most urbanized, proletarianized, and well-educated colonies in Africa. No less than 22 percent of the population lived in the cities, 40 percent of the active male population worked for an employer, and 60 percent of the children attended primary school.[13] This situation was both new and precarious. Wages had risen spectacularly during the early 1950s, but from 1956 on, that growth had stagnated; there had even been a major reversion. The fall of raw material prices on the international market (due to, among other things, the end of the Korean War) put a brake on the economy. Unemployment began to appear in the cities.[14] Soon there were some twenty thousand jobless people in Kinshasa.[15] Those who had lost their jobs moved in with family members who still had an income.

The houses and yards of the *cité* became overcrowded.[16] Little bars began popping up all over. Alcoholism and prostitution increased proportionately, for when life is hard, morals become easy. It was in this atmosphere of unrest that the first elections were held.

That only adult men were allowed to vote did not mean that the women and young people were politically apathetic. It is precisely in these circles that one saw around that time the rise of alternative displays of social involvement: the *moziki* and the *bills*. The former were women's associations in which successful women gathered to save money and talk about the latest fashions. That might sound banal. At their parties the members of these associations all dressed in the same new, luxurious materials. But these customs also constituted a form of social commentary. The *moziki* had names like La Beauté, La Rose, and La Jeunesse Toilette, in French, for that was the language of social prestige. It was their way to respond to the gap between the sexes. They adopted the idiom of the male *évolués* and affirmed their own social progress. The members were social workers, teachers, or merchants. Victorine Ndjoli, the woman with the driver's license, and a few of her friends set up La Mode: "We were influenced by the European fashions we picked up from the mail-order catalogues. Those French names proved that we had been to school, that we were civilized. Women were only given the right to learn French quite late, so speaking it was a way to place ourselves on the same level as the men."[17] The radio announcer Pauline Lisanga also belonged to a *moziki*.

Many of these clubs aligned themselves with one of the city's popular orchestras. The word *moziki*, by the way, comes from *music*. Victorine Ndjoli's La Mode was an unqualified fan of OK Jazz, the Orchestre Kinois led by François Luambo Makiadi, nicknamed Franco. Makiadi is still considered the greatest guitarist and composer of Congolese rumba and, in a less Anglocentric history of black music, would take his rightful place beside the likes of B. B. King, Chuck Berry, and Little Richard. Franco de Mi Amor was what they called him, *le sorcier de la guitare* (the wizard of the guitar), Franco-le-Diable. Victorine and her lady friends went to his shows (later he even married one of them) where they drank *mazout*, beer mixed with lemonade. It had to be Polar beer, though, because that was made by Bracongo, the brewery where around that same time a young man by the name of Patrice Lumumba became an employee. "I was for Lumumba, we supported his MNC," Victorine said. In a city where the Abako ruled the roost, that

choice was hardly self-evident. "When he died, we all went into mourn-
ing."[18] Women were not allowed to vote, but fashion, music, nightlife, drinking, and dancing took on political portent. They voted with their glasses.
Primus, the beer brewed by the competition, was the one Kasavubu's supporters drank.[19]

And then there were the young people. After half a century of birth deficit, population figures began rising significantly from the 1950s. Between
1950 and 1960, 2.5 million babies were born in Congo. On the day of independence the country had some 14 million inhabitants. Congo was getting younger all the time: in the mid-1950s, 40 percent of the population
was under the age of fifteen.[20] Young people became a category of major
importance, not only demographically, but also in society and politics. The
bills were the colony's first youthful subculture.[21] What the *nozems* were to
Amsterdam, the *zazous* to Paris, and the rockers to London, the *bills* were
to Léopoldville. They took their inspiration from the Westerns screened in
the *cité*. As the name suggests, their hero was Buffalo Bill. They spoke an
argot of their own, *hindubill,* and had their own dress code: scarves, jeans,
and turned-up collars that were a reference to the Far West and mocked the
impeccable *évolués*. This latter group, in turn, voiced great concern about
how young people were going to the dogs. Unwholesome movies were to
blame:

> Restraints must be imposed on the movie theaters. Detective and cow-
> boy movies are extremely popular. All of those scenarios demonstrate
> to the audiences, who are largely young people and often even chil-
> dren, how to go about stealing, killing, and, in a word, doing wrong.
>
> When one sees the posters and billboards, one sometimes feels as
> though one has entered the realm of boorishness and sensuality.
>
> How are we to teach our sons and daughters modesty, goodness,
> charity, self-respect, and respect for others? The great evil is housed in
> the movie theaters.
>
> What else does one see in those alcoves but the most erotic films
> consisting of the most lustful scenes, to which extremely sensual music
> is then attached?
>
> One evening I attended a showing. There were, altogether, ten
> adults in the theater. The rest? . . . Children between the ages of six and
> fifteen. The theater was full of these "lads." A hellish noise . . . The boys

bounced with impatience. The screen lit up . . . A cowboy movie . . .
Applause . . . Shouts of joy . . . A love story . . . Kissing going on everywhere and shouts of "ha" from every corner . . . Then came the fistfights
and the gunfire that elicited indescribable rapture. . . . Two ugly movies
. . . After the show began the reenactment of what we had been watching on the screen for the last two hours. Young girls leaving the theater
were accosted and kissed on the cheeks. . . . The boys followed each
other with sticks and imitated the sound of a pistol, in emulation of
the cowboys. . . . See here the moral lesson provided by that evening's
showing . . .

Abominable!

Let us cherish no illusions. The movie theater will become a school
for gangsters in the Belgian Congo, unless the screening of certain
films is banned in the *cité* or the *centres extra-coutumiers*.[22]

The *bills* were seen as hooligans who took to thievery, debauchery, and
the use of marijuana. Juvenile delinquency in the cities did increase during this period, but almost never involved anything more than the theft
of a basket of papayas or in the worst cases a bicycle, as opposed to serious
crime.[23] Still, this was something new. Parental authority was crumbling,
the prestige of the *évolué* was being mocked, the influence of the traditional
chief had vanished long ago. The *bills* created a world of their own. They
split up into gangs, each with its own territory in the city, and those territories were rechristened with names like Texas or Santa Fe. Explicit political
interest was quite foreign to the *bills*, but they generated a volatile atmosphere of rebellion and resistance.

On Sunday, June 16, 1957, sixty thousand spectators crowded into
Raphaël de la Kéthulle's Stade Roi Baudouin to watch a historic soccer
match: F.C. Léopoldville, forerunner of the first national team, against
Union Saint-Gilloise of Brussels, one of the most successful teams in the
history of Belgian soccer.[24] This was something new. For the first time, a
Congolese team would play against a Belgian team in the colony. It was to be
a fierce match with a rowdy ending. The referee was a Belgian army officer
and his calls caused resentment. When he blew off two Congolese goals for
offside violations, the crowd reacted furiously. The final score was 4–2 in
the Belgians' favor. The supporters shouted that the match had been rigged.
Upon leaving the stadium, *bills*, workers, the unemployed, hoodlums, angry

mamans, and schoolchildren vented their rage on the surroundings. There
was screaming, blows were dealt out. Youth gangs and onlookers rushed in
to join the fracas. Stones hailed down on the cars of white colonials trying
to leave the stadium. They had never experienced anything like it. Soccer
was supposed to keep the masses docile, wasn't it? The police should have
intervened. At the end of the day, the toll was forty persons wounded and
fifty cars damaged.

This mounting tension resounded loudly in the elections held on
December 8, 1957. It was an enormous popular success: 80 to 85 percent
of all eligible voters turned out. The Abako did an outstanding job in Léopoldville and succeeded in winning the votes of many who were not even
Bakongo. It took 139 of the 170 seats on the municipal council. Of the eight
native mayors, six belonged to Abako. In Elisabethville, the migrants from
Kasai, the largest population group in the city, won a large portion of the
votes. The Union Congolaise, a Catholic, pro-Belgian association of *évolués,*
also achieved sound results. Nine white candidates were elected as well.[25]

For Brussels, the successful and orderly elections rang in the start of the
controlled democratization of the colony. Local elections were now to be
held in other places as well, followed by provincial and later national ones.
But it was too late for such gradual change, Kasavubu felt. In his acceptance
speech as mayor of the borough of Dendale in Léopoldville, he did precisely
what Lumumba would do in 1960 at his inauguration as prime minister: he
gave a fiery speech.

> Democracy is not in place as long as people, in order to contain demo-
> cratic action, still appoint officials rather than elected representatives
> of the people. Democracy is not in place when the police includes no
> Congolese constables. The same goes for the army: we have neither
> Congolese officers nor Congolese supervisors in the medical service.
> And what then of the top levels in education and the inspectorates?
> There is no democracy as long as suffrage is not universal. The first
> step, in other words, has not yet been taken. We call for general elec-
> tions and internal autonomy.[26]

Those words earned Kasavubu a government reprimand, but that
hardly fazed him. The office of mayor, in addition to a comfortable salary,
also earned him a great deal of respect from the local population. And so he

went on campaigning. The elections did not serve as oil on troubled waters, but in fact fueled the fires of unrest.

AND THE TIME BOMB WENT ON TICKING. 1955: the Abako turns to politics. 1956: publication of the manifestoes of *Conscience Africaine* and the Abako. 1957: the municipal elections and the start of the malaise. But the year of the great turnaround was 1958. The immediate cause, however, was cheery and took place in an atmosphere of heartfelt fraternity: the world's fair in Brussels. There was nothing to indicate the potentially revolutionary effect of an easy-going tour of the pavilions at Expo 58. But that is how it turned out. The world's fair left Belgium with the protomodern monument of the Atomium, and Congo with an acute hankering for autonomy.

Jamais Kolonga confirmed that. For several years already, small groups of *évolués* had been allowed to take educational trips to Belgium, but now hundreds of Congolese, including a large group of soldiers, were invited for a few months' stay to visit the Expo. It seemed like a sort of *Wiedergutmachung* (reparation) for the three hundred Congolese who had been put on display at Tervuren in 1897. There was a Congolese village this time too, in the shadow of the Atomium, but most of the guests from the colony were there as visitors. "My father was allowed to go to Belgium in 1958," Kolonga told me. "He was very impressed by what he saw. Europeans who washed dishes and swept the streets, he didn't know that existed. There were even white beggars! That was a real eye-opener for him."[27] What a contrast with the image of Belgium he knew only from the missionaries' stories and his superiors' attitude! The white man was not an unapproachable demigod. That did not come as a disappointment; on the contrary, it gave him hope. This meant there was room for social development, in Africa too. What's more, the Congolese saw that they were welcome in the restaurants, cafés, and movie theaters of Brussels—yea, even in its brothels, it was whispered.[28] That too was very different from the daily segregation they experienced in the colony.

The visitors to the Expo not only discovered a different Belgium, but they also discovered each other. People from Léopoldville spoke for the first time with citizens from Elisabethville, Stanleyville, Coquilhatville, and Costermansville. Their country's vastness and the restrictions on travel had meant there was little contact between the various regions. Farmers migrated to the city, but urbanites seldom or never moved to other cities. During those months in Belgium, however, the visitors exchanged anecdotes, talked about the situation at home and dreamed of a different future. During the Expo, a number of *évolués* were also approached by Belgian politicians and trade union leaders, from both the Left and the Right. That too fostered a growing political awareness.

But Longin Ngwadi, nicknamed "The Rubber Band," the star soccer player for Daring who had become boy to Governor General Pétillon, had less luck. When I interviewed him in Kikwit he told me that he had been allowed to go along to Belgium in 1958, but had never seen the Expo. "We went by plane. I went along as Pétillon's houseboy. I stayed in Namur and had to cook and do the laundry. Pétillon went to the world's fair to look at all the *merchandise*. Copper, diamonds, everything from Congo, everything from every country." But while the governor general was dining in Brussels with the duke of Edinburgh and the Dutch minister of foreign affairs, Longin remained behind in the kitchen in Namur. "I ate well there. I used a knife and fork. I had watched the others to see how that worked. *Madame de gouverneur* used to burst out laughing when I ate the wrong way. Things were very good in Belgium. I got lots of presents. I heard about trains that disappeared under the ground and about the big seaport. Namur was an intelligent village, just like Kikwit."[29]

Pétillon found the whole Expo idea ill-advised. Send three hundred Congolese to Brussels and expose them for months to the indoctrination of some of these Belgian types? "In the jumble of the crowds and the exuberance of the Expo, they could do exactly as they pleased. They succeeded in their hideous task of undermining and poisoning, even among the soldiers of the Force Publique. It is horrible to think that this happened under the very eyes of the Belgian government, which did not seem to realize that Congo was more or less descending into a prerevolutionary state."[30] As a man of action, he raised vociferous objections. Which was precisely why, during this same official trip, he was asked to remain in Belgium and become the new minister of colonies. His predecessor, Auguste Buisseret, one of the rare liberals to occupy that post, had followed an all-too-idealistic course—among other things, by introducing secular education in the colony. That had breached the hitherto-closed ranks of white authority, according to all those who stood to profit from a subjugated Congo. A technical supervisor was needed: better a fieldworker than a quibbler. King Baudouin applied his influence, Pétillon accepted the job, but after only four months he threw in the towel. Longin never got to see the Atomium.

One Congolese visitor who *was* able to admire the structures of steel and pre-stressed concrete at the Expo was a twenty-eight-year-old man from Équateur. The son of a cook at the Capuchin fathers' mission, he had attended primary school with the Scheutists in Léopoldville. After one year of secondary education he joined the Force Publique. He became a secretary—bookkeeper—a typist, and in 1954 he was promoted to the rank of sergeant. The typing appealed to him. Working on his military typewriter, he began writing articles under a pen name for colonial publications like *Actualités Africaines*. In 1956 he left the army to become a full-time journalist. Two years later he was chosen to go to Brussels. At the Expo he cut a rather nondescript figure; a lanky, timid man who peppered his conversations with Europeans with the stopgap "n'est-ce pas?" (isn't that so?). He was courteous enough, to be sure, but otherwise only rather awkward. His name: Joseph-Désiré Mobutu.

The final months of 1958 were exceptionally turbulent. The Expo visitors returned to Congo, the war of independence in Algeria was coming to a head, and Morocco and Tunisia had already thrown off the colonial yoke. Closer to home, neighboring Sudan was transformed from a British colony into an autonomous state, and in Brazzaville French president Charles de Gaulle spoke the historic words: "Those who want independence must come and get it!" It was intended as a provocation (for anyone responding to the invitation immediately lost all support from France), but the Belgians across the river choked on their coffee when they heard that on the radio.[31] In the working-class districts, however, a cheer went up.

On October 10, 1958, the Belga wire service in Léopoldville received a press release announcing a new political party. That in itself was nothing special. Other parties had been set up in Congo that same month: the Cerea (Centre de Regroupement Africain) in Kivu and Conakat (Confédération des Associations Tribales du Katanga) in Katanga. Every region suddenly seemed to want a party of its own: the electoral success of the Abako had escaped no one's notice. What was new, however, was the communiqué's radically national approach. That was reflected even in the organization's name: Mouvement National Congolais (MNC). The new party's platform included the resolution to "fight vigorously against all forms of regional separatism," as being "irreconcilable with the greater interests of Congo." The Abako had bemoaned only the fate of Bas-Congo, but the MNC resolutely struck a national chord. Congo had to be liberated from "the grasp of imperialistic colonialism, with an eye to the country's independence, within a reasonable period and by means of peaceful negotiations."[32] For the first time, there was a native political movement that viewed Congo as an entity. The list of names at the bottom of the communiqué included people from various tribes and different parts of the country. There were Bakongo, Bangala, and Baluba, people of the Catholic, liberal, and socialist persuasions, trade union members and journalists. The name of the self-appointed party chairman was Patrice Lumumba.

Lumumba was born in 1925 in Onalua, a village in Kasai. In ethnic terms, he belonged to the Batetela, the tribe that had led the great mutiny during the Arab campaigns in the late nineteenth century. Lumumba's father was a poorly educated Catholic known for his volatile temper and stubborn nature. A man who brewed his own palm wine and drank it himself. Lumumba went to school at Protestant and Catholic mission posts and, after a time spent crisscrossing the interior during the war years, moved to the big city: Stanleyville. There he found work as a minor administrative official, then as a postal clerk. The colonial post office sent him for training to Léopoldville, where he improved his paltry French and acquired an insatiable hunger for knowledge. Back in Stanleyville he became a fervent reader, working at the library as a volunteer and never missing a reading or cultural evening. In 1954 he acquired the much-coveted registration card. His self-confidence grew by leaps and bounds. He became extremely active in the town's club life, and seemed to have no trouble juggling a whole series of board positions. He was chairman of the association of postal workers, led the regional branch of the APIC trade federation, maintained contacts with Belgian liberal parties and became chairman of Stanleyville's Association des Évolués.[33] He had a reputation for being able to get by on two or three hours sleep each night.[34] In addition to his busy schedule of meetings, he wrote political analyses. He began submitting articles to newspapers such as *Le Croix du Congo* and *La Voix du Congolais,* and even set up his own periodical: *L'Écho Postal.* All who met him in those days in Stanleyville were impressed. Lumumba was quick and acute, zealous and energetic. He had the gift of the word and the power of his convictions. With his spectacles, bow tie, and—a rarity among African men—his beard, he made an intelligent and attractive impression on many. The fact that he was fairly bursting with ambition was camouflaged by his charm and glibness, though he had the tendency at times to say what he knew the listener wanted to hear. At some moments, this made him seem a bit like a chameleon.

In 1955, the year that Kasavubu became chairman of the Abako, Lumumba steered Stanleyville's Association des Évolués in a more political direction. This made him the city's most influential Congolese. During a reception in the provincial governor's garden at the time of King Baudouin's visit, he succeeded in talking to the king for ten whole minutes. Amid the bougainvilleas at the river's edge he explained to the young king, his compeer, a few of the problems encountered by the native population. Baudouin listened attentively and asked questions. A true conversation took place. The rumor of that meeting spread like lightning through the streets of Stanleyville. Lumumba's popular status was established for good. Soon afterward he was invited to join a study trip to Belgium by young, promising Congolese, a trip during which he praised the benefactions of Leopold II and Belgian colonialism without a hint of irony.[35] But after his return, after eleven years of faithful service at the post office, he was prosecuted for forgery and embezzlement. Later, he would say: "Did I do anything but take back a little of the money that the Belgians had stolen from Congo?"[36] After serving a twelve-month prison term he left for Léopoldville. He took a job with Bracongo, the brewers of Polar beer, and became the brewery's commercial director, a position that provided him with a salary better than that earned by many whites. He led Polar beer into the fray with its competitor, Primus. Lumumba passed out bottles of beer in the working-class neighborhoods. Here too, his *flux de bouche* (eloquence—or glibness) worked wonders. He brought beer and promised freedom. He quenched the thirst of the masses, but left them longing for more. Emancipation began with a free pint. Polar prospered and Patrice became famous. As time went by he became acquainted with a whole slew of young intellectuals. Unlike these acquaintances, however, he was familiar with large parts of the colony; before arriving in the capital he had lived in three of its six provinces. For Lumumba, therefore, the ethnic frame of reference was of lesser relevance. There were, after all, not many Batela living in Léopoldville. He preferred to "struggle on behalf of the Congolese people," as that notorious press communiqué had said.[37]

In Kisangani, the former Stanleyville, I had the privilege of speaking with several people who had supported Lumumba from the very beginning.

Eighty-year-old Albert Tukeke came from the same area; their mothers were even related. Like Lumumba he had worked for the post office and had received his schooling in Léopoldville. He became a counter clerk in Elisabethville, where he learned a great deal about colonialism, the hard way. "Whenever a European would come into the post office, he would never wait in line. He simply said: 'Clear the counter!' They always had that rude way of putting things. We were young and couldn't talk back. If they needed something they would say: 'Isn't anyone here?' By 'anyone,' they meant a white person. That hurt." Colonialism was not only a huge international system, it also consisted of thousands of little humiliations, of telling turns of phrase and subtle facial expressions. Lumumba denounced that resolutely, Albert Tukeke remembered. "Lumumba was a man like everyone else, who simply demanded rights for black people. But his personality, his insight and vision were very different. He had traveled a hundred kilometers by the time others had covered only one. And I'm not just saying that because I'm a Batela."[38]

Jean Mayani was a warm supporter who spoke just as fervently of Lumumba in 2008 as he had in 1958. I listened to him for one whole morning at his house in Kabando, a borough of Kisangani. As early as 1959 he had served as MNC party secretary for his district; one year later he was Lumumba's chief deputy during the municipal elections. Mayani spoke clearly and analytically:

> Listen, there was no extreme racism back then, but there was a clear division. In the shops, in the neighborhoods, in the schools, and even in the graveyards there was a form of apartheid. We were very impressed by the évolués who had a certificate of civil merit or a registration card. They enjoyed social advantages, they went to the European schools. But still, what a difference with the colonial policies of the French! The blacks in the French colonies could go to France to study. [Léopold Sédar] Senghor [later president of Senegal] was a member of parliament in Paris and became a deputy minister. So I was very interested in the MNC's arguments. In 1958 I was one of the first supporters here in Kisangani. I still remember the first meetings in the cité. We met at bars and sports parks. Lumumba talked about the history and the outrages of colonization. He was truly, unbelievably courageous. He told things the way they were: the suffering, the banishment of Kimbanguists, the

racial hatred, the lack of humanity, the forced labor in the mines, road construction, and the railroad. The crowds became completely enraptured by a leader like that.[39]

Old Raphaël Maindo agreed completely. He thought back to those days nostalgically. "When Lumumba spoke, no one wanted to go away. Even when it was raining, even at night, the people stayed and listened." Unlike Jean Mayani, he had not been a party leader, but a grassroots militant: he sold membership cards. "That was very easy. Everyone wanted one. Even women joined. I held membership card number 4. They cost twenty francs, the same price all over the country. We traveled everywhere, sometimes seven hundred kilometers (430 miles) away. We had cars."[40] For many Congolese, buying one of those cards was more than a political act, it was an impassioned form of self-confirmation and pride.

In December 1958 Lumumba went to Ndijili, Léopoldville's airport. He was on his way to the Ghanese capital of Accra. One year earlier Ghana had become the first country in sub-Saharan Africa to gain independence. President Kwame Nkrumah enjoyed a heroic status that extended from Senegal to Mozambique. He was the embodiment of Pan-Africanism, the dream of a free and peaceful Africa joined in solidarity, which was why he had called together leaders and thinkers from all over the continent. Kasavubu went to the airport too, but the customs service—probably with forethought—balked at accepting his vaccination card: the colonial government had not forgotten his incendiary speech at the mayoral inauguration. Lumumba and two of his confidants were the only Congolese representatives in Ghana. The Accra Conference made a deep impression on him, more than any book he had read. He spoke there with intellectuals and activists and saw that they were interested in what he had to say. He encountered Julius Nyerere and Kenneth Kaunda, the future presidents of Tanzania and Zambia respectively, and [Ahmed] Sékou Touré, the first president of Guinea-Conakry. The social-climbing évolué of yore became a self-aware African, proud of his roots, his country, and the color of his skin. The Belgian Congo seemed to Lumumba more and more like an anachronism that kept the people under its thumb for no valid reason. He would liberate his country from fear and shame.

IT IS JANUARY 4, 1959, and it is bitterly cold in Brussels. A quiet Sunday morning, frozen solid. The streets are covered in a treacherous layer of sleet. There is almost no traffic. Down the chic tree-lined streets of Elsene, close to the abbey of Ter Kameren, a car rolls slowly past the stately homes. Behind the wheel is Jef Van Bilsen, the man whose thirty-year plan has—according to many—unleashed the hounds of hell. But he is also the Belgian with the best contacts among the Congolese elite. Almost no one is better informed about what is going on among the *évolués* than he. Early that same morning, he had received a call from Arthur Gilson, the Belgian minister of defense, with the urgent request to come to his home. Gilson has spent the entire weekend after New Year's fretting over the text of a government white paper. During the final months of 1958, a government task force had traveled throughout Congo, taking inventory of the population's wishes. A laudable initiative, except for the fact that there was not a single Congolese in the group. Their report was nonetheless intended to result in a forceful statement that would outline the foundations for a new colonial policy. A number of cabinet ministers had reviewed the text during the Christmas holidays, but were stymied. As was the defense minister. Perhaps Van Bilsen could take a look at it? On that peaceful Sunday morning in the minister's study, Van Bilsen tries to explain that such a crucial statement is useless unless it speaks of independence and proposes a concrete date on which that is to take place. The minister is flabbergasted. "What followed was a discussion, or something that more closely resembled a mummery, concerning what was deemed desirable from a Congolese point of view and what was feasible from a Belgian point of view," Van Bilsen says later.[41] The deadlock remains unresolved. Without having achieved what he came for, he shuffles back to his car.

It is January 4, 1959, and it is blazing hot in Léopoldville. The rainy season is not over, not by a long shot; the air is sticky and close. At the governor general's residence, everything is being prepared for the annual New Year's reception in the garden.[42] Glasses are being polished, tasks handed out. The new governor general's name is Rik Cornelis, he does not yet know that he will be the last. Some Belgians are still in bed, sleeping in after a late night of dancing at the Palace or De Galiema. Others are enjoying a breakfast of bread and strawberry jam. The more adventuresome among them have already been for a swim, or played tennis at the *cercle sportif* (country club). It is going to be a stylish reception. A few Congolese have also been invited, as is fitting within the philosophy of the Belgian-Congolese Community. A few of the native mayors will be there. In his New Year's address, the governor general will undoubtedly talk about the major challenges of the coming year. The champagne will flow, the crystal will gleam. Speakers will "express hope" and "underscore faith," and many will speak of "mutual understanding," all of course "in a spirit of friendship."

It is January 4, 1959, and a few kilometers across town, in Bandalungwa, a housing tract recently built for for *évolués*, Patrice Lumumba has been invited to dine at the home of a new friend. During his time in prison, he had regularly read articles in the newspaper *Actualités Africaines* by a certain Joseph Mobutu, the soldier who became a journalist and visited the Expo. After his release, Lumumba makes friends with him. He goes to visit his home regularly and enjoys the splendid meals Mobutu's wife makes for them. During this Sunday lunch, the two make plans for that afternoon. At two o'clock, at the YMCA in the center of the *cité*, they know that the Abako has planned a meeting. Last week, Lumumba spoke to a crowd of seven thousand listeners about his trip to Accra. It was his best performance yet. The people were wildly enthusiastic. Afterward, the crowd had chanted "Dipenda, dipenda!" a corruption in Lingala of the word "independence." That is probably why the city's first mayor, the Belgian Jean Tordeur, has decided at the last minute to ban today's meeting. A security measure; he doesn't feel like dealing with rabble-rousers. Lumumba and Mobutu decide to go and take a look anyway. They don't have a car, but Mobutu has a scooter. That is the image on which we will freeze: Mobutu and Lumumba, together on the scooter, two new friends, the journalist and the beer marketer; one is twenty-eight, the other thirty-three. Lumumba is sitting on the back. They ride together in the muggy afternoon air and talk loudly, to be heard above the sputtering of the exhaust.[43] Two years later, one of them will help to murder the other.

It is January 4, 1959, and the crowd is flowing into the Stade Roi Baudouin for a major match in the Congolese soccer competition. The huge stadium is only a few hundred meters from the YMCA. Twenty thousand supporters have come from far and near.[44] They are wearing colorful shirts and *pagnes*. Some of them have feathers in their hair and stripes painted on their faces, like in the old days, broad white stripes of clay that glisten in bright contrast to their cheeks and foreheads. They dance defiantly, wide-eyed. It is frightening to see. The steep concrete grandstands around the field fill with people and rhythmic beats. There are tom-toms and drums, whooping and screeching. It's as though there's a war on. It's like the banks

of the Congo in the 1870s, when Stanley and his men first sailed by in their
metal boat. The throb of the war drum, the thousands of enraged voices,
the dancing that grows wilder all the time, the eyes of the warrior. In the
catacombs of the stadium the players tighten their shoelaces and slide shin
guards into their socks. Elsewhere in the city, at the governor's residence,
the champagne bottles have been removed from the cooler and are lined
up, sweating in the sun.

It is still January 4, 1959, and on Avenue Prince Baudouin, close to the
YMCA, Kasavubu tells the drummed-up crowd that the meeting, unfortunately, has been canceled. There are loud murmurs and protest, pushing and
shoving. As a pacifist and admirer of Gandhi, he urges his supporters to remain
calm. That seems to work, even without a microphone. He is the leader, he is
the chief, he is the mayor. Relieved and reassured, he returns home.

But it is January 4, 1959, the day that everything changes, although you
wouldn't say so yet. Congo is going along with the times, it seems. Léopoldville is the second city in the world with a gyrobus, an electric bus with
antennae on the roof that charges its motor at the stops. The first city with
such futuristic public transport was in Switzerland, but now these buses
zoom around the *cité* too.[45] A few thousand Abako supporters remain mop-
ing around the spot where their meeting was to have been held. A white
gyrobus driver gets into an argument with one of them and raises his fist.
Futurism meets racism. Right away, the blows rain down on him. The genie
has left the bottle. There are fistfights, there is shoving. The police arrive,
black constables, white commissioners. It's because of New Year's, the police
think, the people are still drunk or already broke, one of the two. Two commissioners deal out punches. That is not a good idea. "Dipenda!" the cry
goes up. "Attaquons les blancs! Let's get the white men!" Panic breaks out.
The police fire warning shots in the air. Farther along, one of their jeeps
is overturned and set on fire. At that moment the soccer stadium empties
out—commotion, ecstasy, frustration, sweat—and the supporters join
in with those who had been waiting to attend the Abako meeting. Soccer
is gunpowder. In 1830 Belgium became independent after an opera performance; in 1959 Congo demands independence after a soccer match. A
scooter comes racing up, with two young men on it. They can't believe their
eyes. In the last few years, both of them have worked their way up by edu-
cating themselves, but now they see the rage of the masses from which they
have withdrawn. They no longer look down on them, as *évolués* are wont to

do, but feel a sense of solidarity. The elite and the masses have found each other at last.

Léopoldville at that moment has four hundred thousand inhabitants, twenty-five thousand of whom are Europeans. There is a bare-bones police force of only 1,380 officers.[46] There is no national guard. The very next level of law enforcement is the army. The city's barracks house some twenty-five hundred troops, but they have been trained to wage war abroad, not to repress uprisings among their own people. The police do their best to handle things, but within a few hours the entire *cité* is in an uproar. Cars belonging to white people are covered by a downpour of stones. Windows shatter. Fires break out everywhere. The police turn their guns on the demonstrators. Puddles of dark blood gather on the asphalt, reflecting the flames. Thousands and thousands of young people begin looting. All things Belgian become their targets. Catholic churches and mission schools are vandalized; neighborhood centers where sewing classes are held are stripped. Around five o'clock, a few youth gangs descend on the shops belonging to the Greeks and the Portuguese, places where the people otherwise do their shopping. The looters strike ruthlessly and run off with meters of floral fabrics, bicycles, radios, salt, and dried fish.

At the governor general's New Year's reception, the telephone rings. "Ça tourne mal dans la cité. Things are getting ugly down in the *cité*." Heavy rioting has broken out in a zone ten or twelve kilometers (about 6.2 to 7.5 miles) long. The city's European district has been locked down. The army moves in, first with tear gas, then with guns. Demonstrators are being mowed down. "That was like using a hammer to kill a mosquito," people realized afterward.[47] Some of the colonials, however, are so furious that they take their hunting rifles down off the wall to go out and "help." Years of piled-up contempt and fear, especially the latter, burst loose. At around six, when darkness falls, relative calm descends on the city. The fires smolder on. At the European hospital, dozens of white people show up for treatment. Outside, before the door, their elegant cars are parked in the darkness, dented, scratched, and ruined. At the villas, for the first time in years, the women have to do their own cooking: the boy has disappeared completely.

The next day, many of the Belgians feel more resigned than outraged. "We completely lost face," they tell each other on Monday morning.[48] Some of them begin stocking up on canned sardines and vegetable oil, others book one-way tickets with Sabena for Brussels. The army takes three or four days to get the city back under control. The final toll is unbearable: forty-seven fatalities and 241 wounded on the Congolese side, according to the official figures at least. Eyewitnesses speak of two hundred, perhaps even three hundred people killed.

It was January 4, 1959, and things would never be right again.

"A few days later I flew to Brussels aboard a DC-6," Jean Cordy told me in the fall of 2009 at his service flat in Louvain-la-Neuve. In 1959 he had been the principal private secretary to Governor General Cornelis. "My directives were clear: I was to convince the Belgian government to include the word *independence* in their long-awaited policy paper. The governor general had said this was an opportunity we should absolutely not miss. I also visited the king and told him that Belgium had to refer to independence."[49]

On January 13, 1959, more than a week after the riots, both the policy paper and a royal statement were publicized. The ministerial text was fuzzy, technical and incoherent, but Baudouin's speech was both apt and crystal clear. A tape recording of his message was sent to Congo and immediately broadcast on the radio. Fishermen on the beach at Moanda, farmers amid the sugarcane, workers covered in the dust of the cement factory, seminarians immersed in their books, nurses washing their hands, village chieftains in the interior, helmsmen on the riverboats, nuns weeding their gardens, the elderly, and the adolescent listened to their transistor radios and heard their beloved king pronounce the historic words: "Our decision today is lead the people of Congo in prosperity and peace, without harmful procrastination but also without undue haste, toward independence."[50]

People could hardly believe it. This was too good to be true! As they drove through the villages of Bas-Congo, the truck drivers honked their horns and sang loudly out the window:

> *Independence is coming.*
> *Independence will soon be ours.*
> *Mwana Kitoko [Baudouin] has said so himself.*
> *The white chiefs have said so too.*
> *Independence is coming.*
> *Independence will soon be ours.*[51]

But this exuberance did not mean that Congo was back to business as usual. Unrest continued and extended far into the countryside. In regions with a long tradition of protest, like Kwilu and Kivu, things were rumbling once again. In Kasai a conflict arose between the Lulua and the Baluba, and there were mass demonstrations in Bas-Congo. After the riots on January 4, the Abako was disbanded by decree, and Kasavubu, along with two other leaders, was sent to prison for a time (they would later be released by Maurits Van Hemelrijck, the new Belgian cabinet minister charged with overseas affairs). This only increased Kasavubu's fame in the interior, while the attitude toward the colonizer was becoming increasingly grim. Kasavubu had issued a call for civil disobedience and peaceful resistance. Secretary Jean Cordy, one of the only white, card-carrying members of the Abako, traveled through the province in July 1959 with interim governor general André Schöller. "Suddenly, the people's support for Kasavubu had become absolute. No one talked to the authorities anymore. 'Kasavubu is our leader, negotiate with him,' is what they told us. They simply didn't react, not even when I spoke to them in Kikongo. I had never run into anything like it before, and I had been in Congo since 1946. The bridges had been torn down, despite the statements on independence from the king and the government, despite Van Hemelrijck's visit. The dialogue was over. Their silence felt very, very strange."[52]

The prospect of a political turnaround aroused in many the ambition to govern. New parties arose everywhere. In late 1958 there had been only six; eighteen months later there were a hundred. Each week saw the birth of a new movement, with names like the Union Nationale Congolaise, the Mouvement Unitaire Basonge, and the Alliance Progressiste Paysanne. It rained abbreviations (Puna for the Parti de l'Unité Nationale, Coaka for the Coalition Kasaïenne, Balubakat for the Baluba of Katanga); the acronyms sometimes had more letters than the party did members.

Who were these political leaders? Time and again one saw that they were relatively young men with a secondary school education. They formed the country's intellectual upper crust and and lived in the cities, to which they had moved as young people. Often they were active in alumni or cultural associations and nurtured their interest in politics by means of readings and debates. Admittedly, their tone was often more acute than their insight, and their knowledge of actual developments came in second to their drive. With a few exceptions, their party platforms were meager.[53]

One characteristic, however, cannot be overemphasized. Despite their urban surroundings, their tender years, and modern lifestyle, this budding political generation cherished ties with something that seemed to come from long ago and far away: the sense of tribe. That seems contradictory, but it is not. The sense of ethnic identity was an urban feeling par excellence. Only in contrast with others did one start to think about one's own origins. The young and upcoming politicians hooked up with the existing ethnic organizations and modernized them. Following the tribal tack was, in terms of political strategy, a smart move: it allowed you to reach the masses. There was gain to be found in hammering on the fact that you were a proud Tshokwe, Yaka, or Sakata. Besides a larger constituency, it also guaranteed a greater chance of being heard by the various levels of colonial government. Kasavubu spoke for the Bakongo, Bolikango stood up for the Bangala, Jason Sendwe for the Baluba from Katanga, Justin Bomboko for the Mongo people, et cetera. Tribal rhetoric allowed a young elite to step forward as spokesmen for their communities.[54]

Understandably enough, this *jeunisme* (leadership by the younger generation) did not please the chiefs in the interior, some of whom still exercised a certain degree of influence over their migrant communities in the cities. And what was happening here was, indeed, quite revolutionary. In large parts of Central Africa, authority was traditionally based on age. Age meant respect. Now, suddenly, there was a generation of twenty- and thirty-year-olds competing for power and, in so doing, also competing for the people's favor. They had little choice, for the Belgian government had decided to introduce universal suffrage. "With the introduction of universal suffrage," the chief of the Bayeke in eastern Congo said, "the traditional authority will be completely undermined and is doomed to disappear." And he was right: after 1960, a relatively young generation assumed the reins in Congo. They had proved to be the only ones capable of understanding the game of democracy and playing it successfully. The great chief of the Lunda, the inhabitants of a former kingdom along the border between Katanga and Angola, called universal suffrage an "unforgivable aberration."[55]

But the most famous Lunda of those days, and in fact in the entire history of Congo, was someone else: Moïse Tshombe. In 1959—he had just turned forty, he lived in the city and had studied bookkeeping—he accepted leadership of a new political party, the Conakat (Confédération des Associations du Katanga). A family fortune had left Tshombe well-to-do, but he himself was a not particularly successful businessmen, with a look that was often misinterpreted as brooding. He came from a prestigious Lunda family, his father was a rich trader, he himself married one of the daughters of the great Lunda chieftain. Tribal pride was not foreign to Tshombe (for a time he had led the most important Lunda association in Elisabethville), but he did not oppose universal suffrage. The Conakat was a political party that used democratic means to obtain more rights for the original inhabitants of Katanga, such as the Lunda, the Basonge, the Batabwa, the Tshokwe, and the Baluba (although not the Baluba from Kasai: they were "newcomers"). Due to the decade-long import and immigration of workers, primarily from Kasai, the original population felt threatened; in Elisabethville, the Baluba from Kasai had even won the 1957 elections. Tshombe wanted more power for the "true" Katangan tribes. His Conakat in that way greatly resembled Kasavubu's Abako; both movements advanced the interests of the city's original inhabitants (although the Abako was monoethnic), both desired a return to far-reaching regional autonomy and both dreamed—unlike Lumumba—of a federal, highly decentralized Congo. Bas-Congo and Katanga, if need be, might even become independent states. But when it came to the future role of Belgium, they also entertained fundamental differences: the Abako was radical and anticolonial, particularly after the January riots; the Conakat, on the other hand, was not out to burn any bridges. Tshombe, who was surrounded by Belgian advisers, dreamt of a calm and orderly independence, but continued to believe in the idea of a Belgian-Congolese Community. "If we call for independence, that is not to chase away the Europeans: on the contrary. We want to continue working together with them, hand in hand, to build this country's future."[56]

Amid the profusion of political parties there ran only two major fault lines. First of all, was one radical or moderate? Radical meant that you were in favor of rapid decolonization and a total rupture with Belgium. And, second, did one think in federal or unitary terms? The Abako (Kasavubu) was radical and federalist; the MNC (Lumumba) was radical and unitary; the Conakat (Tshombe) was moderate and federalist. All the other parties could also be characterized along these same lines.

LUMUMBA REALIZED that political sectarianism was not a good idea. So in April 1959 he called together eight political parties in Luluabourg (Kasai) for the purpose of joining forces. It was Congo's first political congress, a sort of mini-Accra. Jean Mayani, the man who had been one of Lumumba's earliest supporters, was present. In his living room in Kisangani he told me: "I went there as party secretary for my borough. All the nationalist parties were there. The Cerea from Kivu, Sendwe's Balubakat from Katanga, the Parti Solidaire Africain (PSA) from Kwilu, Kasavubu's Abako. Really, everyone was there. Lumumba had almost three-quarters of the population backing him."[57] The Cerea opposed white supremacy in eastern Kivu province. The Balubakat stood up for the rights of the Baluba in Katanga, in direct opposition to Tshombe's Conakat. The PSA was active in Kwilu, but would soon gain a national reputation with such major figures as Cléophas Kamitatu and Antoine Gizenga.

Lumumba wanted the parties to jointly propose a date for independence. In his speech, King Baudouin had promised that it would arrive "without harmful procrastination but also without undue haste," but when did procrastination become harmful and when was haste undue? It would be a huge step forward, Lumumba realized, if the Luluabourg congress could agree on a date. What's more, it would also be a major triumph for him: the initiator's kudos would be bestowed on him and he would be recognized as the country's most important political figure. His own suggestion was: January 1, 1961. Did anyone object to that? "Why such a hurry?" one of the delegates remarked. "Is the world going to come to an end on January 1, 1961?" Whereupon Lumumba snapped back: "You speak like a colonial."[58] Two years seemed like plenty of time to prepare the switch to the new system. That was how it had gone in Ghana too. In a time of weak party platforms and budding political leaders, there was little time for nuance and reflection. Anyone who called, even apologetically, for more gradual change was laughed off the platform as a colonial flunky. The parties became entangled in a symbolic, unparalleled game of ante-up. Rhetorical bravura was valued more highly than pragmatic sense. Rapid and unconditional independence became a goal in itself, even an obsession; people were ready to throw away the baby with the bathwater, if need be. "Better poor and free than rich and colonized."[59] Slogans like that were extremely popular. But what else could one expect? None of those present, with the exception of a few shantytown mayors, had ever been given a political mandate. Administrative experience, realism and an eye for planning were completely lacking. They were all just muddling along. And no one wanted to lag behind. This, however, happened to be about the future of a country the size of Western Europe.

It was not only sucking up when the great chief of the Luanda welcomed the governor general and the Belgian minister to his district with the words: "We do not want you to make decisions under pressure from loud-mouthed minorities. We do not understand the hurry many are in to achieve independence. We solemnly confirm that we, too, desire independence, but not yet. We need a great deal of help and support to arrive at normal development. All needless haste could once again plunge our country into the poverty and misery of the past."[60]

What seemed like a reactionary standpoint at the time was a widely heard lament in Congo in the year 2010, a lament prompted by all the recent misery. Many young people blamed their parents for having demanded independence at all costs. On the street in Kinshasa, someone once asked me: "How long is this independence of ours going to last, anyway?" As a Belgian, I had heard it countless times: "When are the Belgians coming back? After all, you're our uncles, aren't you?" That was often meant as flattery, but sometimes there was more to it. Even Albert Tukeke, the man from Kisangani who was a distant relative of Lumumba's, said at the end of our conversation: "We shouldn't have become independent so quickly. But after the war, you know . . . there was that urge. If it hadn't all happened in such a hurry, we wouldn't have been faced with all these shortcomings."[61]

Head-over-heels decolonization was the result of symmetrical escalation with the colonial government and a symbolic and continuous raising of the stakes among the various political parties. The killing of a few dozen of Lumumba's supporters during riots in Stanleyville did nothing to improve the situation. The proud Lumumbist Jean Mayani said of that: "After the congress, the colonial powers interpreted the MNC's demands as a form of racial hatred and xenophobia that was being turned against the Belgians." It took a bit before I realized that, in the colonial vocabulary, xenophobia was a trait ascribed to the Congolese. "The Force Publique dealt repressively with Lumumba's partisans. Twenty people were killed in Mangobo, a borough in Kisangani. Lumumba was arrested and thrown into prison. It was just like the January 4 riots in Kinshasa."[62]

Municipal elections were held in late 1959, but boycotted by the Abako, the MNC, and the PSA. The parties were no longer interested in transitional measures and slow processes. The important thing now was immediate independence, and nothing else. Belgium hoped that gradual democratization would win the people's favor, but things turned out differently. The tensions had risen too far. The first elections had been organized in 1957, in the hope that that would placate the men of *Conscience Africaine* and the Abako. But it had the opposite effect. After the January 1959 riots, the Belgians promised independence, but not even that could smooth the feathers that had been ruffled. The colonizer thought it was doing the right thing, but struck out each time. That resulted in 1959 in the loss of a great deal of valuable time and goodwill; assets that could have been used to prepare well for independence. Rather than try to slap together an improvised, well-intentioned policy, perhaps the time had come to finally ask the Congolese themselves what they wanted.

ON JANUARY 20, 1960, a group of some 150 men in winter coats gathered at the Palais des Congrès in Brussels—about sixty Belgians and some ninety Congolese. The idea was to spend one month discussing, frankly and on an equal basis, a number of touchy issues. Hence the name: it was to be a "round table conference" (even though the tables were actually arranged in a rectangle). The Belgian Socialist Party, part of the parliamentary opposition at the time, was pleased with the initiative. The Belgians were represented by six cabinet ministers, five members of parliament, and five senators, accompanied by a few dozen advisers and observers. The politicians had little on-the-ground knowledge of the colony; "dry-season pilgrims" was how the Belgians in Congo itself referred to them mockingly. But many of them were rather smitten with the United Nations' new-fangled ideology of decolonization. The Congolese delegates came from the major political parties (Kasavubu, Tshombe, Kamitatu among them) and included a dozen tribal elders to represent the traditional authorities. Just before the conference began, the Congolese delegates gathered to form a common front that would bridge the interparty rivalries, ethnic tensions, and ideological fault lines. They did not want this conference to turn into a messy game of Ping-Pong; they wanted to act as a single player. *L'union fait la force*—united we stand, divided we fall: Belgium had at least taught them that much. This sudden coalition came as a great surprise to the Belgian politicians, divided as they were between Catholic, Liberal, and Socialist sociopolitical blocs, between cabinet and parliament. Many of them were ill-prepared. There was no agenda, no government standpoint. After all, this meeting was not meant to decide anything, was it?

During the first five days of the round-table conference, however, the common Congolese front achieved three crucial victories. First, they were able to convince the Belgians that Patrice Lumumba, who had been imprisoned after the Stanleyville riots, should not be absent. Without him, they stated, the conference was not representative and might merely fan the flames in Congo. Deciding to play it safe, the Belgians had Lumumba released from prison and flown to Brussels. The second major victory: the Belgian delegates had to promise that the resolutions of the conference would afterward be molded into draft bills that would then be sent to both parliament and the senate. The Congolese knew all too well that they had no legislative power, but this gave them the guarantee that the decisions made would not end up at the dead-letter office. It would be hard to overstate the importance of this particular victory: what had begun as an informal colloquium in this way became a summit meeting of far-reaching portent. The third victory was even more conspicuous: the date! The Belgians had hoped to knock around a few ideas about the political structures of a Congo that would become independent at some point in the future, but for the Congolese delegates there was one question that went before all others: when?

On the fifth day of the round-table conference, even before Lumumba arrived, a discussion took place between Jean Bolikango, leader of the common front, and August De Schryver, acting minister for Congo, that most closely resembled the process of haggling and undercutting at a Kinshasa street market. January 1, 1961, the date only dreamed of back in 1958, had meanwhile become superseded. Things could not go quickly enough. In accordance with the old Flemish motto "you never know until you ask," Bolikango made the first bold move and proposed June 1, 1960. The Belgians were astounded: but that was barely four months away! What could they say to that? Their counterproposal was July 31. A two-month respite. Given, it wasn't much, but it was all right. Shall we make it June 30 then? Split the difference? Going, going, gone! On June 30, 1960, Congo would become independent. The die was cast. In the Palais des Congrès, an applause went up from the Congolese *and* Belgian delegates. No one in the Congolese delegation had thought it would be so easy; they were all flabbergasted.[63]

What was going on here? Had the colonizer, in an unguarded moment, given away independence? No. The round-table conference had indeed gained more momentum than first intended (as was the case with almost every initiative in Belgian colonial politics after 1955) and the Belgian delegation was indeed badly prepared, but this was no rash decision. In the context of the moment, Belgium had only two options: either to reject the common front's demand, which would almost surely have led to massive rioting, or to agree to the request and hope that things would not get out of hand.[64] There was no time for calm negotiations. The choice, in other words, was obvious enough. There were enough Belgian soldiers stationed at the military bases at Kitona and Kamina, but Belgium was not at all in favor of a conflict model. A bloody struggle for independence had been raging in Algeria for the last six years. There was absolutely no majority to be found in the Belgian parliament for a military show of force. The United Nations Charter and the anticolonial standpoints of the United States and the Soviet Union also gave Belgium little room to maneuver on the international scene. Fight off independence? That was possible, but only at the cost of a risky undertaking in the colony and moral isolation from the international community. In 1960 no less than seventeen African countries were to gain independence; Belgium could not lag behind. The only European countries with no intention of releasing their large African holdings were the southern European dictatorships: Salazar's Portugal, which refused to surrender Angola, Mozambique, Guinea-Bissau, and the Cape Verde Islands, and Franco's Spain, which still clung to Equatorial Guinea. Apartheid South Africa had no plans to let go of Namibia either. Belgium could agree to the date of June 30, 1960, because it knew that, even after that, it would continue to be involved in policy making, the army, and the economy. Top officials would act as ministerial advisers, white officers would remain in service, the big companies would remain Belgian, and missionaries would carry on teaching.

At the Plaza Hotel in the heart of Brussels, the atmosphere was euphoric. All kinds of things still had to be talked about, of course (that Congo would become a republic, that the ties with the Belgian royal family would be severed, that it would become a unitary state, that the provinces would receive powers of their own: none of this had been established yet), but the loot had been dragged in, the cat was in the bag! Joseph Kabasele's African Jazz orchestra, which had had such success with its song about Jamais Kolonga, had accompanied the delegates to Brussels. Even negotiators in three-piece suits have to be able to dance after the plenary sessions.

Charly Henault still remembered it clearly. He was African Jazz's drummer for years, but a Belgian nonetheless. "I was white, but what did I care? I was a drummer in a land full of drummers," he told me when I found him one drizzly day at his home in eastern Belgium. He was deathly ill and stayed in bed; the memories were becoming washed out. "The round-table ball was held at the Plaza, yeah. . . . The joy, the euphoria . . . Kabasele called the politicians by their first names. They all loved him. . . . A man with class, in his powder-blue tuxedo with black piping. Very chic . . . He loved women and he loved making jokes. . . . One time I even hid his pajamas!"[65] Besides all the fooling around, at the Plaza the band made a start composing the song that would soon become the biggest hit in Congolese music: "Indépendance Cha-Cha." The lyrics, in Lingala and Kikongo, celebrated the newly won autonomy, praised the cooperation between the various parties and sang of the great names in the struggle for independence: "Independence, cha-cha, we took it / Oh! Autonomy, cha-cha, we got it! / Oh! Round table, cha-cha, we won it!" After 1960 Congo would adopt a number of national anthems, under Kasavubu, under Mobutu, under Kabila: pompous compositions with pathetic lyrics, but throughout the past half-century there has been only one true Congolese anthem, one single tune that right up until today makes all of Central Africa shake its hips: that playful, light-footed and moving "Indépendance Cha-Cha."

JUNE 30 IT WAS. The round-table conference ended on February 20, 1960, with only four months left in which to knock together a country. The to-do list was impressive. A transitional government had to be formed, a constitution written, a parliament and senate established, ministries expanded, a diplomatic corps appointed, provincial and national elections organized, a cabinet put together, a head of state chosen . . . and that was only the country's political institutions. A national currency also had to be created, and a national bank, in addition to postage stamps, driver's licenses, license plates, and a land registry office.

A great many Belgians in the colony were leery of this mad rush. They were afraid that the colony, which had been worked on so carefully for seventy-five years, would go down the tubes in a few months' time. Many of them began sending their savings, their belongings, their families home. Others migrated to Rhodesia or South Africa. During the first two weeks of June, four times as many passengers left from the airport at Ndjili than in the same period the year before. Sabena had to organize seventy extra flights, the boats to Antwerp were brimming over.[66]

The run-of-the-mill Congolese, on the other hand, was enjoying it immensely. He believed that a golden age was on its way, that Congo would

become prosperous from one day to the next. That, after all, was the promise made him in the dozens of pamphlets circulating around the country.
Almost all the parties were making promises that could never be kept, promises that were sometimes grotesque, sometimes downright dangerous.[67]
"When independence arrives," an Abako broadsheet read, "the whites will
have to leave the country." That was definitely not one of the conclusions
of the round-table conference. "The goods left behind will become the property of the black population. That is to say: the houses, the shops, the trucks,
the merchandise, the factories, and fields will be given back to the Bakongo."
Little wonder then, with such inflammatory texts, that farmers in Bas-Congo
expected nothing short of boundless liberty: "All laws will be abolished, we
will no longer have to obey the traditional chieftains, nor the elders, nor the
officials, nor the missionaries, nor the bosses. . . . ." In that longing for a sudden, radical turnabout one heard echoes from the days of Simon Kimbangu.
Independence itself became a sort of messianic moment that would bring with
it "life, health, joy, good fortune and honor." Kasavubu and Lumumba, both
of whom had spent time in prison, grew to become prophets and martyrs. In
Kasavubu people saw the resurrection of the king of the old Kongo Empire,
while dynamic Lumumba was compared to the Sputnik satellite! Simple
people looked forward to nothing less than a cosmic turnabout. Employment and taxes would disappear. Some of them even assumed that, from
then on, "the black will have white boys" and that "everyone will be allowed
to pick out a white woman for themselves, because they will be left behind
and redistributed, just like the cars and other things."[68] A few hucksters took
advantage of that naïveté and began selling white people's homes for the
trifling sum of forty dollars. . . . Gullible souls, not realizing they had been
swindled, knocked on the doors of white villas to ask whether they could
come in and take a look at their new property. Some of them even asked to
inspect the woman of the house, because they had just paid twenty dollars
for her as well.[69]

On a macroeconomic scale, a number of things had to be arranged too.
Colonial industry, after all, was intertwined in numerous ways with the
colonial state, which would soon cease to exist. To deal with that, a second
round-table conference was held in Brussels. This time, the political parties
in Congo attached far less importance to the meeting. Independence was
the most important thing, they figured, and they had secured that. Besides,
it was already late April and everyone was busy campaigning for the upcom-

ing elections in May. None of the heavyweights had time to leave Congo for any period. Young party members went to Brussels in their stead, where they were assisted by a few Congolese who had studied in Belgium.

One of those delegates was Mario Cardoso. Today he is the deputy vice president of the Congolese senate. He invited me out to lunch in Kinshasa at the restaurant of the stately Memling Hotel.

I was the third student from Congo allowed to study in Belgium. Every year, Raphaël de la Kéthulle would send one of the Scheutists' students to the university at Louvain. The Jesuits felt that they should educate people in their own country, but the Scheutists wanted to show that they had pupils who could stand comparison with Belgian students. The first one to go was Thomas Kanza, in 1951. He studied psychology and education. In fact, he had been hoping to study law, but the governor general had forbidden that, out of fear for subversion. The next year it was Paul Mushiete. He studied psychology and education too, and sociology alongside that. My turn came in 1954. What I really wanted was to attend the military academy, but that wasn't allowed, so I also went for psychology and education. In 1959 I came back to Kinshasa and became an assistant at the University of Lovanium. I was planning to become a professor, but Lumumba asked me to go to the economic round table. I was the head of the MNC delegation, the Lumumba caucus.

The party had split in the meantime: there was Lumumba's MNC-L, which was unitarian, and the MNC-Kalonji, which advanced the interests of the Baluba in Kasai. "There was an awful lot of suspicion at that conference. The Belgian delegation included gentlemen who had been our professors. We had to negotiate with them. That was no mean feat. The talks were about the future status of the colonial companies, but everything seemed to have been decided beforehand."[70]

The economic round table was, above all, an attempt on Brussels' part to save the furniture. Belgium wanted to safeguard its business interests in Congo and felt that Belgian companies should be free to decide where their registered office would be after 1960.[71] Cardoso was still bitter about that: "The companies were allowed to decide whether to continue under Congolese or Belgian law. That measure was forced down our throats as a foregone conclusion." Most companies chose for Belgium, fearing as they did fiscal instability in Congo or, even worse, nationalization. From the time of Leopold II on, Congo had been a test plot for the free-market economy. Companies there enjoyed a generous fiscal regime with almost no government interference. Huge conglomerates, with the Generale Maatschappij in pole position, had experienced a period of unbridled capitalism. Even in those cases where the colonial state was the major shareholder, for example the influential Comité Spécial du Katanga, the government left the actual running of affairs to the businessmen. With independence on the horizon, many business leaders now feared that their days of autonomy and excellent relations with the government were numbered. They remained active in Congo, but chose for a registered office in Belgium, effectively placing their company under Belgian rather than Congolese governance. That transfer cost the Congolese treasury a vast amount of tax revenue.

During this second round of talks, the status of the "colonial portfolio" came up as well. The term referred to the huge package of shares the Belgian Congo held in many colonial companies (mines, plantations, railroads, factories). What was to be done with that? As soon as the Belgian Congo became Congo, those shares would obviously become the property of the new state. The Belgian politicians and business leaders didn't think that was a good idea. They convinced the Congolese delegates that it would be better if those government participations were taken away from the state and transferred to a new Belgian-Congolese development company. It was a sly way to keep a hand on the purse strings.[72] Here too, Congo paid for its delegation's lack of economic experience. People who had been allowed to study only psychology were being asked to make crucial macroeconomic decisions. "Second-rank figures," then-prime-minister Gaston Eyskens opined.[73] One of them was the journalist Joseph Mobutu. He had been sent to negotiate by his friend Lumumba, and the experience was to haunt him for the rest of his life. He said of it later:

> And there I sat, a silly, unmannered journalist, at the same table with the great white sharks of Belgian finance! I'd had no financial training whatsoever, and the other members of my delegation, who represented the other Congolese movements, hadn't either. It is not one of my fondest memories. From April 26 to May 16 we negotiated inch by inch, but I became like one of those cowboys in a western who lets himself be bamboozled time and again by professional con men. We talked until

late at night, and the next day we discovered that the Belgian parliament
had meanwhile made decisions that rendered the negotiations obsolete.
We had to fight for everything. . . . Of course we let ourselves be rolled.
Our partners in the discussion used a whole series of legal and technical
ruses to successfully safeguard the hold which the multinationals and
the Belgian capitalists had on the Congolese pocketbook.[74]

The worst was yet to come, but only a few weeks later. On June 27,
1960, three days before independence, the Belgian parliament—with
the endorsement of the Congolese government, no less—disbanded the
Comité Spécial du Katanga.[75] Congo could not have made a worse blunder!
With that, the new state lost control over mining giant Union Minière, the
motor of the national economy. How could that have happened? The CSK
was essentially a public enterprise that awarded concessions in Katanga to
private companies in return for shares. It held a majority share in Union
Minière and therefore the power of control. Historically, that right to a
government say in the company's dealing had rarely been used: the colonial
state had always relied on the competence of the business world. Now that
Congo was on the point of becoming independent, however, there was a
chance that the new state would actually involve itself in the activities of
Union Minière and all its subsidiaries. By disbanding the CSK, that possibility was effectively blocked. In all their disaffection with the Moloch of
Western capitalism, the Congolese delegates to the economic round table
had no problem with that, and Lumumba's new government would soon
adopt that same line of reasoning. . . . Congo remained a part owner, but as
minority shareholder had far less power and received far less of the profits
than the big Belgian trusts, such as the Generale. In that way it not only
missed out on many millions of dollars, but also on the opportunity to let
the industry work in the service of the country itself.

Dancing with ignorance, the country moved toward the precipice of
independence. The political keys were already in its pocket, but the economic ones were now safely tucked away in Belgium. Nevertheless, one day
after this unbelievably cunning move, the two countries signed a "pact of
friendship" that spoke of aid and assistance.

FINALLY, IN LATE MAY, the long-awaited national elections were held. The
turnout was huge, the results predictable. After Patrice Lumumba's MNC,

the biggest winners were the regional parties, with or without separatist tendencies. The Abako won in Bas-Congo, Conakat in southern Katanga, and Balubakat in the north, Kalonji's MNC in Kasai, Cerea in Kivu, and the PSA in Kwilu. The latter two were not true tribal parties, but provided the ethnically highly splintered regions of Kivu and Kwilu with a sort of super-tribal élan. The electoral map of Congo in 1960, therefore, was largely identical to the ethnographic maps drawn up by the scientists half a century before. This tribal reflex should not be seen as atavistic. Were pan-European elections to be held in Europe today, after all, there is a great chance that most of the French would vote for a Frenchman and most Bulgarians for a Bulgarian. In a vast country like Congo, where the greatest part of the population had no more than a primary school education, it should come as no surprise that many voted for candidates from their own region. The three strongest figures to come out of the elections were Kasavubu, Lumumba, and Tshombe. Kasavubu held sway over the western part of the country, Lumumba over the northwest and center, and Tshombe over the far south. That corresponded with the major cities: Léopoldville, Stanleyville, and Elisabethville. The smaller parties divided among themselves the countryside that lay between.

This fragmentation made it no easier to form a representative coalition government. No one party had an absolute majority (Lumumba's resounding victory only secured about one-third of the parliamentary seats from five of the six provinces; he made absolutely no headway in Katanga), and even a rudimentary coalition with only a few partners was ruled out. The negotiations were going to take a long time. What's more, the Belgian government was quite disappointed to see that Lumumba, whom they considered a seditious demagogue, had been able to win over so many voters. This concern went so far that Brussels even appointed a new minister resident, W. J. Ganshof van der Meersch, and sent him to Congo to supervise the formation of a new government. In his wake, new Belgian troops were also sent to the colony. Lumumba had little patience with these demarches and made no effort to disguise it. The two men irritated each other no end. Kasavubu was the first to be appointed to try and form a new government, but when that failed the job was given to Lumumba. He was faced with the almost impossible task of bringing together all the widely diverse individuals into a single political team. Up until one week before independence, the new minister resident still had hope that Lumumba would not become prime minister.

But on June 23 the first Congolese cabinet became a fact. It numbered twenty-three ministers, nine deputy ministers, and four ministers of state, posts that were divided among no less than twelve political parties. The way things go with difficult compromises, this feat produced more shouts of pain than of joy. Bolikango, the gray eminence from Équateur who had led the common front in Brussels, saw the office of president slip through his fingers at the last moment. Lumumba, after all, needed the support of the Abako, and received it by means of a compromise: if Kasavubu would repress his separatist urge, he could become the head of state. Lumumba, the big winner, therefore, did not become president himself, but only prime minister, even though his party had won 33 of the 137 parliamentary seats and Kasavubu's only 12.

Tshombe realized at last that he had missed the boat and that his party would have to make do with one ministerial post and one deputy minister. His Katanga accounted for the lion's share of the nation's income, but was receiving little in return: that stung. Sooner or later, it would have to have repercussions. The parliament, too, was hesitant: the new cabinet was only barely ratified by the elected representatives.[76] In the early days of the Lumumba cabinet, therefore, there was nothing like the collective effort of a government team providing unified support for a political project.

The team that was installed was not only heteroclite and petulant, but also extremely youthful. Seventy-five percent of them were under the age of thirty-five. The youngest was only twenty-six years old. That was Thomas Kanza, the first Congolese to obtain a university diploma. He became the new ambassador to the United Nations, certainly no sinecure in the first months after independence. The oldest cabinet minister was Pascal Nkayi, but he was only fifty-nine. He was made finance minister, after a lifetime as clerk to the post office administration. A new elite also held primacy in the parliament: only 3 of the 137 seats went to traditional chiefs.[77]

The first government of Congo inherited from Belgium a country with a well-developed infrastructure: more than fourteen thousand kilometers (nearly 8,700 miles) of rails and more than 140 kilometers (about eighty-seven miles) of highways and streets had been built; there were more than forty airports or airfields and more than a hundred hydroelectric and power plants and there was a modern industrial sector (Congo was world leader in industrial diamonds and the world's fourth largest copper producer). In addition, a start had been made with general health care (three hundred hospitals for natives, plus medical centers and birth clinics) and the country enjoyed an extremely high degree of literacy (1.7 million primary school pupils in 1959)—achievements that were truly striking in comparison with other African colonies.[78] What's more, the army had had major successes in both world wars. But there is more to life than infrastructures. Thomas Kanza, the fresh-faced cabinet minister who had studied psychology, knew that for many Africans those successes were only relative: "Unlike what most Europeans were willing to admit, they had suffered more under the lack of sincere sympathy, respect, and love from the colonizers than from any lack of schools, roads, and factories."[79] Besides, what were you supposed to do with a fully appointed country if no one knew how to run it? On the day of its independence, the country had sixteen university graduates. And although there were hundreds of well-trained nurses and policy advisers, the Force Publique did not have a single black officer. There was not one native physician, not one engineer, not one lawyer, agronomist, or economist.

"BELGIUM HAD NO EXPERIENCE WITH COLONIZING," Mario Cardoso said during our elegant lunch at the Memling, "but it had even less experience with decolonization. Why did it all have to go so quickly? If they had waited five years, the first batch of Congolese officers would have finished their training. Then there would have been no mutinies in the army." Between 1955 and 1960 the colonial regime had searched feverishly for reforms that could stem the tide of major social unrest, but it was too little and too late. And so the process of decolonization became a runaway locomotive and no one could find the brake. By bowing too late to the understandable demands of a frustrated elite, Brussels released a play of forces that far exceeded its own ability to control. But the same applies to that same young elite, which not only pinpointed and canalized the social dissatisfaction of the lower classes, but also whipped it up and magnified it until it took on proportions that it, too, was unable to handle. The chronology of events brought to light a paradox that could be noted at best, but not resolved: the decolonization had begun much too late, independence came much too early. Disguised as a revel, the breakneck emancipation of Congo was a tragedy that could only end in disaster.

CHAPTER 7

A THURSDAY IN JUNE

J AMAIS KOLONGA CLIMBED OUT OF BED THAT THURSDAY MORN-
ing at four o'clock. He had slept at his tailor's the night before, just to
make sure nothing was left to chance.[1] The ceremony was not until eleven,
but this was not a day like any other. The city of almost half a million
inhabitants was still dark and silent. The houses and huts were covered by
a heavy blanket of heat. Nothing moved. The laundry: hanging in deathly
stillness on the line. The fire: brittle cinders. Out of sight, the children
slept in awkward poses. Out of sight, men and women nestled together—
comfort for a single night, or for a lifetime. Along the empty boulevard, the
traffic lights sprang from green to yellow to red. In the European neighborhoods, the water in the swimming pools was without a ripple.

The birds were still silent on their roosts. Further along, past the gardens and the villas, the lawns and the bougainvillea, the black water of the
powerful river flowed by in silence. Little islands of vegetation were still
being carried along, clods of earth and grass and plants, torn from the jungle
hundreds of kilometers upstream, tree trunks that rolled in the darkness
and soon, at the first rapids, would rise up and collide in the foaming river.
That is how it had gone for thousands of years. Nature paid no heed to this
auspicious day.

Jamais Kolonga turned on the light. He prayed and bathed. His brand-
new suit was on its hanger. Carefully, he drew the trousers out from under
the coat. His tailor had made a beautiful tuxedo for him, cut to size. The
trousers' smooth material felt cool, the shirt was wonderfully stiff and

starched, the coat fit his little form to a tee. He looked at himself in the mirror. Who would ever have thought that he, Jean Lema to the registrar's office, Kolonga to the rest of the world, would play such an important role on this day? Until just a few years ago he had worked only at a desk job in the interior, in Équateur. As a clerk for the Otraco, he was charged with the administration for the cargo ships plying the big river. But even then there had been change in the air. At his next promotion, he assumed a position formerly held by a white man, Monsieur Eugène, a Belgian from Verviers. In 1958 he came back briefly to Léopoldville and, as he put it, caught a whiff of "the odor, the perfume, of independence." Joseph Kasavubu was still coming to his father's home regularly; he heard the exciting conversations and sensed the unbelievable opportunities. He didn't want to go back to the interior again, despite his employer's repeated exhortations. On the boulevard, in the center of town, he had run into the great Jean Bolikango. Bolikango had gone to school at Tata Raphaël's as well, he was one of the few Congolese who—with an eye to impending emancipation—had been given a high administrative position, as deputy commissioner at the Ministry of Information. Bolikango knew, of course, how eloquently Kolonga could speak in public, and remembered his father's status as "über-évolué." After all, King Baudouin had even visited his home! Bolikango had rolled down the window of his car and invited him there and then to become an editor/announcer/translator for the government's information service. Jamais Kolonga agreed on the spot. From desk clerk for river transport he became a radio journalist for the public broadcasting system. From then on he would be able to inhale the perfume of independence each and every day. As a reporter he not only went from fashion show to soccer match, but he also saw his country's great political turnabout from close up. During the round-table conference in Brussels he reported the goings-on each day from the studio. And on June 26, 1960, when Kasavubu was sworn in as first president of soon-to-be-independent Congo, it was his scoop. With his TEAC, the leaden tape recorder of that day, slung over his shoulder, he was the one who had done the interviews.

His new black shoes were buffed to a mirrored shine, their soles were still a virginal white. Kasavubu's inauguration had been held only four days ago. Kolonga had done a good job on that. Two days ago they had asked him to do the live reporting on the solemn independence ceremony as well. He agreed. But it meant that his tailor would have to work around the clock.

June 30, 1960. Officially, Congo had become independent at midnight, but the ceremony at the Palais National would be the actual confirmation. King Baudouin flew in specially from Belgium; after fifty years of Belgian colonial rule, seventy-five years after his predecessor Leopold II had established the Free State, he would hand over the reins to President Kasavubu. And Jamais Kolonga, the reporter, would be at that historic event.

The history of the Belgian presence in Central Africa had deeply affected his own family history. By means of study, his father had become one of the colony's most prominent *évolués*, while his grandfather had still been a hunter in his native village. Kolonga knew the stories about him. "When the whites arrived in Bas-Congo he carried their baggage on his head. He wasn't afraid of the white men, but he did what they said. He was polygamous, but when he was baptized he sent away two of his three wives." No single individual life, not even in the depths of the interior, had been left unaffected by the great course of history. It had all gone very quickly.

At a quarter past six there was a briefing from the commissioner general of information. The press kits were prepared. A text had just come in from Prime Minister Patrice Lumumba and could now be handed out to the journalists. Kolonga was shown to his seat, up at the front of the hall. Everything was to proceed in a dignified and orderly fashion, that was emphasized again. There had already been an embarrassing incident yesterday, while the king and Kasavubu were being driven around the city in an American convertible.

As he had in 1955, Baudouin had waved to the people, who turned out in great numbers to wave back at him from the side of the road. But then suddenly, from somewhere in the crowd, a man had wormed his way to the front and grabbed the king's sword. The incident had been filmed and photographed. Baudouin was standing upright in the car, wearing his white ceremonial uniform. Kasavubu was standing to his left, in a black, custom-tailored suit. Baudouin saluted the troops of the Force Publique, who were standing on the left side of the road, holding up a banner with the Belgian tricolor. When he felt something at his right hip, the king did not realize at first what was going on. A man with a high forehead and an oblong face raced away, holding aloft the royal sword, one of the regalia of the Belgian monarchy. More than merely a weapon, it was an object that symbolized the power of the royal household.

The incident provoked loud commentary. "That man wasn't in his right mind," Kolonga said, "he was a *feu-follet*, a misguided, restless soul with a mild form of psychosis. People had always said he was crazy." They had little choice in the matter. Many Europeans considered it an idiotic display, a stupid, sophomoric prank that made a mockery of the change of power, but for many Congolese in the working-class neighborhoods this was no joke. For them, it was pure foolhardiness. To touch and then take away a sacral object belonging to a chief? This man would die that very night, they said. If a mask, an ancestor image, a leopard skin, or a monkey tail already had magic powers, what about the sword of a European king? Among the *évolués*, the rebellious gesture also met with disdain. Victorine Ndjoli, the photo model with a driver's license, said: "We were so embarrassed when some idiot stole King Baudouin's sword. We only heard later on that he was out of his mind."[2]

If only things proceeded calmly today, Kolonga thought. The ceremony had to go without a hitch. But people had such strange expectations about independence. There were many who had buried boxes of pebbles in the hope that they would change to gold after independence. There were also many who believed that the dead would rise again.[3] Some people had even laid clothes on the graves of their ancestors, as a sort of gesture of welcome. The graves of those less well-loved were sometimes covered in corrugated iron sheeting, to prevent them from crawling up out of the ground. Some villagers in the interior locked themselves up for four days in their huts, out of fear of the risen dead. Pregnant women refused to leave their homes.[4]

In the cities, the fever of independence assumed more socialized forms. In Stanleyville, a few native inhabitants built unauthorized huts on land belonging to Europeans. Adherents of the Kitawala religion, who had lived in secrecy for years, moved into the abandoned villas of Belgians, where they performed their rituals and sang their songs by torchlight. In Léopoldville, during the run-up to the great day, a clear rise was seen in the number of thefts and acts of vandalism. Boys laughed in their boss's face and sat on the hood of his car, stubbornly refusing to get off.[5]

Around nine that morning, Kolonga watched as the huge rotunda of the Palais National began to fill with dignitaries. There were members of parliament and senators from Belgium, high-ranking officers and civilians. There were delegations from friendly African nations; Prince Hassan of Morocco was there, beside President Fulbert Youlou of Congo-Brazzaville and King

Kigeri of Rwanda. But above all were the newly elected members of the Congolese parliament and senate. The Palais National, built only a few years earlier as residence of the governor general (at the time, people had thought that position would remain intact for decades to come) had now become the new house of parliament. Most of the guests seated beneath the big dome were dressed in dark, Western-style suits, but others wore traditional head-dresses decked out with seashells, feathers, and skins, headdresses every bit as impressive as the white pith helmet with vulture feathers worn by the governor general.

When everyone was seated, Prime Minister Lumumba came in. A few moments later, the audience rose to its feet to greet King Baudouin and President Kasavubu. Baudouin was the first to address the auditorium. The charming king gave a speech that seemed more like something writ-ten in 1900 than in 1960. He praised the work of Leopold II as though no investigative committee had ever condemned his predecessor's regime: "The independence of Congo constitutes the completion of the work that arose from the genius of King Leopold II, that was undertaken by him with undaunted courage and set forth by the determination of Belgium." Nor did the young king eschew a certain paternalism: "It is now up to you, gentlemen, to show that we were right to have confidence in you. . . . Your task is immense, and you are the first to realize that. . . . Do not hesitate to turn to us, if need be. We are prepared to stay by your side and to assist you with our counsel."[6]

When he was finished, the audience applauded politely. At that moment, thousands of people glued to their transistor radios in the villages and working-class neighborhoods heard the clear voice of Kolonga, announc-ing in French, Lingala, and Kikongo: "Ladies and gentlemen, you have just heard the speech given by His Majesty the King of Belgium. As from this moment on, Congo is independent."[7] He, that little Kolonga with the twin-kle in his eye, was the first Congolese to call his country independent.

After that came President Kasavubu, the man Kolonga had seen so often in his parents' living room, in enthusiastic conversation with his father, the man who had leveled blistering accusations at the colonizer dur-ing his first mayoral speech. This time, however, his address was restrained and conciliatory. Little wonder, really: the text was written by Jean Cordy, the Belgian who had once been Governor General Cornelis's private sec-retary. "I wrote Kasavubu's text, or at least the initial version of it. I had also written the text for him when he became president."[8] According to the protocol, the part of the day's ceremonies dedicated to speeches had now come to an end.

But they had overlooked something.

Throughout the president's speech, Lumumba had been busily making corrections. He had a pile of paper balanced on his knees and was scribbling comments here and there. Lumumba had seen Kasavubu's mild-mannered speech days before the new president gave it, and felt that he couldn't let things go at this. He was bound and determined to talk back to the colonizer one last time. Doing that would also put him back in the limelight, for it disturbed him greatly to see that it was not he, but Kasavubu, doing the honors. As the big winner of the election, he could only watch powerlessly as his archrival Kasavubu, the regionalist who did not even carry Congo in his heart, stood there showing off beside King Baudouin.[9] Lumumba had written his speech the night before: he was still able to get by with only a few hours' sleep. Rumor had it that his Belgian adviser and faithful supporter Jean Van Lierde had worked on the text as well. Today it is seen as one of the great speeches of the twentieth century and a key text from the decolonization of Africa:

> For if today Congo's independence is being announced in agreement with Belgium, a friendly nation with which we operate on an equal basis, then still no Congolese worthy of the name can ever forget that this independence was gained by struggle, a daily struggle, a fiery and idealistic struggle, a struggle in which we spared neither our efforts nor our hardships, neither our suffering nor our blood.
>
> That struggle, which was one of tears, fire and blood, fills every fiber of our being with pride, for it was a noble and a just struggle, an inevitable struggle to end the humiliating slavery that had been imposed on us by force.
>
> The fate that befell us during eighty years of colonial rule is not something we can eradicate from our memory, our wounds are still too fresh and too painful. We have known grueling labor, demanded from us in return for wages that did not allow us to eat decently, to clothe ourselves or have housing, nor to raise our children as loved ones.
>
> We have known mockery and insult, blows that we underwent in the morning, in the afternoon and evening, because we were Negroes.

Who can ever forget that a black man was addressed as *tu,* not out of friendship, but because the honorable *vous* was reserved only for whites?

We have seen our raw materials stolen in the name of documents that were called legal, but which recognized only the right of the most powerful.

We have seen that the law was never equal when it came to black and white: accommodating for the one, cruel and inhuman for the other.

We have seen the terrible suffering of those exiled for reasons of their political convictions or religious beliefs, banished in their own country; their fate was worse than death itself.

In the cities we have seen magnificent houses built for the whites, and hovels for the black, that a black was not allowed into the so-called "white" movie theaters, restaurants and shops, that a black man traveled in the hold of the riverboats, beneath the feet of the white man in his luxury cabin.[10]

It was, indeed, a memorable text. Like all great speeches, it clarified the abstract course of history with the use of a few concrete details, and he illustrated the great injustice with a host of tangible ones. But Lumumba's timing was highly unfortunate. This was the day on which Congo won its independence, but he spoke as though the elections were still in full swing. Too focused on attaining immortality, too blinded by the romanticism of Pan-Africanism, he who was, after all, the great advocate of unity in Congo forgot that on this first day of autonomy he should be leading his country to reconciliation rather than divisiveness. He professed to be the voice of the people—that fit with the exalted rhetoric of the day (the People, the Yoke, the Struggle, and, of course: the Liberation)—but the people did not stand unanimously behind him. After all, he had won a little less than a third of the votes. Lumumba's speech was therefore a great one in terms of import, but a problematical one in terms of its effect. And compared with the truly grand speeches of history—Abraham Lincoln's Gettysburg Address from 1863 ("a government of the people, by the people, for the people, shall not perish from the earth"), Winston Churchill's first speech as prime minister on May 13, 1940 ("I have nothing to offer but blood, toil, tears, and sweat"), the speech given by Martin Luther King in 1963 ("I have a dream"), the words with which Nelson Mandela lectured the judges on democracy in 1964 ("It is an ideal which I hope to live for and to achieve. But if needs be, it is an ideal for which I am prepared to die"), or the acceptance speech with which Barack Obama thrilled the world in 2008 ("Change has come to America")—Lumumba's address contained more of a look back than a look forward, more rage than hope, more rancor than magnanimity, and therefore more rebellion than statesmanship.

JAMAIS KOLONGA WITNESSED IT ALL from the front row. He heard how the Lumumba supporters in the audience interrupted the speech eight times with a hail of applause, but he also saw the "chilly looks of the invitees and the king's paleness." He saw Baudouin lean over to Kasavubu to demand an explanation, but Kasavubu didn't move a muscle: neither of them had been informed of Lumumba's initiative. His text had been handed out to the journalists under embargo, but neither the king nor the president had seen it. Afterward, Baudouin was furious and deeply offended. For him it must have been a painful replay of his own coronation. Then, at the height of the ceremony ten years earlier, the Communist senator Julien Lahaut had shouted out "Vive la république!" That too was intended to be a festive day, a confirmation of his royal dignity, but then too the ceremony had been ruined by a leftist firebrand who had butted in and claimed all the attention. One week later Lahaut had been mowed down in his doorway by a group of unknown assailants, under circumstances as vague and violent as the fate that awaited Lumumba six months later.

Baudouin wanted to return to Belgium immediately. He no longer had any desire to visit the Pioneer Cemetery or the equestrian statue of Leopold II. But Belgian Prime Minister Gaston Eyskens intervened and during lunch demanded that Lumumba give a second, more friendly speech. And so it came about: Eyskens wrote the text, Lumumba read it aloud drily, Baudouin stayed to the end of the day.

It would be a mistake to assume that all of Congo rejoiced at its prime minister's daring words. Fourteen million people rarely share the same opinion. Kolonga, in any case, found it troubling: "Lumumba was no diplomat, he was far too categorical. Kasavubu, now that was a gentleman. He wanted some of the whites to stay on as deputy director in the provinces, for agriculture, for finance. But our constitution gave too much power to the prime minister. It also made our president after the image of the Belgian king: he ruled, but did not govern." As a native of Bas-Congo, he felt more sympathy for Kasavubu. For many Bakongo, Lumumba was no hero. "Kasavubu was calm, cultivated, and respectful," old people in Bas-Congo say even today. "Lumumba was empty-headed, affected, and rude. He was the source of our problems. The way he talked to the king, that was irresponsible! He should have said: 'You people are independent now, so come on, get to it!' instead of pointing out the minor problems of the past."[11] Almost all older citizens in Boma, Matadi, and Mbanza-Ngungu (former Thysville) can still get wound up about this. "It all started then. Lumumba's speech angered the Belgians. The king didn't even want to stay for the banquet. Kasavubu didn't want to chase away the Belgians, but Lumumba wanted to wipe the slate clean. It was a very bad start. And I say that truthfully, not just for ethnic reasons."[12]

Even Lumumba's fervent supporters had their misgivings. Mario Cardoso, who came from Stanleyville, Lumumba's home town, and who had been his personal representative during the economic round table in Brussels, told me: "I was in the audience and I was struck dumb. Lumumba acted like a demagogue. I was a member of the MNC, but our campaign hadn't been about what he was saying. Some of the deputies applauded, but I didn't. He's committing political suicide, that's what I thought."[13]

In other parts of Congo, however, the incident received little attention. In Elisabethville, the day was calm and festive. Moïse Tshombe, who'd had to be satisfied with the position of provincial governor, reemphasized the importance of warm and friendly ties between Belgium and Congo. During the independence day celebrations in the mining town, a children's choir sang a few hymns. Colonials, who still had to get used to the fact that they were suddenly ex-colonials now, joined the party in the native districts and were welcomed.[14] Elsewhere in the country, too, mass was celebrated, cantatas were sung, and tribute was paid. The news about Lumumba's speech was heard only later. Very few people disagreed with what he said, but many wondered whether it had really been necessary. One capital city inhabitant said: "A birth is always accompanied by painful contractions. That's the way it goes. But once the child is born, it is smiled upon."[15]

And so went the first day of a liberated Congo. There were parades and games, folk dancing and fireworks. The party was to last for four days, from Thursday to Sunday. Congo began its existence with a long, free weekend. There were sports contests at Stade Baudouin (Kasavubu was supposed to hand the trophy to the winners, but Lumumba grabbed it away from him and did it himself).[16] There was a bicycle race through the streets of the city (the most Belgian of all sports, but the first three places went to Congolese cyclists). And above all there was beer, lots of beer, a great deal of beer. It was the end of the month and everyone had just been paid. The walls of the bars were lined with crates. After a few days the new government ordered that all points of sale for alcohol be closed between six in the evening and seven in the morning. The partying got a bit out of hand, but it was innocent enough. There was some rioting in Kasai, but there were no attacks on Belgians, no lynchings, no raping, no looting of European homes.

But on that first day of independence, there was one man who—by his own account—laying groaning in pain on the floor of a prison cell: Longin Ngwadi! The man from Kikwit, the believer who had wanted to become a priest but was not allowed, Élastique, the star player for Daring Club, the former houseboy to Governor General Léon Pétillon, the man who had traveled to Belgium *not* to see the Expo; he, of all people, was the new state's first dissident. "My belly was swollen like a balloon. I was bleeding from my nose and anus. I peed blood, I passed terrible gusts of wind. I was handcuffed, as though I had stolen something." At four in the morning, while Kolonga was busy gussying up for the big day and Lumumba was still working on his speech, Longin had already been lying there for hours, bemoaning his fate. The day before he had been arrested by the provincial governor, Jean-Baptiste Bomans. "They came to get me with two jeeps full of soldiers. 'You're insane,' Bomans told me. 'I'm not insane,' I said, 'I'm normal. King Baudouin is my brother. Do whatever you like, I am a prophet, like Elijah or Jeremiah.'"

After months of searching in 2008, when I finally found Ngwadi in Kikwit, he was washing himself in the river. To welcome me, he put on his most cherished piece of clothing: a shirt with a leopard-skin pattern to which he had pinned a photograph of Lumumba standing beside Antoine Gizenga. Gizenga was his big political hero, a man from his region who had been deputy prime minister under Lumumba and who, at the moment we met, was sitting out his final days as prime minister under Joseph Kabila. Papa Longin Ngwadi was one of the most colorful Congolese I had ever met, and not just because of his amazing life story. Even his gaudery was breathtaking. Around his neck, on that first day we met, he wore a big crucifix, alongside a medal of St. Theresa with the Infant Jesus, a medal of the archangel Michael, a blue cross of Lourdes, an old ICSA door key bearing the stamp "made in Italy," which he described as "the key to heaven," a hammer that was his allusion to the name "Jean Marteau, the nickname for Kamitatu," that other great politician from his region, and a whistle, "because when I have a vision, I call everyone together to pass on the message."

Ngwadi's fantasy knew no bounds. He claimed that he was the man behind that stunt half a century earlier: "Yes, I am the one who took Baudouin's sword." For a long time I thought that he was telling the truth. His high, prominent forehead and oval eyes, after all, strikingly resembled those of the man in the famous picture. But meanwhile we know that there are many stories in circulation about that incident. Any number of elderly Congolese claim to know who stole the sword, and why, while the actual culprit died long ago. Those stories, even if they are usually only that, form a rich source of information about the memories of decolonization. "Baudouin was an icon," Ngwadi said, "a *chouchou*; he was straightforward, very young and very handsome."

After Ngwadi returned from his Belgian adventure and Pétillon was no longer governor general, he too became caught up in the fever of emancipation. He had an eye for its mystic dimensions in particular. He wandered the streets of Léopoldville and went each day to the Église Saint-Pierre, in the borough of Limete. Monseigneur Joseph-Albert Malula celebrated mass there. Malula, an extremely intelligent man who had witnessed the struggle for independence from close at hand and had even been involved in the manifesto issued by *Conscience Africaine*, was enthroned as bishop in 1959. Later he would become the first cardinal from Congo and a direct opponent of Mobutu.

"I went to his church every day. When I prayed, everything became light. I had the power of the spirit and the vision of history. All the prayers came as though I'd known them beforehand; I sang all kinds of new hymns, I broke through all the secrets, I saw flowers, lots of flowers. *Tiens*, I said, so God has given me peace. I went and told Malula about that. He gave me a ballpoint pen and a little notebook and asked me to write down my visions."

Today, Ngwadi is still a deeply religious man. His whole life is saturated with spirituality. He prays constantly, never fails to start a conversation by blessing his visitors with hairspray or perfume, and raises his hands to heaven to ask for protection. For him, religion and politics are joined at the hip. One day, still woozy from the cloud of cheap women's perfume, I walked with him along the street market in Kikwit, a long ribbon of merchandise that extends along the main street of the lower city all the way to the bridge over the Kwilu. Every five minutes he would stop, blow his whistle, and shout in Kikongo to anyone who would hear: "Children of Kikwit, if you still don't believe in my powers, look at this visitor. I asked Gizenga to send me a white man, and here he is!" Half an hour later, his son had to ask him to edit this particular vision, because not everyone was an adherent of Gizenga's and that could compromise my safety. On the market, just before the bridge, was a sinister stand selling fetishes, herbs, masks, and monkey skulls. No one stopped there. "Don't look at it," his son said to me, "that brings bad luck." But Ngwadi examined the wares attentively, obviously feeling more powerful than all this sorcery. At home he had a magic sword he'd made himself. He had decorated an old umbrella stick with artificial flowers, bits of copper wire, a picture of Christ with flowers, and a banner bearing the acronym of the Palu, the Parti Lumumbiste Unifié, Gizenga's current party. The reference to the magic sword of half a century earlier was loud and clear. In his "junk art," memory and mysticism mingled effortlessly.

I found a good spot and waited for Baudouin at the station, close to the railroad workers' monument. Everyone wanted to see him, he was a handsome boy, but there were soldiers with rifles everywhere. It was impossible, but my power allowed me to slip past. I wanted to give the king some flowers, to show my love for him, but then I saw that long, shiny sword and I took it *pour la folie*, just for fun. I got five meters away, then I heard the soldiers loading their weapons. King Baudouin said: "No guns." I walked back to him and said: "I wish you a fine visit to Congo. The Lord was the one who urged me to take your sword. We will travel to parliament together as good acquaintances. It is time that we become independent. The European women are like the Virgin Mary, but later the good Lord will grant us the peace to be able to marry white women. Belgium is far away, as far as heaven, a common property where there must be black people as well. A common market. The blacks will go to Belgium. I am not insane, I am normal. I give you your sword back." Baudouin replied: "No one may hit you! I am going to give you a gift. Don't forget me. It's true, later you will marry a white woman, on condition that you learn French." But he left the same day, without giving me a gift. He never kept his promise.

Whether this remarkable conversation actually took place is very much in question. In it, mysticism and eroticism collide ingeniously with European current events (the common market!) and Belgian linguistic rivalry (learning French!). But that a man with a rather idiosyncratic way of thinking remembered independence, fifty years after the festivities, as a promise never kept says a great deal in itself. Today, through the fissures in his bizarre fable, there shines the light of a profound truth: independence should have been a gift, but it remained an empty promise.

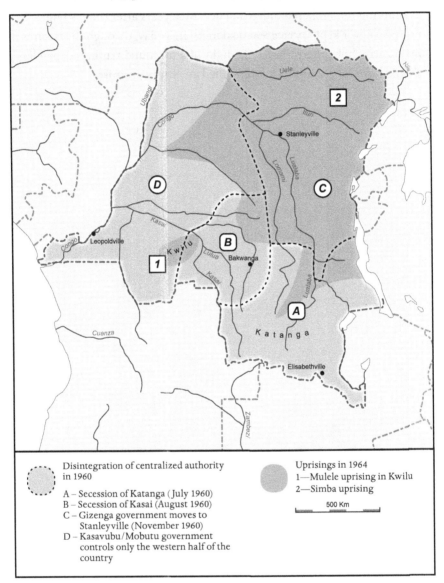

Disintegration of centralized authority
in 1960

A – Secession of Katanga (July 1960)
B – Secession of Kasai (August 1960)
C – Gizenga government moves to
   Stanleyville (November 1960)
D – Kasavubu/Mobutu government
   controls only the western half of the
   country

Uprisings in 1964
1—Mulele uprising in Kwilu
2—Simba uprising

500 Km

CHAPTER 8

THE STRUGGLE FOR THE THRONE

The Turbulent Years of the First Republic

1960–1965

E VERYONE KNEW THERE WOULD HAVE TO BE A GOOD DEAL OF improvising during that first period after independence. That things would not run smoothly as burnished silk was only to be expected. But that Congo, during the first six months of its existence, would have to deal with a serious military mutiny, the massive exodus of those Belgians who had remained behind, an invasion by the Belgian army, a military intervention by the United Nations, logistical support from the Soviet Union, an extremely heated stretch of the Cold War, an unparalleled constitutional crisis, two secessions that covered a third of its territory, and, to top it all off, the imprisonment, escape, arrest, torture, and murder of its prime minister: no, absolutely no one had seen that coming.

And it would take a long time for things to get better. The period between 1960 and 1965 is known today as the First Republic, but at the time it seemed more like the Last Judgment. The country fell apart, was confronted with a civil war, ethnic pogroms, two coups d'état, three uprisings, and six government leaders (Patrice Lumumba, Joseph Ileo, Justin Bomboko, Cyrille Adoula, Moïse Tshombe, and Évariste Kimba), two—or perhaps even three—of whom were murdered: Lumumba, shot dead in 1961; Kimba, hanged in 1966; Tshombe, found dead in his cell in Algeria in 1969. Even Dag Hammarskjöld, the secretary general of the United Nations, the man who headed a reluctant world government, lost his life under circumstances that still remain unclear—an event unparalleled in the history of postwar multilateralism. The death toll among the Congolese population itself during this period was too high for meaningful estimates.

Congo's First Republic was an apocalyptic era in which everything that could go wrong did go wrong. Both politically and militarily, the country was plunged into total, inextricable chaos; at the economic level, the picture was clearer: things simply went from bad to worse. Yet Congo had not fallen prey to wild irrationality. The misery of the first five years was not the product of a renaissance of barbarism, of the revival of some form of primitivism repressed during the colonial years, let alone of any opaque "Bantu soul." No, here too the chaos was a result more of logic than of unreasonableness, or rather of the collision of disparate logics. The president, the prime minister, the army, the rebels, the Belgians, the United Nations, the Russians, the Americans: each of them wielded a form of logic that seemed consistent and cogent within the confines of their own four walls, but which often proved irreconcilable with the outside world. As in theater, tragedy in history here was not a matter of the reasonable versus the unreasonable, of good versus bad, but of people whose lives crossed and who—each and every one of them—considered themselves good and reasonable. Idealists faced off with idealists, but when believed in fanatically all forms of idealism lead to blindness, the blindness of the good. History is a gruesome meal prepared from the best of ingredients.

The turbulent first five years of Congo can be divided into three phases. The first ran from June 30, 1960, to January 17, 1961, the day on which Lumumba was murdered. During the first six months, the house of cards of the colonial state collapsed, and "the Congo crisis" dominated world news week after week. The second phase coincided with the years 1961–63, and was marked primarily by the Katangan secession. It ended when the rebel province, after forceful UN military intervention, rejoined the rest of the country. The third phase started in 1964, when a rebellion broke out in the east and spread across half the country. The central authorities regained control of the territory only with the greatest of difficulty. The year 1965 was to have witnessed a return to normalcy, but ended unexpectedly on November 24 with Joseph-Désiré Mobutu's coup d'état, a putsch that defined the country's history. Mobutu remained in power for the next thirty-two years, until 1997. That was the so-called Second Republic, a regime that, strictly centralized at first, ultimately developed into a dictatorship.

The First Republic was characterized by a jumble of the names of Congolese politicians and military men, European advisers, UN personnel, white mercenaries, and native rebels. Four of those names, however, dominated the field: Joseph Kasavubu, Lumumba, Tshombe, and Mobutu. In terms of complexity and intensity, the ensuing power struggle between them was like one of Shakespeare's history plays. The history of the First Republic is the story of a relentless knockout race between four men who were asked to play the game of democracy for the first time. An impossible mission, all the more so when one considers that each of them was hemmed in by foreign players with interests to protect. Kasavubu and Mobutu were being courted by the CIA, Tshombe at moments was the plaything of his Belgian advisers, and Lumumba was under enormous pressure from the United States, the Soviet Union, and the United Nations. The power struggle among the four politicians was greatly amplified and complicated by the ideological tug of war taking place within the international community. It is hard to serve democracy when powerful players are constantly, and often frantically, pulling on the strings from above.

What's more, none of these men had ever lived under a democracy in their own country. The Belgian Congo had had no parliament, no culture of institutionalized opposition, of deliberation, of searching for consensus, of learning to live with compromise. All decisions had come from Brussels. The colonial regime itself was an executive administration. Differences of opinion were kept hidden from the native population, for they could only undermine the colonizer's prestige. In his seemingly unassailable omnipotence, the highest authority, the governor general with his white helmet decked out with vulture feathers, seemed more like the chieftain of a feudal African kingdom than a top official within a democratic regime. Is it any wonder then that this first generation of Congolese politicians had to struggle with democratic principles? Is it strange that they acted more like pretenders to the throne, constantly at each other's throats, than like elected officials? Among the historical kingdoms of the savanna, succession to the throne had always been marked by a grim power struggle. In 1960 things were no different.

And in fact, wasn't it all about who was going to take over from King Baudouin? Kasavubu was the first and only president of the First Republic. The dress uniform he had designed for himself was an exact copy of Baudouin's. Léopoldville and Bas-Congo supported him en masse. Only rarely was his position as head of state openly called into question, but in 1965, Mobutu—whose own ceremonial uniform later proved to be a copy of Baudouin's as well—shoved him aside.

Lumumba's power base lay to the east, with Stanleyville as its center. He was the most popular politician in Congo, but he resented having to bow to Kasavubu as president. He would only survive the first six months of the First Republic, but after his death his intellectual legacy continued to play a major role in Congolese politics.

Tshombe was perhaps even more resentful. His party had received the short end of the stick during the formation of the new government. He himself had no choice but to settle for the position of provincial governor general of Katanga in Elisabethville. And even though that position—in terms of square kilometers and industrial importance—was comparable in weight to that of a German chancellor in a united Europe, he had to face the fact that the true center of power lay elsewhere, in Léopoldville.

On the day of independence itself, Mobutu, finally, was the least significant of the four: he was Lumumba's private secretary. He had no major city backing him, as the other three did, let alone a powerful people like Kasavubu (among the Bakongo) or Tshombe (among the Lunda). He came from a small tribe in the far north of Équateur, the Ngbandi, a peripheral population group that did not even speak a Bantu language like the rest of Congo. At twenty-nine he was also the youngest of the group (Kasavubu was forty-five, Tshome forty, Lumumba thirty-five). But five years later he was lord and master. He would develop into one of the most influential persons in Central Africa and one of the richest men in the world. The classic story of the errand boy who becomes a Mafia kingpin.

DURING THE FIRST ACT OF CONGO'S INDEPENDENCE, Patrice Lumumba was incontestably the pivotal character. All eyes were turned on him after his inflammatory speech during the transfer ceremony. When the curtain went up on the Congolese drama, he was a dynamic people's tribune, adored by tens of thousands of common folk. Only a few scenes later he was despised, spit upon, and forced to eat a copy of his own speech.

July 1960. The dry season. A cobalt blue sky. The independence party had lasted four days. The army, the Force Publique, kept order as always. The newly independent Congo may still have been sailing against the current— the political institutions may still have been in their infancy, governmental experience may have been null, the challenges were perhaps enormous—but the armed forces were solid as a rock. The officers' corps was still Belgian: a thousand Europeans maintained command over twenty-five thousand Congolese. The chief commander was still General Émile Janssens, the man who had so rigorously quashed the 1959 riots. Without a doubt the most Prussian of all Belgian officers, he was a great soldier with a rigid mind: discipline was sacred to him, protest was a defect, chaos the sign of weak character. He had to put up with being answerable to Lumumba, who was not only prime minister but had also been made minister of national defense. Concerning him, Janssens would later write: "Moral character: none; intellectual character: entirely superficial; physical character: his nervous system made him seem more feline than human."[1] That was how things lay: Congo was independent, true enough, but the Belgians not only ran things economically, they also maintained a total grip on the military apparatus.

Fireworks had graced the night sky on Thursday, June 30, but by Monday, July 4, things were already awry. Congo's existence as a stable country lasted only a few days. During the afternoon parade at the Leopold II barracks, a few soldiers refused to obey orders. General Janssens intervened and did what he had always done in such cases: demoted the recalcitrant elements on the spot. This time, however, that move backfired. The next day some five hundred soldiers gathered in the mess hall to express their dissatisfaction. The soldiers were tired. For the last eighteen months they had been zigzagging across the country, putting down minor insurrections. They yearned for opportunities for advancement within the military hierarchy, for better pay and less racism. Shortly before independence, they had written:

> No one has forgotten that within the Force Publique we, the soldiers, are treated like slaves. We are punished arbitrarily, because we are Negroes. We have no right to the same advantages or facilities as our officers. Our two-person rooms are extremely small (7.5 m² [about 79 square feet] of floor space) and have no furnishings or electricity. We eat very little and our food in no way complies with the rules of hygiene. The wages we are given are insufficient to meet the current cost of living. We are not allowed to read newspapers published by blacks. One need only be caught with a copy of *Présence Congolaise, Emancipation, Notre Congo* . . . to receive two weeks in the brig. After this unjust punishment one is then transferred to the disciplinary camp at Lokandu, where one is taught to live in military fashion. . . . In the Force Publique our officers live like Americans; they have better housing, they

live in big, modern houses, all furnished by the Force Publique, their standard of living is very high, they are arrogant and live like princes; all this in the name of prestige, because they are white. Today it is the unanimous desire of all Congolese soldiers to have access to positions of command, to receive a respectable salary and to put an end to every form of discrimination within the Force Publique.[2]

Radical military reforms were needed to counter so much frustration, but General Janssens had no intention of countenancing them during the tumultuous months before and after independence. The first batch of Congolese officers was in training at the Royal Military School in Brussels, and a school for noncoms had been set up at Luluabourg. Within a few years those men would be on active duty, but until then everything would remain the same. On Tuesday morning, July 5, Janssens went to the Leopold II barracks and gave his troops an unambiguous lesson in military discipline: the Force Publique was there to serve the country, that's how it had been in the days of the Belgian Congo and that's how it would be now. To underscore his message, he wrote in big letters on the chalkboard: *"Avant l'indépendance = après l'indépendance"* (before independence is the same as after independence). That was not a good idea. The slogan stuck in the soldiers' craws. They had watched as Congolese civil servants, from one day to the next, were assigned top administrative positions, they had seen how well the politicians did by themselves during the big turnover. One of the new parliament's first acts, after all, had been to decide that they had a right to a 500,000 franc honorarium, almost twice the amount earned by their Belgian colleagues.[3] The soldiers awoke with a start to the fact that independence was doing them very little good.

The mutiny within the army is often explained by referring to Lumumba's inflammatory speech. But that remains questionable: the soldiers were as angry with their own fresh-faced politicians as they were with their white superiors. They wanted to vent their rage not only on General Janssens, but also on Lumumba himself! To them he was not so much a hero as a defense minister who had never served, an intellectual in a dress suit and bow tie who was out to cut a dashing figure while their fate remained unchanged, despite all his glorious promises.[4]

That very same day, July 5, the mutiny jumped the gap to the garrison town of Thysville, barely a two-hour drive from the capital. Things there took a much more violent turn. Hundreds of soldiers rose in revolt. They beat up their officers and forced them, with their wives and children, to take refuge in the mess hall. Meanwhile, the soldiers occupied the munitions dump. Outside the barracks, along the road to the capital, heavy rioting was seen in the Madimba-Inkisi district. This time the soldiers did not turn on their white officers, but on white civilians. A number of European women were subjected to sexual violence. One of them was raped sixteen times within a five-hour period, in the presence of her husband, mother, and children.[5] The rumors reached the capital only a few days later.

Meanwhile, Lumumba did all he could to stop the mutiny in his army. He took three successive measures, each with the best of intentions, but also with consequences far beyond what he could oversee. On July 6, in the company of General Janssens, he inspected the troops at the Leopold II barracks. On that occasion he promised to promote each soldier in rank. "The private second class will become a private first class, the private first class will become a corporal, the corporal will be a sergeant, the sergeant will become a sergeant first class, the sergeant first class will be sergeant-major, and the first sergeant-major will become adjutant."[6] It did not have the desired effect. "Lokuta!" the soldiers shouted, "lies!"[7] They weren't about to be appeased that easily. For them, it was all about the officers' corps.

Two days later, Lumumba took things a step further. He dismissed General Janssens and appointed Victor Lundula to replace him as chief commander of the armed forces, with Mobutu as his chief of staff. The Africanization of the army top brass, that should boost the troop's morale, shouldn't it? Then he moved on without hesitation to his third measure: the accelerated and drastic Africanization of the officers' corps. The soldiers were allowed to nominate their own candidates. In this way, at one fell swoop, sergeants and adjutants became majors or colonels. And to emphasize this break with the past, the Force Publique was now given a new name: the Armée Nationale Congolaise (ANC).

These decisions did help to calm things down a bit, but the final result was disastrous: after only one week, the newborn Republic of Congo no longer had a functional army. The new state's most solid pillar had toppled. In today's demilitarized Europe, where the NATO invisibly safeguards its members, it is hard to imagine the importance of a standing army for a nascent state. The state can only become a state when it assumes the monopoly on violence (be that social, tribal, or territorial). In the turbulent Congo of the 1960s, the army was vitally important. But the Force Publique, the colonial army that could look back on crucial victories in the first and second world wars, was reduced in the space of one week to an unruly mob. The supreme command was now in the hands of two reservists: Lundula, the mayor of Jadotville, who had served as a sergeant-medic fifteen years earlier, and Mobutu, a journalist who had worked for a spell as a sergeant-bookkeeper and had recently become Lumumba's confidant. Once the two men had driven together through the streets of Léopoldville on a scooter, now they were the prime minister and chief of staff of a vast country with a ragtag army. That Mobutu might also be the confidant of the Belgian and American intelligence services was a suspicion Lumumba refused to entertain. It was a refusal that would soon cost him his life.

Lumumba's attempts to mollify the mutineers remind one of Belgium's attempts to pacify social unrest in the 1950s: confronted with a rebellious element in society, he also made too-hasty decisions that consisted of important concessions meant to buy social stability. But once again, the result was the exact opposite of what was intended. The resentment was not dammed, but actually continued to spread.

"OUR WOMEN ARE BEING RAPED!" The rumor spread like wildfire through Congo's European community. On July 7 a train full of Belgians who had escaped Thysville arrived in the capital. For many, their stories went beyond even the worst nightmare scenarios. Some of them had been spit upon, humiliated, and jeered at; many of them felt threatened. But it was the rumor of sexual violence that caused the most panic. In colonial society there was no greater gap than that between the African man and the European woman (the reverse, contact between a European man and an African woman, was a matter of course). Jamais Kolonga had become a national celebrity by dancing with a white woman. Longin Ngwadi had told King Baudouin that he wanted to marry a European. Before *30 juin,* naive souls had believed that they could buy a Belgian home *and* a Belgian wife. The white woman was inaccessible, and it was for that very reason that she generated such intense curiosity. In the late 1950s a Belgian colonial was privy to a humorous, yet telling, incident:

> The post office at Katana had a native postmaster. One day the postmaster came to me and said: "Sir, they have cheated me." And I replied:

"Tell me what you mean." "Well, sir (all this was said in Swahili), look here, I have a catalogue from the Au Bon Marché in Brussels and look at this picture here. (The picture showed a lovely girl with a beautiful bra.) I ordered it, and do you know what they sent me? An empty bra." Our postmaster told me later that he had thought he would get the girl along with it; the price was much more reasonable than that for the dowry of a native woman.[8]

White females in colonial Congo were almost always married women or nuns. Their sexual availability was negligible. Sexual violence after independence was a brutal way to nevertheless claim the most unattainable element in colonial society and to deeply humiliate the former rulers. Clichés abounded on both sides: if the white woman was a semimythical being for many Congolese men, then many Europeans still had semimythical conceptions of African sexuality. The clichés influenced the events. The rapes were hideous, but their frequency stood in no proportion to the panic they caused among the Europeans. Everyone was goading everyone else with horror stories.

Not a single European had been killed, but the result was a large-scale exodus. An estimated thirty thousand Belgians left the country within a few weeks.[9] Between Léopoldville and the Beach, cars were backed up for kilometers to catch the ferry to Brazzaville. Lots of Volkswagen beetles, lots of pickups, lots of Mercedes with with the CB sticker (for *Congo belge*) still on their bumpers . . . Elsewhere the cars were simply left behind. Before independence, Brussels had asked as many Belgians as possible to remain at their posts in the colony—young Congo would be badly in need of their expertise—but two weeks later Belgium was advising its citizens to return home, or at least to bring their wives and children to safety. Sabena organized an airlift that within three weeks took tens of thousands of Europeans out of Congo. It was a hallucinatory withdrawal. Some ten thousand civil servants, thirteen thousand private-sector workers, and eight thousand colonists (plantation owners) left the country.

We know today that this mass psychosis bore no relation to the actual danger. It was like a movie theater emptying out after someone has shouted "Fire! Fire!," while in reality the flames are limited to an overfull ashtray. "See what I mean, look at that fire!" the moviegoers shout on their way to the exit, apparently not realizing that the fire is being fed precisely by the draft they are creating themselves. The situation was serious, without a doubt, but there was no reason for a general evacuation. But reason had gone out the window. At a certain point, every wave of panic achieves an energy that can no longer be tempered. Just as the barracks at Luluabourg had been vacated in 1944 due to irrational fear of a vaccination campaign, so too did the European inhabitants of Congo leave the country due to a misjudged security risk.

But there were also those who kept a cool head. In the village of Nsioni in Bas-Congo, I spent a few days in 2008 with the old physician Jacques Courtejoie. As a child in Stavelot (in the Belgian province of Liège), he had witnessed the Ardennes Offensive (the Battle of the Bulge) in 1944 as it passed within three hundred meters (about four hundred yards) of his parents' home. A lesson in level-headedness. He had lived in Congo since 1958, always on his own, always unmarried, as a missionary of science, a one-man repository of humanism, dedication, and optimism. He had educated and trained half a dozen people from that same area; he gave them responsibility and self-assurance. The booklets and posters with medical information they made together were distributed all over Congo: books about tapeworms, eye disease, and domestic rabbit breeding; posters with information about washing one's hands, tuberculosis, and breastfeeding. Rarely had I seen a man serve the cause of human dignity so straightforwardly and under such difficult circumstances. An unsung Dr. Albert Schweitzer. From the very first day of his stay in Congo, Courtejoie had been averse to colonialism. "In July 1960 I heard the reports on the radio. Panic was breaking out everywhere, everyone was running away. I tried to stay calm and rational. I really saw no reason why I should leave." He was one of the few who stayed. After three months of independence, Congo had only 120 physicians.[10]

There was so much irrational fear at the time. For example: two months after independence, I went to dinner at the home of a white regional administrator. He came home late, because he'd been to a political meeting of the Abako. When he got home, his wife said: "I certainly hope you didn't shake Kasavubu's hand!" I can still hear the way she said that. Even by that time, people still thought Africans were dirty! And two months later that man became the president of Congo! That's what the mood was like back then. A black person was never allowed to go along in the car, at most in the back of a pickup, even when the person in question was sick or pregnant. One time I even saw them make the elderly mother of a black priest travel in the back of the truck, even though she was seriously ill. Here in this area, the whites never sat down at the table with black people."[11]

Courtejoie still combats prejudice every day. Whenever he goes out with his staff members, everyone piles into the jeep until there's no more room. During lunch breaks he shares their manioc loaf and eats with them from the same can of sardines.

Many Europeans ran away with the idea that they would come back a few months later, when everything had calmed down. But they did not. Among Belgian ex-colonials, proud as they were of their own achievements, this caused a great deal of bitterness. Many of them sincerely felt that they, as subjects of a little country, had outdone themselves and displayed great dedication and boundless drive. In the 1980s Vladimir Drachoussoff, the agronomist who kept such a fascinating diary during World War II, remembered "the joy of helping to develop a huge country that is today foreign but which we experienced intensely as our own."[12] The colony had offered many opportunities that would have been beyond their reach in Europe, it was their most cherished homeland. Now it had become a foreign country. Thomas Kanza, the first man with a university diploma in Congo and a very young cabinet minister under Lumumba, displayed amazing insight into their state of mind when he wrote: "Almost all of them had achieved more in Africa than they would have in Europe, because the opportunities to take initiatives, to display their skills, their energy, in short to confirm their personality, were greater abroad than they were there."[13] Leaving Congo therefore also meant giving up a dream, a dream of self-fulfillment that, for many of them, went hand in hand with a paternalistic ideology. Drachoussoff, once again, was completely frank about that: "Our paternalism was solid and serene: we were deeply and sincerely convinced that we were not only the bearers of a more modern civilization, but of civilization as such, of the rule and standard for all of the peoples on the earth. . . . Almost all of us were proud of being European and we approached the world around us as builders and designers, with the will to mold and to transform, and the conviction that we had the right to do so." Of course that calm self-confidence also had a dark side, he realized. The sudden hostility between white and black had not appeared out of thin air: "An understandable but dangerous sense of superiority had influenced the daily practice of colonization. . . . The 'civilizers' dearly wanted to protect and to educate, as long as it went from top to bottom and the pupils remained respectful and obedient. None of us escaped completely from that God-given hierarchy that expressed itself among the mediocre in the form of straightforward racism and provided the more magnanimous with a good conscience."[14]

If that exodus was frustrating for the whites, for the young country itself it constituted a second heavy blow. To put it simply: after one week Congo was without an army; after two weeks it was without an administration. Or, to put it more accurately: it was without the top layers of an administration. Of the 4,878 higher-ranking positions, only three were occupied by Congolese in 1959.[15] Suddenly, people with a simple education now had to assume important roles within the bureaucracy, roles that were often far beyond their ability. The army was crucial for maintaining order, the administration for the operations of the state. In Kisangani, I talked about this with the very colorful Papa Rovinscky, the nickname of Désiré Van-Duel, which was in turn also a Belgian-sounding alternative for his true African name: Bonyololo Lokombe. When one's country changes names four times in one's lifetime, why not adapt your own from time to time? Papa Rovinscky welcomed his visitors with music. He played the slit drum and the gong, and was still able to broadcast messages in the language of his tribe, the Lokele, over great distances. "The white man has arrived and is sitting in the easy chair," he drummed out on his bush telegraph, as soon as I had pulled out my ballpoint and notebook. On his living room wall he had hung the hand-written story of his life and his curriculum vitae. He had noted the names of the thirty-five children he had sired by nine different women, "dont 8 cartouches perdues" (including eight near misses). He described himself as an "independent journalist and deacon, a born national and international historian, an external staff member of the communicational class [I have no idea what he meant by that, but it sounded good], peace artist and multidimensional *griot*." But today, at the age of seventy-three, he mostly lived from building coffins, primarily for children, which were in great demand. In Congo, one out of every five children dies before the age of five. Before independence he had worked as a stenographer and typist for the colonial administration. He could touch-type ("My fingers had eyes"), but after independence he was suddenly pitchforked into the job of first municipal secretary of Tshopo. "There were only a few whites, the rest of the city managers were black. None of them were ready for it. The mayor put together a team. Because I could take stenography and type, I became the municipal secretary. I had to take the minutes of the city council meetings. That was very difficult for me! I'd had no training at all!"[16]

The Belgian exodus had major economic consequences too. During the second half of 1960 the export-oriented farming sector suffered a drastic dip. Cotton, coffee, and rubber, ready for harvest, were no longer being exported. The crops stood rotting in the fields. Exports of cacao and palm nuts fell by more than 50 percent.[17] Other sectors highly dependent on European know-how suffered as well: forestry, road construction, transport, and the service sector. Mining was the only industry that remained more or less stable. Unemployment rose sharply. Those who had served as boy, cook, or maid to a white family were suddenly out of work. Tens of thousands of employees on the plantations, at the sugar refineries, soap works, and breweries lost their jobs. In the long run, industrialized agriculture made way for more traditional forms. People once again began raising manioc, shucking corn, and collecting locusts, they once again turned to family when they became hungry. The nuclear family, the *évolué*'s ideal and the object of tireless promotion by the missions, would gradually make way for the extended family, the broad network of uncles, nephews, and nieces on whom one could fall back in times of scarcity.

THE UPRISINGS OF 1960 affected not only the army, the administration, and the economy; they also led to armed conflict. On July 9 the first casualties fell in Elisabethville: five Europeans, including the Italian consul, were murdered. This was the bloody limit, Belgian Defense Minister Arthur Gilson decided that same evening. Going against the advice of Foreign Affairs Minister Pierre Wigny and without informing the Belgian ambassador in Léopoldville beforehand, he gave the green light for military intervention.[18] The lives of countrymen were at stake, he reasoned. Early in the morning of July 10 Belgian planes took off from Kamina airbase with troops for Elisabethville. Paratroopers were dropped over Luluabourg that same day to free the Belgian nationals.

It was, in every way, an ill-fated move.

A few weeks before independence, Belgian soldiers had already been stationed at the Kitona and Kamina military bases. According to the "agreement of friendship" signed by both countries, Belgium was to provide military support for an independent Congo, but only at Léopoldville's express request; that is to say, at the request of Defense Minister Lumumba. That was absolutely not the case here. Brussels hid behind the argument that the intervention was meant only to protect Belgian nationals, yet the liberation of Belgians soon made way for the occupation of large parts of the former colony. Now that the Congolese army was in disarray, Belgium decided to maintain order (and the economy) on its own; what had taken three-quarters of a century to build must not be razed in a month. That was understandable, but foolish. Belgium should have limited itself to protecting its own citizens and then turned to the United Nations to handle the rest. As it was, its self-willed intervention now boiled down to nothing more than the military invasion of a sovereign, independent country. In Katanga, Belgian soldiers forcibly disarmed Congolese troops who had not even been mutinying! Seemingly without much awareness of the fact, the kingdom of Belgium was carrying out an offensive of its own on foreign soil for the first time since 1830.

Kasavubu and Lumumba were inclined, at first, to turn a blind eye to the Belgian actions—there were, after all, Belgians in danger—but one day later reversed their well-disposed stance. Which was entirely justified. On July 11 the real story made itself known—two times, in fact. First, on that day two Belgian naval vessels shelled the port city of Matadi. That had nothing more to do with the protection of Belgian nationals, almost all of whom had been evacuated, but with the taking of a strategic harbor. Second, and vastly more important, on that same day Tshombe declared the independence of Katanga and immediately received Belgian support. At that same time Kasavubu and Lumumba were traveling around the country, dealing in a diplomatic way with uprisings. They were just as concerned about their country's disintegration as Belgium was. In Bas-Congo, individuals including borough mayor Gaston Diomi and Charles Kisolokele, one of Simon Kimbangu's sons, carried out brilliant and courageous work in containing the mutiny. Successful domestic initiatives, therefore, were already being taken. When the president and the prime minister heard about the Katangan secession they flew to the province, but the Belgian commander, Weber, refused them permission to land at Elisabethville. That, of course, created a lot of bad blood: the numbers one and two of the democratically elected government were being denied access to their country's second largest city! By a foreign officer who had entered the city only the day before![19]

Kasavubu and Lumumba inferred right away that Belgium was behind the Katangan secession. An understandable assumption, but not entirely correct. The Belgians and Katangans had long maintained excellent contacts, but it would not be right to claim that Brussels had helped plan the province's secession.[20] In fact, the Belgian government had been unpleasantly surprised by Tshombe's rash deed. On the ground, however, great rapport immediately arose between the Katangan leaders, the Belgian soldiers, and the management of Union Minière. Belgian soldiers disarmed Lumumba's troops and immediately helped to form a new, Katangan army, the Gendarmerie Katangaise. Brussels never formally recognized the Katangan state, but in actual practice Tshombe could count on massive Belgian support. The Belgian national bank even helped to set up the central bank of Katanga.[21] The Belgian court, too, was well disposed toward the rebel province. King Baudouin held Tshombe in much higher esteem than he did Lumumba. He wrote to him: "An eighty-year association, like that which unites our two peoples, is far too fervent and fond a link to allow it to be disbanded by the hateful policies of one single individual." In the definitive draft, that word "hateful" was scrapped. That he was referring to Lumumba was clear enough already.[22]

With its military intervention, Belgium meant to restore order, but the move resulted in total escalation. The history of Congo between 1955 and 1965 is nothing but a series of attempts by various governments to contain unrest; attempts that resulted again and again in even more unrest. But this time the Belgian authorities had added an inordinate amount of fuel to the fire.

In July 1960 four Belgian Harvard fighter planes began patrolling the skies above a restless Bas-Congo, picking out specific targets for strafing and missile attacks. Within six days, one had crashed and another had been shot down. The other two had bullet holes in wings and fuselage.[23] The badly wounded pilot of the plane that was shot down was murdered by Congolese soldiers; his body was thrown into the Inkisi.

Deputy regional administrator André Ryckmans, son of the former governor general, was shot and killed as well. One of the brightest minds of that day's administration, he was a man who felt very much at ease in the villages.[24] Anyone hearing him speak Kikongo would have sworn he was African. His feeling for the Congolese perspective was unparalleled. Old Nkasi remembered him as one of the few truly amiable whites. But when

Ryckmans went to negotiate with the mutineers over the release of a number of white hostages, he was murdered before the eyes of an angry mob. The lynching of one of the most brilliant and empathic minds in the administration by a furious mob can only be seen as an indication of how badly the Belgian military intervention had ruined things.

"Monsieur André, oh yes, I knew him," Camille Mananga said with a smile when I met him in Boma. "He was a real Congolese. He considered himself Congolese too. But they killed him, at the bridge over the Inkisi." I asked what he remembered of the Belgian military operation. Without missing a beat, he replied: "I was in Boma. The Belgian soldiers from the Kitona base had come to disarm the army. The airfield was full of tanks. It was early in the morning, I was on my way to work. I was a government clerk back then, a minor civil servant. The town was full of soldiers. A Belgian stopped me. 'Where are you going?' he asked. 'I work for the regional administration,' I said. 'Go back home,' he said, 'the Belgians have occupied the city.' But I just kept walking, I was too curious to go home. It was the first time in my life that I had seen a tank. I went to take a look. The Belgians didn't stay long, but it was an occupation, nothing more and nothing less."[25]

Peace, in other words, did not return. All over the country, violence against Belgians increased. Civil servants and plantation owners were beaten with clubs, whips, and belts. Some were forced to drink urine or eat spoiled food. Catholic nuns had to undress in public and were tied up. Soldiers asked them why they weren't members of Lumumba's party, and whether they slept with the priests. Others suggested putting a hand grenade in a white woman's vagina. Humiliation was an end in itself. In the period between July 5 and July 14, approximately one hundred European men were assaulted, an equal number of women were raped, and five whites were killed.[26] Belgium had granted Congo independence in order to avoid a colonial war, but got one anyway. And it was its own stupid fault.

THE GOVERNMENT OF THE REPUBLIC OF CONGO REQUESTS UNO ORGANIZA-
TION URGENTLY TO SEND MILITARY ASSISTANCE STOP OUR REQUEST JUSTI-
FIED BY DETACHMENT OF BELGIAN TROOPS FROM MOTHERLAND TO CONGO
IN VIOLATION OF TREATY OF FRIENDSHIP SIGNED BETWEEN BELGIUM AND
REPUBLIC OF CONGO THIS JUNE 29 STOP ACCORDING TO TERMS OF TREATY
BELGIAN TROOPS ONLY TO INTERVENE AT EXPLICIT REQUEST OF CONGO GOV-
ERNMENT STOP THAT REQUEST NEVER FORMULATED BY GOVERNMENT OF RE-

PUBLIC OF CONGO STOP CONSIDER UNSOLICITED BELGIAN OPERATION AS ACT
OF AGGRESSION AGAINST OUR COUNTRY STOP TRUE CAUSE OF MOST UPHEAV-
ALS ARE COLONIAL PROVOCATIONS STOP ACCUSE BELGIAN GOVERNMENT OF
DETAILED PREPARATION OF KATANGAN SECESSION TO RETAIN GRIP ON OUR
COUNTRY STOP GOVERNMENT SUPPORTED BY CONGOLESE PEOPLE REFUSES TO
SUBMIT TO FAIT ACCOMPLI POSED BY CONSPIRACY BY BELGIAN IMPERIALISTS
AND SMALL GROUPS OF KATANGAN LEADERS STOP . . . INSIST EMPHATICALLY
ON EXTREME URGENCY OF SENDING UNO TROOPS TO CONGO FULLSTOP[27]

Signed: Joseph Kasavubu and Patrice Lumumba. With this telegram, the
president and prime minister of Congo called in the support of the United
Nations on July 12, one day after the Katangan secession. At that point the
United Nations was a relatively young organization, with only four short-
lived observer missions to its name during its fifteen-year existence. Its secretary general was Dag Hammarskjöld, the son of a former Swedish prime
minister and a man imbued with a Protestant sense of duty. Kasavubu and
Lumumba had all their hope fixed on the United Nations. Their country had
been a member for less than a week.

That same evening Hammarskjöld called an emergency meeting of the
UN Security Council. In the austere meeting room in New York, the delegates spent the whole night discussing the recent developments in Congo.
The Soviet Union called for total compliance with Kasavubu and Lumumba's
request. The other members agreed to the need for intervention, but were
hesitant to reprimand Belgium. The secretary general felt that an international task force should serve primarily to keep the peace and not so much to
carry out the Congolese government's orders. He also refrained from passing judgment on the Belgian invasion of Congo. Poland and Russia felt that
the Belgians, as aggressors, should leave the country immediately. A little
before 4 A.M. UN Resolution 143 was approved. The Security Council called
on "the Government of Belgium to withdraw its troops from the territory
of the Republic of Congo" and decided to send in peacekeeping forces.[28] The
operation, known under the name ONUC (Opération des Nations Unies au
Congo) was at that point in history the biggest UN mission ever.

But the UN resolution did not please Lumumba. It contained no denun-
ciation of Belgium and the text said nothing about the Katangan secession.
He had expected a much more assertive stance from the Security Council.
He had hoped that the UN "blue helmets" would take over the work of his

hobbled army, that they would drive out the Belgian soldiers and bring about
the reannexation of Katanga. The resolution did not provide for that. It was
like calling for the police during a major riot and having the fire department
show up. Useful, but not enough. That was why he, along with Kasavubu,
asked for help from the country in the Security Council that had shown the
most sympathy for his cause: the Soviet Union. On July 14 Congo severed all
diplomatic ties with Belgium and contacted Moscow:

COULD BE INDUCED TO REQUEST INTERVENTION BY SOVIET UNION IF WEST-
ERN CAMP DOES NOT TERMINATE ACT OF AGGRESSION AGAINST SOVEREIGNTY
REPUBLIC OF CONGO STOP NATIONAL CONGOLESE TERRITORY CURRENTLY
OCCUPIED BY BELGIAN TROOPS AND LIVES OF PRESIDENT OF REPUBLIC AND
PRIME MINISTER IN DANGER FULLSTOP.[29]

It would be hard to overstate the importance of this move. At a single
swoop, this telegram opened a new front in the Cold War: Africa. Until then,
the tension between East and West had been played out largely in Eastern
Europe and Asia (Korea and Vietnam). Now, suddenly, Africa was the focal
point of attention. The telegram had barely been sent to Russia before it was
leaked to the CIA. Its contents caused great nervousness in Washington: was
Congo actually asking the archenemy for assistance?

In 1960 seventeen African countries had gained independence. The
result was a new scramble for Africa. Unlike in the nineteenth century,
this was not about Western European powers in search of overseas colonies, but about the victors of World War II trying to expand their spheres
of influence around the globe. Economic interests still played a major role,
but ideological, geopolitical, and military factors were much more decisive.
Congo was the first African country to become involved in the tug of war
between the two new superpowers. Not only was it a huge and strategically
located country from which all of Central Africa could be controlled, but it
also had crucial stores of raw materials for the production of weapons. The
Americans knew all too well that it had won World War II with the help
of uranium from Congo, and that cobalt, an ore used in making missiles
and other weapons, was found in only two spots in the world: Congo and
Russia itself.[30] To leave Congo to the Russians would seriously compromise
America militarily.

Did Kasavubu and Lumumba realize the impact their telegram had?

Most probably not. Inexperienced as they were, they were simply trying to obtain foreign assistance in solving a conflict concerning national decolonization; in doing so, however, they had opened the Pandora's box of global conflict. A great deal of ink has been spent on Lumumba's supposed communist sympathies. The contacts with Russia in that regard are often seen as proof of his Bolshevist disposition. But that is not correct. Economically, Lumumba leaned more toward classical liberalism than communism. He held no truck with the collectivization of agriculture or industry; he counted more on private investments from abroad. What's more, Lumumba was a nationalist and not an internationalist, as would have behooved a good communist. Despite all the Pan-Africanism, his frame of reference was Congolese through and through. The notion of proletarian revolution was foreign to him as well. As an *évolué* he was part of the newborn Congolese bourgeoisie; he had no desire to overthrow his own social group. What's more, he had also turned to America to help solve his country's problems. And it is often forgotten that he wrote his request to Nikita Khrushchev along with Kasavubu, who was anything but a communist. Even Khrushchev realized that: "I could say that Mr. Lumumba is as much a communist as I am a Catholic. But if Lumumba's words and actions overlap with communist ideas, I can only be pleased."[31]

Nor was the request to Moscow prompted by Lumumba's fickle nature, his suspicious turn of mind, his unreasonable behavior or any other personality trait that people thought they detected in him. Lumumba did indeed have a reputation for being irritable and capricious, but reading the telegrams to the United Nations and Russia today, one feels a very different psychological register: panic. Panic accompanied by total outrage, a great fear of losing control and the fear of being murdered. We should not forget that Kasavubu and Lumumba had occupied no major political positions before being placed at the helm of their country. Kasavubu had been mayor of a borough in Léopoldville; Lumumba's very first political appointment was that of prime minister. After two weeks of independence, they lost their grip on events. It was as though they had just received their driver's licenses and suddenly found themselves in the cockpit of a jet fighter that was about to crash. Confronted with Belgium's unsolicited military intervention, they did what they thought best at that fearful moment: quickly called for assistance from whoever was ready to help. And Russia was more than ready. One day later, in an extremely enthusiastic letter, Khrushchev let them know that, should the "imperialist aggression" of Belgium and its allies continue, the Soviet Union would "not hesitate to take resolute measures to end that aggression." His country, after all, could only sympathize with "the heroic struggle of the Congolese people for the independence and integrity of the republic of Congo." To which he added: "The Soviet Union's demand is clear: hands off of the republic of Congo!" Saying that, he conveniently forgot how the Russian army had ground Hungary beneath its heel four years earlier.[32]

Hammarskjöld understood the threat of a global conflict and succeeded in getting peacekeeping forces to Congo within the next forty-eight hours: on July 15 the first Moroccan and Ghanaian contingents arrived, followed by other African troops from Tunisia, Morocco, Ethiopia, and Mali. Meanwhile, Russia sent ten Ilyushin transport planes to Congo with trucks, food, and weapons. America considered bringing NATO forces into play, but that could have unleashed a second Korean conflict or even a new world war. Washington therefore chose to exercise influence through two more discreet channels: the United Nations and the CIA: the path of diplomatic lobbying in New York and that of clandestine influence in Léopoldville. Larry Devlin, head of the American intelligence service in Congo, had access to huge funds for the purpose of nudging Congolese politicians in a direction favorable to America. Kasavubu and above all Mobutu were to become his minions.[33]

Through my talks with Jamais Kolonga, I gained a picture of those tumultuous days. One of his anecdotes was very telling. In late July, Lumumba decided to go to America to negotiate with the United States and the United Nations. The usual protocol, under which such an official state visit is carefully arranged by top officials on one side and diplomats on the other, was thrown to the wind. A member of Lumumba's staff went to the American embassy in Léopoldville and demanded on the spot that twenty-four visas be issued for the prime minister and his retinue. More than one eyebrow was raised at that. There was no program, no protocol, no appointments had been made.[34] "I went to Ndjili airport to wave goodbye," Kolonga said. Since June 30 he had been working for the prime minister's press department. The people he met there included Mobutu, Lumumba's secretary.

A brass band played, the door of the plane closed, the stairs were rolled away. But inside the plane, Lumumba realized that he lacked a press attaché. The door opened again and Lumumba pointed to our little group. Who was he pointing at? At me? At the person beside me? None of us could figure it out. *"C'est vous!"* he shouted, pointing at me. I walked over to the plane. I had to go along. All I had with me was a Parker pen and a notebook. No clothes, except for the green suit I was wearing! No passport, no visa, I went on board without any baggage. But when it was over I came back with two full suitcases and a shoulder bag. And in the meantime I had seen Dag Hammarskjöld at work at the United Nations.[35]

This nonchalance was characteristic of the spirit of improvisation that reigned within the young Congolese government. It was one of the reasons why Lumumba did not make a good impression during his visit. With no appointment having been made, President Eisenhower refused to receive him. At the United Nations, officials were annoyed by the way Lumumba "made impossible demands and demanded immediate results."[36] C. Douglas Dillon, U.S. deputy secretary of state at the time, complained about his "irrational, almost 'psychotic' personality": "He never looked you straight in the eye, he looked up at the sky. And then came this huge flood of words. . . . His words were never related to what we were trying to talk about. You got the feeling that he, as a person, was possessed by a fervor I can only describe as messianic. He simply wasn't rational. . . . The impression he made was extremely negative, this was someone you couldn't work with at all." His asking a top State Department official to arrange a blonde call girl for him did not make a good impression either.[37]

AFTER ONE MONTH this was the situation in Congo: the army had been tossed topsy-turvy, the administration decapitated, the economy was on the blink, Katanga had torn itself away, Belgium had swept down on the country, and world peace was being threatened. And all this because, at the outset, a few soldiers in the capital had demanded better pay and a higher rank.

Meanwhile, Lumumba had burned many of his bridges. After his speech against Baudouin and his dismissal of General Janssens, Belgium had had it with him. After the telegram to Khrushchev and his trip to America, the United States was finished with him. The United Nations' patience was running out as well, while in his own country his high-handed dealings had estranged him from Kasavubu. Western diplomats, advisers, and intelligence personnel drove a wedge between them. Each and every one of them chose Kasavubu's side and recommended that he drop Lumumba. In August 1960 Lumumba was a lonely man, supported only by the Soviets.

What's more, his wrath had only grown. On two occasions the UN Security Council had called upon Belgium to withdraw from Congo (on July 22 that was to happen "quickly," on August 8 even "immediately"), but Belgium refused to budge as long as the blue helmets could not guarantee its subjects' safety.[38] It was not until late August, none too early, that all ten thousand Belgian soldiers had left Congo. In Lumumba's eyes the United Nations was toothless, at best. Perhaps even pro-Western.

On August 8, to top it all off, the southern part of Kasai province declared independence as well. After Katanga, the diamond province was Congo's most important mining area. Albert Kalonji had himself crowned king. A former supporter of Lumumba, with whom he'd had a falling-out before the elections, he had missed out on a ministerial post in the new national government. His secession, however, was ethnically motivated as well. Kalonji stood up for the Baluba, the inhabitants of Kasai who had gone to work in the mines of Katanga in great numbers and were hated there as immigrants and fortune hunters. In Kasai itself, the Baluba faced off against the Lulua; violent clashes had become commonplace. By proclaiming a new nation, Kalonji hoped to create a homeland for the Baluba. Tshombe supported the initiative and he and Kalonji even decided to establish a confederation.

Together with Katanga, newly seceded South Kasai accounted for one-quarter of Congo's territory, and the wealthiest quarter at that. For a unitarian like Lumumba, that was unacceptable. What's more, Jean Bolikango was also thinking about withdrawing Équateur from the republic. That was no coincidence: Tshombe, Kalonji, and Bolikango considered themselves the ones duped most badly during the government's formation, because they had not received a ministerial post. Lumumba wanted to act but could not count on the UN emergency forces, seeing as they had done nothing to stop Katangan independence. As defense minister, therefore, he sent the renovated Congolese army to the rebellious diamond province. But the government army was broke and led by officers who had been promoted two months earlier without any preparation.

The results were horrific. Kasai in late August of that year was the scene of senseless confrontations that led not to victories, but to massacres that claimed thousands of civilian lives. During an attack on a Catholic mission where noncombatant Baluba had gone for refuge, more than fifty people were slaughtered, including women and children. In addition to the machine gun, the government soldiers also wielded the machete. UN Secretary General Hammarskjöld expressed his abhorrence and suggested that the Baluba were the victims of genocide. He called it "one of the most flagrant violations of rudimentary human rights, [which has] the earmarks of a crime of genocide."[39] Lumumba had now completely blown his chances with the United Nations as well.

ALL THIS TIME, Kasavubu had remained pretty much in the background. But on September 5, 1960, he seized the opportunity to do what many Western advisers had been prompting him to do: he removed Lumumba from office. Article 22 of the *Loi fundamentale*, the new country's provisional constitution, gave him the power to do that: "The head of state appoints and dismisses the prime minister and the cabinet ministers."[40]

For those listening to the national radio station, it must have been one of the strangest evenings in the history of the government broadcasting service. Just after eight o'clock that evening, the normal programming—a radio course in English—was interrupted and they heard the high voice of President Kasavubu saying that he had just removed the prime minister from office. All around the *cité*, in the working-class neighborhoods and in inland villages, the Congolese people were hearing that Lumumba was no longer their prime minister, that he had been replaced temporarily by Joseph Ileo, a political moderate who had written the 1956 manifesto in *Conscience Africaine*. Then, to their amazement, less than an hour later, the listeners heard Prime Minister Lumumba announce in his staccato French that he, in turn, had just dismissed President Kasavubu! So much confusion—the rules of English grammar were nothing in comparison! As if Congo didn't have enough on its hands already with a military, administrative, economic, ethnic, and global crisis, it now received a constitutional crisis to boot.

Lumumba appealed to Article 51 of the provisional constitution, which stated that "only the Parliament and the Senate can provide authentic clarification of these laws."[41] It was a wise gamble, for on September 13 the parliament confirmed its faith in Lumumba and refused to recognize Ileo as the new prime minister. President Kasavubu was put to shame so badly that the next day he sent the parliament into recess for a month.

The imbroglio was now complete. Congo was being ruled not by government, but by arguments. National interest was made subordinate to power struggles. And in the midst of this chaos, Colonel Mobutu, the army's chief of staff, stepped forward to put an end to the squabbling. That very same day, September 14, 1960, he carried out his first coup d'état, with the approval and support of the CIA. He told the press that the army would be taking over the reins until the end of the year. Lumumba and Kasavubu were "neutralized." But whereas Kasavubu was ultimately allowed to stay on as a sort of figurehead president, Lumumba was placed under house arrest in his capital city residence. The friendship between Mobutu and Lumumba was over for good.

Mobutu placed national policy making in the hands of a team of young university students and graduates, a move intended to counter the lack of expertise in Lumumba's government team. Mario Cardoso, who had attended the economic round-table meeting and was popular among the Congolese students in Belgium, told me the following: "Colonel Mobutu asked the students and academics to come back from abroad and apply their knowledge in the service of the country. We would not be given the title of minister, but of commissioner general. We were to become apolitical administrators, we would not represent any party, tribe, region, or village. We had a diploma, and that was enough." Within that council of commissioners general, Cardoso was charged with education. Justin Bomboko, charged with foreign affairs, was the chairman and served as de facto prime minister. This situation was to last only a few months. "We were a transitional government. Mobutu only wanted to restore order, because the fighting between Kasavubu and Lumumba just wouldn't stop."[42]

This government of academics did not please everyone, not by a long shot. Lumumba repeated his claim to be the only democratically elected prime minister of Congo. The Belgian government, on the other hand, was only too glad to see him removed and maintained warm relations with the young commissioners. Many of them had studied in Brussels or Liège. Any return to the political arena by Lumumba was to be blocked at all costs, even physically if need be. Two Belgian military men, operating under the protection of Minister of African Affairs Harold d'Aspremont Lynden, made preparations to kidnap or murder Lumumba.[43] In addition, U.S. president Eisenhower personally ordered the CIA to liquidate Lumumba. In true James Bond style, the Congolese prime minister was to be poisoned with a tube of hypertoxic toothpaste.[44] There were also many people in Congo who would have been pleased to see him go.

Aware that attempts might be made on his life, Lumumba asked the UN for protection. He received a contingent of Ghanian blue helmets, who camped in his garden to keep any attackers at bay. That proved necessary; on October 10, Mobutu sent two hundred soldiers to Lumumba's residence to take him into custody. The United Nations stopped them. The resulting standoff lasted for weeks. Lumumba's house was under a twofold siege: by a ring of blue helmets, to protect him as long as he stayed inside, and by Congolese ready to arrest him as soon as he came out. His telephone was cut as well. Lumumba was silenced. Deputy prime minister Antoine Gizenga therefore took on the role of representative of the Lumumba government. Gizenga came from Kwilu, and even today he is adored by older people, including Longin Ngwadi, the swordsman from Kikwit. As Mobutu's coup gathered momentum, however, Gizenga realized that there was no place in Léopold-ville for him and other Lumumba supporters. In early November, therefore, he left with the remnants of the first government for Stanleyville, the cradle of Lumumba's movement, to govern and retake the country from there.

THE SITUATION WAS GROWING MORE COMPLICATED all the time. Congo was now four months old and already had four contiguous governments, each with its own army and foreign allies. In Léopoldville Kasavubu and above all Mobutu enjoyed unconditional American support. Thanks to the massive funding supplied by the United States, Mobutu was able to reorganize the national army. Around him there rose up the "Binza group," named after the residential neighborhood in the capital where they met. It was an informal group with a great deal of power, generously supported by the CIA. In Stanleyville Gizenga was keeping alive the Lumumbist body of ideals. He was backed by a portion of the armed forces and his government received support from the Soviet Union, although that was never as systematic and substantial as the American support for the capital.[45] In Elisabethville Tshombe stood at the helm of a self-proclaimed, independent country. Belgium was very generous with its logistical and military support. The Katangan military police included a great many Belgian officers. Union Minière financed the secession on a large scale. In Bakwanga, Kalonji led Kasai, an independent Baluba state where Belgian diamond delvers were active. The necessary means were provided by Forminière.

Tshombe and Kalonji were only regional leaders, but Kasavubu and Gizenga both claimed the legitimacy of a national government. Who would be proved right? Both went in search of international recognition, and their battle was fought out before the UN General Assembly in New York. Congo showed up there, divided into two camps: Kasavubu/Mobutu versus Lumumba/Gizenga. Thomas Kanza, the twenty-six-year-old psychologist, represented the Lumumba government at the United Nations, but President Kasavubu traveled to New York himself to convince the world that he, and only he, embodied the legal authority of the republic. He argued that his dismissal of Lumumba was allowed under the constitution, a claim with which the Americans, Belgians, and many UN officials had little problem. On November 22 the verdict came in: fifty-three countries recognized Kasavubu, twenty-four voted against him, nineteen abstained.[46] Cardoso, who worked for Mobutu at the time, remembers it as a triumph: "That's when we won the seat in the U.N. Kasavubu was the head of our delegation, and Lumumba lost internationally."[47] With that international marginalization, Lumumba's swansong began.

He was still locked up in his home in the capital. When news of the vote in New York reached him, he realized that his days in Léopoldville were numbered. Would the blue helmets in his garden still protect him, now that the United Nations had voted against him? He was bound and determined to join up with his political friends in Stanleyville. It was nighttime, it was November, the rainy season was in full swing. On November 27 an unusually heavy tropical storm forced his Congolese besiegers to seek shelter. Their attention lagged. Lumumba crawled into the back of a Chevrolet and was driven out of the house in the pelting rain.

The Congolese roads at that point were still in excellent condition. Had his chauffeur driven on steadily for two days, they could have reached Stanleyville. But on the night of his escape, Lumumba hung back in the capital to speak to the people. Along the way as well he stopped in the villages and enjoyed the locals' warm welcome.[48] But it was the rainy season. In the capital, Mobutu found out about Lumumba's escape and vowed to keep him, at any price, from reaching Gizenga. A successful reunion there could only mean a political rebound, and Mobutu's Belgian advisers and the CIA wouldn't like that. The United Nations refused to help search for the fugitive, but a European airline supplied Mobutu with a plane and a pilot accustomed to carrying out low-altitude reconnaissance. It did not take them long to find the convoy, which now consisted of three cars and a truck. On December 1 Mobutu's troops arrested Lumumba and his retinue as they tried to cross the Sankuru River close to Mweka. Lumumba was flown to Camp Hardy at Thysville, the base where the army mutiny had begun a few months before. From that moment on Lumumba could no longer count on UN protection, but was a prisoner of the Léopoldville regime. When he arrived, without glasses and his hands tied, someone stuffed a piece of paper in his mouth: the text of his famous speech.

What were Kasavubu and Mobutu going to do with him? Hold him in custody forever, like a sort of Simon Kimbangu of the First Republic? Wouldn't it be better then to have him taken to Katanga. Or Kasai? Hostile provinces, to be sure, but that was exactly why it might be a good idea. He would have no supporters there. Where he was now, the trouble was starting all over again. On January 12 the soldiers at Thysville started another mutiny. The situation grew restless. The Belgian government, in the person of Minister of African Affairs d'Aspremont, endorsed the plan to take Lumumba to Katanga, come what may, as long as it was far from the capital and somewhere no mutineers could come to his rescue. D'Aspremont's support for the plan also meant a strengthening of the ties with Kasavubu, and Belgium was interested in reestablishing diplomatic relations with Léopoldville. The former colonizer wished to avoid the impression that it sympathized only with Katanga. Reluctantly, Tshombe accepted the arrival of Lumumba and two other political prisoners. At the last minute, d'Aspremont had applied his influence to that end.

At 4:50 P.M. on January 17, 1962, the DC-4 carrying Lumumba and his two confidants, Maurice Mpolo and Joseph Okito, landed at Elisabethville. During the flight the men were beaten. A force of about one hundred armed troops was waiting for them; the soldiers were led by the Belgian captain Gat. A convoy took them immediately to Villa Brouwez, an isolated, vacant mansion belonging to a Belgian, close to the airport. The security inside and outside the villa was in the hands of the military police, led by two Belgian officers. There they received a visit from at least three Katangan cabinet ministers—Godefroid Munongo, Jean-Baptiste Kibwe, and Gabriel Kitenge, charged with internal affairs, finance, and public works respectively—who beat them as well. Tshombe was not with them. At that moment he was sitting in a movie theater, watching a film with the, in this context, preposterously cynical title *Liberté* (Freedom), from the Moral Re-Armament movement. When the movie was finished, he met with his ministers. There were no Europeans present. The meeting lasted from 6:30 to 8:00 P.M., but all practical measures for the rest of the evening seem to have been taken beforehand. The decision to send Lumumba to Katanga was taken jointly by the authorities in Léopoldville, their Belgian advisers, and the authorities in Brussels; the decision to murder Lumumba, however, was made by the Katangan authorities themselves. Munongo, the grandson of Msiri, the nineteenth-cenutry Afro-Arab slave trader who had taken the Lunda empire by force, played a particularly decisive role.

After the meeting, a ministerial delegation once again left for Villa Brouwez. There the prisoners were loaded into the back of a car. Along with a few other vehicles and two military jeeps, they drove off. Darkness had fallen by then. The convoy drove to the northwest, over the level road through the savanna toward Jadotville. In the glow of the headlights, to the left and right: bushes, the silhouette of a termite mound. After about forty-five minutes the vehicles left the main road. A few moments later they stopped at a secluded spot. The prisoners had to get out. In the wooded savanna beside the dirt road they saw a shallow well that had been dug only hours before. There were a few uniformed black policemen and guardsmen, but also a few men in suits: President Tshombe, the ministers Munongo and Kibwe, and a few of their colleagues. Four Belgians also took part in the execution: Frans Verscheure, police commissioner and adviser to the Katangan police force; Julien Gat, captain in the Katangan national guard; François Son, his subordinate police sergeant; and Lieutenant Gabriël Michels. One by one, the prisoners were led to the edge of the hole. They had been in Katanga for no more than five hours. They were beaten and mishandled. Only four meters away from them stood the firing squad: four Katangan volunteers with machine guns. Three times, a deafening salvo sounded through the night. Lumumba was the last to be dealt with. At 9:43 P.M., the body of Congo's first democratically elected prime minister tumbled back into the well.[49]

LUMUMBA'S MURDER WAS KEPT QUIET for a time. Shortly afterward, to wipe out all traces, Gerard Soete, the Belgian deputy inspector general of the Katangan police, dug up the three bodies. Rumor has it that a hand, possibly Lumumba's, was still sticking out of the ground.[50] Soete sawed the bodies into pieces and dissolved them in a tub of sulfuric acid. He pulled two gold-lined teeth from Lumumba's upper jaw. He cut three fingers off his

hand.[51] For years, at his house in Brugge, he kept a little box that he sometimes showed to visitors. It contained the teeth and a bullet.[52] Many years
later, he threw them into the North Sea.

The world received the news of Lumumba's murder with total dismay. From Oslo to Tel Aviv, from Vienna to New Delhi, people marched
in the streets. Belgian embassies in Belgrade, Warsaw, and Cairo were
attacked. While a university in Moscow was named after him, in the West
the "Lumumba"—a popular cocktail made with brandy and chocolate
milk—became popular. Gizenga's Lumumbist government was promptly
recognized by the Soviet Union, Poland, East Germany, Yugoslavia, China,
Ghana, and Guinea-Conakry. In no time, the murdered prime minister was
elevated to a martyr of decolonization, a hero to all the earth's repressed, a
saint of godless communism. He owed that status more to the grisly circumstances of his death than to any political successes. He had been in power for
less than two and half months, from June 30 to September 14, 1960. His track
record read like a pile-up of blunders and misjudgments. His abrupt Africanization of the armed forces was sympathetic but disastrous; his appeals for
military assistance to the United States and the Soviet Union were understandable but frighteningly frivolous, his military offensive in Kasai took the
lives of thousands of his countrymen. During his lifetime, Fulbert Youlou
and Léopold Sédar Senghor, the first presidents of Congo-Brazzaville and
Senegal respectively, already considered his actions quite doubtful.[53] On the
other hand, here was a man who was barely prepared for his task, who was
forced to deal with a rash domestic exodus and a Belgian military invasion,
and who watched as the United Nations hesitated about forcefully condemn-
ing the Belgian aggression. But with his unfortunate way of responding to
true injustices, Lumumba systematically cultivated more enemies than
friends. The tragedy of his short-lived political career was that his greatest trump card from before independence—his incredible talent for rousing
the masses—became his greatest disadvantage when, once in power, more
cool-headed behavior was expected from him. The magnet that had first
attracted now repelled.

A number of players share responsibility for Lumumba's demise. Less
than two weeks after independence, Brussels had already indicated that it
wanted a different prime minister. After only one month, the United Nations
and the United States were eager to get rid of him too. At first the intended
ousting was purely a political one, but American and Belgian authorities

gradually began thinking about eliminating him physically as well. In fall 1960 the CIA was behind Mobutu's coup and was charged by sources in the White House with liquidating Lumumba. The Belgian minister of African affairs also provided cover for covert actions aimed at taking him out. All these attempts failed. But when Lumumba was transferred in January 1961 from Thysville to Katanga, it was not merely at the initiative of authorities in Léopoldville and Elisabethville: the logistical and operational planning was carried out by Belgian advisers in Léopoldville (who, among other things, drafted the blueprint of the transfer during a meeting at the offices of Sabena) and received active support from certain government offices in Brussels, particularly the Ministry of African Affairs. That ministry was not unaware of the potentially fatal consequences for Lumumba, yet took no precautions. The same goes for the CIA: when he heard about it, the chief of station in Léopoldville entered no protest against Lumumba's transfer to Katanga, even though he knew it could have drastic consequences. The actual execution was the work of the Katangan authorities. The role played by their Belgian advisers remains shadowy: we know at least that on the evening of January 17 they were informed that Lumumba had landed at Elisabethville. In any case, they made few attempts to prevent the murders, even though they knew that their influence could have made a difference. A few Belgian military men, who were in charge of the Katangan guardsmen, took part in the killing itself.

The first act in the play of an independent Congo was over. It was characterized by an absolute centripetal force, a nonstop flow of events and complications. And it ended with a few teeth from an inspired African swirling in slow motion to the sandy bottom of a gray, European sea.

IN APRIL 2008 in a beautiful garden in Lubumbashi, I met Mrs. Anne Mutosh Amuteb. At ninety-one she was the oldest Congolese woman I had the honor to interview during my study. She was still an impressive sight. Anne Mutosh was a princess; her grandfather had been the *Mwata Yamvo*, the traditional king of the Lunda empire. That made her a member of Moïse Tshombe's clan; in the African sense of the word, she was his "aunt." To talk with her was to talk with the history of Katanga. She told me that her parents had already learned to read around the year 1900, taught by American Methodists. She herself had been a midwife, but her business talent proved greater than her obstetrical skills. I asked what she considered the best period in her life. She didn't even have to think about it. *L'époque Belge* and the Katangan secession," she said in her deep voice. "During the Belgian period, everything was well-organized. There was no corruption, commerce went the way it should. I imported textiles from the Netherlands, but also flour and grain. I once placed an order for fifty sacks. That was easy to do back then. During the secession, imports were no problem either. Only when Mobutu came along did things become so difficult."[54]

Considering her pedigree, it was little wonder that she favored Katangan independence. The Lunda mourned the loss of their empire and had for a long time dreamed of regional autonomy. In that, they were supported by those Europeans who remained behind. Many former colonials were for the secession. That rhymed with the tendency seen throughout southern Africa to perpetuate white rule. There were great differences between the apartheid in South Africa, Rhodesia, Southwest Africa (later Namibia), and the Portuguese colonies of Angola and Mozambique, but while the rest of the continent was becoming independent, white, right-wing regimes in the south were tightening their grip on power. Katanga fit that context.[55]

The Katangan secession constituted the second act of the First Republic. It was proclaimed on July 11, 1960, and came to an end on January 14, 1963. After the murder of Lumumba, on January 17, 1961, it assumed an entirely different complexion. After Tshombe had stood at the edge of that man's grave, he became the dominant player. Of the four pretenders to the throne of independence, only three were left. Kasavubu and Mobutu had as much blood on their hands as Tshombe, but Lumumba's death did not drive them closer together. From now on, the power struggle would take place between the three of them.

It is rather amazing that Tshombe became such a central player. After Lumumba's murder, after all, his Katangan state was the pariah dog of the international community. The Communist bloc expressed its abhorrence; the United Nations decided to act more forcefully. Not a single state ever recognized Katanga, not even Belgium or America. Tshombe's ability to stay on top for so long, however, had everything to do with the Belgians. Union Minière funded the new state by no longer paying taxes to Léopoldville, but to the local regime. Belgians manned the military, administrative, and economic infrastructure. Behind each Katangan minister stood a Belgian adviser. Professors from Liège and Ghent wrote the Katangan constitution. Key institutions like the Katangan national guard, the state intelligence service, and the central bank were led by Belgians.[56] In the lobbies of the Elisabethville hotels one frequently saw white men with a Katangan flag pin attached to their lapels.[57]

In addition, Tshombe remained in place with the help of a small army of white mercenaries. These "volunteers"—there were never more than five hundred of them—came from South Africa, Rhodesia, and England, but also included Frenchmen who had fought in Indochina and Algeria, veterans of the Foreign Legion. Ragtag types, roughnecks, rabid right-wingers, machos, Rambos, tough guys who drank till they couldn't remember their own names, let alone the name of the whore they'd ended up in bed with. They came for the money, for the adventure, and for vague ideals of white supremacy. Belgian officers took active part in their recruitment, training, and deployment.[58] They formed the creepiest contingent of the Katangan armed forces.

Their adversaries were the UN blue helmets, the Congolese national army, and the Baluba from the north of the province. That sounds more impressive than it was. The United Nations was hesitant about acting on its more forceful mandate. The ANC was still a shambles. And the Baluba waged war with poisoned arrows and machetes.

EXACTLY ONE YEAR after Lumumba was murdered, a twenty-two-year-old Fleming arrived in Elisabethville for the first time. He had never been outside Europe before. He came from a farming village in West Flanders and had just graduated from the polytechnic in Ghent with a degree in technical engineering; his specialism was low-voltage electrical engineering. He had been recruited by the Nouvelle Compagnie du Chemin de Fer du Bas-Congo au Katanga, the BCK. Working on the railroad was not really his boyhood dream. He had applied for jobs with Sabena airlines and with Union Minière, the showpieces of the Belgian economy. He wanted to become a pilot, but years of diligent study had ruined his eyesight. His name: Dirk Van Reybrouck. Ten years later he would become my father.

The country in which he arrived was called Katanga, not Congo. To him, the rest of Congo was a foreign country. All he had seen of Léopoldville was the Sabena guesthouse, where he had spent the night during a layover. The Katanga where he landed had its own flag, its own currency, its own postage stamps. His registration card made that clear as a bell. It is here before me as I write. The bilious green card was still printed in two languages, French and Dutch. "Congo Belge/Belgisch Congo" is written at the top. Someone had scratched that out with a ballpoint and struck it with a big rubber stamp: *État du Katanga*.

My father was based in Jadotville, present-day Likasi. He was responsible for the electric locomotives, overhead wiring, and substations along a six-hundred-kilometer (roughly 375-mile) stretch of rails leading to the Angolan border. For independent Katanga, that east-west stretch of tracks was a lifeline.[59] Ores and raw materials could no longer be taken north and shipped by way of Léopoldville and Matadi, for that was enemy territory. Everything, therefore, went by rail to the Angolan coast. That Benguela railroad, a single track still served in Angola by steam locomotives, was crucial for Katanga's imports and exports. My father was often "out on the line," as they called it. Aboard a *draisine,* a diesel-driven railroad car that served as his mobile workshop, he would go into the interior for two or three weeks at a time, checking transformers and replacing switches. BCK was a hierarchical company, but during those years the old guard placed a great deal of responsibility in the hands of young employees. "They had already sent their families back to Belgium," Walter Lumbeeck, one of my father's former colleagues, told me. "They just wanted to sit out their term and let others do the work. Your father was timid. His job was demanding for someone so young, and at first his French wasn't too great. But after a while he was able to communicate well with the blacks."[60] He also took Swahili lessons. Years later, at home, our dog was called Mbwa (Swahili for *dog),* and sugar and tobacco were still *sukari* and *tumbaku.*

The warring parties were well aware of the strategic importance of the Benguela line. While still alive, my father—a poor storyteller, unfortunately—would recall how he had been awakened in the middle of many a night "because a bridge had been blown up somewhere." Then he would head out with his *drezzine,* at daybreak, in the fragile morning light, while the world slowly took on color. A few of his African employees would pilot the wagon over the rails so that he could try to sleep a little longer. At the sabotaged bridge, they had to repair the overhead lines and rebuild the rails.

"In Katanga we still ruled the roost," Lumbeeck said, "that was the dominant way of thinking. We've kept things together here, let the rest go to hell, as long as things go well here, people figured. Copper was commanding a good price. Union Minière was still rolling." Congo may have been independent, but in Katanga, colonialism was in fact still in force. The Belgian employees had whisky and fruit from South Africa; fresh mussels were even flown in from Belgium. Young Belgians led a sunny existence, far from their parents, village, and church. Those were the days of barbecues and parties, when absolutely everyone smoked: stylish young women with beehive hairdos, men in white shirts and narrow ties. Those were the days of Adamo, Juliette Gréco, and Françoise Hardy. On Sundays people went to the *cercle*, a country club for sports and recreation. People lay there at pool's edge, drinking white Martini and Rossi vermouth, while the plop and thud of tennis balls came from a little farther away.

In July 2007 I walked the grounds of the Cercle de Panda, the club to which my father had belonged. The swimming pool was empty, the playground equipment rusty. The diving boards were em-dashes with no text between.

"Your father had a Ford Consul convertible," said Frans and Marja Vleeschouwers, a couple he had befriended at the *cercle*. "That car burned more oil than it did gasoline. Dirk always had to carry liters of motor oil around with him."[61] They took trips together to the waterfalls on the Mwadingusha. They visited the mission post at Kapolwe and drank beer brewed by the Flemish priests. Or they went fishing in the *brousse*, at spots where the old natives still paid with currency from the days of the Belgian Congo. Ties of friendship were more important than family. When Frans and Marja had a baby daughter, they asked my father to be her godfather—in Flanders, an honorary job reserved strictly for family.

But it was a closed world. "Everyone was allowed to join the *cercle*," Frans and Marja recalled, "but the membership fee was so high that no black person could afford it. The whites couldn't either, actually, but Union Minière automatically transferred the sum back to our accounts in Belgium. Unbelievable, isn't it?" There were also other things that made one think. "We let our daughter play with black children. 'Aren't you worried about her coming down with something?' some people asked then. It wasn't really apartheid, but at the butcher shop the blacks were still served by blacks and the whites by whites."

Walter and Alice Lumbeeck, his other friends, were in complete agreement about that. In the pictures made at their parties, you never saw an African.

Contact with black people was avoided back then. If you took a black person along to a party, you lost your friends. A white man with a black woman, people really looked down on that. That was something for the older generation. At BCK or Union Minière you still had a few older white men with a black wife, but not in our circles. That was beneath one's station, that wasn't chic. It would be comparable to a managing director visiting prostitutes these days. White men tended more to have affairs with their colleagues' wives. Your father was single then, he enjoyed being in the company of people who spoke Dutch. If he had brought a black person to a party, he wouldn't have been invited anymore.

KATANGA WAS AN ANACHRONISM. After Lumumba's death, the United Nations decided to deal firmly with Tshombe and his neocolonial secession. During the first half of 1961, this took place through diplomatic channels. Conferences were held in Tananarive (Madagascar), Coquilhatville and Léopoldville. The United Nations was pressing for a federal or confederal Congo, a reunited country with major powers for the provinces. Belgium, too, favored that option, but the Belgian advisers to the Katangan ministers systematically boycotted the search for a compromise. This obstinacy led to great friction. In August 1961, things came to a head. The United Nations mediated in a final conference, held at Lovanium University in the capital. Congo was to have a new prime minister. Not Ileo, who was pushed forward by Kasavubu; not Mobutu, who was pushed forward by himself; not Bomboko, who had led the government of academics—but Cyrille Adoula, a moderate and competent trade unionist who was acceptable to all parties. What's more, national reforms were in the offing: less centralism from the capital, more power for the provinces. A consensus seemed imminent, but at the last moment Tshombe withdrew from the discussions.

Then it would have to be by force, the United Nations decided. In August, September, and December 1961, the blue helmets launched heavy offensives to regain Katanga, to disband the local army and drive out the foreign mercenaries. They did not succeed. The mercenaries withdrew to Rhodesia and continued the fight from there. The UN campaign caused much suffering among the civilian population. Ambulances were fired upon, hospitals bombed, innocent civilians killed. More than thirty Europeans lost their lives. In addition, the UN offensive resulted in a dismal premiere: the first refugee camp in Congolese history. More than thirty thousand Baluba took to the road, out of fear for reprisals from Tshombe. They were not in favor of the secession and felt threatened. At the edge of Elisabethville they settled down in little huts of cardboard, leaves, and cloth.

Anne Mutosh, too, had bad memories of the UN offensive. "At the road-blocks, the Moroccan blue helmets raped a lot of women, even pregnant women. I was chairperson of the Union des Femmes Katangaises at the time, and I sent letters to Dag Hammarskjöld and President Kennedy. I actually met Hammarskjöld once."

The UN secretary general himself was determined to put an end to the neocolonial state of Katanga. He engaged in intensive mediation between Léopoldville and Elisabethville. On September 18, 1961, he flew to Ndola airport in North Rhodesia for a meeting with Tshombe. But shortly before landing, the plane went down under circumstances that have never been cleared up. No one survived the crash. "Pray that your loneliness may prompt you to find something to live for, great enough to die for," he had once written.[62]

The conflict seemed endless. Congo was like a broken vase that could not be glued back together. Still, around this same time (December 1961–January 1962), Mobutu's government forces succeeded in putting an end to the secession of Kasai and overthrowing Gizenga's government in the east of the country. With that, two of the four governments had been eliminated. Katanga would hold out for another year. The new secretary general of the United Nations, U Thant, spent all of 1962 trying to reach a solution through negotiations, but in late December of that year the United States decided that enough was enough. President John F. Kennedy provided considerable American support for a final UN offensive, known as Operation Grand Slam. Within two weeks, Katanga had been taken.

IT WAS JANUARY 3, 1963; my father was standing at the window on the second floor of his house in Jadotville. BCK had given him one of the apartments for single men. Not a villa with a garden, but a spacious flat with a downstairs garage and a modernistic stairway. It was just outside the city, beside the main road to Elisabethville. He knew that the UN blue helmets had already taken the capital. "Liberated" said some, "occupied" said others. The international force was now moving north toward Jadotville, the second largest town in Katanga province, following the same road along which Patrice Lumumba had made his final journey by car two years earlier. At the Lukutwe and Lufira rivers they had encountered resistance, but moved effortlessly into Jadotville around noon on January 3.

My father was looking out the window. He saw a white Volkswagen Bug coming up the road from the direction of town. Now that the troops had passed, apparently, the road to Elisabethville was open. Life was resuming its normal course. Suddenly, a volley of shots rang out. Across from his house, the Volkswagen careened to a stop. There were three people in it, a man behind the wheel and two female passengers. Three Belgians and a dog. My father had seen them before. The driver climbed out. Albert Verbrugghe worked in a cement factory. He raised his hands to show that he was unarmed. Blood was running from a cut under his eye. He screamed, wailed, staggered. The two women—his wife Madeleine and her friend Aline—did not get out. Across the fronts of their floral dresses, huge red spots were spreading. Only when their bodies had been pulled out of the car and laid in the grass beside the road, did Verbrugghe realize what had happened. The Indian blue helmets had apparently taken them for white mercenaries.[63] The dog, too, was dead.

"That dog lay there for a whole week," my father told me, sometime in the early eighties. I was sitting with him in the dentist's waiting room. On the coffee table was a well-thumbed copy of *Paris Match*. The front page bore a black and white photograph of the scene with the Volkswagen. I was ten or eleven at the time and could see the mortal fear in the man's eyes. My father looked at it for a few minutes, then said: "That photographer must have been more or less standing beside me. That happened right in front of my door." Later I learned that the images had been shot by an American cameraman and sent around the world. *Time* magazine published them in January 1962; today they can be found on the Internet.[64] My father had been an eyewitness to the most famous photograph of the Katangan secession.

On a Sunday in July 2007 I found myself standing where my father had stood then. I looked out his window at the spot where it happened. The road was dusty; there was a big sign advertizing CelTel. Someone was pushing a bicycle along, heavily laden with charcoal. My father's apartment still existed. It was inhabited these days by a friendly young magistrate, his lovely wife, and two darling children. On Sundays the man was a preacher in the Armée de l'Eternel, one of Congo's many Pentecostal denominations. The windowless garage where my father had parked his Ford Consul for five years was now an impromptu house of worship. I attended a service there. Some thirty believers were packed in close together on shaky wooden benches. Light came through the half-opened garage door. In the semidarkness I saw the glowing colors of the people at prayer. I thought about black and white photos. The years 1963 and 2007 faded into one. My father had died one year earlier, in 2006. The people sang beautifully.

WHILE THE DOG LAY THERE ROTTING, Katanga was retaken. Ten days later, on January 14, 1963, Tshombe announced that the secession was over. His Katangan gendarmes and white mercenaries fled across the border to Angola. Tshombe himself fled to Franco's Spain. Among the Belgians, the mood was extremely anti-American: Kennedy was blamed for that final UN offensive. "It's over and out," my father's colleague, Walter Lumbeeck thought at the time: "Everything became Congolese again. That came as a great disillusionment. A lot of people left."⁶⁵ The ANC moved in, young soldiers who spoke no Swahili, only Lingala, and who acted with the arrogance typical of victors. The administration was placed in the hands of people from Léopoldville. "During the secession, our boy's pension plan was still in effect," said Frans and Marja Vleeschouwers, "but under the new administration that all disappeared. People went back to cooking on *makala*, charcoal fires. You couldn't get anything in the shops except for milk and meat."⁶⁶

For his cigarettes and shaving soap, my father had to turn to the Ethiopian blue helmets. The United Nations remained in Katanga for another eighteen months, to keep the peace. In the mid-1980s, when I first started shaving with a razor, my father produced a big, old-fashioned tube of Palmolive. He himself had switched to an electric razor long ago. As an electrical engineer, he couldn't see the charm of brush and foam. "Don't use it all up right away," he said nonetheless, "I bought it more than twenty years ago from the UNOC soldiers." I still have that tube. The soap is half a century old, but it hasn't lost its foam.

The Katangan secession was over, but the white enclave lived on. They held "Bavarian Alps" festivals in Jadotville, where they walked around in lederhosen and waved steins of beer. In the middle of the savanna . . . Alice Lumbeeck recalled that, on November 22, 1963, there had been a celebration with dancing at the home of her Belgian neighbors. What was going on over there? They had once had a Katangan politician as neighbor. A frantic household. The crates of beer had been stacked to the ceiling. The Katangans had used the garden to barbecue rats. A white mercenary had also lived next door for a while, one of the *"affreux,"* the terrible ones. But now they heard the ruckus coming from their Belgian neighbors, normal citizens. "I asked them what the party was about. 'Kennedy has been killed! Kennedy has been killed!' they cheered."[67]

THEN THERE WERE JUST THE TWO OF THEM, Kasavubu and Mobutu. At the start of the third and final act of the First Republic, it was Kasuvubu who had triumphed: Lumumba was dead, Tshombe was in exile, and Mobutu hadn't exactly distinguished himself with the liberation of Katanga. That had been the work of the blue helmets. For the first time since independence, Kasavubu could rule over the entire territory of Congo. The country was reunited and he traveled around in it. The ties with Belgium had been restored and those with America had been strengthened. As a token of appreciation, Washington had sent a free package of enriched uranium to Léopoldville for research in the nuclear reactor at Lovanium.[68] Kasavubu owed the stability of those years in equal part to his prime minister, Cyrille Adoula, who remained in office for three years. That was far and away the longest term of office during the First Republic, where a prime minister's tenure tended to be expressed more in months than in years. Adoula was a hardworking, intelligent, but introverted bureaucrat who, due to his irresolute nature, never constituted a threat to Kasavubu.[69]

During this third act, Kasavubu was able to considerably fortify his position. Now that peace and quiet seemed to have returned, he became a zealous advocate of a new constitution to replace the provisional *Loi fondamentale*. In the course of 1964, a commission turned its attention to drafting the country's future rules of play. The result was the Constitution of Luluabourg, a text submitted to a national referendum, and a two-base hit for the president. The new constitution transformed Congo into a decentralized state, something Kasavubu had dreamed of for the last decade. The provinces were given more power, but were also greatly reduced in size. In 1962 the six gigantic provinces of colonial times had already been split up into twenty-one *provincettes*, miniprovinces, which reflected more closely the ethnic reality and the historical territories.[70] What's more, the new constitution granted the head of state a lot more power. From then on he would reign supreme over the prime minister and his government. The parliament, too, was hobbled. If it voted in favor of a law that did not appeal to the president, he could send it back with a request for a recount; anyone, after all, can make a mistake. And to avoid the same mistake being made twice, a two-thirds majority was needed to overturn the presidential alternative. In cases of emergency, the president could even draft legislation himself. Dickering with a recalcitrant prime minister became a thing of the past. "Accordingly, he may at his own initiative terminate the carrying out of the duties of the prime minister or of one or more members of the central government, particularly in those cases where he finds himself in grave opposition to them," Article 62 read.[71] Kasavubu was sitting pretty: the country had been broken up into little blocks, Katanga had dissolved into three innocuous miniprovinces, and he held the reins more tightly than ever. Divide and conquer is what they called that. He was safely out of range. Or so he thought.

On November 19, 1963, two Russian diplomats were arrested as they returned from Brazzaville. They were found to be carrying extremely compromising documents. In the capital on the far shore they had met with Christophe Gbenye, former interior affairs minister under Lumumba. Brazzaville had become a haven for those who had been Lumumbists from the very start, close enough to Léopoldville, yet safely out of reach of Kasavubu. The documents spoke of the setting up of a revolutionary movement, the Comité National de Libération, led by Gbenye. Delegations had already visited Moscow and Peking. In the documents, the committee asked for Russia's support in training young soldiers. It asked for radio units, small tape recorders, miniature cameras and photocopiers, "or other, similar material for espionage." It sounded like *Mission Impossible*. The group also asked for "20 miniature pistols (with silencer) in the form of lighters or ballpoint pens" and several "carryalls with a double bottom." Kasavubu's peace of mind had been premature.[72]

The first eruption of defiance took place in Kwilu. The instigator was Pierre Mulele, former minister of education and the fine arts under Lumumba and one of Gizenga's confederates. Mulele had nothing to do with the conspirators in Brazzaville, but he was on the same track. After the debacle with the first government he had fled abroad and ended up in China. There he grew acquainted with the ideology and practices of Mao's peasant revolt and gained experience with guerrilla warfare techniques. With this training, he returned clandestinely to his native region. Gizenga was a Pende, the tribe that had fought against the colonial authorities in 1931; Mulele belonged to the neighboring Mbunda tribe. He tried to reignite the fires of rebellion among the local farmers. The enemy this time was not the white colonizer, but the first generation of Congolese politicians who had murdered Lumumba. After all, weren't they more concerned with power than with the public interest? Wasn't their lifestyle bloated and their grab for financial gain despicable? Weren't they simply depraved, bourgeois pigs? Rather than serving the people, he claimed, they abused their power and stuck their fingers in the till. Their pro-Western attitude only aggravated their greed. The farmers listened to Mulele and nodded in assent. It was true, they really hadn't noticed much of the benefits of independence. In fact, their lives had become harder. Wasn't it about time for "a second independence," they wondered. The expression was an authentically popular one; a new *dipenda,* but this time the right kind.

Mulele started in on his peasant revolt, the first major rural uprising in Africa since independence. He displayed remarkable idealism and great selflessness. He became a sort of Congolese Che Guevara, a leftist intellectual hooking up with the common people. In villages and huts, he instructed them in revolutionary ideology. Time and again he underscored the importance of self-discipline during the revolt. His precepts were based explicitly on the writings of Mao.[73] Those who took part in the revolution were to show respect for everyone, including prisoners of war. Stealing was forbidden, as was prayer.[74] That which was destroyed had to be reimbursed. "Respect the women and do not toy with them as you might be inclined." No, the revolution needed daughters as well. In Mulele's maquis, women received training too.

The weapons available to them, however, were very limited. Mulele did not want to be dependent on foreign powers; the revolt had to support itself. And so the peasants went to war with only obsolete firearms, knives, and poisoned arrows made from bicycle spokes. Schools were torched, mission posts destroyed, bridges sabotaged. The death toll was in the hundreds. Despite the precepts, massacres took place. But the revolution did not catch on. Mulele's Chinese doctrine was not received enthusiastically everywhere. It was probably too secular. Why weren't the combatants allowed to pray? The simple farmers from Kwilu did not know what opium was, and were not interested in stories about false consciousness. Their reflexes remained extremely religious and tribal. Mulele's power base, therefore, never extended outside the tribal territories of the Pende and the Mbunda. The cities were beyond his grasp. The revolt lasted only from January to May 1964, but it was of great symbolic significance. For the first time since Tshombe, Kasavubu's authority had been openly challenged and Lumumba's ideology

proved very much alive. If Lumumba was a martyr, then Mulele was his
new prophet.

IN THOSE DAYS, along the broad streets of Stanleyville, amid the modernistic showpieces beneath a burning sun, one could sometimes see a very old
woman. She was eighty, perhaps ninety. Mama Lungeni was the widow of
Disasi Makulo, the man who Stanley had bought out of slavery. Her illustrious
husband had died in 1941, but she lived on for more than twenty years without
him. In 1962 she went to Stanleyville for her granddaughter's wedding, but
poor health kept her from returning to her native village in the rain forest.[75]

As a young girl she had been the victim of tribal violence, and now, old
and stiff, she could only watch as the war returned. She did not know, of
course, that the revolutionary comrades-in-arms in Brazzaville had decided
to go into action, but she would notice that soon enough. Gbenye's Comité
National de Libération was planning to invade the eastern part of the country. In Burundi, which along with Rwanda had been independent since 1962,
the rebels-to-come were being trained by Chinese specialists in guerrilla warfare. The Soviet Union, too, was ready to help. In southern Kivu the rebellion
was led by a man by the name of Gaston Soumialot, in northern Katanga
by someone named Laurent Désiré Kabila. Their soldiers were very young,
mostly boys of sixteen or seventeen, but some of them were barely even in
their teens, children too at times. They were more susceptible to magic than
to all the Maoist and Marxist-Leninist rhetoric put together. They called
themselves *simbas* (lions) and they believed strongly in martial rituals.

Kabila and Soumialot's army of liberaton employed a powerful *féticheur*,
Mama Onema, a woman in her sixties. Every young soldier was personally initiated by her. With a razor, she made three little incisions between
his eyes. From a matchbox she then produced a black powder—the ground
bones and hides of lions and gorillas, mixed with smashed black ants and
crushed hemp—and rubbed it into the cuts. He was given a *grigri*, an amulet
he wore around his wrist or neck, meant to give him strength. Each time he
went into battle she sprinkled his chest and weapons to make him immune
to the enemy. The warriors were expected to abide by a strict code of conduct. They were never to shake the hand of a non-Simba and they were
not allowed to bathe, to comb their hair, or to cut their nails, for otherwise
they would become vulnerable again. Many of these rules were less bizarre
than they might seem. Most of the Simbas had no uniforms and there were

almost no firearms. They entered the fray with their chests bared, decked out in twigs and animal skins, and were armed only with spears, machetes, and clubs. That was all they had to bring to bear against Mobutu's government troops, troops that may still have been a chaotic mess, but a chaotic mess nevertheless armed with machine guns. The magic rules forced the Simbas to abide by a form of military discipline. Sex was forbidden, because otherwise the warriors would go off on a rampage of sexual violence. Panic was forbidden, because otherwise they would run away. Looking over one's shoulder was forbidden, hiding was forbidden. The Simba warrior had to run straight at the enemy, loudly screaming *"Simba, Simba! Mulele mai! Mulele mai! Lumumba mai! Lumumba oyé!"*(Lion, lion, water of Mulele, water of Lumumba, long live Lumumba!). If they shouted that, the government forces' bullets would turn to water as soon as they touched their chests. Anyone shot and killed had apparently violated one of the precepts.[76] Nonsensical? Yes, but no more nonsensical than some charges in World War I when soldiers were driven into enemy fire. And the weird thing was that not only did the Simbas believe in their magic powers, but the government soldiers did too. Mobutu's men were scared to death of these drugged, hysterical madmen who came rushing at them, screaming and wide-eyed.

In May 1964 the Simbas took Uvira and Albertville, two major cities on the western shore of Lake Tanganyika. For Kasavubu and Mobutu, it was a humiliating defeat. The government soldiers tied twigs to the barrels of their own rifles in the hope of neutralizing the Simbas' magic, but much more frequently they turned and ran. Screaming and ranting, the rebels conquered eastern Congo. They confiscated automobiles and plundered shops. They picked up the guns the government army had left behind in a panic. Soumialot and his boys advanced from Uvira to Stanleyville, a few months' journey on foot through the jungle. Everywhere they went, in towns and villages, young people joined up. These were people who hated the new independence. The mess created by political intrigues at the top meant that thousands and thousands of young people in the east of the country could no longer go to school.[77] Their teachers were paid poorly or not at all. All across the country, teachers went on strike.[78] Secondary education, the prime means of social promotion, was a mere shadow of its former self. These were students without teachers. The word *révolution* for them contained more promise than the word *indépendance*. They were too young to have a wife, a house, or a plot of land, but not yet old enough to surrender all their dreams. They had nothing to lose. They were rebels without a cause, young lions, the ones who had lost most from independence. And they were horrifying killing machines.

Mama Lungeni saw the rebels come to town. In early August of 1964 they took Stanleyville. The stronghold of Lumumba and Gizenga was theirs once more. They went in search of those who had squandered independence. *Évolués*, intellectuals, and the rich became the brunt of their attacks. Around the statue to Lumumba, some 2,500 "reactionaries" were murdered. The Simbas cut out their hearts and ate them, to keep the dead from coming back. Other cities, too, witnessed their extreme cruelty. "Butter, butter!" they shouted in Tshombe when a machete split an enemy's skull and the brains ran out.[79] Babies and children were taken from their parents and laid in the burning sun for days, until they died.[80] In Kasongo they disemboweled a few elderly people and forced bystanders to eat their intestines.[81] In addition, they were pronouncedly anti-American, anti-Belgian, and anti-Catholic. The American consul at Stanleyville was forced to tread on the American flag and eat a piece of it.[82] Anyone carrying an object with "MADE IN USA" on it ran the risk of being slaughtered. It became a game to set the beards of Belgian missionaries on fire and then extinguish the flames with a beating. Many of the Simba had a background in the secret Kitawala cult, which had always been prominent in eastern Congo.[83] They bitterly hated white people. Any number of nuns at the missions were raped and murdered; missionaries were sometimes tortured and then butchered.[84]

Mama Lungeni was afraid she would never be able to return to the Protestant mission at Yalemba where Disasi was buried. That was where she hoped to die and be laid to rest beside him. But on September 5, 1964, the rebels announced the formation of a new state. Their territory was to be called the République Populaire du Congo, in analogy to the People's Republic of China. The various militias were fused to form the Armée Populaire de la Libération (the People's Liberation Army). Gbenye, the man from Brazzaville, became the new republic's president; Soumialot became defense minister; the post of commander in chief of the armed forces went to General Nicholas Olenga. One-third of Congo was in their hands. Mama Lungeni could not get away.

FOR KASAVUBU, this was a complete affront. It made Mobutu look like a fool, with his troops that kept turning on their heels and running. In his attempts to modernize the army, he received assistance from Cuban fighter pilots, men who had fled from Castro's regime and were determined to obstruct left-wing revolutionary uprisings wherever they could. But even that could not turn the tide. Was Congo about to fall prey to communism after all? The Americans would not be happy about that. What if Katanga were to be retaken? What if Tshombe came back from Spain and joined up with the rebels? He had enough means and troops to do so. Then two-thirds of Congo would be in the hands of the revolutionaries.

What happened then was one of those unlikely twists of fate so exclusive to Congolese political history: Tshombe did come back and . . . he sided with Léopoldville, the regime he had fought against for two and half years! It was an about-face to beat all about-faces, but not, when one ruled out things like integrity, entirely illogical. Mobutu and his comrades in the Binza Group (most particularly Foreign Affairs Minister Justin Bomboko, intelligence service chief Victor Nendaka, and national bank director Albert Ndele) realized that Tshombe could still mobilize his Katangan guardsmen and mercenaries.[85] All he had to do was bring them in from across the Angolan border. If they were to align themselves with the rebels, Léopoldville would be lost. Better to have a troublemaker in the house who pissed out the window, they decided, than a troublemaker in the garden who pissed in.

Tshombe, in turn, had always longed for a power base in the capital. The offer made him by Mobutu and company was the perfect opportunity to end his exile in Madrid and add a new entry to his political curriculum vitae. Obsequiously, he wrote to Kasavubu: "During this difficult period that stands before us, from which the country must emerge stronger than ever in order to deal with the enormous tasks that lie before us, I renew my offer to place myself fully at your disposal in the service of the fatherland."[86]

For the first time since independence, Lumumba's three enemies formed a troika: Kasavubu as president, Mobutu as commander in chief, and Tshombe as prime minister. In July 1964 he replaced Adoula and promised the people "a new Congo within three months." At the huge soccer stadium in Léopoldville, he was cheered on by thirty to forty thousand spectators. In Stanleyville, shortly before it was taken by the rebels, he even laid a wreath at the monument to Lumumba, the man he himself had helped to murder.[87]

Tshombe had two aces in the hole: his mercenaries of yore and the American military. The soldiers of fortune included Colonel Mike Hoare, a South African of Irish descent, nicknamed "Mad Mike"; Colonel Bob

Denard, a Frenchman who was without a doubt the twentieth century's most notorious mercenary; and Jean Schramme, otherwise known as Black Jack. This latter man was not your classic mercenary, but a plantation owner in Katanga who had decided to dedicated himself to "saving" Congo. In disreputable cafés in Brussels, Paris, and Marseille, new troops were recruited. They signed contracts that stipulated how much they would receive in damages for the loss of a toe (30,000 Belgian francs), a big toe (50,000 Belgian francs), or a right arm (350,000 Belgian francs). Or how much their widow would receive (1 million Belgian francs).[88]

The Americans placed a fleet of aircraft at Léopoldville's disposal: thirteen T-28 fighters, five B-26 bombers, three C-46 cargo planes, and two small twin-engined passenger planes. All World War II surplus, but good enough to wage war against bare-chested boys who believed in their own invulnerability.[89] While the mercenaries started in on a ground offensive, along with Katangan guardsmen, Congolese government troops, and Belgian officers, the Americans harassed the Simbas from the air. Their strongholds fell, one by one.

The Simbas reacted furiously. Appalled to find that they could actually be killed, they blamed their losses on the season's rain, which washed away their magic powers.[90] Like men possessed, they went in search of broadcasting equipment among those whites who had remained behind, for they believed that this was how the enemy was receiving its information. Anyone found to possess a transistor radio or even a ballpoint pen became suspect. They took hundreds of Europeans from what remained of rebel territory and held them hostage in the Victoria Hotel in Stanleyville. They threatened to murder them all. That was the starting sign for a large-scale military operation by the Belgians and Americans. It consisted of an offensive on the ground (Operation Ommegang) and one in the air (Operation Dragon Rouge). On November 24, 1964, 343 Belgian commandoes were dropped into Stanleyville and seized the airport. Meanwhile, ground troops moved into the city. Two thousand Europeans were freed and evacuated aboard fourteen C-130s; about one hundred were killed during the operation. In the days that followed, the Simbas retaliated by murdering ninety clerics and laypeople in the interior.[91] The number of Congolese killed has never been established.

Mama Lungeni escaped by the skin of her teeth. At 5:30 P.M. on the day Stanleyville was liberated, she heard the roar of aircraft engines and locked herself in her house along with her family. "A little later one of the planes flew over our neighborhood, Tshopo," one of her sons recalled. "Just above our home it fired a missile that landed about ten meters from the house. A section of the projectile disappeared into the ground, while the shrapnel blasted against the front door and blew out all the windows." At that moment, Mama Lungeni was sitting in the parlor, across from the door. She fell into a swoon. "Everyone, the children and the grandchildren, began shouting: Mama is dead! Grandma is dead! We carried her out to the yard, and soon she began breathing again and opened her eyes."[92]

After Stanleyville was taken, the rebels scattered across the interior. Two of Mama Lungeni's daughters, who lived beside the river, came to get her in a canoe. But the mission at Yalemba was still no safe place to be. Terrified as they were of the American bombers, the people fled their villages.

The people ran away into the jungle or to the islands. Mama Lungeni and her children were among the refugees in the forest, but conditions there were terrible. They had to keep building temporary huts to stay out of the rain, and moved from place to place. Mama Lungeni was exhausted and had to be carried. Her daughter Bulia and her granddaughters Mise and Ndanali took turns with her on their back, while the little ones, Naomi, Toiteli, Maukano, Moali, and their little nephew Asalo Kengo walked along behind and carried their baggage.

Because of the bad conditions and the dangerous situation, they decided to leave the forest and take shelter on the island of Enoli, in the middle of the river, where Uncle Anganga and his family lived.[93]

The old woman ended where she had begun: amid the misery of war. One day after evening prayers, she went to sleep. A heavy thunderstorm rolled in. At three in the morning, her eldest daughter, who slept next to her, lit the lantern. Mama Lungeni had passed away. It was May 1, 1965. Her body was taken by canoe to Bandio, the place where Disasi was abducted in 1883. The gong sent news of her death to the surroundings. People came out of the equatorial jungle to attend her funeral. She was buried beside her husband.

AND THE CIVIL WAR RAGED ON. Léopoldville was slowly gaining ground. But just as the rebels were reaching the end of their rope, they received help in the east from an unexpected source. The badly organized revolution had never paid serious attention to diplomacy, and the support from sympathetic countries like Egypt, Algeria, China, and the Soviet Union remained at a minimum. But suddenly, in April 1965, on the shores of Lake Tanganyika, no one less than Che Guevara himself stepped onto dry land! He had been flown over from Cuba and brought more than one hundred well-trained Cuban soldiers with him. In order to avoid detection, those soldiers were all of African origin, descendants of Central African slaves. Now they had come to help Kabila and his Simbas retake Congo. El Che noticed soon enough, however, that the flame of revolution no longer burned so brightly among his Congolese charges. The sound of loud dance music echoed from their secret camps in the bush, where women and children loitered. The Congolese comrades loafed about and had no training whatsoever. They had no desire to dig trenches, because holes in the ground were for corpses. Target practice didn't interest them at all, because they were unable to close only their right eye. They preferred shooting from the hip.[94] "One of our comrades said jokingly that in Congo all conditions were unripe for revolution," Che Guevara sneered in his diary.[95] The few times they actually made it to the front, the Cubans witnessed "the pitiful spectacle of troops that advanced but, once the fighting began, scattered in all directions and tossed aside their costly weapons in order to run faster."[96] Kabila himself stayed in Tanzania the whole time and appeared briefly only two months later, after which he disappeared again quickly. Che admitted that Kabila was the only one with leadership ability—but a true revolutionary commander, that was a different thing altogether. "He must also possess a serious attitude concerning the revolution, an ideology that serves to channel his actions, and a willingness to make sacrifices that is expressed in deeds. So far, Kabila has not shown himself to possess any of this. He is still young and may perhaps change someday, but I am not at all reluctant to state here, in writing that will see the light of day only many years from now, that I seriously doubt whether he is capable of winning out over his shortcomings."[97] Kabila would continue to hang around in the maquis for more than thirty years. By the time he ousted Mobutu in 1997, El Che was long dead.

After seven months, Che Guevara and his soldiers left Congolese territory. The rebellion had been fruitless. Bitterly, he noted: "During the final hours of my stay in Congo I felt alone, more alone than I had ever felt before, neither in Cuba nor in any other place where my wanderings around the world had taken me."[98]

TSHOMBE TRIUMPHED. The rebellion had been quashed, thanks to "his" mercenaries and "his" guardsmen. On the heels of this military triumph, he also achieved an extremely important diplomatic victory. He had gone to Brussels to negotiate about the notorious "colonial portfolio." That was the term used to refer to the sizable package of shares that Belgium had appropriated shortly before independence. The discussion concerning the return of the securities became known as the Belgian-Congolese dispute. Tshombe was able to convince the Belgian negotiators that the packet of shares actually belonged to the Congolese state, thereby effectively bringing the goose that laid the golden egg back to the farmyard. When he returned to Congo, he waved a leather attaché case everywhere he went.[99] The portfolio! The people laughed and beamed. The war was over, the money had come back. "Now we can eat *makabayu* again!" they sang, that lovely salted cod that had been prohibitively expensive for so long.

During the First Republic, the average Congolese had suffered financially. Inflation had skyrocketed: a kilo (2.2 pounds) of rice that had cost only nine francs in 1960 was up to ninety by 1965.[100] Buying power had withered.[101] Unemployment became a great burden. Anyone who still had a job was forced to feed more and more mouths with less and less money.[102] Hunger was widespread.[103] Diseases that had been under control, such as sleeping sickness, tuberculosis and river blindness, once again claimed countless lives.[104]

In 1965 Tshombe was by far the most popular politician in Congo. For the first time since the decolonization, the country held parliamentary elections. Tshombe won by a landslide. With his supercartel of parties, he took 122 of the 167 parliamentary seats. Kasavubu realized that Tshombe posed a threat to his position as president. At that point, he already held the combined powers of prime minister and those of minister of foreign affairs, minister of foreign trade, and minister of employment, planning, and information.[105] On October 13, therefore, Kasavubu did exactly what he had done with Lumumba: he dismissed the prime minister and pushed forward one of his lackeys as alternative: Évariste Kimba, a man in whom the parliament had no confidence. The move was allowable under the new constitution, but it seemed like everything was starting all over again.

DURING ONE OF OUR CONVERSATIONS, Jamais Kolonga produced a brightly tinted photograph, rumpled but remarkable. It showed a little group of young men, grinning broadly around a table. In their midst, I immediately recognized the young Mobutu. Even then he had looked like an African remake of King Baudouin. "This was taken on Mobutu's thirty-fifth birthday. The party was held in the restaurant at the Léopoldville zoo, the best restaurant in town." It was October 14, 1965, one day after Tshombe was fired. "The one on the left is Isaac Musekiwa, trumpet player with OK Jazz, the one next to him is Paul Mwanga, the singer. That's me, Jamais Kolonga, standing beside Mobutu! The men on the right are from African Jazz. First the singer, Mujos, then the great Kabasele himself. This one here is Roger Izeidi, from OK Jazz. And the one all the way on the right is no one less than Franco!" The entire *fine fleur* of Congolese music had gathered that evening around the chief military commander, as though the Beatles and Stones had had their picture taken with the supreme commander of the British armed forces. Jean Lema, alias Jamais Kolonga, was still tickled by it. "Do you know what Mobutu let slip to me that evening? I had worked with him for three months, in 1960, under Lumumba. 'Jean,' he said, 'within a month I will be president of the republic.'"[106]

And so it was. At 9 P.M. on November 24, 1965, a date every Congolese knows by heart, Mobutu summoned the nation's entire military brass to his residence in the capital. His desktop was covered in folders, newspapers, and magazines. He had spent the whole day in meetings, and his mind was made up: he was going to be the new head of state. The First Republic had been an utter catastrophe, it was up to him to set things aright. If Kasavubu was going to repeat the same tricks he'd pulled five years ago then he, Mobutu, would repeat his coup, this time not for five months, but for five years. He dictated a communiqué to one of his staff members. A sublieutenant was ordered to read the text on the radio, while a major went to sabotage Kasavubu's telephone lines. Everyone promised his support. The beer flowed freely. Mrs. Mobutu treated the guests to fish with fried plantain. Still, she was not at ease with the plans: "Stop this nonsense, would you? If they catch you, you'll all be killed," she whispered to her brother-in-law. But at two thirty that next morning she handed them all a glass of champagne. Three hours later, news of the coup was broadcast on Congolese radio.[107] The programming for the rest of the day consisted of martial music. The First Republic was over and done with. Not a single shot had been fired. The struggle for the throne had been decided. Each of the four protagonists had had his own finest hour, but it was Mobutu who would walk away with the goods.

CHAPTER 9

THE ELECTRIC YEARS

Mobutu Gets Down to Business

1965–1975

S EPTEMBER 1974. ZIZI KABONGO SHOOK HIS HEAD IN AMAZE-
ment when the letter arrived. He received mail often enough here
in Paris, but delivered personally by the director of his school? That was
something new. Since when had the rector of the celebrated Institut National
de l'Audiovisuel (INA), become a sort of honorary courier? Didn't the head of
the world's leading school for radio and TV journalists have better things to
do than play postman to a handful of African students?

But this letter had been sent by someone with clout, Zizi saw. It came by
way of the embassy and in those days that could only mean: from the president's office. Mobutu had handed the letter over to his cabinet minister, the
minister had handed it to the ambassador, and the ambassador to his staff.
That's how things went these days in Zizi's distant homeland. Since Mobutu
had taken power nine years ago, he was the one who pulled the strings. All
the strings.

September. The academic year had just begun. Paris was reawaken-
ing: the French had come back from vacation, the subways were full again,
people trundled hurriedly along the boulevards. "Ambassade de la République du Zaïre," Zizi read on the envelope. Even after three years, it still
took some getting used to. In 1971, the rounded vowel-sounds of *Congo* had
made way for the hiss of *Zaïre*. Mobutu found that more authentic than the
old colonial name. The Father of the Revolution had based his choice on
one of the earliest known historical documents: a sixteenth-century Portuguese map. But shortly after the name was changed, Mobutu discovered

that it was all a foolish mistake: Zaïre was a slipshod spelling of *Nzadi*, a run-of-the-mill Kikongo word for river. When the Portuguese had landed at the estuary and asked the locals what that huge, roiling mass of water was called, they had simply replied: "River!" *Nzadi*, they repeated. *Zaïre*, that was what the Portuguese thought they heard. For thirty-two years, Zizi's country would owe its name to the sloppy, four-century-old phonetics of a Portuguese cartographer.

And so Zaïre it was. That was what the country was called, and from then on the river as well, and the currency and the cigarettes and the condoms and all manner of other things. A bizarre name, with its atypical Z and that troublesome diaeresis. When you typed the name, you came up with one of those holy trinities of dots over the *i*. Mobutu's American allies could never quite wrap their mouths around it. They spoke of the one-syllable *zair*, as in *air* with a *z* up front.

À l'attention du Citoyen Kabongo Kalala, that was the address on the embassy envelope. The French found it charming, that use of *citoyen* as a form of address. At least one country, two hundred years after the storming of the Bastille, still held aloft the tenets of revolutionary etiquette.

Zizi, of course, wouldn't have expected them to address it to "Zizi." There weren't very many people who knew his real name, but in official correspondence he remained plain old Isidore—especially in France, where *zizi* is another word for *weenie*. But the address didn't mention Isidore either. Kabongo Kalala, that had been his official name for the last two years. Born in 1940 as Isidore Kabongo, he had gone through life since 1972 as Kabongo Kalala. With no first name. Christian names were now forbidden as being, once again, too colonial.

The people's minds, Mobutu felt, had been bent beneath the old yoke for too long. His plan was to liberate them mentally as well. A whole host of name changes would help in that process. Léopoldville was to become Kinshasa, Stanleyville Kisangani, and Elisabethville Lubumbashi. Lesser towns also received a new, indigenous name: Ilebo for Port Francqui, Kananga for Luluabourg, Moba for Baudouinville, Mbandaka for Coquilhatville, Likasi for Jadotville. Lake Leopold was renamed Maï Ndombe, the black water. Lake Albert became Lake Mobutu. And, in order to puncture local pride, Katanga was now to be called Shaba.

But a different toponymy was not enough, according to Mobutu. People's names had to reflect the change as well, for there were some who still looked up far too much to Belgium. Individuals bearing the name Lukusa continued to corrupt that to De Luxe. Kalonda sometimes became De Kalondarve. The singer Georges Kiamuangana preferred the Flemish-sounding Verckys as his stage name. And Désiré Bonyololo, the stenographer from Kisangani, liked to call himself Désiré Van-Duel. This was an affront to the ideologists of the Second Republic: the new Zaïrian should be proud of what he was, rather than ridiculously try to flaunt what he wanted to be. From now on, only native names.

And so Christian names were axed as well. They had been introduced by missionaries who had christened each baptized child with the name of a European saint: Joseph, Jean, Christophe, Thérèse, Bernadette, Marie. Shouldn't the true Zaïrian, the president said, prefer to describe himself in relation to his ancestors rather than to some remote saint? That is why he banned Christian names and prescribed the use of ancestral ones. The *prénom* vanished, the *postnom* (an unintentionally comic Mobutian neologism) took its place. It was a sly attempt to undermine the power of the church. Isidore Kabongo became Kabongo Kalala. Under Mobutu, everything, but then everything, was different.

"AT FIRST WE WERE QUITE PLEASED with Mobutu's coup," Zizi Kabongo told me during one of our many conversations in Kinshasa. There were few informants with whom I met as often as I did Kabongo.[1] He spoke about his country's complex history with great lucidity and finesse. He had attended seminary for a time, like so many of his generation, but was stranded halfway in his calling as a teacher of Latin and Greek in Katanga. He would ultimately choose the path of journalism. Today, at sixty-nine, he is a manager for the national radio broadcaster. "'Whew!' we said back then. At last, a little organization! The First Republic had been a huge mess. All that sniping back and forth between Kasavubu and Tshombe. . . . [I]t was a great disillusionment. The trains had stopped running, prosperity had been crushed, unemployment was on the rise. And meanwhile you saw the politicians being driven around in limousines, and sending their children to study in Europe. Mobutu abolished the political parties for five years, and everyone was quite satisfied about that."

Mobutu did, indeed, introduce a sudden change in style. Shortly after his coup d'état he spoke to the masses in Kinshasa's big soccer stadium. Here one had a slim young orator who wore no extravagant tuxedo, but a

khaki uniform and a beret.[2] He railed vigorously against "the sterile conflicts between politicians who sacrificed the country and their countrymen
to their own interests." His listeners could only agree to that. "For them,
the only thing that counted was power and what they could do with it. Fill
their own pockets, exploit Congo and the Congolese, that was their motto."
Mobutu called it as he saw it. His language was robust, his reasoning clear. "I
will always speak the truth to you, no matter how hard that may be to hear.
It is over, the time of assurances that all is well when all is not well. And now
I will tell you right away: in our beloved country, everything is going very
badly indeed."

He went on to treat the packed stadium to a lecture on national economics. He produced sobering statistics. The production of corn, rice, manioc,
cotton, and palm oil had fallen drastically. State spending had grown exponentially. Buying power had plummeted. Corruption was alive and kicking.
Things could not go on like this. "Special circumstances, special measures,
and that in every area." Mobutu announced a five-year moratorium on political parties. During that period he planned to get the country back on track and
to do that he needed the help of every man and woman. "To achieve this plan
of recovery, we need hands, a great many hands." Mobutu rolled up the sleeves
of his own uniform, to set the right example. "We will see each other again
here in five years' time. In five years' time you will see the difference between
the first and the second legislature. I am certain that you will notice then that
the Congo of today, with its misery, its hunger and its adversities, will have
changed into a rich and prosperous country where the living is good, the envy
of the world."[3]

Since Patrice Lumumba, no politician had spoken in the capital with
such passion. Mobutu employed Lumumba's vigorous idiom and supplemented it with a concrete plan of action. He radiated confidence and conviction. Congo was going to become a modern country.

What Zizi really wanted was to go to Europe and write his thesis about
Charles Baudelaire, but Mobutu felt that the young intelligentsia should
serve their country in a more tangible fashion. Along with a few other of his
countrymen, Zizi was therefore sent to Paris to learn the business of television. State television was to become an essential instrument in Mobutu's
attempts to get the country back on its feet. On November 23, 1966, exactly
one year after the coup, the first Congolese TV program was broadcast. One
year later, the first Lingala-language programs began.

"Antennae and relay stations began popping up everywhere," Zizi told me. "Congo had color television long before large parts of Eastern Europe. An entire generation of journalists received excellent training. We went to Paris and from Mobutu we received student grants that were two times the French minimum wage. I had my own apartment, I went to the movies. I earned more than a French worker!" When Mobutu came to Paris once to visit his students, they were all taken out to the Champs-Élysées to buy five suits, at his expense.[4] When they went to Brussels to shoot some footage, his chief of protocol came by to check the camera crew's baggage and make sure their clothes were up to snuff. Even the cameraman had to wear a bow tie. In the end, the per diems were so exorbitant that Zizi was able to build his own house.

And then came that letter, in September 1974. In it, Zizi read that he was to come to Kinshasa at the end of the month for a visit of no longer than forty-eight hours. All the Zaïrian students at the INA were summoned because if your studies are being paid for by Mobutu, it is only normal that you do something in return. The reason for such great urgency? An important boxing match was going to take place, and it had to be broadcast live. A boxing match featuring Muhammad Ali.

THE FIRST DECADE of Mobutu's thirty-year reign was a time of hope, expectations, and revival. "Mobutu was electric," the writer Vincent Lombume told me once.[5] And not only because he brought in television and built hydroelectric power stations, but also because he himself delivered a moral jolt to a nation in disrepair. The period 1965–75 is remembered as the golden decade of an independent Congo. And indeed, Kinshasa was hopping as never before; the beer flowed, the nights never stopped. "Kin-la-Belle," the city was called. From 1969, beer production rose by 16 percent annually. In 1974, the year of the heavyweight bout, a total of 5 million hectoliters (about 132,086,026 gallons) was brewed.[6] But the first five years, as Mobutu set about consolidating his own power, were also marked by extremely grim moments. Moments surrounding the euphoria like shards of glass cemented to the top of a concrete wall.

It was early in the morning of a gloomy Thursday in Kinshasa that the first people arrived at the big, open field in the cité, the wasteland beside the bridge west of Ndolo airport. Was it really going to happen? Young women carrying baskets of sugarcane on their head slowed for a look. Mothers with babies on their back stopped and stared. Civil servants in suits departed from their usual route to the office. Urchins in torn T-shirts came running up. Was it really true? Hundreds, thousands of feet crossed the big field. Chic Italian loafers stepped through the dust beside bare, callused feet. Spike-heeled slippers poked little holes in the sand. Trucks full of soldiers were waiting. In the midst of the military detachment stood proof enough for everyone to see: a wooden podium had been built, topped by a gallows.

It was Thursday, June 2, 1966, and Mobutu had been in power for six months. On Monday he had read a radio statement saying that a plot against him had been thwarted. Four days earlier, on Pentecost Sunday, the people all heard, four members of the old regime had been caught planning a coup. They were Alexandre Mahamba, a former cabinet minister under Lumumba, Joseph Ileo, and Cyrille Adoula; Jérôme Anany, defense minister under Adoula; Émmanuel Bamba, minister of finance in that same government and also a prominent Kimbanguist leader; and above all Évariste Kimba, the man who had briefly served as prime minister at Joseph Kasavubu's request, just before Mobutu staged his coup. Had they really been planning to overthrow the regime? Most probably, they had walked into a trap. Army officers pretending to be turncoats had asked them to draw up a list of candidates for a new government. The trial that followed was a mockery. None of the officers involved were called to the witness stand; the four civil defendants didn't stand a chance. When one of them tried to come to his own defense, the court-martial chairman said: "Gentlemen, we are here for the military tribunal, not for a debate. We are here to mete out punishment, the court-martial therefore won't take long."[7] A few moments later the verdict was handed down: the four were to be hanged. None of them had ever committed an act of violence, been in possession of a weapon, or even started to implement a plan against the regime.

The people converged. Many tens of thousands of them. The French AFP wire service spoke of no less than three hundred thousand.[8] It was the biggest crowd in Congolese history. Kinshasa's population had doubled in recent years and now numbered more than eight hundred thousand souls.[9] More than half of them were under the age of twenty.[10] Due in part to the civil war going on in the interior, migration to the city had picked up again after independence. Kinshasa was bursting at the seams. Across a fifteen-kilometer (about 9.3-mile) zone there now stretched out an endless sea of corrugated iron and makeshift huts, most of them only with a ground floor, all of them overcrowded. The only tall buildings were in the city center. All of these old and new inhabitants of Kinshasa, the "Kinois," now thronged together on that one Thursday morning after Pentecost. In the 1930s, the colonizer had held executions publicly, as a deterrent. Would Mobutu dare to go that far? To execute four former cabinet ministers, no less?

Mobutu, as the people had found out by now, was no mama's boy. His former opponents had no choice but to seek refuge elsewhere immediately after the coup. Kasavubu fled to his native region, Tshombe returned to exile in Spain. They were taking no chances. Kasavubu had written to Mobutu to say that he would accept the coup as "being in the country's greater interests." As an elected official he could perhaps claim his seat in the parliament, but he "considered it more useful at this point to leave his post." Kasavubu had always had something monastic about him, but he had never spoken this meekly before. "What I want most is to take a bit of a rest in Bas-Congo," he went on to note. He wanted to go back to his village, exchange his European clothes for native garb, and tap palm wine for visiting friends and guests.[11] And as though it were not clear enough already, he added: "I have no desire to stir up any agitation whatsoever."[12] Exit Kasavubu, in other words. Four years later, at the age of fifty-two, he died of cancer.

But Tshombe was cut from very different cloth, the people knew. Many of them had voted for him at some point. After his resounding electoral victory, he, as the nation's savior, continued to nurture major political ambitions. He shuttled back and forth between Paris, Madrid, and Palma de Mallorca, brooding over a possible return. Mobutu was dead set against that. Hadn't he publicly stated that he would apply himself to "l'élimination pure et simple de la politicaille" (the elimination, once and for all, of the political weasels)?[13] None of those gathered around the gallows now could imagine it, but one year later Tshombe—even though democratically elected, like Lumumba—would be condemned to death, in absentia, for "subversive activities." In June 1967 a shady French businessman with contacts in the highest Congolese circles invited him for a airborne jaunt from Palma to Ibiza. On the return flight the man suddenly drew a pistol, fired two shots, and ordered the pilots to fly to Algiers. Upon arrival, Tshombe was thrown straight into prison. Congo demanded his extradition but, going against the recommendation of the Algerian supreme court, President Houari Boumédienne refused to let him go. He originally hoped to extradite Tshombe in exchange for Congo's breaking diplomatic ties with Israel, but French President Charles de Gaulle personally called on him to forgo any such exchange. An extradition would most certainly have resulted in another murder, like that of Lumumba.[14] Two years later, on June 29, 1969, three months after Kasavubu, Tshombe died in his Algerian prison cell. A heart attack, the physicians said. Murdered, according to many in Congo. He was only forty-eight.

Mobutu had won the struggle for the throne, but during the first years of his regime he systematically eliminated his rivals from the First Republic. Even Lumumba, five years after his death, had to be neutralized. His backers were still far too numerous, and not only in the east of the country. Mobutu responded with a masterful move that displayed both strategic brilliance and bottomless cynicism: he, Mobutu, the man who had played a key role in Lumumba's murder, now pronounced that same Lumumba to be . . . a national hero! During the celebration of the national holiday, the Congolese people heard Mobutu say, without a waver: "All honor and fame to this illustrious Congolese, to this great African, the first martyr of our economic independence: Patrice Emery Lumumba."[15] Boulevard Léopold II, one of Kinshasa's main arterials, was promptly redubbed Boulevard Patrice Emery Lumumba. And it bears that name still. At the top of the boulevard, a huge statue of Lumumba stands waving to the mass of honking cars.

The move bore witness to a craftiness beyond compare. Just as Mobutu had neutralized Tshombe in 1964 by employing him in his fight against the Simbas, he now neutralized the person of Lumumba by means of posthumous rehabilitation. The Lumumbists had the wind knocked out of them: their hero had suddenly become the enemy's hero too! Mobutu had, as it were, dragged him onto the back of the scooter of his coup d'état. Neutralization by encapsulation would, in the next thirty years, become one of the favorite tricks of his dictatorship.

Neutralization was also a key theme in the first months of his regime. After banning the political parties, he now put the parliament out to pasture as well. To the MPs and senators he said: "Go now and get some rest, take a five-year break!"[16] He, in the meantime, would see to the country's legislation. The provinces, too, had their turn. The proliferation of miniprovinces was a waste of money, Mobutu felt. He preferred to keep things simple and so reduced their number from twenty-one to nine, all ruled over now by Mobutu adepts. This centralization was intended to counter the centrifugal forces of secession and tribalism. And the ball kept rolling. Congo was transformed from a federal, civil democracy into a decentralized military dictatorship. At the time of the takeover Mobutu had appointed General Léonard Mulamba as his prime minister, but after a time saw the need to neutralize that function as well. Zizi Kabongo knew the real reason: "The people loved Mulamba more than Mobutu. That's why he sent him away. Mulumba became the new ambassador to Japan. That's how it always went. Ostensible promotions, tassels and fringe, money, all kinds of favors just to keep people quiet." As a result, Mobutu—in addition to legislative and military authority—now assumed executive power too.

But a public hanging? That was altogether different from treating a rival to a remote ambassadorial post in a sumptuous villa. "No one thought it would really happen," Zizi said. "Mobutu's power base was still fairly shaky. All he had was the army, and the four condemned men all had tribesmen in that army. They could have mutinied." Mobutu hesitated. For a few days he avoided his wife, fearing that she would talk him out of it. Archbishop Joseph-Albert Malula, too, had requested that the men be pardoned. Even the pope had called. But to give in now would be a sign of weakness. . . . Mobutu's favorite book in those days was Machiavelli's *Principe*.

That same day, a military brass band played on the gallows field. The sea of people looked on as a jeep drove up. The four condemned men were in it! When they reached the platform, two women began to scream in despair. The "conspirators" were family of theirs. They had to be removed from the grounds, along with their children. The women were hysterical, their torsos were bared, their hair hanging down over their faces. All eyes were now fixed on the gallows. The first to climb it was the executioner, a burly man dressed all in black, with a black hood over his head. Immediately afterward the crowd saw a tall, blindfolded man on the steps. He was wearing only a pair of blue soccer shorts with white and red stripes. It was Évariste Kimba, former prime minister of the Democratic Republic of Congo. At the foot of the platform a priest had heard his confession, beside the four coffins that lay in wait. The executioner read the verdict. Kimba stood straight. The noose was placed around his neck and the trapdoor fell open. Horrified cries went up from the crowd, followed by a deathly silence. Kimba's death throes lasted more than twenty minutes. While the crowd looked on in silence, the former prime minister's body continued thrashing about. It took an eternity. From the jeep, the three remaining condemned men could see the fate that awaited them.

During the final hanging, panic broke out in the crowd. The people began running, knocking down the soldiers as they went. Children and adults stumbled and fell in the stampede. Within only a few minutes, tens of thousands of people ran away. When it was over, the field was dotted with groaning bodies and lost shoes. A little farther away a fourth coffin was being nailed shut. On that day, June 2, 1966, the people no longer cheered for Mobutu, but trembled in fear of him.

"Because it is difficult to unite love and fear in one person, it is much safer to be feared than loved, when—of the two—either must be dispensed with," Machiavelli had written.

"FROM THEN ON, everybody lived in fear," Zizi told me. "The state intelligence service became very powerful. No one dared to dine at the restaurant at the zoo anymore, the place where politicians and diplomats had always met, because they were afraid of being spied on by the waiters. Even at memorial services, we were afraid of the little boys who sold peanuts. Maybe they were spies. Mobutu used those hangings to set an example. 'No one toys with my power.' He wanted to strike fear into people's hearts and affirm his own status."

In an interview two days later, Mobutu said: "With us, respect for the chief is sacred. A striking example had to be set." The whole business of secessions, rebellions, and dismissals could not be allowed to start all over again. "When a chief decides something, it is decided, period."[17]

Mobutu made sure that the cornerstone of his authority, the army, lacked for nothing. There would no mutiny against him. Unrest was immediately counteracted with money. The army underwent drastic modernization. New waves of recruits received new opportunities. In addition to an officers' academy, he also organized specialized military training. Kisangani had been liberated by Belgian paratroopers. His army, Mobutu decided, would have parachutists as well.

In Kinshasa I spoke with Alphonsine Mosolo Mpiaka. She had been the first female paratrooper in the Congolese army. In 1966 she was twenty-five. "We received our ground training here in Ndjili. A center for paratroopers was set up there. Our instructors were Israelis." America backed Mobutu, so Israel did too—to the great annoyance of the Arab world. "For the jumps themselves, we had to go to Israel. I did twelve. I was the first woman; after me, Mobutu recruited another twenty-four girls. The team had to be mixed, ethnically too. A couple of Bakongos, a couple of Balubas, a few from Katanga." Detribalization here as well. Mobutu wanted an army that no longer thought along tribal lines. Loyalty was something he paid for. "We were highly respected and extremely pampered. My billet-money was enough for me to buy a plot of land with a house on it. But I never had to jump in the course of combat, only for the parades here in Kinshasa."[18]

Her expertise, however, still came in handy. The rebellion in the east of the country was not yet fully under control, but Mobutu preferred to leave that to the white mercenaries. Bob Denard and Jean Schramme did most of the work and were decorated afterward. Schramme later turned against him and tried to "save" Congo singlehandedly, but that adventure met an ignominious end.[19] And once the rebellion was over, the national army was able to shake off its white soldiers of fortune once and for all. In late 1967 Gaston Soumialot and Christophe Gbenye took to their heels, and all of Congo once again fell under the authority of the capital city. All of Congo? In its eastern reaches, a mountainous region close to Lake Tanganyika, Laurent-Désiré Kabila continued to rule the roost. But after Che Guevara's departure, his "revolutionary" pocket of resistance between Fizi and Baraka came to resemble the comic-book village of Asterix and Obelix: autonomous, to be sure, but above all harmless.

Congo was pacified and from 1968 on Mobutu began restoring civil authority.[20] He himself even began appearing in public out of uniform. For the first time he was seen sporting the accessories that would become his trademark: the characteristic leopard-skin hat, and in his hand a carved ebony walking stick. The traditional symbols of chiefdom.

Pierre Mulele figured it was safe to return home. After his peasants' revolt in Kwilu in 1964 he had fled to Brazzaville. In 1968, however, Mobutu granted him amnesty. Justin Bomboko, the minister of foreign affairs and an insider of the Binza Group, assured him that he would be welcomed like a brother. In September of that year Mulele crossed the river and received a festive reception on the near shore. He was invited to stay at Bomboko's house. Three days later, a contingent of soldiers came to pick him up for a big appearance at the soccer stadium. The impassioned and self-willed freedom fighter would be allowed to speak to the people there. Instead, the soldiers drove him to an army base where he was tortured gruesomely that same evening. They cut off his ears and nose. They gouged out his eyes and chopped off his genitals. While he was still alive, they then cut off his arms and legs. A few hours later, a sack filled with his remains was plunged into the big river.[21]

KASAVUBU, TSHOMBE, KIMBA, Gbenye, Soumialot, Mulele: one by one, within a few years' time, Mobutu's former rivals left the scene. To consolidate his new-won power, though, he also had to make sure that no new rivals arose. In 1967, therefore, his absolute sway became firmly embedded in a new constitution. "The Congolese people and I," he once told parliament, "are one and the same."[22]

The days that followed were bitter ones. Just outside the capital, on a leafy green hillside, lay the University of Lovanium. While Mobutu was busy establishing absolute sway, the student movement continued its incredibly brave attempts to knock the pins out from under him. The May 1968 student revolts in Paris, Louvain, and Amsterdam, so crucial to Europe, seem like little more than frivolous happenings when compared with the dedication and intensity of the Congolese student movement. Mobutu had succeeded in silencing all countermovements. The trade unions had been bound and gagged; the church was keeping its head down. Only the students still dared to make their presence known.[23]

In April 1967 Mobutu and his staff set up the Mouvement Populaire de la Révolution (MPR); the primary text was written on May 20. The MPR was called a popular movement, but was in fact merely Mobutu's political party. Its meetings were held outside the capital, in the little town of Nsele. Within a few years this riverside village would expand into a vast conference center with white, modernistic visitors' quarters and impressive meeting halls. It became a sanctum of Mobutism. The text drafted on May 20 was sent into the world under the title "Manifeste de la Nsele" and became familiar to all in the course of time. In analogy to Mao's *Little Red Book,* it was published and distributed widely in the form of a little green book, to serve as the catechism of the new regime. From now on, the text stated, every inhabitant of Congo belonged to the MPR. "Olinga olinga te, ozali na kati ya MPR," people sighed. "Whether you like it or not, you're a member by definition."[24]

At first Mobutu seemed to be making room for an opposition party, but he quickly abandoned that idea. Like so many African countries shortly after independence, Congo became a single-party state. The abrupt transition from a monolithic, colonial administration to a democratic, multiparty system had included no intermediate steps, which was precisely why it resulted in a fiasco. The MPR was out to reunite the people. "More than the class struggle, the union of all is the guarantee for progress," the manifesto said.[25] The entire nation had to be made enthusiastic about the country's reconstruction. The inner core of the MPR consisted only of a group of young Mobutuist volunteers, but soon the party's power reached astronomical heights. The MPR became the country's supreme institution—so much so, in fact, that the line between party and state was obscured. "The MPR is the designation of the state," Mobutu's house ideologist went so far as to say.[26] At the top one had the president and his cabinet, the extremely powerful Bureau du Président. Beneath that was the MPR Congress and the Political Bureau, followed by a legislative, executive, and judicial council. All titles had been changed. A cabinet minister was henceforth to be called a state commissioner, a governor was a regional commissioner, and a member of parliament a people's commissioner. Every citizen was a party member, even the ancestors and the embryos.

The students were not at all keen about this. Mobutu was out to do away with organized politics, they noted rightly. In doing so, he was turning back the clock: during colonial times, too, there was only a bureaucracy, an administrative leviathan that maintained statistics and spewed reports but made no allowance for public participation. Congolese academic circles had welcomed the coup at first, but their enthusiasm quickly dried up. The most important student movement took a resolutely anti-imperialist stance. Lumumba became their hero, Mobutu their enemy. When American Vice President Hubert Humphrey visited the country in January 1968 and wanted to lay a wreath at the Lumumba monument, the students viewed that as a provocation. During the ensuing demonstration, many people were arrested.

Clashes between students and the new regime escalated in the course of 1968 and 1969. The students demanded greater say, less interference from the MPR, and a fairer distribution of scholarships. A big demonstration was planned for June 1969, but Mobutu sent his troops to the campus. Lovanium was sealed off from the outside world for days. Still, a few hundred students succeeded in slipping around the guards and made it to the city center by bus. There they entered into heavy confrontations with the army. The soldiers fired tear gas, but the students made masks out of wet handkerchiefs and threw the canisters back. More and more local people joined them. The army opened fire. According to official figures, six people were killed and twelve wounded; the students said there were fifty casualties and eight hundred arrests. MPR? *Mourir pour rien,* die for no reason, the students said in disgust. Mobutu vowed to eradicate the student movement root and branch. Each campus was to have it own MPR youth association, the "Manifeste de la Nsele" became required reading and everyone was ordered to return to their books. The resistance was quashed. The leaders of the student revolt received harsh prison sentences of up to twenty years. This critical voice was now silenced as well.

HANGINGS, TORTURE, MASSACRES. The first five years of Mobutu's presidency read like a catalogue of horrors, but that is only part of the story. Many older people in Congo today look back on that period with a certain nostalgia. "Things were orderly," Zizi Kabongo said when I expressed amazement at that. "The soldiers went back to their barracks. There were commodities to be had, prices went down, industry received a boost. For me, too, it was the start of the most prosperous period of my life."

For the first time since independence, major infrastructural projects were under way. Mobutu began work on the first hydroelectric plant on the Congo: the Inga Dam, which produced 351 megawatts. Kinshasa's new neighborhoods received drinking water and electricity. A sewer system was built. The city's central hospital had fifteen hundred beds and received four thousand patients a day. Ten thousand operations were carried out each year, and 1.6 metric tons (about 1.76 U.S. tons) of laundry were processed each day.[27]

Mobutu was no democrat, but he did change the course of the nation. All able-bodied men had to spend a few hours each Saturday afternoon doing volunteer work for the state, a taxation in kind like that seen in colonial times. *Salongo* it was called now. People were required to pull weeds, repair bicycle paths, and sweep the streets. In addition, to help boost agricultural production, everyone was encouraged to cultivate a plot of ground. Even army generals went out harvesting manioc. Work, work, work. Mobutu himself set the good example. He got up every morning at five. He read piles of newspapers, had breakfast with diplomats, attended a constant flow of meetings, and put in days of eighteen hours or more. In 1969, barely thirty-nine, he had a minor heart attack. "How would *you* lead this goddamned country?" he asked his personal physician.[28]

Mobutu was nothing like the flabby figure he would become later. After the total debacle of the First Republic, he put Congo back on the map. He won respect and gave the country new élan. Had the Americans landed on the moon? He invited the crew of Apollo 11, making Congo the only African country to welcome the moon travelers.[29] Were the Europeans organizing a Miss Europe contest? He convinced the organizers to hold the finals in Kinshasa, and to give them a native twist. The winner, including in the category "African Costume," was a ravishing blonde from Finland. Were Congolese women still seen as the most beautiful on the continent? He backed Maître Taureau in organizing the first national Miss Congo contest. "The winner was Elisabeth Tabares from Katanga. She had lovely heels and not those stubby little toes."[30]

In short, Mobutu made good on the promises that independence had awakened but been unable to keep.

But it wasn't all circuses: there was also bread. In January 1967 a cheerful funeral procession moved through the streets of Kinshasa. Mobutu was there and young people from his corps of volunteers held aloft a cross topped by a pith helmet. The banner draped across it read: "Requiescat In Pace, UMHK, born 1906, died December 31, 1966." The Union Minière du Haut-Katanga was being carried to its grave! The big coffin had been made to fit the dimensions of Louis Walaff, then chairman of the board of directors. In order not to rile the ancestors, the mining Moloch's "remains" were tossed into the river.[31]

This lampoonery, however, represented an extremely important matter. Mobutu made no bones about his displeasure with the way Tshombe had wangled the notorious colonial portfolio away from Belgium. Along with that, of course, there was the humiliation to which Mobutu had been subjected at the economic round-table conference in 1960. Congo, he said, was politically independent, but economically far from that. The figures hardly proved him wrong. Only 5 percent of the employees in Katanga were foreigners, yet they took home 53 percent of all wages paid.[32] The amount they paid for a good bottle of whisky equaled a miner's monthly salary. In 1967, therefore, Mobutu set about nationalizing Union Minière, a move greatly deplored by the Generale in Brussels. The company was rechristened Gécomin, Générale Congolaise des Mines, but later also became known as Gécamines, Générale des Carrières et des Mines. The copper revenues would now flow directly into the national treasury. And those revenues were considerable. With the Vietnam War raging in the background, the price of copper on the international market had gone sky high. The Congolese economy had always profited from wars in other parts of the world: that had been the case in 1914–18, 1940–45, and during the Korean War, but the Vietnam War put even more money in the till.

To underscore his new economic regime, Mobutu also changed the country's currency. At independence, 1 Congolese franc had equaled 1 Belgian franc, but by 1967 it was worth only one-tenth of that.[33] Mobutu introduced the zaïre as the new monetary unit: 1 zaïre was worth 1,000 old Congolese francs, and equaled 100 Belgian francs and 2 U.S. dollars. The first banknote showed Mobutu and a few dignitaries resolutely rolling up their sleeves: "*Retroussons les manches!*" was the slogan. Getting back to business!

For many people, those were golden years. In Lubumbashi I met Paul Kasenge, a former Gécamines employee. "We had everything we wanted. I was twenty-six, and after studying commercial economics I became a manager. I was one of the first blacks to do that. The foreign managers had left, the Congolese took over. We were paid well. Copper commanded a good price. We had a house and a garden. There were schools and hospitals for our children. We even received a loan in order to buy a car, and could pay it off in installments."[34] A bicycle had once been the ultimate dream; now it was an automobile.

For others, the MPR offered new opportunities. André Kitadi, the cautious World War II veteran who had crossed the desert and gone out to dinner in English after the war, told me: "Through the MPR, I became a city councilman in Ngaliema. For the first time I had access to a higher position. I'd been waiting a long time for that." The people were not dissatisfied. When Mobutu had himself reelected in 1970, he received 10,131,699 votes, with only 157 votes against him. Those all came from a single polling place, in Kinshasa's student district. Also worthy of note is the fact that more votes were cast in his favor than there were registered voters, even though poll attendance was not compulsory.[35] André Kitadi had changed his mind only much later on: "The dictatorship brought about the nation's fall, but we didn't know that back then."[36]

SEPTEMBER 1974. Zizi Kabongo is getting ready to go home for the boxing match. During the first five years, Mobutu consolidated his power; during the next five he ruled through generosity. The world-championship heavy-weight bout between Muhammad Ali and George Foreman was to be the absolute high point of that flamboyant bonhomie. The match would go down in history as "the rumble in the jungle"; in Congo itself they called it "le combat du siècle" (the battle of the century). And it was, indeed, one of the great sporting events of the twentieth century. By refusing to enter the U.S. Army ("No Vietcong ever called me a nigger"), Ali had lost his title, but after a three-and-a-half-year suspension he was out for revenge. Foreman was seven years younger, only twenty-five, the Olympic champion, the world champion, unbeatable. He had knocked boxing legend Joe Frazier to the mat a total of six times in only two rounds before the fight was stopped. But Ali wanted his title back.

Promoter Don King was demanding a $10 million purse, an insane amount by all standards. No one was prepared to lay down such an astronomical sum for a slugfest that was bound to last no more than twelve times three minutes. No one but Mobutu. The Zaïrian economy had just gone through six years of uninterrupted growth and it was time for a party. Ali was euphoric about the decision, but probably unaware that the prize money Mobutu was coughing up came indirectly from the war in Vietnam. For him, the match in Kinshasa was his ultimate chance for revenge; for Mobutu it was the ultimate opportunity to do some "country marketing."

That the founding president of the MPR chose boxing as his public relations tool should come as no surprise. Boxing had always been part of the black struggle for emancipation. Fists made possible what the law ruled out: the black man's triumph. In 1910 the American Jack Johnson became the first black heavyweight champion of the world; after he gave Jim Jeffries a thumping, race riots broke out all over America. The Senegalese Battling Siki had beaten the Frenchman Georges Carpentier with a well-placed uppercut in the 1920s: until then, it had been unheard of for a colonial subject to so humiliate a superathlete from the mother country. In 1938 world heavyweight champion Joe Louis beat the German Max Schmeling with a technical knockout. "Heil Louis!" people shouted in the streets of Harlem that night. The fight in Kinshasa was between two black men, but Ali was the Zaïrians' favorite from the start, Zizi said. "The people saw Ali as the good black man. He was very smart, he went into the cité. Ali, boma ye! the people shouted: Ali, kill him! Foreman was considered a white black man, just another American, not one of us."

Muhammad Ali and Mobutu: the two had more in common than might have appeared at first glance. They came together in their distaste for white arrogance; both men wore their blackness as a source of pride. Both had shed their Christian names for politico-religious reasons: the Christian Cassius

Clay had become a militant Muslim; the Catholic Joseph-Désiré now bore
the ancestral-sounding name Mobutu Sese Seko Nkuku Ngbendu wa Za
Banga, "the powerful warrior whose stamina and willpower carry him for
victory to victory, leaving behind only fire" (but also "the rooster that leaves
no hen unruffled," depending on the translator). The American sportsman
and the African dictator were both young, strident voices that challenged
the dominance of the white West. And what voices they were: virtuosic,
voluble, humorous, and razor-sharp. With words, too, one could deal out
blows. The agile French that Mobutu employed with such bravura was the
equal of Ali's cascade of English. Pokerfaced, shortly after the public hangings, Mobutu had told two Belgian journalists: "We Bantus can administer
democracy, but not to the letter, not like you." To a flatterer he once thun-
dered: "I didn't ask you to come here to hear your angelic voice or your
evangelical message. Speak your mind, man! What is your problem?" But
to someone who did dare to speak his mind, he said: "So you're saying that
you feel caught in a game of cat and mouse?" "Yes, that's right." "Well then
tell me: who is the mouse?" "We are, papa!" "And who is the cat?" "Um . . .
we're that too." "All right, so what's the problem?" Ali enriched the English
language with one-liners like: "I'm so bad, I make medicine sick" and "My
toughest fight was with my first wife." During his stay in Kinshasa, he came
up with the immortal quip: "I've seen George Foreman shadowboxing, and
the shadow won."

That latter claim proved close to the truth as well. During a workout
with his sparring partner, Foreman suffered a torn eyebrow and the fight
had to be postponed for five weeks. Zizi Kabongo could stay in Paris a little
longer. But the cultural component of the rumble in the jungle got off to
a start anyway. Mobutu had brought the world's greatest black musicians
to Kinshasa. From Latin America there were performances by Celia Cruz
and Johnny Pacheco, from the United States came B. B. King, the Pointer
Sisters, Sister Sledge, and James Brown. The Cameroonian saxophonist
Manu Dibango and South African singer Miriam Makeba shared the stage
with the big stars of Zaïrian music. Old Wendo Kolosoyi, godfather of the
rumba, was there, along with Franco and his OK Jazz. Tabu Ley, the man
once known as Rochereau, gave a show, and the younger generation was
represented by the funk-driven *soukous* band Zaïko Langa Langa, the most
influential Congolese group of the 1970s. The three-day music festival in
Kinshasa was a powerful, intercontinental expression of African pride.[37] It

was a sort of black Woodstock. What the slave trade had driven asunder, Mobutu brought back together.

At last Zizi was able to leave. He took the opportunity to visit his father in Kasai; he had bought a flour mill for him in Europe. A former railway official with the BCK, his father too had become a part-time farmer under Mobutu's new agricultural plan. An electric mill made it so much easier to grind manioc. "When I saw him, my father was very upset. Mobutu had just said that the American artists were the descendants of slaves, and that slaves had not been sold into captivity by the whites, but by native chieftains. He said: 'Mobutu claims that the blacks sold our brothers to the white man!' 'He's right, Papa.' 'But that's unbelievable!' It distressed him greatly. I suspect that Mobutu was spreading those ideas intentionally. It helped him to break the power of the local chiefs."

Mobutu did all he could to combat ethnic reflexes. A strong nation was incompatible with tribal logic; the younger generation had to be given a new frame of reference. The national soccer team had to include players from all over the country. Girls from each province took part in the Miss Zaïre contests. The army was to be inclusive as well: even Pygmies were allowed to join the military.[38] To boost Zaïrian awareness, Mobutu also implemented reforms in higher education. The country's three universities were fused to form one huge, national super-university with three campuses. You could go to Kinshasa to study law, economics, medicine, the natural sciences, or polytechnics. Kisangani was for psychology, teaching, or agricultural engineering. And Lubumbashi, close to the mines, was for earth sciences. There too, far from the capital, one could follow the "risky" curricula such as social sciences, philosophy, and literature.[39] This reform weakened the student movement and resulted in an obligatory tribal mixture among young academics. The most striking example of that was one I came across in a yard in Bukavu one evening, just before dusk. I had been invited to visit Adolphine Ngoy and her family. Her daughter was preparing dinner over a charcoal fire. Adolphine came from Moanda, a seaside town on the Atlantic Ocean. How had she ever ended up two thousand kilometers (almost 1,250 miles) to the east, close to the Rwandan border? "Dodo and I met in Kinshasa. He was doing the polytechnic, I was studying linguistics. He was a Mushi from Bukavu, I was a Mukongo from Moanda. As the eldest son he was supposed to marry within his tribe, but he chose me. I moved here. His family objected strenuously. It took years for the neighborhood and the family to accept me."[40]

Just as the Erasmus Program was intended to instill young people with a greater love for Europe, by means of a foreign sweetheart if need be, so too did Mobutu's education reform create greater Zaïrian awareness. Mobutu liked to surround himself with young, enthusiastic Zaïrians who were completely taken by his national project. The two most influential people from his entourage were Citoyen Sakombi Inongo and Citoyen Bisengimana Rwema.

In April 2008 I traveled from Goma to Bukavu by ship across stunningly beautiful Lake Kivu, which forms the border between Rwanda and Congo. On board I was introduced to a reserved, extremely distinguished young man: the sort of gentleman one would never find out on the windy rear deck of a passenger ship, but who prefers to remain below deck and make phone calls. He was the son of Bisengimana, who had been the number-two man in Zaïre for years. "My father started working for Mobutu in 1966, but in 1969 he was promoted to director of the Bureau du Président de la République. Mobutu trusted him highly. My father was even allowed to disagree with him. They called him *le petit léopard* (the leopard cub). He wore a leopard-skin hat too. He remained Mobutu's cabinet chief until 1977, when they had a falling out. After my father left, no one ever had as much power again under Mobutu."[41]

The most unusual thing about that appointment, however, lay outside the ship's window. The boat roared across the water. On our port side rose up the contours of the island of Idjwi, with Rwanda just behind. Bertrand Bisengimana was from that island and he was on his way home. Idjwi had been a German possession at first, but passed into Belgian hands even before World War I. The population consisted largely of Tutsis from Rwanda. Like him and his father. The Tutsis were an ethnic minority that had for centuries formed the social and political upper crust of the Rwandan empire, a position they owed to cattle breeding. Cows were to the Tutsi what coal had been to the industrial barons: everything. As early as the nineteenth century, Tutsi cattlemen had left overcrowded Rwanda to settle on the lake's far shore. They moved onto the plateaus of South Kivu, to the volcanic region of North Kivu, and to the island of Idjwi. To the Congolese they were, in every way, "different." They looked different and spoke differently. Their Kinyarwanda was a highly specific Bantu language, spoken only in Rwanda and the south of Uganda and related to the language of Burundi. The archetypal Tutsi was tall to extremely tall (1.95 meters—about 6 feet 5 inches—was not unheard of), with a pointed noise, a high forehead, and thin lips. A cliché, of course, but every bit as true as the clichés concerning the Irish, Italians, and Swedes. According to that same cliché, they had a reputation in Zaïre for being arrogant and devoid of humor, yet Mobutu still appointed one of them to be his cabinet chief.[42] "In the beginning, Mobutu didn't want to give preferential treatment to his own tribe," Bertrand said, "otherwise my father, a Tutsi from Idjwi, could never have become the regime's second man." For Mobutu, of course, there were advantages to the fact that his direct associate came from a small tribe of migrants that could pose no threat to him. . . . Little did he know then that, in 1997, Rwandan Tutsis would depose him.

MOBUTU HAD GIVEN THE PEOPLE greater prosperity; now his aim was to give them a dream. That dream was Zaïrian nationalism. And that dream's architect was named Dominique Sakombi, better known—in accordance with the tenets of the day—as Sakombi Inongo.[43] Sakombi was an intelligent, extremely eloquent young man, a greater Mobutist than Mobutu himself. In spring 2008 I spoke to him briefly on the phone: his voice had grown thin as rolling paper, in no way reminiscent of the vocal barrage of yesteryear. He was very ill and could not bring himself to grant me an interview.

In the early 1970s Sakombi's achievement was a particularly ingenious one: he did not ban tribalism, but raised it to the state level. The Zaïrians were still allowed to love their tribe . . . as long as that tribe was called Zaïre. He said: "For us, the ancestral village extends all the way to the borders of the national territory."[44] The arbitrary domain established by nineteenth-century European politicians now had to feel like a natural phenomenon. More than the head of state, Mobutu was to become the national village chieftain, the headman de luxe. And the citizens were his villagers, his children.

Sakombi was the state commissioner of information. His ministry had fourteen hundred staff members, its budget second only to that of the ministry of defense. Mobutu knew where his priorities lay: in a former lifetime he had been both a soldier and a journalist. If his dictatorship had at first relied on the power of the army, from 1970 on it relied on propaganda.

Sakombi designed a sweeping cultural policy that was marketed to the people under the slogan *Recours à l'authenticité!* Resume authenticity! The changing of the name of the country, of the cities and even of the citizens themselves was a part of that, but it went much further. The resumption of authentic living impacted almost every aspect of daily life. When a Zaïrian got up in the morning he knew what to wear. A ban had been imposed on Western clothing. Men were no longer to wear a suit and tie, but were obliged to put on an *abacost,* a high-necked outfit based on Mao's own, with an upright collar and cravat. (*Abacost* was yet another Mobutuist neologism: it came from *à bas le costume* [down with the suit]. The language, too, was being changed.) Women were no longer allowed to wear miniskirts, only the traditional *pagne,* an elegant outfit in three parts—skirt, blouse, and headscarf. Only natural hairstyles were allowed. Extensions and the "conking" or straightening of hair was forbidden. Even more strongly forbidden were preparations for lightening the skin. The authentic Zaïrian was the diametrical opposite of the *évolué,* a person who no longer aspired to be what he would never become anyway, but who drew strength from his or her own identity, culture, and traditions.

If the Zaïrian happened to live in the city, he saw on his way to work new monuments being erected everywhere. The statues of Stanley, Leopold II, and Albert I were pulled down. As Sakombi drily stated at the time: "As far as I know, there is no statue of Lumumba in the center of Brussels either."[45] On squares and in front of government buildings arose stylized figures in concrete, their arms raised to the sky or toting baskets. Two hundred sculptors were active in Kinshasa alone.[46] Their style was strikingly modern (works influenced by Ossip Zadkine, Pablo Picasso, and Constantin Brancuşi were legion), but that was all right, for those Europeans had themselves been strongly influenced by African art. The authenticity policy was no exercise in nostalgia, but a complex admixture of tradition and modernity. Of it Sakombi said: "We respond as our forefathers would have done, had their culture not been interrupted by colonial acculturation."[47] He was not out for a *retour à l'authenticité,* a going back, but a *recours,* a resumption. From out of the old visual idiom, a new art was to be born. And so Mobutu had art treasures brought together from all over the country. Tens of thousands of masks and fetishes found their way into the national museums, just as all manner of artifacts had made their way to Tervuren during the colonial days.[48] The national ballet was expected to study traditional dances in the interior and reinterpret them. A national theater company was set up, a national literature prize established.[49]

When a Zaïrian listened to the radio during the day, Zaïrian music was what he heard. Western music was banned. Mobutu set himself up as the great champion of popular music. Franco, the leader of OK Jazz, was placed at the head of a new government institution intended to support the music business. Wasn't this the man who had stood beaming beside Mobutu at the leader's birthday party, just before the coup? Tabu Ley toured the country. With Mobutu's support; he even became the first black man to perform at the Olympia in Paris. Docteur Nico experimented with traditional percussion. Franco brought the old accordionist Camille Feruzi back into the limelight. *Recours à l'authenticité,* one heard him sing. Kinshasa's music industry experienced its busiest years. Music recorded at five in the evening was in the shops the next morning at nine. There were recording artists everywhere. In Matonge, the absolute heart of the city's nightlife, the central square of the Rond-Point Victoire was redubbed Place des Artistes. A huge statue was erected there to the pioneers of Congolese, no, Zaïrian music.

When the Zaïrian came home from work in the evening, he ate authentic cuisine. *Pundu, fufu, makayabu:* manioc loaf, grubs, all seasoned with the mother of all peppers: *pilipili.* Before taking a sip of your beer or palm wine you first spilled a few drops on the ground. Making libation to the ancestors, that was part of it too. When you turned on the TV after your wonderful meal, you saw the *animation politique,* huge groups of people in geometrical formation, all dressed in the same outfits (usually of green cloth with the national flag on it), dancing and singing the praises of the MPR. Day in, day out, songs went up to the benevolence of the illustrious leader. This went on for six, sometimes even twelve hours a day.[50] Then, at six o'clock, there began the high point of state television: the news. It opened with one of Sakombi's ideas. The president's face appeared against a sky full of fluffy clouds and grew larger and larger, until it looked as though Mobutu were floating down from heaven, right into your living room. The children thought that he was God the Father. "Everything the president and his wife did was shown on the news," Zizi said, "and also everything done by the members of the Political Bureau and the central committee. It became a real personality cult. Sakombi called Mobutu 'the African pharaoh.' That kind of thing."

Even when one crawled into bed at night one could not leave the state's propaganda behind, for Mobutu had called on the people to be fruitful and multiply—the revolution required many pairs of hands. Even at the most intimate moments of one's private life, one heard the supreme leader calling. The joke went that, during lovemaking, he himself never cried out "Ça va jaillir!" (I'm going to come!), but "Ça va zaïre!" . . . Just as the missions had dictated their view of the "good" colonial body (the use of soap, the covering of one's nakedness, the practice of monogamy), so too did the dictator worm his way into the intimacy of personal life and subject it to a new, all-inclusive regime. There was no getting around it. To have an orgasm was to serve the nation.

And it worked. The Zaïrian began feeling Zaïrian. With Sakombi's help, Mobutu accomplished within a few years what the European Union has failed to achieve after more than half a century: people truly began feeling like part of a greater whole. The British and the French still refused to become Europeans, but the Bakongo and the Baluba were proud to be Zaïrian.

BUT WAS THERE NO RESISTANCE? Of course there was, but only discreet resistance. Zizi: "Not being allowed to wear a necktie, that was difficult. In Katanga you sometimes saw men walking down the street wearing a suit and an ascot, out of protest. The police would stop them right away: 'What's with the colonial outfit? What are you, a foreigner?' 'Yes, from Zambia,' they would say then. After all, you could be executed for that!" While in Europe the necktie became the symbol of bourgeois values and repression, in Congo it developed into a statement of resistance and the desire for freedom. "Some people would put on a necktie, just to sit in the living room."

The mandatory name change also prompted sly protest. "My father sent me a list of nine names from our family, from which I could choose my *postnom*. But one of my colleagues was named Gérard Ekwalanga. He was a great sports journalist and very religious, so he was quite attached to his Christian name. In protest, he named himself Ekwalanga Abomasoda. That *postnom* was not an ancestral one at all. In Lingala it means: 'He who kills soldiers'! Or Oscar Kisema, who chose the name Kisema Kinzundi. That sounds like a normal name in Lingala, but in Swahili it means the 'big vagina.'"

The ban on Christian names came as a blow to the church. "Mobutu wanted to destroy the power of the Catholic Church," Zizi said. "He wanted to replace the saints with the ancestors." At first the church had shown itself loyal to the new regime. One month after the coup, Cardinal Malula solemnly stated: "Mr. President, the church recognizes your authority, because authority is God-given. We will faithfully abide by the laws you see fit to pass."[51] But six years later, on January 12, 1972, this same Malula delivered a cutting speech against the regime. Mobutu was furious. He immediately expelled Malula from the Order of the Leopard, sent him into exile abroad, and forbade Christians to pray for their archbishop. To little avail. The church long remained one of the regime's most vociferous critics. The bishops enjoyed the backing of an international network and they also ran the schools. States usually have two means for molding their citizens: the schools and the media. Mobutu had only the media. He therefore did all he could to curb the power of the church (mission schools had to have a native headmaster, crucifixes were burned, seminarians had to join the MPR youth movement, Christian young people's organizations were banned, Christmas became a normal working day, even all religious gatherings, with the exception of mass and confession, were taboo at a certain point). And when none of this had the desired effect, he simply offered the bishops top government positions or bought them jeeps and limousines.

Mobutu's cultural policies did not explicitly stipulate what the Congolese were to believe, and ancestor worship received no detailed national theology, but Kimbanguism, the religion persecuted so heavily under the Belgians, flourished as never before. It was seen as an authentic African religion. The Kimbanguists' own organization developed into a miniature version of the state: hypercentralistic and hierarchical. The religious leader was venerated in song and dance, just like Mobutu. The underdogs of the colonial era now became the heralds of Mobutuism.[52]

RELYING ON ONE'S OWN IDENTITY was a lovely idea, but of course also fraught with catches. Why did Mobutu promote the native kitchen, when his own favorite dish was still *ossobuco alla romana*? What was so authentic about that gruesome *animation politique*, which he had only copied from Kim Il-sung of North Korea? What was so Zaïrian about the notorious *abacost*, which was really nothing more than a Mao outfit with more color to it, the finest examples of which came from Arzoni, a textile plant in Zellik, close to Brussels? What was so typically African about the *pagne*—made from Indonesian batik and praised by the nuns for the way it covered the breasts—the most colorfast variations of which (the famous *wax hollandaise*) came from the Vlisco plant in Helmond, the Netherlands? What made Camille Feruzi an authentic musician? He played the accordion, for God's sake, and he had obviously listened to a lot of Tino Rossi.

Was this *recours à l'authenticité* then simply a ruse? A charming ideology meant to disguise a deeper reality? Yes, it was. And that deeper reality was: Mobutu had started caring less and less about his people. He was so busy safeguarding his position that he neglected major governmental duties. He was so caught up with handing out cars, appointments, honorariums, and ambassadorial posts that the state treasury was drained. Yes, one could speak of economic recovery, but that was due more to Vietnam than to prudent policy making. It was a chance period of economic boom on which Mobutu was able to surf along in comfort, but it in no way served to combat poverty. He used the wealth of revenues to keep his own power base intact. In essence, he owed his power to an extreme form of pork-barrel politics. Mobutu perched atop a pyramid of clientalism where, directly or indirectly, thousands ate from his hand. He and his retinue were bound to each other by a network of mutual debts and favors. In exchange for benefits, his followers gave him the loyalty he needed to remain in power. Mobutu needed them and they needed Mobutu. A monstrous alliance. Mobutu was a slave to his own thirst for power.

In Zaïre, therefore, a true state bourgeoisie arose, a large group of individuals who owed their prosperity to the regime.[53] In the most literal sense, the state served as economic base for this new middle class, which did not hesitate to diplay its newfound wealth in the form of expensive cars, lovely homes, and a luxurious lifestyle.[54] Those who drove around in a Jaguar or Mercedes received the nickname "Onassis." "And anyone who felt a nasty cough coming on flew to his family physician in Brussels," Zizi said.

This clientalism went well as long as there was money. The nationalization of Union Minière had produced huge revenues for Mobutu, but his attempts to retain power were consuming more of that money all the time. "I used to have, as it were, no family at all," he moaned once, "no one cared a whit about me! But since I have become president, it seems that half of Zaïre has discovered that they could very well be related to me in one way or another, and therefore have a right to my assistance."[55] All of this took place, of course, to the disadvantage of the common Zaïrian, who was unable to recall any family ties with the head of state. To keep his growing clientele satisfied, Mobutu had to keep on finding new sources of income. Foreign investments, bilateral agreements, and international loans came in handy.[56] The more needy his country was, the more he was able to rake in. Poverty pays. It was an economic jackpot.

But it was still not enough. On November 30, 1973, he made a drastic decision. He had just returned from a tour of China, where he had seen the country's planned economy. "The peril is more white than yellow," he said upon his return. "Politically we are a free people, culturally we are becoming that, but in economic terms we are not at all the masters of our fate."[57] Mobutu began a process of "Zaïrianization": those small- and medium-sized businesses, plantations, and trading companies still in the hands of foreigners, a few thousand enterprises in all, were expropriated and given to his faithful followers.[58] From one day to the next, Portuguese restaurant owners, Greek shopkeepers, Pakistani TV repairmen, or Belgian coffee growers saw the work of a lifetime disappear. At the head of their company came a Zaïrian from the president's circles who usually had no sense of how to run a business. In the best of cases he allowed the original owner to work on as manager and came by each month to collect the profits. In the worst cases, he immediately emptied the till and sold all the stocks on hand.

The consequences were grotesque. An elegant lady who never left the capital might suddenly be running a quinine plantation on the other side of the country. Gentlemen who couldn't tell a cow from a bull became heads of a cattle company. Generals were allowed to run fisheries, and diplomats soft-drink factories. Minister of Information Sakombi became the owner of a series of newsstands and movie theaters, but also of a few sawmills. Bisengimana received the Prince de Ligne plantations on Idjwi, which comprised one-third of the island itself.[59] Our friend Jamais Kolonga, a small fish in the network around the president, became head of a lumberyard in his native district. The party animal from the capital was now suddenly required to manage a stock of tropical hardwood. Some made a mess of it, others rose to the occasion. In one fell swoop, pop star Franco became the new owner of Willy Pelgrim's recording empire, a sector with which he was indeed familiar.[60] Thanks to Zaïrianization, Jeannot Bemba became the country's wealthiest businessman. He was made chairman of the employers' association and even started his own airline, Scibe Zaïre. Finally, Mobutu treated himself to fourteen plantations spread all over the country. He controlled a quarter of the production of cacao and rubber, had twenty-five thousand people on the payroll, and so became the nation's third biggest employer. Thanks in part to the mining revenues, he was now estimated to be the world's eighth richest man.[61]

But Mobutu saw his country, and it was not good. In late 1974 he switched to "radicalization." Ailing companies were now taken over by the state. That way they could continue to yield revenue and with those yields he could stay on friendly terms with his friends. It had not been a good idea, letting them run the companies. But this new economic reform worked out badly as well. Without asking for it, Mobutu, that close friend of the Americans, suddenly found himself stuck with a communist economy. By means of a third reform, this one dubbed "retrocession"(rhetoric was the only branch of business still solidly on its feet), he tried to give the plucked and dressed companies back to their original owners, but they were no longer interested in the least.[62]

The social consequences were disastrous. As brilliant a communicator as Mobutu was, he was an equally great flop as an economist. The fiasco of Zaïrianization caused unemployment to rise. Those who still had a job, for example as civil servant or teacher, could no longer get by.[63] Everyone moonlighted, as bricklayer, chauffeur, or beer vendor. Their wives tried to earn a little through microcommerce. They would spend the whole day on the market, sitting beside a little pile of charcoal or onions. They bought bread at the wholesale bakery and carried it on their heads around town until it had all been sold. They stayed at home with the children and started a little shop where the neighbors could come to buy tea bags, matches, and soap. They let their homes be used as depots for a brewery or a cement factory, and sold soft drinks or bags of cement at a tiny profit. Everyone tried to make ends meet. Even if that meant turning to their families for help.

In 1974 things came to a head. The end of the Vietnam War resulted in a dramatic fall in copper prices. What's more, the start of the oil crisis was being felt in Zaïre too. Prices shot up. The entire process of Zaïrianization contributed even more to inflation, for now that a class of the super-rich had arisen, shopkeepers pumped their prices up as well. For the average citizen, however, this meant a further decline in spending power. In 1960 an unskilled worker had to work one day in order to pay for a kilo (2.2 pounds) of freshwater fish; by the mid-1970s that same worker needed to work ten days to do so.[64] Food became unaffordable and consumed the entire family budget. Farming in the interior had been neglected. Why should a farmer cultivate his land when there were no more roads to bring his goods to market? Zaïre, one of the world's most fertile countries, therefore became highly dependent on expensive, imported food. Cans of tomato paste were unloaded in the harbors, while in the interior, tons of beefsteak tomatoes hung rotting on the plant.

Mobutu's promise of economic recovery had ended up in catastrophe.

One of the MPR's early slogans had been: *Servir et non se server* (Serve, but not to serve yourself), but Mobutu and his clan served themselves very well indeed. His popularity declined. The bread was running out. And what about the circuses?

AFTER HIS ADVENTURE WITH BAUDOUIN'S SWORD, Longin Ngwadi returned to Kikwit. He began work as a salesman for Bata, the international shoe chain that had shops in Africa. One day he saw a pretty girl come into the shop. She looked at a few models, then went off to buy fish. A few minutes later Longwin closed the shop for lunch and went after her. She was just finishing her shopping. Fish was still affordable in those days. He spoke to her the immortal words:

"I'll pay for your fish, if you become my fiancée."

"Really?"

"Yes, really?"

"Then I'll give you my address."

That evening he visited her at her home. She called in her father and her uncles. The family first wanted to see who this oddball was.

"I am ready to take this girl as my wife," Ngwadi said.

"Do you have money?" the family asked.

"Yes."

That was not entirely true, but his European boss at the Bata shop was willing to advance him the dowry. It wasn't the first time he'd done that for an employee. Ngwadi had to pay back a little each month. Bata had a good reputation, it was a serious shop. The father and uncles gave their approval.

Ngwadi worked for Bata for many years. Meanwhile, his wife, because that happened to be the tradition, worked the land: she grew corn, manioc, and peanuts. The young couple had everything they needed. The first of six children was born in 1969. A few years later Ngwadi bought a big lot, thirty meters by forty (nearly 100 by 130 feet), and built a spacious clay house, the same house where I interviewed him. "That was the wealthiest period of my life."

But then came Zaïrianization. "My European boss left. Bata was taken over by a Zaïrian. He ran the company. That was not good. Bata went bankrupt." Hard times came. Increasingly often, Ngwadi went to pray at the grave of Kuku Pemba, a dangerous spot, a mythical spot. Kuku Pemba had been the first man from that region to see a white man. In times of famine, people turned to something higher. Kuku Pemba was seen as a powerful ancestor; even Mobutu was afraid of him.

In 1974, for the first time in years, Ngwadi went back to the capital. "I went to Kinshasa to see the fight. I saw Ali praying too. He was a Muslim and wore a little chain around his neck. Foreman had a big dog with him, like a European. I was sitting in the stadium. The match was held in the middle of the night. Foreman was stronger. Ali went to the ropes. He did that throughout the whole match. Foreman was swollen up like a pig. It was a fantastic fight, fantastic!"

How could you stay angry at a president who treated you to a wonderful party like that?

Because the American viewers needed to watch the fight during prime time, the match did not start until 4 A.M. The air in the city was hot and humid, the rainy season had arrived. The stadium already began filling up the morning before the match. "The children had a day off from school. Companies had to give their employees a day of paid leave. The bars had to sell beer at half price. Even the flour was free," Zizi Kabongo recalled. Spectators came from all over, even as far away as Angola and Cameroon. Seventy thousand people had a seat in the stadium. A few thousand of those seats were reserved for VIPs, most of them Mobutu's yes-men. A huge crowd milled around outside the stadium. Because of the unusual hour, Mobutu had had lighting installed. Around the grandstands, four gigantic flyswatters rose up out of the darkness. They were equipped with a battery of blinding lamps that, thanks to the current from the Inga Dam, bathed the entire stadium in bright white light. Mobutu was truly electric.

In the middle of the soccer pitch stood the ring where it was all going to take place. The American TV crews had brought an impressive array of equipment. The children sitting on the concrete steps beamed with pride. Theirs was the only country in the world that could organize this match! Even the ring had been brought over from America! The Americans even had their own water with them! And their own toilet paper!

The Zaïrian television crew, too, was well-equipped. To make sure nothing could go wrong, the state broadcasting company had bought five brand-new Arriflexes, heavy cameras you could carry on your shoulder. In addition, the reporters had a few Bell & Howells, lighter cameras for details and close-ups. Everything in color, of course. There were two directors, two

commentators in French, and one in Lingala. All of them received a great
deal of extra pay for working at night.

Zizi Kabongo was put on the camera that was to record the crowd's
reactions. A brass band marched around the track, playing traditional Congolese music. A huge cheer went up when Ali came out of the catacombs and
moved to the ring, dancing and shadowboxing as he went. He removed his
cape. A god's body shone in the spotlights. *Ali, boma ye! Ali, boma ye!* Zaïre
chanted.

But the most amazing thing of all was: Mobutu himself wasn't there.
The stadium where the people had welcomed him in 1965 he now avoided.
Was he afraid he might be less popular than Ali? Did he fear for his own
safety? Did he consider himself, as president-founder, more emphatically
present when he was absent? Kabongo wasn't sure. He did know, however,
that Mobutu was viewing his shots live in his own palace. The chief, it so
happened, had the country's only closed-circuit television network. Kabongo
let his camera glide over the sea of faces. On his monitor he saw the colorful
celebration of a cheering crowd reduced to a mute scene in bluish gray.

He was only able to catch an occasional glimpse of the match itself. He
did not see it when, in the very first round, Ali tried to knock out Foreman
with a series of brutal right leads to the body and face. He did not see how
enraged Foreman became and how Ali forgot to dance. "Float like a butter-
fly, sting like a bee," that was what he'd promised. Dance he would, dance
he must, but it didn't happen. Through his camera Kabongo saw only the
crowd, the crowd that cheered at first, then cowered in fear. He did not see
how, starting in the second round, Ali kept to the ropes and leaned far back
to avoid Foreman's blows. Ali concealed his face behind his black gloves and
soaked up a relentless tempest of punches to the body. "Everlast" was the
word printed on the cushions in each corner of the ring, but the question
was whether this could last forever. Foreman had one of the hardest punches
in the history of heavyweight boxing. Ali's plan was to beat his opponent by
exhausting him. The "rope-a-dope" was how he would refer to the technique
later. Kabongo did not hear how Ali kept shouting around the white grimace
of his mouthguard: "George, you disappoint me." "Come here, sucker! They
told me you could punch." "You're not breaking popcorn, George."

Kabongo filmed and filmed. His shots were not meant to go around
the world. The Americans were taking care of that. This was for domestic
consumption. He saw the dignitaries: the state commissioner of sports, the

provincial governors, the diplomats, the members of the Political Bureau and the central committee: the whole clique that ate from Mobutu's hand. Sycophantic spectators came up and handed him money, asking him to get a good shot of them, so the president would see. Especially women. A woman wearing a red *pagne*, a lady in white . . . Could he zoom in on them just a little?

Every once in a while he turned and looked. Each time he saw that giant of a Foreman pounding Ali, who was hanging terribly over the ropes. Kabongo did not see how, in the eighth round, thirteen seconds before the bell, Ali suddenly bounced off the ropes and struck out with a lightning right-left-right combination. The final punch was a crushing sledgehammer blow to Foreman's jaw that distorted his face into a lump of modeling clay. Foreman's arms, which had rocked like the steel crossbars on a locomotive for eight whole rounds, suddenly milled helplessly in thin air. Foreman bent over, he couldn't believe it. He had never been knocked out before. The floor of the ring rushed up to meet him.

IT WAS A CRAZY NIGHT. Right after the match, an extremely heavy thunderstorm broke loose. The nightclubs of Kinshasa filled to the brim. The drinks were on the house. Everyone partied, everyone laughed, everyone went on a binge. But on his way home, Kabongo couldn't help but wonder how Mobutu had viewed his shots. Alone in his palace with a few family members? Reveling in the spectacle he had given his country? Curious about the woman in the red *pagne*? Or peering restlessly at the crowd's reactions, apprehensive of every face that did not smile broadly enough?

CHAPTER 10

TOUJOURS SERVIR

A Marshal's Madness

1975–1990

I N THE LONELINESS OF HIS OMNIPOTENCE, MOBUTU CONTIN-
ued to stare at the screen. But, fifteen years after the historic bout,
he saw things that knocked him for a bigger loop than any footage he'd seen
before. It was Christmas Day 1989 and on a foreign channel he saw a turtle
poke its head out, slow, helpless, with the fear of death in its eyes. No, this was
no turtle, it was a man who came crawling—or who was squeezed, rather—
from a compartment beneath an army tank. Amid the grayish-green steel
his upper body moved so clumsily—his arms were pressed against his sides,
his hands still in the compartment—that it made him look like a turtle. A
soldier waiting on the street supported the man and pulled him out, like a
midwife.

The video footage was yellowed and grainy, the scene had something
wintery about it. But Mobutu recognized the man right away. It was Nicolae
Ceaușescu. He and his wife had been arrested shortly before, after days of
protest in his country. Mobutu watched the Romanian president stumble
to his feet and take off his black astrakhan cap to arrange his hair. The cap
looked like a wintertime variation on his own leopard-skin model. That was
not the only similarity. Like him, Ceaușescu had come to power in 1965
and Mobutu greatly admired the way he had kept Romania on a course
independent from the Soviet Union. And like Mobutu, Ceaușescu had been
able to count on great Western support. Both men derived their power from
faithful allies abroad and an obedient clique at home, which allowed their
presidencies to grow into a sort of monarchy. Both were fond of the same

nickname: Ceauşescu had people call him the *Conducător*, the leader, while Mobutu liked to be called *le Guide*. Surrounding the "Genius of the Carpathians," another one of those nicknames, there had grown a personality cult as remarkable as that surrounding the "Great Helmsman" in Kinshasa. In Zaïre, the philosophy of authenticity had meanwhile been transformed into "Mobutuism"; in Romania, "Ceauşesism" reigned supreme. With so much authority on their sides, neither of them were good at dealing with criticism. They curbed the freedom of the press and when it came to dissidents, they were pleased to see the back of them. Let them spew their rancor over full ashtrays in some grimy Parisian café, blind as they were to the blessings these men had brought with them. The security of the state deservedly took pride of place. Ceauşescu's Securitate displayed striking resemblances to Mobutu's DSP, the Division Spéciale Présidentielle. The ties between Kinshasa and Bucharest were extremely warm and were topped off by a close personal friendship between Mobutu and Ceauşescu. Mobutu looked to America for money and to the East for methodology. He had learned a lot from Mao and Kim Il-sung, but the only Communist head of state with whom he was still on friendly terms was Ceauşescu. Their wives got along well too.

Mobutu saw the footage. Only one month earlier their two parties had held a bipartisan summit in Bucharest.[1] Now he watched as Nicolae and Elena took their seats in a dismal classroom. How worn-out they looked, all of a sudden . . . Nicolae was a gray-haired senior citizen in a long winter coat, Elena an elderly lady with a big fur collar. An old Eastern European couple. They were sitting at a desk with thin metal legs. Nicolae waved his arms, raised his voice. The camera swung to the right. Shots of a few ranking officers covered in medals. Soldiers who sprang to their feet. A man reading aloud a printed statement.

It had been an extremely turbulent year in Europe. Glasnost, perestroika, the Wall . . . Mobutu followed it all vigilantly. Mikhail Gorbachev's impetus for a political thaw had prompted a chain reaction no one could stop now. Least of all Gorbachev himself. The democratization of a huge, single-party state seemed a totally reckless venture to Mobutu:

> Look at what is happening in the Soviet Union; even without a multi-party system installed, allowing the mere possibility of it was enough to cause regionalism and separatism to rear their heads. I pass no judgment whatsoever on the Baltic, Armenian, Georgian, or Belorussian

movements; I limit myself to noting that the mere thought of a multi-party system works in favor of centrifugal forces.[2]

Democratization, Mobutu was leery of that. He remembered all too well the debacle of the First Republic. The fall of Communism in Eastern Europe in many ways resembled the decolonization of Africa: an abrupt process in which a latent hope suddenly found itself caught up in an uncontrollable momentum. With true sophistry, he reasoned: "Were we to forcibly impose here a Western-style democratic system, only then would we fall into dictatorship."[3]

The end of the Communist era had arrived all over Central and Eastern Europe, without bloodshed. During the last few days, Mobutu had seen squares in Bucharest where tens of thousands of people had braved the cold to demand that the Conducător step down. But this shaky footage from a little village outside the capital gave him the real shivers. Suddenly Nicolae and Elena were no longer sitting in that classroom but standing on a empty playground, in front of a yellow wall. Mobutu saw a cloud of dust. Heard a rattling. Like someone shaking a can full of pebbles. The home video of world history. Faded colors. Muffled voices. Eternal winter. The camera then swept over two wax figures. Elena lying on her side, indifferent to the flow of blood trickling from her head. Nicolae on his back, his calves folded back unnaturally beneath his body, like a jumping jack. Mobutu looked, and kept looking.

ZOOM OUT. Dolly back. Reframe. New focus: more than ten years earlier, 1978. Bright sunlight. Mobutu, brimming over with self-confidence. Shots of his figure. He's grown a bit fatter since the coup; the presidency has obviously served him well. In 1970 and again in 1977 he was reelected as head of state. The term of office had been extended to seven years and there was no longer a statutory limit on the number of consecutive terms. Mobutu had always been the sole candidate. At the polls, the voters had only to deposit a red card or a green one in the ballot box. Red, as one was told by an MPR official in the polling place itself, stood for chaos, bloodshed, foreign ideologies. Green was the color of hope, of manioc, and the MPR itself. Everyone could see you vote. Mobutu received 98 or 99 percent and governed more comfortably than ever. He walked a little more slowly, spoke a little more slowly too. Dignity had become more important than energy.

The rocket was ready to be launched. At the edge of a plateau overlooking the valley of the Luvua stood a slender juggernaut, twelve meters (thirty-nine feet) high. It was supported by a double framework of steel. It was 11:30 A.M. on Monday, June 5, 1978. A beaming Mobutu had invited a gaggle of friends and journalists to witness another of his stunts: the launching of a rocket from Zaïrian soil. He had come to an agreement with a German firm a few years earlier. That company, OTRAG (Orbital Transport- und Raketen Aktiengesellschaft), was given free run of a huge stretch of savanna to experiment with the construction and launching of inexpensive rockets. OTRAG received German government funding in its attempts to find an alternative to the costly projects run by NASA and the European Space Agency (ESA).[4] In the long run, these German *Billigraketen* (bargain rockets) were meant to launch satellites into earth orbit for a mere pittance. A private company that built rockets: that was a first in the history of space travel. And a company that received support from an African dictator: that was an absolute first. Driving force behind the project was Lutz Kayser, but the most striking name on the payroll was that of Kurt Debus, a German who had worked during World War II on the development of the V-2 and who after the war had headed the Kennedy Space Center for many years, where he was in charge of the Apollo program.

OTRAG had gone looking for a large, empty spot along the equator, and had already taken a look at Indonesia, Singapore, Brazil, and Nauru, all countries bordered by an ocean. Zaïre entered the picture only late in the game. The savanna of Shaba, former Katanga, was thinly populated enough too. Within ten days in 1977 a deal was signed with Mobutu: the arrangement was stunning in every way. OTRAG became lord and master over an area of one hundred thousand square kilometers (thirty-nine thousand square miles), one and a half times the size of Ireland. It was reminiscent of the nineteenth-century rubber companies with their huge concessions that allowed them to "do business" unobstructed. For a period that extended to the remote year 2000, OTRAG leased almost 5 percent of Zaïre's territory under extremely favorable terms. The company was exempted from paying import duties and was not to be held responsible for any environmental damage. Its employees paid no taxes and enjoyed legal immunity. And because the savanna was not quite as empty as the ocean, they were even allowed to relocate native settlements if they got in the way of the launch. Mobutu, the man who had fought against secessions and rebellions, was now effectively handing over the control over a substantial part of his country. In return he asked for no more than 5 percent of the net profits, if profits were ever made, and the launching of an observation satellite for domestic security, if such a launch should ever take place.[5] But things never got to that point. In anticipation, however, he pulled in $25 million dollars' rent each year, which immediately disappeared into his own pocket.[6]

Mobutu, beaming with pride, stood with his cronies and awaited liftoff. The countdown was in German. The first two tests had been successful. One year earlier, in deepest secrecy, the company had fired a six-meter-high (twenty-foot-high) rocket twenty kilometers (about 12.5 miles) into the air. Two weeks before, a heavier projectile had actually reached an altitude of thirty kilometers (about nineteen miles). Today, nothing could go wrong. This colossus was going to make it to a hundred kilometers (sixty-two miles).

Mobutu loved such spectacles. Wasn't he the man who had invited the moon walkers to Kinshasa? Hadn't he organized the match of the century in Congo? Hadn't the public hanging been a spectacle too? But performances alone were not enough. He also wanted to treat the country to a series of megalomaniacal infrastructure works. The Inga Dam on the Congo he had rebuilt into one of Africa's biggest hydroelectric plants. Upon completion in 1982, the new dam, Inga II, was to produce 1,424 megawatts instead of the former 351. Soon afterward Mobutu began dreaming of Inga III, a station with a capacity of no less than 30,000 megawatts, enough to supply energy for all of Africa and a part of Europe too. Before things got to that point, however, he had a high-tension line stretched from Inga all the way to the mining province of Shaba, an 1,800-kilometer-long (1,100-mile-long) extension cord straight through the jungle. Shaba itself was already well-equipped with power stations, but that line was Mobutu's way of keeping a finger on the main switch of the rebellious province. The project required 10,000 pylons. In Maluku, on the Congo north of Kinshasa, he had a foundry built that could produce an annual 250,000 metric tons (275,000 U.S. tons) of steel.[7]

All these prestigious projects bore identical earmarks: they were built by foreign companies and equipped with the very latest gadgetry, they were delivered as turnkey projects—and they never worked as they were meant to. As soon as payment was received, the French, Italian, or American contractor would leave the country, abandoning all the high-tech equipment to people who did not know how to use it or had not had the time to learn. Inga II cost $478 million, but Zaïre continued to be plagued by blackouts.[8] Maintenance of the turbines was neglected and the two (of the original eight) that still work today generate only 30 percent of the intended yield. The high-tension line to Shaba cost a dizzying $850 million, but often carried no more than 10 percent of its capacity.⁹ In addition, the project included no trunk lines to serve the cities and villages along the way. A $182 million price tag came attached to the steel mill at Maluku, but the company never turned a profit: it was unable to process local iron ore, only imported scrap metal.¹⁰

All that wasted money . . . That never became clearer to me than in 2007, the first time Zizi Kabongo showed me around the national broadcasting building. Mobutu's construction craze was not limited to heavy industry; Kinshasa was to be beautified too, just like Brussels in the days of Leopold II. In the borough of Limete a huge traffic cloverleaf was built, with broad exits and entrances and daring overpasses; in the middle of the rotunda there arose a modernistic replica of the Eiffel Tower, a pointed steel-and-concrete structure 150 meters (about 485 feet) high. A panoramic restaurant was to be built at the top of it, but the complex was never finished. Along the banks of the Congo he had built the CCIZ, Zaïre's international trade center, a high-priced structure that has stood in disrepair for decades. Shortly after the official opening, when the air-conditioning broke down, it turned out that the building's windows could not be opened—a bit of a nuisance in a tropical country. In the center of the city there arose a chic, multilevel shopping mall called the Galéries Présidentielles. And a few kilometers farther away came the media park for the RTNC, the national broadcasting company, Kabongo's new place of employment. Cost price: $159 million.

"The French built this," he said as he showed me around. "They were determined to get the contract. In exchange for the commission, they gave Mobutu free Mirage jet fighters." He showed me the dilapidated recording studios. Two of the original nine were still in use: massive, unequipped hangars. During live broadcasts, a little band of intrepid journalists availed themselves of two old cameras and a few microphones, at least if the electricity hadn't gone out. I saw it happen once myself. As part of an artists' exchange program between Brussels and Kinshasa, I took part in a morning talk show with a few other guests. The ceiling sagged. In the light from the spots we could see the asbestos floating down ceaselessly. Power cables were exposed; mixing consoles were lashed together with rope. I couldn't understand how they could produce live television here. Before the talk show came a news report. The anchorwoman had no autocue, not even notes, but she presented the items perfectly, by heart, without the slightest hesitation and with amazing presence. The only thing was: after the news had been going for a few minutes, a technician realized that there was no microphone on her table. The broadcast had to be interrupted. While the crew feverishly went in search of a mike that still worked, the Congolese viewers were treated to a long stretch of test pattern. I saw the elegant anchorwoman sitting there at her brightly lit table, in the vastness of a darkened, rundown studio.

"The complex was originally built for six thousand employees," Kabongo said. "Two thousand people still work here." The central building was a nineteen-story phallus. The reception desk in the entryway had a switchboard that could accommodate hundreds of incoming calls. It had all been out of order for years, just like the elevators. These days everything went by way of the emergency stairwell, a dark labyrinth like some sketch by Escher that stank terribly of urine because the plumbing on the top floors was broken as well. In the old days the managing director had his office on the building's top floor, from where he had a majestic view of the whole city. Today no one feels like clambering up to that eagle's nest. The current director enjoys the great privilege of a ground-floor office. The higher you work in the building, the lower your status. "What a waste of money," Kabongo sighed as we climbed to his fifth-floor office, "the RTNC, the CCIZ, all those projects . . . and all of it at a point when there was so much poverty elsewhere in the country."[11]

It is truly amazing, the way Mobutu kept throwing money around. Ever since 1975 and the start of an endless war of decolonization in neighboring Angola, Zaïre had been unable to use the Benguela Line—the stretch of railroad on which my father had worked and that connected the Katangan mining basin with the Atlantic. It became much harder to export ore and Mobutu missed out on a lot of foreign revenue. The country was crumbling, but he seemed hardly aware of that.

Vier, drei, zwei, eins . . . a burst of flames lit the surroundings. The roar swelled. Slowly, the rocket rose from the launching pad. A hundred kilometers into the atmosphere, that's where it was headed, a new step forward in African space travel. A lavish lunch was waiting for the guests. But before the projectile had left the pad, even a child could see that something was going wrong. The rocket listed, cut a neat arc to the left and landed a few hundred meters away, in the valley of the Luvua, where it exploded. As a thick cloud of smoke rose up from the savanna, Mobutu turned away in silence. Against the sky, the spectators could briefly see a dark vapor trail describing the curve the rocket had made.[12] A parabola of soot. It looked like a graphic representation of Mobutu's regime: after the steep rise of the first years, his Zaïre toppled inexorably and plunged straight into the abyss.

AND THERE WERE MORE THINGS TO COME DOWN out of the blue in those years. Between 1974 and 1980, two of the Zaïrian army's C-130 transport planes, two Macchi fighters, three Alouette helicopters, and four Puma helicopters went down.[13] Not a single one of those crashes took place during combat. The reason for so much bad luck? The soldiers were so badly paid that they had started selling the spare parts for their aircraft. Pierre Yambuya, a helicopter pilot in the national army, saw it all happen. His testimony provides a unique glimpse of the state the armed forces were in at the time. "Anyone with a private plane knew that Kinshasa was the world's cheapest market for spare parts. The soldiers sold them for twenty times less than the factory price."[14] Mobutu showed off with his prestigious projects, but began neglecting the institution that had made his coup possible: the army. Air force pilots supplemented their incomes by selling, wherever they landed, a part of their kerosene to the local population, who used it as lantern fuel. It became such a common custom that children would run with their yellow jerrycans to the landing strip as soon as a government plane arrived. Yambuya knew what he was talking about: "A sergeant-major earned 280 zaïres, a bag of rice cost 1,200 zaïres back then. An adjutant got 430 zaïres. But a school uniform cost 850 zaïres, and with the 5 zaïre allowance he received for each child, you couldn't even buy a pencil." That suddenly makes corruption much more understandable. The soldiers did not protest "up through the ranks," for that could cost them their jobs or even their lives, but repeated at lower levels that which went on over their heads. "To lead a reasonable life, for example, I sold the fuel from my helicopter. My superior stuck the funding intended for my mission in his own pocket and said: 'If you land somewhere, just sell some fuel. After all, what you do is your own business.' "[15]

Zaïre became sick. The deeper cause was a shortage of revenues (due to the copper crisis, the oil crisis, failed Zaïrianization, and grotesque public spending), and the worst symptoms were the withdrawal of the state and the spread of corruption. It was in the army that that first became visible. Soldiers took military vehicles away from the base and used them to run their own taxi services. Radios and record players disappeared from the mess halls, bulldozers and trucks from the garages. Officers even took their subordinates home with them and used them as servants. Absenteeism in the barracks was high, sometimes more than 50 percent. The few soldiers who did show up for roll call were not highly motivated. Discipline was something from long, long ago. An internal document, the "Mémorandum du Réflexion," did not shrink from self-criticism when it came to a concise summary of the troops' morale: "Everyone wants to command, but no one wants to obey."[16]

Meanwhile, on the other side of the Angolan border, Moïse Tshombe's troops—the veterans of the Katangan secession—were increasingly active. Many of them belonged to the Lunda tribe, a people whose traditional territory reached into Angola. Mobutu had driven them into exile many years ago, after they had defeated the Simba rebels. But now, along with their sons and new recruits, they were out for revenge. These notorious Katangan guardsmen had followed a remarkable course. During the Katangan secession (1960–63) they had fought for a rightist, European-run Katanga, but in Angola they had taken sides since 1975 with the Marxist MPLA, the Movimento Popular de Libertação de Angola. The reason for the ideological turnaround was simple enough: the MPLA, like them, held a grudge against Mobutu.

After the Carnation Revolution in Portugal, Angola started in 1975 a violent struggle for decolonization. As in Congo the contest was one for the throne, but in Angola the conflict was far bloodier. There were three factions. Agostinho Neto's left-wing MPLA faced off against the FNLA of Holden Roberto and Jonas Savimbi's UNITA. The superpowers got involved. Angola was the spot where the Cold War experienced its most heated African episode. The MPLA received massive support from Russia and Cuba; the two other militias had American backing. The U.S. support went by way of South Africa and Zaïre: Pretoria backed Savimbi in the south; Kinshasa supported Roberto in the north. Because Roberto also happened to be Mobutu's brother-in-law, the former Katangan guardsmen chose to join up with the MPLA. Their leader's name was Nathanaël Mbumba, their new nom de guerre the FLNC (Front pour la Libération Nationale du Congo), their nickname *les Tigres Katangais* (the Katangan Tigers).

The rebels invaded Zaïre on two occasions. In 1977 and again in 1978 they crossed the border and seized a large part of western Shaba (the so-called Shaba I and Shaba II wars). In numbers and logistics they were far inferior to the national army, but the local population received them joyfully; not only were they fellow Lundas, but the people were also tiring of Mobutu. The rebels won ground easily and in 1978 even took the important mining town of Kolwezi. For the first time in a decade, Mobutu had to deal with a military uprising. Dissidents who had fled to Brussels and Paris hoped the dictatorship would topple and saw in the invasion the "embryo of a people's army."[17] Mbumba, they felt, could breathe new life into the dreams of Lumumba and Mulele. The sun king's empire seemed to be tottering.

Mobutu, however, did everything in his power to portray the rebellion as a foreign, Marxist intervention. According to him, Mbumba was merely a pawn of the MPLA and therefore of Cuba and Russia. With this line of reasoning he hoped to draw support from abroad, for his own army was now virtually worthless. And it worked. After eight days, Shaba I was quashed by Moroccan troops flown in in French military aircraft. Shaba II was put down after only a few days by French Foreign Legion troops and Belgian paratroopers. Mobutu's allies sprang into action after the rebels had slaughtered thirty whites in a villa in Kolwezi. What these foreign friends did not know was that the white people had probably not been murdered by the rebels at all, but by Mobutu's own troops. Helicopter pilot Yambuya was in Kolwezi, and was clear about what he had seen:

> On Sunday, May 14, Colonel Bosange [of the national army] suddenly orders all those Europeans locked up in the villa to be executed. According to him, they are all mercenaries. Bosange will tolerate no objections, and General Tshikeva does not try to dissuade him. Only old Musangu raises his voice in protest. Bosange commands the head of the intelligence and security services, Lieutenant Mutuale, and three other soldiers to carry out his orders. Mutuale and his firing squad go to the villa, where the doors and windows are hermetically sealed. They fire their machine guns through the closed metal blinds. The volleys echo like the sound of a car crash. Five minutes later, Mutuale and his men are back: mission accomplished.[18]

Mobutu knew his history. In 1960 Belgium had invaded the country after the murder of five whites in Elisabethville. In 1964 Stanleyville was retaken by Belgian paratroopers after hundreds of whites were taken hostage. Kill a few Europeans, Mobutu knew, and you've got a Western army on your side. That is: as long as you can put the blame on someone else.

The two Shaba wars were short-lived, but the lessons they taught were of great importance. First, Mobutu was capable of doing absolutely anything to maintain his position. Second, his army was worthless. Third, he survived by dint of foreign support. America had been a faithful ally ever since 1960 (regardless of occasional tensions), but now France came along as well. President Valéry Giscard d'Estaing implemented a very explicit policy of increasing the French sphere of influence in Central Africa. As the world's largest Francophone country, Zaïre of course received special attention. Back in 1960, with decolonization in full swing, France had even tried to acquire Congo from Belgium with a reference to the historical *droit de préemption* from 1885![19] Giscard, however, was much more interested in financial gain. Trade with Zaïre received a strategic boost. It was against that background that the deal had been made to build TV studios in exchange for Mirage jets. The main contractor was Giscard's cousin, while another of his cousins was one of the project's biggest financers.[20] Nepotism was hardly a Zaïrian invention.

In Kinshasa and Brussels I spoke a number of times to Colonel Eugène Yoka, who had been one of the Zaïrian army's few fighter pilots. He was the son of the last surviving widow of a World War I veteran and came from a soldiering family. His father had fought against the Germans; his grandfather was one of the first soldiers in the Force Publique. He himself had put in more than two thousand flight hours. In 1961 he was among the first batch of Congolese pilots to complete his training; he had been taught to fly in an SV4-BIS—a propeller biplane—at Tienen, in Belgium. Afterward he had flown Dakotas, T-6 aircraft, P-148s, you name it. He was there when the Concorde made its first flight to Africa in 1973; Mobutu would soon charter the supersonic plane for jaunts on a regular basis, including trips with his family to Disneyland Paris.[21] Yoka also became one of the select circle of pilots able to fly the Mirage. He had been trained in France. I asked him about his memories of the Shaba wars. "I was there," he said, "for both Shaba I and Shaba II, but not as a pilot."[22] I'd received a similar answer from Alphonsine Mosolo, the first female parachutist, who had received her training in Israel. "The wars in '77 and '78, I never had to jump then." Both soldiers had received extensive training abroad, both of them had to show up for the annual parades in Kinshasa, but neither of them had to apply their expertise when the time came. The armed forces seemed to have fallen into disuse. Mosolo said: "Instead of jumping, I cooked for Mobutu aboard the *Kamanyola*. That was his private yacht, the one he used on the river. One evening there was a party on board. I was finished cooking. Mobutu liked it when spirits were high, he was a real partygoer. I was sitting in a chair, but he wanted me to dance. He even took my shoes off to get me to dance. Really! The president himself! Down on his knees! Even though my feet stank so badly!"[23]

MOBUTU COULD KEEP ON DANCING for a time in the full conviction that his country's economic recession was only a temporary dip. Copper no longer commanded the price it once had and at the same time oil had become so expensive. But anyone can have a little bad luck, he reasoned, especially with an economy so dependent on a single sector like mining. All right, his country couldn't pay off all its loans right away, true enough, but soon the international demand for ore would rise again. He turned to his French, American, and also his new Arab allies to ask for a little help.

But Zaïre's burden of debt was not merely cyclical. In 1977 the deficit had risen to 32 percent of the total budget.[24] Year after year, GNP dropped by a few percentage points.[25] An annual inflation rate of 60 percent became normal.[26] Between 1974 and 1983 prices rose six times over.[27] This was no longer just a passing problem, the people knew. In 1984 they had to work two days to pay for a kilo (2.2 pounds) of rice, ten days for a kilo of beef. The unskilled Zairian who wanted to buy a cheap forty-kilo (eighty-eight-pound) bag of manioc for his family had to work eighty days to do so.[28] And by the time he could finally afford that bag of manioc, the price had risen again. By 1979, purchasing power had plummeted to 4 percent of that in 1960.[29]

At first, Western and Japanese banks had had no problem with granting loans to the young Mobutu to carry out his program of industrialization—Zaïre was, after all, rich—but from 1975 on they started worrying about their money. Zaïre's foreign debt by that point totaled $887 million, spread over ninety-eight banks.[30] To consolidate their claims, those banks finally joined forces in the "Paris Club." They directed a joint appeal to the International Monetary Fund (IMF), the financial watchdog of the world economy, set up in the wake of World War II to prevent another depression like that in the 1930s. The IMF was asked to provide emergency loans, to make sure Zaïre didn't go completely off the deep end.

Mobutu, however, had no desire to entertain the busybodies of the IMF. After all, all the power he possessed was founded on the conscientious maintenance of a large group of followers. If he let the IMF come in he would no longer be able to pass out goodies. But if he didn't, he would have no money left. This latter option would lead to the immediate collapse of his regime; the former still left him with a few possibilities. The trick was to pay lip service to the IMF, to nod amiably in response to all its demands, and then go on plundering the state treasury behind the scenes.

Mobutu, the man who had been so adamant about his country's "economic independence," now had to accept the IMF, the Paris Club, and later the World Bank as key players in the domestic economy. In 1976 the IMF launched the first of many stabilization plans for Zaïre. In exchange for a first installment of $47 million, Mobutu had to agree to cut public spending, raise tax revenues, devaluate the currency, stimulate production, enhance infrastructure, and improve the country's financial management. Only then were the international banks prepared to talk about a possible extension of payment.

Many capital injections and bridge loans would follow but, in the period 1977 to 1979 alone, Mobutu—by the most conservative estimates—siphoned off more than $200 million for himself and his family.[31] After the stabilization plans of the 1970s came the much more rigorous, structural adjustment programs of the 1980s, but that didn't help either. By around 1990 Zaïre's total national debt had risen to the insane sum of more than $10 billion. Only then did the flow of money stop.

It was, however, not the first time that Mobutu's creative bookkeeping came to light. The first alarm had been sounded by a meticulous German banker as early as the late 1970s. In 1978 the IMF had charged Erwin Blumenthal, for years a top official with the West Germany Bundesbank, with the onerous task of cleaning up the mess called the National Bank of Zaïre. During this period, the IMF had placed the country's major financial institutions under receivership. Blumenthal doggedly tried to pick up the pieces at the central bank: time and time again he unearthed shameless examples of corruption. "There is not a single official at the Fund or the World Bank who does not know that all attempts to impose stricter budgetary control here run into major obstacle: the presidency," he wrote. "Who is going to shout 'stop the thief!'? It is an impossible task to monitor the financial transactions within the president's office. Within that office, no distinction is drawn between personal needs and state expenditures. How can it be that international organizations and Western governments blindly trust President Mobutu?"[32]

The systematic embezzlement of government funds, the discovery of a whole slew of secret bank accounts in Europe, the bald-faced, systematic greed of Mobutu and his clique filled Blumenthal with disgust. After less than two years, he resigned his mission. The confidential report with which he announced that resignation was grimly unambiguous: "New promises will undoubtedly be made by Mobutu and his government, and the payment of the country's foreign debt, which is accruing apace, will once again be postponed; but there is absolutely no chance, I repeat, absolutely no chance that the foreign creditors will ever see their money again."[33]

The Blumenthal Report was so damning for Mobutu and his clan that it had to leak out at some point. Zizi Kabongo still remembered those days: "Mobutu wanted to make absolutely sure that the report didn't appear here. No one in Zaïre knew about it at first, but circulating in Paris was a text on the subject that had been published by Nguza Karl I Bond. Journalists coming back from abroad were frisked at the airport." For eight months, Nguza had been Mobutu's prime minister. After he fell from grace in 1981 he went to Europe, where he continued to fire broadsides at Mobutu's regime in the form of books and pamphlets. For him, the president-founder was the embodiment of the "Zairian sickness."[34]

Blumenthal said aloud what everyone suspected, but his revelations did not lead to a radical turnabout. Zaïre's national debt was already up to $5 billion by 1981; for the French, however, Mobutu was too important a cultural and economic partner to cross. For the Americans he was too valuable as an ally in an Africa in the throes of socialist and communist experiments (in Angola, Congo-Brazzaville, Uganda, Tanzania, and Zambia, to mention only the neighboring countries). "Mobutu is a bastard, but at least he's our bastard," the CIA reasoned. Secret directives stated that "a negative frame of mind on the part of the IMF, or a negative attitude on the part of the U.S., might cause Mobutu to reconsider our extremely close ties. This could endanger a program that the president [Reagan] considers to be of cardinal importance for U.S. security."[35] Republican presidents in particular, like Richard Nixon, Ronald Reagan and George Bush Sr. maintained extremely warm contacts with Kinshasa; after Jimmy Carter's election in 1976, relations cooled for a time.

THIS COLD WAR LOGIC formed a considerable encumbrance for the IMF's recovery plans. Yet the IMF too, was not without sin. Historically speaking, the organization had been set up not to help poor nations back onto their feet, but to avoid global financial crises.[36] Even in the 1970s, its genteel staff members tended to know more about macroeconomics than about anthropology. They preferred examining charts in their Washington offices to talking to the people it was really all about. The consequences of that lack of on-the-ground expertise were highly unfortunate.

On Christmas Day 1979, under the watchful eye of the IMF, one of the most remarkable monetary measures in the country's history took place: the depreciation of the zaïre. To combat inflation, all citizens were summoned to bring their five- and ten-zaïre notes—the highest denominations known till then—to the bank and exchange them for new ones. In late 1976 there were 59,000 five-zaïre notes in circulation; by late 1979 there were 363,000, six times as many. The result was inflation. Currency is to an economy what oil is to a motor: too little of it is not good, but too much is not good either.[37] In addition to inflation, hoarding had also become a problem. In a vast country with a shaky economy like Congo's, almost no one wanted or was able to put money in the bank. People stashed it away in suitcases, pillows, or jugs. Didace Kawang, an actor to whom I once gave a master class in playwriting, told me about his uncle, who had been a successful merchant in Lubumbashi: "He did business with Zambia. The banknotes came in through the big gates. He had piles and piles of them. He bound them together in *brikken*, bundles the size of a brick, wrapped with a rubber band. He had a mattress made of money. Really! He slept on it!"[38]

The IMF bankers knew that it is extremely unhealthy for a national economy when there is more money in circulation (in the form of coins and bills) than there is in the banks. They knew the big theories: money in the bank is used to provide new loans; money used as a mattress doesn't help the economy one little bit. To combat hoarding, therefore, they rolled out a process of currency depreciation. Anyone who turned in his banknotes on Christmas Day 1979 would be given new ones, at least for half the amount brought in. The other half had to be placed in a bank account. It was a clever way to bring a lot of "dead" money back to life and at the same time to deal with inflation. The move was intentionally announced at the last minute and lasted only one day, to keep people from fleeing abroad with their cash. The border crossings were closed and even the country's airspace was shut down. Zaïre was going to freshen up in a monetary jiffy, in order to reappear spic and span in the international footlights. But the country was far too vast for such a lightning operation.

"My uncle had no choice but to put his savings in the bank too," Didace said. "But they had reserved only one day for that. There was a huge line. People came in dragging sacks full of money. The sun went down and my uncle still hadn't been able to turn in his bills. All his piles of banknotes became worthless. . . . In one swoop, he was poor as a church mouse. He died in his native village." And he was not the only one, not by a long shot. Many of those who lived too far from a bank or who did not understand what the operation was about lost all their savings, while in Mobutu's circles everyone had been briefed beforehand and had taken steps long before.

Not only was there something awry with the practical side of the IMF measures, but the basic philosophy was skewed as well. After the stock-market crash of 1929, the fund had dealt with the excesses of unbridled market thinking; by 1975, however, the IMF itself had evolved into one of the great heralds of free enterprise. Almost all its officials were firmly convinced that the creation of favorable market conditions was enough to jump-start a national economy, regardless of the local culture, the state of the economy or the governmental structure. Here too, a form of macroeconomic blindness reigned. As long as the government kept its distance, the invisible hand would do its work; that was the institution's mantra. No one had an eye for the pace and sequence of the needed changes.[39] The whole package was imposed at once, in the form of programs for "structural adjustment." For these zealots of liberalization, all forms of poverty reported to them afterward (for they rarely entered the field themselves) could be blamed on the defective implementation of their infallible, yea, holy formulae.

The Zairian currency was devalued no less than six times: in 1975 it was still worth two dollars, by 1983 only three cents.[40] Those devaluations were intended to stimulate international trade. As part of its "structural adjustments" the IMF demanded a drastic reduction in government spending and far-reaching privatizations. Governmental and semigovernmental enterprises had to be slimmed down and operated with greater autonomy. The infrastructure and production had to improve.

In the early 1980s the IMF's prescription seemed to be taking effect. Inflation was indeed tempered and the economy seemed to be achieving a higher degree of organization. The charts were looking good. The Paris Club creditors breathed a sigh of relief and hoped that perhaps their loans really would be paid back. On nine separate occasions they voted for a program of debt relief. But on the ground, things turned out quite differently. As is often the case with IMF interventions, success was short-lived: inflation resumed after a time, poverty rose. The per capita GNP fell dramatically from six hundred dollars in 1980 to two hundred in 1985.[41] People ate less; infant mortality was high. Onions were cut into quarters before they were sold.[42]

Slim down the public sector? The ranks of the civil service were reduced from 444,000 to 289,000; the number of schoolteachers from 285,000 to 126,000.[43] That was, indeed, one way to combat inflation, but it meant that thousands of families ended up with no income. The civil service and the schools had been the country's last major employers.

Cut back on spending? Government funding for education and health care was reduced, so that those with no money at all suddenly had to pay for their own children's schooling and their visits to the doctor. The charts didn't show it, but it was the poorest of the poor who paid most dearly for the IMF's well-intentioned measures, while the international funding kept Mobutu firmly in the saddle.[44]

Measures to jump-start foreign trade? As long as Mobutu failed to use the available funding to restore the country's infrastructure, Zaïre could only become more dependent on imports. The country had all it needed, for example, to again become a major coffee producer, but in the cities people drank only imported instant coffee. Little wonder: of the 140,000 kilometers (about 87,000 miles) of passable roads that had existed in 1960, only 20,000 kilometers (12,400 miles) were left.[45] The IMF was out to reorganize the country, but in fact dismantled it. Zaïre became nothing more than a sales outlet and would remained that way for decades to come.

In 2008 I once spent an afternoon beside the Congo River in the old port of Boma. Swallows zigzagged across the water. Fishermen paddled out in the canoes to inspect their nets. It could have been 1890—until a huge cargo ship came sailing past. It was on its way from Matadi to the ocean. The ship rode high on the water. At the back, close to the prop, I could even see the keel. The ship was empty, completely empty. With the exception of a few spare containers, it was carrying nothing at all. I was reminded of Edmund Morel, who had watched a century earlier as the ships entered Antwerp's harbor loaded with rubber and ivory from Congo, but left again empty. To Morel, the difference between ships riding high or low in the water was proof that the Free State was not engaged in commerce, but in plunder. The difference in draft I noted suggested that free trade, as roundly promulgated for decades by the prophets of the international economic institutions, could be a form of plunder as well.

IN THE 1980S Mobutu became a tired, somber man who seemed to draw little pleasure from his duties. After the deaths of his mother and his first wife, no one in his immediate surroundings had any control over him. His new wife, Bobi Ladawa, and her twin sister, who was also Mobutu's mistress, never had the same impact as *mama* Yemo or *mama-présidente* Marie-Antoinette, his first spouse. Mobutu had been very fond of his old, mettlesome mother. Her death weighed heavily on him. His wife Marie-Antoinette had also been an outspoken character who had stubbornly refused to give up her Christian name. For a long time she had had a restraining effect on her husband's tomfoolery. But now Mobutu had expelled his cabinet chief, Bisengimana Rwema, and his personal physician, the American William Close, had left the country.

Mobutu became a lonely man who grew more melancholy with each passing day. He seemed to fall prey to the longing for excess that marks all those for whom life holds no more surprises. In Europe he bought one chic property after the other. He owned a dozen castles, storage spaces, and residences in the wealthy Brussels boroughs of Ukkel and Sint-Genesius-Rode. He owned a luxurious, eight-hundred-square-meter (nearly 8,500-square-foot) apartment on Avenue Foch in Paris, Savigny Castle close to Lausanne, Switzerland, a palazzo in Venice, a sumptuous villa on the French Riviera, an equestrian estate in the Portuguese Algarve, and a series of hotels in West Africa and South Africa, not to mention his luxury yacht on the Congo.[46] But the most incredible of all his residences was, without a doubt, Gbadolite. In the middle of the jungle in his native region, close to the border with the Central African Republic, he had a city built, complete with banks, a post office, a well-equipped hospital, a hypermodern hotel, and a landing strip that could accommodate the Concorde. (Zizi Kabongo: "That's right, as a journalist I once took the Concorde from Gbadolite to Japan.") A cathedral was added, with a crypt that was to serve as the family grave, and a Chinese village with pagodas and imported Chinese people. The crowning glory was Mobutu's opulent, 15,000-square-meter (158,400-square-foot) palace. The mahogany doors were seven meters (nearly twenty-three feet) high and inlaid with malachite. The walls were covered with Carrara marble and silk tapestries. Crystal chandeliers, Venetian mirrors, Empire furniture: no expense was spared, no luxury was too excessive. There were Jacuzzis, massage rooms, a swimming pool, and a beauty parlor. Mrs. Mobutu had a walk-in closet fifty meters (over 160 feet) long where she hung her extensive collective of French couture, some one thousand creations in all. Beneath the building itself, thousands of top French vintages lay gathering dust (if not actually going sour in the tropical climate). There was a discotheque for the children and a bomb shelter for the family.[47] The fountains scattered around the grounds splashed around the clock and were lit at night—in a region that had almost no electrical network. Mobutu threw state banquets for thousands of guests where the pink champagne—his favorite beverage—flowed freely and the suckling pigs lay grinning with an orange in their mouths.

"He had the great chefs of France and Belgium flown in," recalled Kibambi Shintwa, a man who still retained his "authentic" name. Shintwa had worked as a reporter for the *présidence* from 1982 and was closely acquainted with Mobutu. "After years of hard work, he started taking it a little easier. He enjoyed good food and good restaurants. But he also derived a lot of pleasure from giving to others. He was extremely generous." That generosity, however, was functional. "He always felt the need to remind others that he was the chief. He wanted to display his power."[48]

Mobutu's corruption was so shocking that it caused a long-forgotten term to resurface in the English language: kleptocracy. The unforgettable Jamais Kolonga had witnessed it firsthand. After his short-lived adventure as sawmill owner, he went to work for Miba, the national diamond enterprise in Kasai. "Oh, but I visited Gbadolite often. I usually went along with the great Mibaas Jonan Mukamba to see the president. Every time we went I had to carry an attaché case and hand it to the president when we met. Here you are! A briefcase full of diamonds, that was."[49] But kleptocracy was only part of the story. It was also a "giftocracy": Mobutu stole in order to share and so ensure his popularity. No one left Gbadolite emptyhanded, or so the saying went. A few hundred dollars, a valise full of zaïres, a cigar box full of diamonds: Mobutu always had a gift ready for his guests.

"Mobutism" and the cult of personality that went along with it had already made clear the boundless nature of Mobutu's vanity. Of the seventy-nine series of banknotes printed during his regime, seventy-one bore his likeness.[50] But in the 1980s his narcissism became nothing short of pathological. No one knew that better than the Flemish tailor Alfons Mertens, whom I met in a well-to-do residential area in Antwerp province. A good-natured family man he had never dreamed that he would become directly involved in world history, but he worked for Arzoni in Zellik (close to Brussels), the company that made the world's chicest *abacosts* and so became a brand name in Zaïre, like Dior or Versace. Mertens was such a skilled tailor that in 1978 he became Mobutu's private couturier. "Between 1978 and 1990 I traveled

to Kinshasa more than a hundred times. I always stayed at the Intercontinental. Mobutu would have me come in to take the measurements of Air
Zaïre's pilots and stewardesses, or of his army generals. When his son was
promoted to the rank of sublieutenant, I had to design a dress uniform and
a ceremonial uniform for his entire class at the military academy: twenty-
seven cadets in all. I often made clothes for Mobutu himself, including his
civilian dress. His wife or mistress would pick the material, my boss would
draw the pattern, I took the sizes. The Mobutus always went for extremely
costly materials, like natural silk, wild silk. His sizes didn't change much.
He was tall, almost one meter eighty [five feet ten inches], but he never wore
anything bigger than a size 54. He was a fine man. You had to meet him a
few times before you won his trust, but after that he was nice person."

In 1983 Mertens received the most ceremonious commission of his
career. "I had to make new uniforms for all the generals, and no less than
four full-dress uniforms for Mobutu himself, two black ones and two white
ones. His generals had decided to confer on him the rank of marshal, and I
went to work." Mobutu, commander in chief of the armed forces, who had
cut a bad figure indeed during the Shaba uprising, was now to be given the
historically rare rank of marshal! The idea, of course, was his own.

Mertens showed me pictures of the ceremony and explained his creations. "Look, that collar, the belt, and cuffs, they were embroidered with
real gold thread. That chain is made of gold too. All of it hand tailored. He
had two times seven stars on his sleeves. All made from solid gold, from
France."[51] His uniform cap bore a cockade with the emblem *Paix Justice
Travail*, even though his country offered no peace, no justice, and no work.
The photographs of his marshal's inauguration bear witness to unparalleled
gaudiness. Mobutu wore white gloves and held a scepter. He was driven
around in a white Mercedes and waved to the people along the way. He
inspected the troops, the magistrates, and the top officials, and gave a speech
from beneath an ornamental canopy. A marshal needs a motto, he told the
nation. His would be: *Toujours servir.* To serve at all times. It wasn't even
laughable anymore. It was the sad low point of madness cast adrift.

BUT DIDN'T ANYONE SPEAK OUT? In December 1980 a group of thirteen
members of parliament had the audacity to send a fifty-two-page open letter to the president, calling for political change. Their leader was Étienne
Tshisekedi, a former member of Mobutu's staff who had written the 1967

constitution and occupied a number of ministerial and ambassadorial posts. Like everyone whose name starts with "Tshi," he was a Muluba from Kasai. His obstinacy was legendary.

> For fifteen years we have obeyed you. Look at all the things we've done during that time, just to please you? Sing, dance, perform political drama . . . we have been through every form of humiliation, all sorts of insults unlike anything even pressed upon us by our colonizers. And we did all that so that you would lack for nothing as you set about achieving, even if only half, of the social model you presented to us. Did you succeed in that? Unfortunately, no.
>
> After fifteen years of government, which you have carried out without any distribution of power, we are now faced with two absolutely divided camps. On one side we have a few scandalously wealthy members of a privileged caste. On the other we have the masses of the people who live in darkest misery and can depend at most only on international charity to survive after a fashion. And if that charity happens to reach Zaïre, these same wealthy few make arrangements to claim it to the disadvantage of the needy masses! . . .
>
> Citoyen Président-Fondateur, this dry-eyed analysis shows that our society is faced with grave problems. You have often said that a real chief is one who admits his own mistakes. You have done so often enough. But the tragedy is that you do not always assume the consequences of those mistakes. And the worst thing is that you take one step forward, and three steps back.[52]

It had been a long time since Mobutu had been spoken to so candidly. The group of thirteen was arrested and sent into exile abroad, but in 1982 a few of its members set up the Union pour la Démocratie et le Progrès Social (UDPS), an illegal opposition party that aimed to challenge the MPR's single-party state. That party would become yet another nail in Mobutu's coffin.

I talked about this with Raymond Mukoka, one of those involved from the very start. He had helped to write the MPs' letter. "The cosigners were sentenced to fifteen years in exile. My name wasn't on it, but as coauthor I was flown first to the Ituri, then on to Kasai. I had to pay for the food and wages of my own guards! We received support from Amnesty International and the Catholic Church, who used its field phones to keep my family informed. *Jeune Afrique* wrote about us. In 1985 I paid a brief visit to the capital. The UDPS arose in exile, just like the Kimbanguists. Some of the party members dreamed of forming a paramilitary wing, but we always remained nonviolent. Tshisekedi said: Our pen and our words, those are our weapons. In 1987 Mobutu invited us to Gbadolite. He said: join the MPR. We said: No! Then he said: Well, then take part in the MPR institutions. He gave us access to the central committee, offered us ministerial posts or management positions in the state-owned companies. A lot of our people took him up on that, but I didn't want to, and Tshisekedi didn't either."[53]

Mobutu mollified his critics by giving them gifts and one of his favorite presents was a ministerial post. A political career was too lucrative for most people to refuse. Between 1965 and 1990 no less than fifty-one cabinets were installed, each of them with around forty ministers.[54] Regular reshuffles, on the average of every six months, made sure no one could gain real power, while providing the next group with a chance to move up to the buffet.

Mobutu was a political schemer par excellence. "He didn't like big meetings," Zizi said, "he always chose for the tête-à-tête, for private consultations that allowed him to play out one politician against the other. He had a frightening ability to generate hatred between individuals." Mobutu had an entire arsenal of techniques for binding people to him. He could be charming, friendly, and funny, but also manipulative, treacherous, and vicious. The emotional yo-yo was a tool he used consciously. He could be hearty and jovial one day, only to treat you with frosty distance the next. Kibambi Shintwa told me about that: "Mobutu was protean, slippery, impossible to figure out. He was fickle. He changed every day. What he was primarily interested in was showing that his power was not to be toyed with. He was jealous, like an animal clutching its prey."

The very people he protected were sometimes humiliated in plain public view. Others, who had seemingly blown it for good, like former prime minister Nguza Karl I Bond, might suddenly be forgiven and allowed to return to Kinshasa—Nguza took Mobutu up on that offer and lost all credibility. Still, for a time, Nguza had been the hope of the clandestine opposition. In this way, critics were made to dance to the pipes of the MPR and Mobutu triumphed as wise and mild village chieftain.

Another privilege of the traditional chieftain of which Mobutu made avid use was his *droit de cuissage* (right to deflower). Zizi Kabongo said:

When he traveled around the country, the local chiefs always offered him a virgin. It was a great honor for the family when a girl lost her virginity to the supreme chieftain. It was an old custom, but Mobutu took it even further. He didn't hesitate to employ women in his power games. He used women from his province to advance his political ends. He slept with the wives of his cabinet ministers, in order to hear their secrets and to humiliate his ministers. When they were summoned to Gbadolite, they never took their wives, they would take a niece instead. They didn't mind that as much. . . . Mokonda was a legal counsel, one of his closest associates. He had a very beautiful wife. One day Mokonda was in a meeting with Mobutu at Gbadolite. What he didn't know was that his wife was sleeping in the room next door. The president had had her flown in on a private jet. Mobutu, we used to say, is multipolygamous. He destroyed a lot of marriages.

Political and sexual intrigues were only the tip of the iceberg. The more Mobutu withdrew to his yacht or his palace, the more he wanted to know what was going on in the country. In the 1980s the intelligence services became as important as the propaganda services had been in the 1970s. The president had half a dozen such secret services, all working at cross-purposes, but here too the motto was: divide and conquer. Spies were everywhere. Men distrusted their wives, mothers their sons, sisters their brothers. Mobutu had informants everywhere, even in Belgium. Paranoia became the emotional bottom line. Cabinet ministers asked to dine with the president feigned a strict diet or intestinal problems, afraid as they were of being poisoned. Others brought their own bag lunches.[55] Rumor had it that an underground canal ran from the presidential palace on Kinshasa's Mont Ngaliema and the river; political opponents were thrown into the canal to feed the crocodiles. Belgian diplomats, even among themselves, no longer pronounced Mobutu's name. During meetings they preferred to speak of "Jefke Van den Bergh": Jef was the Flemish equivalent of "Joseph," Van den Bergh meant "of the mountain," a reference to Mont Ngaliema.

A true reign of terror settled over Congo. The country was ruled by caprice, and there was nothing to be done about it. In 2005, during one of my first visits there, I came in contact with Madame A., an elderly lady and a former newsreader. During dinner she told me her incredible life's story:

My husband was general editor of the daily news programs. We had five children. He was a handsome man. Mobutu's sister-in-law saw him on TV and wanted him, married or not. One evening, while we were sitting around the table, armed soldiers suddenly came to the door. My husband had to go along with them. They told me: you keep your mouth shut, or you and your children will end up in the *fleuve* out at Kinsuka. At work they told me: don't try anything, he hasn't been taken by just anyone. I never saw him again. Mobutu gave him ambassadorial posts in Togo, Argentina, Austria, and Iran. He died in 1995, while serving as ambassador to South Africa. A lot of people in Kinshasa know the story, but not many of them know it's about me.[56]

Mobutu's secret services were so ruthless that even today Madame A. insists on remaining anonymous. The Division Spéciale Présidentielle (DSP) acquired a particularly sinister reputation. The corps consisted of several thousand specially trained and well-paid soldiers from Mobutu's native region. The great unifier of the nation had become so neurotic that he now drew his Praetorian Guard only from among his own tribesmen! It was an army within the army. They were loyal and unrelenting. The hard core consisted of *les hiboux* (the owls), so called because they came at night and silently spirited people away. Opponents or suspected opponents were arrested and held without trial in filthy prisons without food. In Zaïre, like everywhere else in the world, the human mind was extremely creative in devising tortures. There was "the fish," a method in which the prisoners' hands were tied behind his back and he was hung upside down before being dipped in a tub of water. There was "the Boeing," in which the prisoner was raised to the ceiling on pulleys, beaten with sticks, and then allowed to fall through "air pockets." There was the "stenographer," whereby blocks of wood were shoved between the fingers and then tightened to crush the fingers. There was "the nutcracker," in which the feet were clamped into wet blocks of wood and the prisoner was then put in the sun; the wood dried and shattered the tarsal bones.[57] Electric shocks were applied to genitals and cigarettes stubbed out on lips. Amnesty International submitted an official protest and tried to estimate the scope of the human rights violations, but the exact number has never become clear.[58] As in colonial times, people were sent into domestic exile. Others disappeared without a trace.

Pierre Yambuya, the helicopter pilot who had sold his kerosene, flew secret missions on a number of occasions. He was required to fly over the Congo River or a lake, while commandos in the back tossed out dozens of bags; bags containing bodies, he saw. "Between March and October of 1983 I flew four such missions and each time a load was dumped close to the rapids at Kinsuka. To the best of my knowledge, at least one such flight was carried out each week." Sometimes they didn't even bother to kill the prisoners first. One day Yambuya had to land his Alouette on the *Kamanyola*, the presidential yacht. Yambani, one of Mobutu's ranking bodyguards climbed into the helicopter with two bound men and two commandos. Mobutu stood and watched. "When we take off again, Yambani tells me where to fly to. At a certain point he asks me to take it up to a thousand meters [3,250 feet]. He looks around to make sure there's not a living soul in sight—except perhaps a few hunters in the jungle—and orders the commandos to lock in to their harnesses. The commandos obey, then open the right hatch at the back of the helicopter. They throw the first prisoner out, before he even has time to protest. The second prisoner starts weeping and begging for mercy, but then he too is pushed out of the hatch, in a free fall over the jungle."[59]

IN KINSHASA, the repressive climate gave rise to a circuit of rumors in which truth and fantasy flowed together. That grapevine was referred to as the *radio-trottoir* (the sidewalk radio) because the official media spewed only government propaganda. The street became the venue for suspicion and sarcasm. Clandestine comic books and primitive paintings were sold at the crossings where the taxi-buses came together. Kinshasa developed a lively visual culture. Social, political, and moral topics were portrayed on mimeographed sheets or rudimentary canvases, with no explicit opinion expressed for or against. With virtuoso irony, cartoonists and painters depicted life in the big city under the dictatorship. The subjects were often ambivalent: the artists reveled in portraying transgressions and took potshots at all that was sacred. The scenes resembled something from Hieronymus Bosch or Pieter Breughel.

Young people disgusted with Mobutism developed a very unique form of social commentary. They did not protest with words or images, but with clothing. The *évolué*'s suit had been banned, and the mandatory *abacost* they considered old-fashioned. And so they dressed in brand-new, extremely flashy outfits. They saved their money and imported shockingly expensive brand clothing from the boutiques along Louisalaan in Brussels and Place

Vendôme in Paris, or at least that's what they claimed. They christened their movement *la Sape* (Société des Ambianceurs et Personnes d'Elégance, the Society of Mood-Makers and Trendsetters). The musician Papa Wemba, a working-class-boy-become-international-pop-star, was their pope, *le Pape de la Sape*. It was a highly remarkable movement. Seemingly ridiculous at a first glance, a man in Kinshasa during times of crisis in a pair of gaudy sunglasses, a Jean-Paul Gaultier shirt, and a sable coat, but the *sapeurs'* materialism was a form of social criticism, just as punk was in Europe. It displayed a deep aversion to the misery and repression they experienced, and allowed one to dream of a Zaïre without cares. Materialism is one of the most common symptoms of poverty. La Sape was about success, about visibility, about being in the picture and scoring. A disco was meant to be entered with a combination of *chic, choc,* and *chèque*. The true *sapeur* was übercool: he moved and spoke with total control, he treated his friends to beer and picked up girls with a snap of the fingers. He was a dandy, a playboy, a snob. Luxury meant respect. The *sapeur* wasn't looked down on, but admired. For many dirt-poor young people, the extravaganza he put on kept hope alive.

Kabongo had been too old for that. "Mobutu threw a big party. Franco and Tabu Ley came and played there. The guests wore *abacosts* with the MPR logo on the collar. But Mobutu's own sons were big fans of Papa Wemba. They wore baggy pants and shirts with flashy collars. Those were two separate worlds! La Sape was really the young people's music. They considered themselves a new generation and rebelled against their parents. Papa Wemba refused to talk about politics. His music wasn't made to listen to carefully, but to get the feet moving right away. It was music as anesthesia."

An entire generation grew up in a world of poverty and misery. Music provided an escape valve, but going to school remained extremely popular. Even if the university auditoriums were a shambles, even if the professors rarely showed up, and even if the workbooks were missing and the mimeographed sheets worn to a tatter, the college classrooms filled week after week with young people who hoped that a university diploma would pull them out of the morass. The thirst for knowledge and diplomas was enormous and that has never changed. But the level of education was miserable, and corruption was found in all walks of life. For many poorly paid professors, everything was negotiable. Many female students exchanged sexual favors for a good grade. "For many girls, the body is no longer a source of beauty, but has necessarily also become a source of profitability," one worried professor of moral philosophy wrote. The phenomenon even extended down to the secondary schools. Principals, party officials, and magistrates liked to brag about having *une série 7* (a number 7 series), a teenage girl born in the 1970s.[60] "Many girls' schools have been transformed into sexual fishponds for leading figures in political-administrative circles. They leave their offices before the official closing time and mingle with the rows of cars waiting to pick up the children at the school gates. The evenings usually start at a restaurant in a working-class neighborhood, with roast chicken or fish and lots of *pilipili*, and end in the wee hours in some little hotel, ensconced in the darkness of the tropical nights."[61]

To counter the crisis, a new parallel economy arose, revolving around home commerce. Women would get up early in the morning and fry a single chicken, then take it to the market. Well-to-do ladies who had acquired a pair of chic foreign pumps through the informal circuit would sell them in their own neighborhoods. Nurses who worked in hospitals by day would take home a strip of pills and sell them one by one. Pilots would flog a few jerrycans full of kerosene. Civil servants haggled over every document they stamped. Policemen were delighted with every traffic violation. There was always something that could be "arranged." Quid pro quo. *Madesu ya bana*, the people said, beans for the children, in the old tradition of *matabiche* and baksheesh: the Esperanto of the desperadoes.

In response to a state that was withdrawing from its citizens, the citizens withdrew from the state. "Article 15" they called it, in reference to a fictitious article in the Zaïrian constitution that read: "Débrouillez-vous!" (get it while you can!). Often enough, this involved illegal activities (contraband, theft, fraud)—but what does illegal mean when the country itself is criminal? Grassroots corruption was the best way to counter corruption at the top, for faithfully paid taxes would simply evaporate up there anyway. Hadn't Mobutu Sese Seko himself more or less promised to turn a blind eye? During a huge rally in Kinshasa's soccer stadium, he had said: "If you must steal, then steal a little bit and leave a little bit for the nation."[62]

He should not have said that. During those years, 30 to 60 percent of the coffee harvest was smuggled out of the country, good for a sum of $350 million between 1975 and 1979. Seventy percent of the diamonds, 90 percent of the ivory, tons of cobalt, and hectoliters of gasoline crossed the borders unseen.[63] The country was leaky as a sieve, and the state lost out on hundreds of millions of dollars in revenue. It went little by little, just as the president had suggested, until nothing more was left. "The cockroach can finish a whole loaf of manioc, using only his teeth," the people said.[64] There was no other way to survive. Thanks to the informal economy, people were able to pay teachers and nurses. The country was living on borrowed time, but at least it was alive.

This economics of pillage, of course, could not endure. Congo was being cannibalized. No one gave a hoot about the state anymore. The postal service no longer functioned, water and electricity became scarce, there was less than one telephone line for every thousand inhabitants.[65] The country, as Zao said in one of his songs, was becoming *cadavéré*, it was going into rigor mortis. Boats on the river became slowly drifting villages that would perhaps reach their destination someday. Air Zaïre, the national airline and a former source of pride, received the nickname "Air Peut-Etre" (Air Maybe) with as motto "la seule chose au Zaïre qui ne vole jamais," the only thing in Zaïre that won't disappear into thin air. Humor was the best remedy.

"Mobutu and Reagan and Mitterand were flying around the world in the Concorde," began the best joke from the Mobutu era. "Reagan stuck his hand out the window and said: 'I think we're flying over America.' 'How can you tell?' the other two heads of state asked. 'I just felt the Statue of Liberty,' said Reagan. Then Mitterand stuck his hand out the window. 'I believe we are now flying over France,' he said right away. 'How can you tell?' Mobutu and Reagan asked him. 'I just felt the Eiffel Tower.' Finally Mobutu stuck his hand out the window. 'I know for sure that we're flying over Zaïre,' he told his fellow passengers. 'But how can you be so sure?' they protested. 'Zaïre doesn't have any towers, does it?' 'No,' Mobutu said, 'but somebody just stole my watch.' "[66]

THE CRISIS CHANGED THE RELATIONSHIP between the sexes as well. Many men had lost their jobs and felt humiliated because they could no longer support their families, let alone their mistresses. They were poorer than their parents and often had to turn to them for help. The man had once been the breadwinner, the one who came home with a paycheck, but now it was the woman who saw to the family's income. A school principal in Kikwit told me: "We would run through my wages in two days. They didn't add up to anything. And often enough I didn't even get paid. My wife had a stall on the market. She sold soap, sugar, and salt. That was our family's main source of income. During that period, many women earned more than their husbands. Sometimes the women even moved out altogether. Young women started going to college and became more independent."[67]

The informal economy, however arduous and unpredictable, provided some women with new opportunities. A number of them adopted a more fighting spirit. Manioc saleswomen in Kivu, farming women themselves, refused to simply accept the way that local police and officials kept coming up with new taxes whenever they brought their baskets to market. They filed official protests, all the way up to the provincial governor's office.[68] In Bukavu, Régine Mutjima, headmistress at a girl's school, noted that the policy of spending cuts implemented by the IMF and World Bank was leading to abuses.

[Léon] Kengo wa Dondo was prime minister in 1983, and he had plans to effect drastic cutbacks. Even the pregnancy leave for female schoolteachers was scrapped, ostensibly because there was no money for that, while government funds were being stolen by the truckload at the same time. I was the leader of the Association des Femmes Enseignantes de Bukavu [the Bukavu League of Women Teachers]. A Canadian colleague had told me about Gandhi. I read his writings and those of Martin Luther King, and the works of Lumumba and Nkrumah, even though that was forbidden. I also read the banned weekly *Jeune Afrique*. In 1986 one of my colleagues died during childbirth. She had kept working right up to the day itself, her baby weighed less than four pounds at birth, less than a little rabbit. I had never experienced anything like it. I decided to organize a sit-in. We went in little groups to the payroll office for the local schools. Three-quarters of the female school teachers took part. At 10 o'clock sharp we all sat down. We were running the risk of being shot at, we knew that, but we wanted to close down the city. That evening I was arrested. A Landrover full of soldiers took me to town hall. All I had on was my nightdress. They were all there: the mayor, the head of State Security, the MPR, the board of education, the borough council. It was one woman against fifty men. One by one, they started calling me names, but all I could think about was that baby that weighed less than four pounds, that little rabbit, whose mother, Madame Rumbasa, a good colleague, had died because she wasn't given pregnancy leave. I became furious. I exploded. I screamed at the mayor. I had a lump in my throat, it was only the second time I'd cried during my adult life. When my tirade was over, no one spoke a word, that's how furious I'd been. But then I felt calm. Around midnight, the mayor took me back to my house in his Mercedes.[69]

It was a unique act of courage. Mutjima was not prosecuted, but was allowed to go to conferences in Nsele and Gbadolite to talk about the problems faced by young people. Things did not always work out so well, however. On the other side of the country, Thérèse Pakasa worked as a cashier at a little supermarket in Kinshasa. She had once met Antoine Gizenga, Lumumba's deputy prime minister who lived in exile in Brazzaville but came from her native region. She too read Lumumba and the Universal Declaration of Human Rights. She became enraged when she thought about the situation in her country. "I wanted to organization a demonstration, but the people were so afraid! I could only find three women who were willing to go along. A woman who sold bread and two housewives. We made a banner and pamphlets. In July 23, 1987 we walked down the Boulevard du 30 Juin, where the Belgian embassy was. We were carrying the old, blue Congolese flag."

Four everyday women who risked their lives walking down the biggest street in the capital with a forbidden banner and a forbidden flag . . .

"After a couple of hundred meters we were arrested. The intelligence service kept me in detention for six weeks. I was tortured, but not very badly, and my determination grew. One year later I did it again, this time with ten women. We were arrested again. I was beaten, then sent under military guard into exile in another part of the country. When I came back to Kinshasa, I went straight to prison and all my children were arrested too, even my two-week-old baby. They couldn't believe that a woman would do something like that."[70]

THE BLOOD FLOWING FROM THE WOMAN'S HEAD. The man's legs bent back like those of a jumping jack. The wintry footage from Romania continued to haunt Mobutu.[71] The East bloc was collapsing, the Cold War was coming to an end. Soon he would be an expendable ally to the Americans. Mobutu owed his empire to the fear of communism, but Marx had proven to be a colossus with feet of clay. Loyalty in the struggle against the red menace no longer counted; respect for human rights became the new criterion. At a summit of Francophone countries, President François Mitterand announced that France would from now on support only those developing countries that abided by democratic values and honored human rights. The days of Giscard were over.

Early 1990. Along with the new decade, a new political climate also seemed to descend. In February Nelson Mandela was released from prison, a world event that gave new hope to the entire continent. In the Ivory Coast, in Benin, Gabon, and Tanzania, cries for the introduction of a multiparty system grew louder. The military dictatorships in Congo-Brazzaville and Mali shook on their foundations. The flush of freedom reached Zaïre as well. Mobutu realized that he could no longer ignore his people.

And so he consulted them, just as the Belgians had done in 1958 and Leopold II had done in 1905. On those earlier occasions, those consultations had resulted in a dramatic turnabout, but what would happen now? A group of consultants went from city to city and organized public hearings. All over the country, citizens were allowed to express their opinion of Mobutu's regime, yes, even to air their grievances. There was no need to fear prosecution. The first hearing was held in Goma and Mobutu was there. He was quite willing to hear constructive criticism; after all, no one was perfect. But it rained, no, it *poured* complaints. Mobutu found himself standing in a tropical downpour of dissatisfaction. Old women stood up and delivered broadsides against him. His new nickname at that point was Mobutu *Sesesescu*, a clear reference to his late Romanian friend. Press secretary Kibambi Shintwa saw his reaction: "Mobutu couldn't believe it, he was extremely disillusioned. He felt that the country owed him everything, and he withdrew, he was hurt. He didn't want to admit the truth. He refused to attend the rest of the sittings."

The investigators did their job. More than six thousands reports were drawn up, most of them absolutely damning. The chairman of the investigating committee presented a summary to the president. Zizi Kabongo remembered how that went: "Mobutu had withdrawn to his yacht, the *Kamanyola*. He called together the political bureau of the MPR for consultations onboard. They were stuck there for two or three days. The cabinet ministers had to come too." Even the American secretary of state came by to say that Bush Sr. despite all historical ties of friendship, could not go on granting him unconditional support.[72]

But on April 24, 1990, his mind was made up. Generals, magistrates, cabinet ministers, provincial governors, members of parliament, and foreign journalists were summoned to the whited conference grounds at Nsele, to what had been the Vatican of the MPR for the last quarter of a century. Standing behind an array of microphones, dressed in the black marshal's uniform that Alfons Mertens had sewed for him, Mobutu spoke to the auditorium. He had heard the voice of the people; Zaïre was to be democratized. To everyone's amazement, he announced the end of the single-party state. From now on, three parties would be allowed; room would be made for freedom of the press, free trade unions, and, within a year, free elections. "And what will happen with the chief in all of this?" Mobutu wondered aloud at the end of his speech. "The head of state stands above the political parties. He will be the umpire, even more, the highest court of appeal. I hereby announce that, as from today, I am withdrawing from the Mouvement Populaire de la Révolution, to allow a new chief to be chosen. . . ." Mobutu hesitated for a moment, stumbled over his words and looked helplessly into the hushed auditorium. Like a man grown old in the space of a single second, he lifted his thick-rimmed spectacles, dabbed at eyelids hazy with tears and spoke the words that have since become legendary: *"Comprenez mon émotion,* forgive me for being emotional."

The footage was seen all over the country. Had they understood correctly? Was this how the Second Republic came to a definitive end? By means of a simple speech, diluted with a few crocodile tears? A speech as prosaic as the radio broadcast with which Mobutu had seized power in 1965? With no revolution or fighting in the streets? Young people raided their fathers' closets, rummaged through the old clothes, looking for neckties. No one remembered how you tied one, but who cared? This was the emblem of freedom! Girls put on their brothers' oversized dungarees and went out on the street, giggling. This was one revolution that required no throwing of stones or chanting of slogans. "I remember it clearly," Zizi said, thinking back on the most hopeful day of his life. "That night the streets of Kinshasa were filled with badly tied neckties."

CHAPTER 11

THE DEATH THROES

Democratic Opposition and Military Confrontation

1990–1997

R ÉGINE AND RUFFIN BOTH LIVED IN BUKAVU, THAT LOVELY town on Lake Kivu along the Rwandan border—but that was all they had in common. When Mobutu announced the end of the single-party state, Régine was thirty-five and Ruffin was seven. Régine was head-mistress at a Catholic girls' school, Ruffin was just learning to read at a Catholic boys' school across town. Régine had organized the teachers' sit-in a few years before. When she heard that the MPR had lost its primacy, all she could do was dance. Ruffin was too young to understand the historical portent of the turnaround. He played soccer with his friends and began to dream of life as a priest. Yet both of them would become involved in the dictator's fall, in totally different ways and at totally different moments, Régine in 1992, Ruffin in 1997. When it came to falling, Mobutu took his own sweet time.

Régine Mutijima thought it might all go very quickly. "We really wanted Mobutu to step down after the elections and go on living honorably inside the country."[1] But in the period 1990–97, Mobutu kept hold of power with a stubbornness and cunning no one had thought possible. These were the death throes of a dictator taking his country along with him in his fall.

In 1905, when Leopold II could no longer deny the atrocities taking place in the Free State, he tarried for three years before turning over his conquered land to the Belgian state. Mobutu's attitude after 1990 was no different. The results of the public sittings caused him at first to slacken the reins, only to tighten them again afterward. There was, as far as he was concerned, no big hurry with those elections. He had applied his creativity in 1970, 1977, and 1984 to get himself elected, but knew that that trick would no longer work in 1991. The democratic genie was out of the bottle. Still, he succeeded in remaining in power for another seven years, this time without any elections.

Mutijima knew all too well that Mobutu owed his power to two things: money and violence. Money from abroad, violence at home. But how long could he keep that up, now that the Cold War was over? On May 12 and 13, 1990, a few weeks after his emotional speech, Mobutu ordered the student protests in Lubumbashi crushed by military force, for the students were once again the first to take to the streets. For the West, that was the last straw. Reports spoke of hundreds of casualties, but the exact number has never been stated (there were perhaps only three, Mutijima heard later). Belgium suspended all development aid, France froze all relations, America no longer had any need of Mobutu. In the early 1990s, his former foreign allies were glad to see the back of him. Even the IMF expelled Zaïre as a voting member.

What remained was: violence. But the army was unreliable and, now that more and more bans were being lifted, the intelligence services had lost their grip on the people. The official media, too, had stopped monitoring the flow of information. Government newspapers with "authentic" titles like *Elima* and *Salongo* were overtaken by new periodicals with French names like *L'Opinion, Le Phare,* and above all *Le Potentiel.* Those periodicals did not make it all the way to Mutijima in Bukavu, but they were very important in the capital. At the newsstands along Boulevard Lumumba there arose the phenomenon of the *parliamentaires debout* (the standing members of parliament), groups of the unemployed who read the front pages of the newspapers that were hung up on display and discussed them all day long. A public space, peopled with critical citizens. The founder of *Le Potentiel* was Modeste Mutinga: "We were completely independent. We didn't even maintain ties with other opposition movements, like the UDPS or the Church. I bought a second-hand printing press in Strasbourg. It wasn't until later, after that period of détente in the early nineties, that things got worse again. The DSP burned our presses and the presses at other papers too. Everything was demolished."[2]

Besides the newspapers, democratization expressed itself in other ways. In between the powers-that-be and the masses, there arose a full-blown *société civile* (civil society). Hundreds of new associations were set up: for rural women, for taxi drivers, for altar boys . . . associations for agrarian development, for solidarity between laypeople, for health care . . . even associations for the chairmen of associations.[3] Trade union organizations shot up like mushrooms: in 1990 there was only one state-aligned union; by 1991 there were 112.[4] And, just like in the late 1950s, there was an explosion of political parties. Mobutu had advocated a three-party system at first, but soon had to allow a full multiparty system. In no time, the all-powerful MPR had to countenance some three hundred rivals large and small. Some of them had no more than one member. News of the upcoming abdication caused some people to dream of a bid for power. Mobutu watched ruefully: the proliferation of parties confirmed his fears of disintegration and sectarianism. "But if you can't beat them, join them" he must have thought; in an attempt to weaken the power of the opposition, he paid some loyal followers to set up parties that advanced his own views. "Alimony parties" those were referred to mockingly, or "taxi parties," because their members could all fit in one taxi cab. Was this the *mulitpartisme* Mobutu had promised? It looked more like *multimobutisme!*[5]

In the end, the restless political field crystallized around two poles, with Étienne Tshisekedi's UDPS—the so-called *Union Sacrée de l'Opposition* (Holy Alliance of the Opposition)—on one side, and the MPR and Mobutu loyalists—the *mouvance présidentielle*—on the other. Between them one had the mugwumps. The church sympathized with the opposition, but was often prepared to compromise. Mutijima did not feel called to serve. "I had been a member of the UDPS for a while, when it was still clandestine, but I didn't feel at home in politics. For me, the most important thing in 1990 was the birth of the *société civile*."

The opposition gained a major victory when Mobutu agreed to the organization of a national conference. With that, he hoped to cut a good figure abroad and regain Western support. The plan was to bring together representatives to discuss the past and set out tentative lines for the future, analogous to a similar conference in Benin that had recently reformed that country in ten days' time. The Kinshasa conference was intended to give form to the shift from the Second to the Third Republic; today it is best known by the name given it later, the Sovereign National Conference. The participants were to include not only politicians and dignitaries, but also the rank and file, representatives of the associations and the churches. The meeting would be held in the capital, but with delegations from all the provinces. Everything was to be broadcast live on radio and TV, a high mass of town-hall democracy.

In distant Bukavu, Mutijima donned her battle dress: "The other women teachers in Bukavu said: 'You have to go to Kin!' So I ended up in the South Kivu delegation. All the tribes were represented, we didn't want to think along ethnic lines. At the Sovereign National Conference we were going to denounce everything. We were going to depose Mobutu and demand his head on a platter."

Mutijima went to the capital as one of the twenty-eight hundred delegates, no more than two hundred of whom were women. "There were too few women, not even 10 percent. A lot of women were afraid to express themselves. They were badly informed about how such a meeting worked and about the importance of lobbying." But she herself would prove her mettle. The Sovereign National Conference started on August 7, 1991, and was intended to last three months. The opening session was held in the Palais du Peuple, the national house of parliament. That colossal structure, thrown up by the Chinese, was only a few hundred yards from the new soccer stadium. In the parking lot, Citoyen Jacques Tshimbombo Mukuna, one of the big cheeses in the regime, stood passing out banknotes from a cardboard box to anyone interested in starting a little, off-the-cuff political party. The money was free, all you had to do was stick up for Mobutu. . . . [6] Tshimbombo was the man who once, on the president's behalf, had presented the members of the national women's basketball team with twenty-two Mercedes sedans for winning the Africa Cup, and kept eleven of them for himself. . . . [7] Now that the people saw him standing there with his box of money, they jokingly began referring to him as the "guardian of the national treasury." Mobutu was clearly out to thwart the purposes of the conference, by hook or by crook.

"He kept trying to compromise us by offering us hotel rooms, giving us presents or offering to let us stay at the Nsele conference grounds," Mutijima said, "but we refused. The South Kivu delegation was very militant. We even spent two nights sleeping on the ground in front of the doors of the house of parliament! People brought us food. It was the first time in my life that I tasted manioc bread. And in Kinshasa you had these big, fat mosquitoes. We didn't have any of that in the mountains of South Kivu."

Mobutu was prepared, if need be, to agree to an extensive transitional government with a certain amount of room for dissenting votes. A government of national unity with great power for the opposition, however, was too much for his taste. He had stipulated that he would appoint the conference's chairman himself and entrusted the task to an old supporter, a man whose *nom kilométrique* (mile-long name) alone showed how fanatically "authentic" he was: Kalonji Mutambai wa Pasteur Kabongo. The old man's name still makes Mutijima sigh in despair: "He was a complete marionette. He was hard of hearing and didn't even understand what we said! "Pasteur wa Farceur" was what we called him, the Honorable Joker. I remember thinking: did we come two thousand kilometers to let ourselves be jerked around? We told each other: We need to silence this man! But how? Every morning we had to walk past the police guards and be frisked. We started smuggling in whistles, those little plastic ones. I had five of them tucked away in my shoes and in my braids. Every time the chairman took the floor, we started whistling until he stopped."

The first weeks of the conference went agonizingly slowly, with endless quibbling over procedural questions and interminable haggling over who was to take part in the committees. Mobutu, who followed it all from a distance, must have relished the bickering. A failed conference, after all, would serve him well. But there was growing unrest outside the walls of the Palais du Peuple. On September 23 the soldiers at the paratroopers' center at Ndjili staged a mutiny. They went to the nearby airport and shut down the control tower. From there they cut a swath to the center of town, plundering department stores, shops, gas stations, and even private homes along the way. Everything worth anything was up for grabs: the mutineers dragged away television sets, refrigerators, and photocopy machines; entire warehouses were pillaged, trading companies sacked. With the desperation of the hungry and the poor, the people joined in. It was a great rush, a party, the moment for the Big Snatch. At last the people could do what their leaders had been doing for a quarter of a century! A delirium, the reversal of all values. Forbidden and fantastic! The upheaval spread to other cities, and the plundering went on for days. The Belgian and French armies intervened to free their own nationals. Some 30 to 40 percent of all the urban businesses were destroyed, 70 percent of the small retailers were ruined. Some 117 people were killed and some 1,500 injured.[8]

And Mobutu? He didn't react. He simply let his troops go about their business. Many suspected that he had provoked the mutiny himself in an attempt to scuttle the Sovereign National Conference. Even his loyal press officer Kibambi Shintwa, when we spoke later on the balcony of his little apartment, said he suspected the president of opportunism. "Mobutu wanted to break the country. Maliciously. His pride was deeply wounded by Tshisekedi's popularity, and he wanted revenge. It's like someone with a nice cell phone." He held up his own phone by way of illustration. "The kind of cell phone that other people would like to have, but can no longer afford. So what do you do then?" He lowered the hand holding the phone until it was beside his chair. "You drop it, so it breaks and no one else can have it either. That's what Mobutu did. When the Sovereign National Conference started, he moved out to Gbadolite, permanently. He knew that his people despised him. At three in the morning the soldiers sacked the airport and he didn't do a thing to stop them. It was really a case of: *après moi le déluge*. He saw the pillaging as the people's just desserts. I was very disappointed when he ruined the country like that. For the first time, I was more afraid of being killed by the people than by Mobutu."[9]

Once the ransacking stopped, the conference got a new chairman: this time by popular vote. Laurent Monsengwo, the popular archbishop of Kinshasa and chairman of the national synod, was chosen without delay to replace Pasteur wa Farceur. Monseigneur Monsengwo: the very name prompted great expectations. With his purple vestment and moral authority, he seemed poised to become the Desmond Tutu of Zaïre. The opposition liked him: the Zaïrian synod of bishops had often expressed sharp criticism of the Mobutu regime. Under Cardinal Joseph-Albert Malula the church had evolved into the major counterforce in the Second Republic. When the new civil society awakened, many organizations, even the more secular among them, drew inspiration from the grassroots groups and liberation theology of Latin America.[10] Monsengwo was perhaps not the most radically progressive of Catholic clerics, but the church itself enjoyed credibility among the opposition (which referred to itself—not at all coincidentally—as *sacrée*). There was no doubt about it in Mutijima's mind: "Monsengwo was our candidate, but he even received votes from a few Mobutu supporters!"

Mobutu was not pleased. His relationship with the church had always been ambivalent: he had fought and feared it for almost twenty years. On the eve of the pope's visit in 1980, he had quickly arranged a marriage in the church with his mistress, Bobi Ladawa. He built a cathedral at Gbad-

olite—he, the man who had once tried to outlaw the liturgy and liked to
surround himself with West African miracle workers and soothsayers. With
Monsengwo at the head of the conference, the time had come to watch his
step. Having surrendered the power to choose the chairman of the national
conference, he would now make sure he decided about that other central
position in the transitional government: the prime minister. Zaïre had eight
different prime ministers between 1990 and 1997, seven of whom were given
a leg up by Mobutu himself. The longest term of office had been three years,
the shortest three weeks. The latter had been his archenemy, Tshisekedi. In
October 1991, after the pillaging was over, Mobutu appointed him to chair
the cabinet. Had the uprisings forced the Steersman to acknowledge that
he could no longer get around Tshisekedi? Or was it a cunning move to
discredit him with his own following? The standing members of parliament
along Boulevard Lumumba chattered about it for days, but three weeks later
the prime ministership was over. Mobutu immediately replaced Tshisekedi
with Bernardin Mungul-Diaka, another of his old enemies. He remained in
the saddle for one month. Then it was Nguza Karl I Bond's turn; yet another
dissident from the distant past. *Le vagabondage politique* (political merry-go-
round) was once again running at full speed, and meanwhile nothing was
happening. In January 1992 Mobutu declared that the Sovereign National
Conference had come to an end. The game had lasted long enough in his
eyes, and to his relief nothing had been achieved. This reef too had been
skirted and he had kept his firm grip on the wheel of state.

"The delegates had their trips home paid for them," Mutijima said, "but
we couldn't go home empty-handed. The people of my province demanded
results. Those elections had to be held. The government finally withdrew
our travel allowance, but we stayed in Kinshasa, thanks to the people's support." The Congolese were not about to relinquish the hope for a change.

AND THEN IT WAS FEBRUARY 16, 1992, a day as important in Congolese
history as January 4, 1959, when the riots broke out in Léopoldville. Here
too, the immediate cause was a banned demonstration and here too that led
to large-scale protests in Kinshasa and a bloodbath. The churches wanted
to protest against the closing down of the conference, but the government
refused permission. The charismatic priest José Mpundu, a cleric who
stood closer to the masses than to the hierarchy of the church, was directly
involved in the organization of the protests. I spoke to him in his plainly

furnished house just outside the old soccer stadium. He was wearing short pants—a rarity among Congolese men—and—even rarer—he addressed me right away with the familiar *tu*.

> The bishops had already called for the conference to be reopened. The priests had mentioned that during mass on Sunday. A number of laypeople said: well then, let's do something about it. I went along with their initiative and attended their preparatory meetings, where I talked about nonviolence. Within the bishops' conference, you see, I was secretary of the commission for justice and peace. But Cardinal [Frédéric] Etsou, the new archbishop, wouldn't give permission for the march and Monseigneur Monsengwo felt that bishops should talk, not act. . . . Anyway, we mapped out the routes and decided that the banners would say: "Unconditional reopening of the Sovereign National Conference." Later I was kicked out of the bishops' conference for that.

The march began on Sunday, February 16, after the nine o'clock mass. Starting in Kinshasa's more than one hundred parishes, people left their churches and converged along the broken boulevards and avenues of the capital. They were simple believers, not diehard dissidents or dyed-in-the-wool politicians, merely schoolchildren, students, young parents, poor people, people who felt supported by the common clergy, like the nonconformist Father José. They waved fronds and sang songs. The Protestants, the Kimbanguists, and Muslims took part too. Similar marches were held in Matadi, Kikwit, Idiofa, Kananga, Mbuji-Mayi, Kisangani, Goma, and Bukavu. More than a million people took to the streets, it was the biggest mass meeting in the country's history. People referred to it as the March of Hope.

"I was on my way from Limete to Pont Kasavubu," Mpundu told me, "but when we got to Saint-Raphaël we encountered a battalion of heavily armed soldiers. I was up in front. We had agreed beforehand: if anything happens, we all sit on the ground. Sitting beside me was an old woman, looking in disbelief at those soldiers who were maybe sixteen or seventeen years old. One of them looked her right in the eye, and she said: '*Mwana na nga, est-ce que omelaki mabele ya mama te?*' [My son, didn't you ever drink at your mother's breast?] The boy didn't know which way to look. That's the power of nonviolence, of the truth." For a moment Congo resembled the

India of Mahatma Gandhi. "Then they dispersed us with tear gas. We ran away, but regrouped again a little farther along. We marched on, and we kept singing. At Kingabwa we ran into bodyguards, I think they belonged to Prime Minister Nguza. They threatened to kill us. 'Don't sing, just march,' they shouted. But I said: 'If we keep marching, they'll shoot at us.' A burly fellow with a revolver tried to grab me, but the people held on. The buttons on my cassock popped off. My chain broke. One of the parishioners picked it up. White priests were beaten up too."[11]

The March of Hope ended in a bloodbath. At least thirty-five civilians lost their lives that day.[12] The guardsmen shot at anyone they saw, even from very close range, even at children. They not only used tear gas to disperse the crowds, but also a highly inflammable product rarely used outside military operations: napalm. During one of my many conversations with Zizi Kabongo, at a picnic table outside the canteen at the public broadcasting company, he said: "After that march, Mobutu was afraid he would be excommunicated. The Sovereign National Conference was allowed to reopen, and he withdrew even more to Gbadolite. The conference became much more assertive. The fear was gone. 'Did you really think you could kill us all?' people said out loud. During the march, my wife saw bodies lying around. I was burned too." He shifted his legs from under the table and rolled up his pants legs. I had known him for a few years already, but he had never talked about this or shown it to me. On his shins I saw big, pink spots, as though he were a white man wearing camouflage. A long silence descended. "Napalm," he said at last.[13]

THE CONFERENCE RESUMED IN APRIL 1992 and this time it made a great deal of progress. It became truly sovereign: its decisions were no longer lukewarm recommendations, but expressions of popular will with the force of law. With the conference as the supreme state body, the process of democratization accelerated decisively. After the plenary sessions the delegates split up into twenty-three committees and a hundred subcommittees, spread around the city. Amazing work was done in many of those groups. Inventory was taken of existing problems and realistic alternatives were offered. Régine Mutijima ended up in the "Woman, child, and family" committee. "I was the acting secretary. We worked around the clock. Afterward, all the reports were read aloud during the plenary session, so they could be amended and ratified. The negotiations that finally led to consensus were a

formidable lesson in democracy. The Mobutu supporters discussed openly
with the opposition. We wanted to bring the country's true history to the
surface and give a voice to the powerless."

The Sovereign National Conference voted on a provisional constitution;
its most notable clause read that it was not the president who appointed the
prime minister, but the conference itself. That constituted such a radical
break with the past that the symbols of state had to change as well: Zaïre
was to once again be called Congo and the country's motto and national
anthem would revert to those used before 1965.

And then something peculiar happened: Monsengwo left the conference and went to negotiate with Mobutu on his own. That step ran completely counter to all agreements concerning the conference's sovereignty.[14]
Mobutu told the prelate in no uncertain terms that the country would
continue to be called Zaïre; a name change was completely unacceptable
to him. But he also intimated that he might settle for a more ceremonial
presidency. Mutijima still has mixed feelings about that move: "I thought it
was outrageous of Monsengwo to go off to Gbadolite, but I think he did it to
keep more people from being killed." The men of the DSP, Mobutu's private
army, were still well-armed; a civil war could have broken out. "Monsengwo was for gradual change. He didn't want there to be winners or losers,
because he feared that the latter would ultimately take revenge. Tshisekedi,
on the other hand, wanted a fast victory, even at the risk of a serious conflict.
Monsengwo chose for the gentle landing. He did his best to operate tacti-
cally in a complex situation."

Zaïre remained Zaïre, but a new prime minister was elected directly
for the first time in thirty years. On August 15, 1992, the Sovereign National
Conference appointed Tshisekedi prime minister of the transitional government with 71 percent of the votes; his opponent, Thomas Kanza, received
only 27 percent. Change did not come without a struggle, the offices of the
UDPS had been destroyed only a few days earlier, but Étienne Tshisekedi,
the man who had written that daring open letter to Mobutu a decade earlier, now became the first democratically elected prime minister since Moïse
Tshombe in 1965.

Things went quickly after that. A transitional government was formed,
and a transitional parliament: of the 2,800 delegates, 453 were to take part
afterward in the Supreme Council of the Republic. A new constitution was
drafted, based largely on the 1964 Luluabourg federal constitution, the only

one Congo had ever known that was passed by referendum. An agenda was also established for the coming elections.

The democratic momentum seemed unstoppable. But Tshisekedi's fresh new government excelled in neither vision nor strategy.[15] The prime minister made no attempt to gain control over the government apparatus's most essential instruments, the intelligence services and the army. The cabinet ministers wasted their time with visitors and public ceremony. Governing a nation took more than just sitting around in leather armchairs and talking for hours, but that was something these people, who had even less experience with democratic practice than the politicians of the First Republic, did not know. Tshisekedi himself seemed to have come down with Patrice Lumumba's old ailment: charismatic as long as he remained in the opposition, capricious and unpredictable as soon as he came to power. Becoming prime minister seemed more important to him than providing leadership for Congo.[16]

The Sovereign National Conference was drawing to an end, but the reports from the two committees dealing with the most delicate issues still had to be read aloud: the committee on "unlawfully procured goods" (read: theft) and the committee dealing with the political killings. "Monsengwo wanted to hold those sessions behind closed doors," Mutijima told me. "Mobutu sent tanks to the houses of parliament and had the broadcasts of the conference proceedings stopped." The damning report on government corruption was read only in part; the worse of the two reports, about human rights violations, was not read at all. A few hundred printed copies of the report made the rounds, but missed any effect. "There were two other women with me in the South Kivu delegation. Neither of them could read or write," Mutijima said. In a country where more than two generations had done without proper education and where the spoken word had more authority than the printed one, the lack of those moments of public disclosure were more than symbolic.

In early December 1992, after not three but seventeen months, the Sovereign National Conference closed up shop. The final balance was ambiguous. For the first time, Mobutu had been forced to accept a prime minister he had not appointed himself. The historical review had been of great importance, but the crucial reports were not read aloud and the legislative work remained unfinished. The democratic opposition had not always displayed its political maturity. And the long-awaited elections were still up in the air.[17]

Mutijima provided a sublime summary of the results: "We wanted to uproot the dictatorship, that's right, but cutting down a baobab tree is not simple, because it can fall on you. You have to cut the roots one by one, and then work together to pull it down from a distance."

THE PEOPLE WERE CRUSHED beneath the weight of the toppling baobab. The period of transition to the Third Republic was a true ordeal for many. In the period 1975–89 Zaïre experienced an average inflation of 64 percent annually; by 1990–95 that had risen to an average of 3,616 percent.[18] In 1994, inflation actually peaked at 9,769 percent.[19] A wheelchair in Zaïre had cost 750 zaïres in 1981, by 1991 that was 2.5 million zaïres.[20] Pocket calculators did not have enough zeros to add up the bills. A simple stay in a hotel already ran to exponential figures.[21] Salaries became meaningless. Purchasing power was a farce. The old people said: "In the Belgian days we ate three times a day. During the First Republic we ate two times a day. During the Second Republic that was down to once a day. Where is it all going to end?"[22] Children died of starvation. Rather than making furniture, carpenters were now primarily engaged in building coffins, often for children. Child mortality in the cities was around 10 percent, in the countryside around 16 percent.[23]

Many hoped for a miracle. In the Zaïre of the 1990s games of chance became wildly popular. Lotteries and risky investment and pyramid schemes promised instant success, but in fact made many poor people even poorer.[24] They took their money out of the bank, laid it on the line, won a bit, then lost everything they had. They turned to soothsayers and witchcraft to give luck a helping hand, because money and mysticism went hand in hand. Even Mobutu surrounded himself with powerful marabous and a whole array of *féticheurs* (healers). When he lost two sons to AIDS, he blamed that on occult forces.

In response to this mystic revival a new brand of Christianity awakened, one which was not classically Catholic, Protestant, or Kimbanguist, but evangelical and messianic. The congregations were referred to as *Églises du Réveil* (charismatic churches). They were often initiated by foreign missionaries, particularly from the United States.

The most conspicuous reborn Christian in Zaïre was Dominique Sakombi Inongo, who had been Mobutu's main propagandist for years. The man who had thought up the concepts of *authenticité,* Mobutism, and political entertainment had now become God's own spokesman. After a traffic accident on a highway outside Brussels (driving down the wrong side of the road at night, he had killed a Belgian woman), he'd had a mystical experience. In a dream, the Almighty spoke to him, saying: "Dominique, my son, I grant to you both life and death, but I recommend that you choose Life! For I shall save you and use you." That took a little explaining.

"For years you had my people dance for a man, but from now on it is for Me and only for Me that you shall mobilize the people, so that they may praise Me and be set free at last." Sakombi decided to make a clean break with the Mobutu regime, and personally advised Mobutu to do the same: "You must coexist with Tshisekedi. You must absolutely be converted. . . . I urge you, as a brother, to turn away for good from the marabous, the witches, the *féticheurs* [healers], the sorcerers, et cetera. They are liars. . . . Citoyen Président, do not resist the Lord's call. He died on the cross for you as well."[25] To the former PR man, the symbols of the Second Republic were clearly bewitched. The national anthem, the flag, and coat of arms were of satanic origin. During long prayer meetings, Sakombi told his listeners that he had seen the prototype of the national coat of arms carved in stone in a cave, dozens of meters under the ground, in Egypt. There, close to one of the pyramids of Cairo, on the banks of a subterranean river, there had been elders, singing incantations. . . . The national currency, too, was bewitched: "One need only observe the cabbalistic symbols on it to believe that: they are all about magic. With such banknotes, one can never finance a country's development. . . . You will recall that the recent series of banknotes caused troubles and even lay at the root of the conflict between Mobutu and Tshisekedi. And now you know why . . . because they are diabolical."[26]

And indeed, there was something peculiar about that recent series of banknotes. Sakombi's fantastical discourse did not come entirely out of the blue but put a religious twist on well-known public comment. In 1970 the largest denomination in circulation was the five-zaïre note, in 1984 it was five-hundred zaïres. That in itself was a sign of drastic inflation. But in 1990 there appeared a fifty-thousand-zaïre note, and two years later even one representing five million zaïres.[27] At times, macroeconomics can be childishly simple. When the value of the biggest banknote rises that quickly, it means either that a country is becoming breathtakingly wealthy or that the currency is becoming breathtakingly worthless. Unfortunately, it was the latter: the five-million-zaïre note was worth only two dollars. Nevertheless, it showed Mobutu looking as unperturbed as he had on earlier bills. He stood there proudly in his white marshal's uniform, the one tailored for him by Alfons Mertens, thereby making it one of the twentieth century's most oft-depicted outfits. Printing new money on a massive scale, after all, was Mobutu's favorite way to ensure cash for himself, especially now that he could no longer count on his international backers. He placed his order with the German firm of Giesecke and Devrient, printers of currency for customers from Hitler to Mugabe, and had huge shipments of banknotes brought in by cargo plane. In 1995 alone, 830 million new bills were flown in. Mobutu immediately had to exchange almost half that for dollars, in order to pay the printer.[28] Printing money in order to pay the printer: economics can also be childishly tragic.

When Mobutu introduced the five-million-zaïre banknote in December 1992, Prime Minister Tshisekedi declared it unlawful. He wanted to call a halt to this thoughtless monetary policy, but that resulted in the new prime minister's first serious clash with the president. In the streets of Kinshasa the banknote was soon given the nickname Dona Beija, after a lovely but extremely treacherous character in a popular Brazilian soap opera of that day.[29] Mobutu used the money to pay his soldiers. As always, they took their pay right away to a moneychanger; the wages paid on Friday, after all, could lose a third of their value by Monday morning. The phenomenon of the *cambistes* had manifested itself all over the country, moneychangers (almost always women) who sat under a parasol along the side of the road, piles of banknotes beside them on a folding table. In Zaïre even the black market was colorful. One found them along the street behind the Belgian embassy in Kinshasa, soon referred to as Wall Street, but deep in Matonge too one soon found alleyways full of unofficial change offices. On payday the civil servant, traffic policeman, or soldier would take his plastic bag full of newly printed bills to the *cambiste* and exchange them for dollars. The moneychanger would later sell the banknotes to someone else, often to state enterprises that needed them to pay their own employees. In this way, Zaïre gradually became "dollarized."[30] Even today, dollars are the prime currency for all major expenditures; the local currency is used only for smaller purchases.

But the five-million-zaïre note was a catastrophe. After Tshisekedi declared it unlawful, the *cambistes* refused to accept it. The soldiers who saw their monthly pay dwindling away felt betrayed and decided to collect their wages themselves. From January 28–30, 1993, they went on another rampage. The consequences were terrible. In Kinshasa today, people still speak of the First and the Second Plundering, the one in 1991 and the other in 1993, for these were historical events that impressed themselves deeply on the nation's memory. The Second Plundering was the most violent by far. This time it was Mobutu's elite troops, the DSP itself, that mutinied and helped themselves to public and private property. Before the owners' eyes they smashed shop windows and pulled chandeliers from the ceiling. Because the assortment was often meager, they even yanked the copper wiring and sinks out of the walls. Zaïre had now become the country of last things, a lawless, freebooting, hopeless country that had succumbed to banditry and greed. Thousands lost their lives during the Second Plundering, including the French ambassador and a member of his staff. Once again, French and Belgian paratroopers were mobilized. When it was over the city looked as though a horde of locusts had descended. The streets were littered with bits of paper, notebooks, chunks of debris, and shoes. Curtains flapped through broken windowpanes, brushing the sidewalk. Normal people tried to get a piece of the action as well, for in a bankrupt country even the most trivial takes on new value. Old paper, for example, became a costly commodity The archives at the Kinshasa zoo, a pitiful remnant of colonial days where a crocodile hatched in 1938 still lay basking in the sun (and still does today) were pillaged by city dwellers in search of packing paper.[31] In the weeks that followed, anyone buying a handful of peanuts on the market in Kinshasa by way of evening meal received them wrapped in a piece of yellowed paper on which one could read of the wondrous life of the chimpanzee and the okapi.

HIS COUNTRY COULDN'T GO ON THIS WAY, Mobutu felt. A few months earlier he had celebrated the wedding of one of his daughters at Gbadolite. On that occasion she had worn $3 million worth of jewelry from Cartier and Boucheron. But that wasn't the problem. The wedding was attended by twenty-five hundred guests. There had been caviar and lobster. Thousands of bottles of top French wines were consumed. But that was not the problem either. A plane had flown to Paris and back again just to pick up the cake, a four-meter-high (thirteen-foot-high) construction, from the bakery of *chef-patissier* Lenôtre. None of this, however, was the problem, no. In his eyes, the real problem was Tshisekedi, the man who had rejected a banknote with his picture on it and thereby unleashed the plundering. No, there was no way he could work with a stubborn fool like that.

In March 1993, after ten days in conclave with the leaders of the still-vital MPR, Mobutu decided to set up his own government with its own parliament, constitution, and prime minister. Faustin Birindwa, a former opponent, was the new flunky. He would see to a project of monetary reform in which a new currency, the *nouveau zaïre*, would replace three million old zaïres.

Zaïre now had a shadow government. In addition to the institutions established by the Sovereign National Conference, there were now those put in place by the much more sovereign president. The tremendous work fought for by people like Régine Mutijima now fell prey to the avarice of an old dinosaur. The historical irony was clear to all: in 1960 and again in 1965, Mobutu's coups had been staged in response to Joseph Kasavubu's twice appointing a prime minister of his own beside the democratically elected one (Joseph Ileo versus Lumumba in 1960, Évariste Kimba versus Tshombe in 1965); Mobutu was now doing the very same thing. This insufferable impasse was to last one year. Transnational organizations recognized the seriousness of the situation and feared an escalation like the one after independence. To negotiate a compromise, the Organization for African Unity and the United Nations sent emissaries to Kinshasa. That compromise was finally found in the form of a superparliament with seven hundred members, comprising two parallel bodies: the parliament approved by the Sovereign National Conference and the one imposed by the dictatorship. Within the rather clumsily named Haut Conseil de la République–Parlement de Transition (HCR–PT), the Mobutu faction held a majority. In July 1994 the prime ministership went again to Léon Kengo wa Dondo, a man of Polish-Congolese origin who had led two relatively stable governments in the 1980s and implemented the IMF's structural adjustments. That made him acceptable to the international community; the Zaïrians themselves, however, had chilling memories of those years of spending cuts. Kengo never caused their hearts to leap in hope, the way Tshisekedi had. It was now his job to lead the country toward the elections originally projected for 1995; in 1995, however, they were postponed until 1997.

With this new construction (a parliament that listened to him, a prime minister who did not obstruct him, elections that were not going to happen tomorrow), Mobutu seemed to have once again saved his skin. Yet that was only appearance; Zaïre, the country he had unified and made grand, was gradually falling apart. In Kasai, the people refused to use the new currency, the *nouveau zaïre*: within that autonomous currency zone now lurked the threat of a new secession.[32] In Katanga ethnic violence between the original inhabitants and the Luba migrants from Kasai had flared up again, due to the outright racism of provincial governor Kyungu wa Kumwanza, who dreamed of an independent Katanga and took an advance on that dream by running tens of thousands of immigrants out of the region. The worst violence, though, was seen in North Kivu. There the Banyarwanda, the Rwandan-speaking population, were increasingly seen as undesirable immigrants who unrightfully claimed wealth, land, and power. Most of them had settled in Congo between 1959 and 1962, in response to rioting in their own country. As long as Bisengimana Rwema, the father of the young man I spoke to on that boat on Lake Kivu, remained Mobutu's cabinet chief, these Rwandan-speakers (mostly Tutsis) were seen as full Zaïrians and received the Zaïrian nationality quite easily. But a new law in 1981 purposely tightened the criteria for naturalization, and from 1990 the goal was to get rid of these Tutsi immigrants. The Banyarwanda were to Kivu what the Baluba had been to Katanga: undesirable elements, intruders, outsiders, profiteers, foreigners, people who didn't belong. *Rwandais* became a term of derision. Children sang: "All Rwandans should go home, we don't want them anymore."[33] The animosity between Zaïrians and "the Rwandese" grew to such heights that nationalistic militias arose, the Mai-mai. These ad hoc paramilitary groups were in favor of armed resistance against all foreign influences. They drew on the Simbas of 1964 for their bizarre rituals, but this time the enemy was not Mobutu and his Western allies, but the "immigrants" from the east.

"Je suis zaïrois!" a Mai-mai veteran told me proudly in December 2008, eleven years after his country had changed its name back to Congo. "At first we got along well with the Banyarwanda, but then they tried to eliminate the Hunde, the Tembo, and the Nyanga. I'm a Hunde. The Banyarwanda locked the Hunde up in their houses and burned them down." The conflict, in essence, was about land. Rwanda and Kivu are Africa's most densely inhabited agricultural regions. "It started in 1993. We became Mai-mai. To do that you had to belong to the Bantu race, you had to be highly patriotic and baptized with our special water. You received a ritual scar, traditional potions, and medicinal plants. Stealing and raping was forbidden. There was no raping back then. We adapted the rifles we usually used for bird hunting. We had no alternative. The Banyarwanda were foreigners who wanted to annex North Kivu as part of Rwanda." Overcrowding, poverty, and an absentee state made for a deadly cocktail. In 1993 the tensions in North Kivu led to campaigns of ethnic cleansing that killed at least four thousand, by some accounts as many as twenty thousand, people.[34] "Oh, I took part in at least forty battles myself."[35]

In Goma I talked about this with Pierrot Bushala, a man who still looked back in amazement at the events of that day: "In the 1980s no one knew what ethnic group his classmates came from; that only started in the 1990s. My class at secondary school was *un mélange total* [a complete mix]. I had a Tutsi girlfriend at the time, and I didn't even know that. But in the 1990s, when we wanted to get married, her parents wouldn't allow it. I'm sure that ten years earlier they would have accepted me." He was able to place his heartbreak within a historical context. "Look, when Belgium took over the mandate territories in 1918, the border with Rwanda became porous. The Belgians had exported thousands of Rwandan Hutus to the mines and the Tutsis moved spontaneously across the border. Under Mobutu those Tutsis had Zaïrian passports, but in the 1990s, tribalism grew. Suddenly people starting saying that the Tutsis were no longer loyal countrymen, because they supported their brothers' struggle in Rwanda. 'If you're a Tutsi, then you're a Rwandan,' the Zaïrians said. That's where things went wrong. I ended up marrying a Lega woman; her tribe originally comes from South Kivu."[36]

In South Kivu, these Zaïrian Tutsis were increasingly referred to as Banyamulenge, the people of Mulenge, an ethnic moniker invented and applied to them by others. Ever since the nineteenth century, they had lived with their herds in the cold, misty highlands west of Lake Tanganyika, close to the town of Mulenge. With their height, their fine features, and their felt hats they confirmed the clichés of the Tutsi herder sauntering along behind his cows with his stick over his shoulder. And increasingly, they became the objects of abuse and hatred. They were like bats, a Congolese woman told me once, neither bird nor mouse, neither Rwandan nor Zaïrian, scary and slippery. And a bit dirty as well! Yes, another person chimed in, they earned lots of money with their cattle but the Banyamulenge had no culture. They bought the most expensive clothes, but all in bad taste. Their men wore women's clothes. And their women used toilet pots to pound manioc. Ha ha! And the way they grimaced all the time! Was that because of their buck teeth? Or were they just cold?

THE DEATH THROES   413

Mobutu had tried to awaken national sentiment as an antidote to the
tribal reflex; in times of scarcity, however, enmity was just around the corner. The Tutsis in Kivu (both the Banyarwanda in North Kivu and the Banyamulenge in South Kivu) in particular footed the bill for that. The racial
hatred, in turn, caused them to behave increasingly as a group. Those reviled
as Banyamulenge truly began feeling like Banyamulenge. They looked back
on their history, recalled that they were indeed different from all the rest,
that their roots lay in Rwanda, and that in fact, yes, now that you mention
it, they had never been truly welcome in Zaïre. Groups form as soon as they
are threatened. Ethnic identification became more important than national
identification.[37] Even the father of the nation had withdrawn to his native
region and entrusted his safety to men of his own tribe. Mobutu, the advocate of unity, himself became a tribalist. Zaïre was once again transformed
into a crazy quilt of tribes. Poverty led to aggression, hunger to atrocities.

No money, no foreign aid, and no functional army: Zaïre was crumbling and little would have been needed in 1994 to bring the dictatorship to
its knees. But then, in Zaïre's smallest neighbor, a humanitarian catastrophe
took place that destabilized the entire region so badly that the international
community once again came to see Mobutu as a beacon of stability, a wise
elder, a bulwark amid the storm racing across Central Africa. That catastrophe was the Rwandan genocide, a foreign event that would impact the
history of Zaïre like no other.

Like neighboring Burundi, Rwanda had become independent of Belgium in 1962. During the first democratic elections, the centuries-old ruling
class, the Tutsis, a cattle-breeding minority, lost power. Ascendancy went to
the far more numerous Hutus, a traditional farming people. The social and
economic differences between the two groups were real enough, but the
Belgian colonial regime had accentuated them and rendered them categorical. You were either a Hutu or a Tutsi. After independence, the new Hutu
regime displayed great intolerance for its former masters. Many Tutsis fled
with their herds to Burundi, Congo, and Uganda. From there, at the borders
of their fatherland, they stared at the distant hills and vowed to return some
day and seize power anew. In southern Uganda they banded together militarily in the Rwandan Patriotic Front (RPF) and fought beside rebel leader
Yoweri Museveni in his campaign to oust Milton Obote. Museveni became
president of Uganda and the RPF learned for the first time how one went
about conquering a country. That military experience would serve them

well. Paul Kagame, the current president of Rwanda, became their military leader. Starting in 1990, the RPF began crossing the Rwandan border and initiated a civil war with the Hutu regime. An estimated twenty thousand people were killed in that war between 1990 and 1994, and 1.5 million civilians became displaced. The attacks created so much bad blood among the Hutu population that the hatred toward anything Tutsi grew even further, even toward those Tutsis who had remained in Rwanda and behaved as good citizens. "Cockroaches" is what people called them.

On April 6, 1994, when Hutu president Juvénal Habyarimana's plane was shot down, all hell broke loose. Kagame's RPF had to be behind the attack, the Hutus reasoned, and they began murdering Tutsi citizens on a massive scale. This was no battle fought by soldiers with firearms, but by civilians with machetes. Civilian militias had been trained beforehand by the Hutu regime and equipped with machetes. These militias often consisted of teenage boys weaned on racial hatred, the infamous Interahamwe. They set about the business of genocide, egged on by the broadcasts of hate-radio Mille Collines, which kept repeating that the graves were not yet full and that there were still cockroaches scuttling around. Within three months, eight hundred thousand to one million Tutsis and moderate Hutus had been slaughtered. Meanwhile, from the north, Kagame's RPF continued to press on toward Kigali, the capital.

The international community was not on the ground. At the start of the genocide, the Rwandan government army had murdered ten Belgian blue helmets in order to chase the United Nations out of the country and clear the way for ethnic cleansing. Reporters and foreign journalists fled the country's violence. The eyes of the world in those weeks were turned much more on South Africa, where Nelson Mandela was elected president. Few people knew exactly what was going on and France's President François Mitterand was no exception. He saw the Hutus as victims of the Tutsi invasion and sent French troops to Rwanda to help them. The French support was unconsciously prompted in part by the fact that the Hutus were Francophone while the Tutsis in Uganda spoke English. What Mitterand did not know was that he was in fact protecting the perpetrators of the genocide. Under the name Opération Turqoise, the French troops established a safe haven to which Hutus could flee in the southwest of the country, away from Kagame's advancing RPF, away from the reprisals that were sure to follow.

The genocide was intended to make Rwanda Tutsi-free, but those same

Tutsis were now coming in to conquer it from the neighboring countries. The RPF's military might had been sorely underestimated. The French soldiers took in hundreds of thousands of Hutu refugees and helped them across the border. Here it was not only a people fleeing, but also a regime: the government army, the country's ordnance, the administrative apparatus, and even the state treasury left the country. Some 270,000 people fled to Burundi and 570,000 to Tanzania, but the lion's share of the refugees— approximately 1.5 million—ended up in eastern Zaïre.[38] Mobutu had put his airports at the disposal of the French offensive and granted permission to lodge the refugees in his country. Most of them arrived in North Kivu, in and around the city of Goma (850,000 refugees), and to a lesser extent at Bukavu in South Kivu (650,000).

Along with Pierrot Bushala, the man who had lost his Tutsi girlfriend, I drove in December 2008 to Mugunga, west of Goma, the biggest of the former Hutu camps. It was still being used as a refugee center; since 1994, calm has never returned to Kivu. In the 1990s, under the auspices of the UNHCR, the United Nations' refugee organization, Bushala had been involved in trying to maintain hygiene in the camps. "Can you picture it? This whole area was full of refugees, and there was nothing at all," he said as his jeep bounced through a sinister lunar landscape overrun with garish green vegetation. The earth's surface in this place consisted of black lava from the imposing Nyiragongo volcano a little farther along, and suddenly 850,000 people had been dropped here. Bushala was responsible for the sanitary conditions in one of the camps. "At first the people relieved themselves wherever they could. But then the UNHCR and the Red Cross brought in tents, and quicklime to sprinkle around. It was only later that toilets were built, over a hole in the ground." As we walked around Mugunga itself, I realized how grim a task it must have been to dig toilet holes in that volcanic rock. Pierrot looked out over the desolate landscape of clotted lava covered with little huts and tents. "We combated flies, mosquitoes, we walked around with spray guns, we had teams to empty the toilets, we collected garbage." But it was to no avail. Cholera and dysentery broke out in the camps. At least forty thousand people died. Their bodies were piled along the road. The stench was unbearable. The clouds of flies were so thick that drivers could barely see through their windshields.

The misery that came after the genocide restored Mobutu's international respectability. The French were grateful to him for his assistance and soon invited him to an international summit at Biarritz. The United Nations recognized his role in taking care of the refugees. When the camps were struck by epidemics, dozens of NGOs and international aid organizations were allowed to descend on the country. The outbreak of the highly contagious Ebola virus in Kikwit one year later lent Mobutu the aura of victim, rather than villain. Now that the world was taking a milder view, Prime Minister Kengo wa Dondo could easily go about delaying and sabotaging the election process. There was no rush.

But providing shelter for a million and a half refugees within one's own territory was of course a high price to pay for rehabilitation, particularly in a region that was already overpopulated and where hatred toward Rwanda had been growing for years. Just as the people tried their luck with risky games of chance, Mobutu too was playing for high stakes with those camps. At first his gamble yielded some returns, but in the end it was to be his downfall.

ON THAT SAME SATURDAY IN 1996, Ruffin Luliba was playing soccer with the locals. A sunny day. The sound of children's voices calling for a pass, the dull thud of soccer slippers against the ball, a few screaming spectators, the arbiter's whistle. Ruffin was thirteen by then. After elementary school in Bukavu he had gone to the Marist Brother's boarding school in Mugeri and began studying at the minor seminary there. On that particular Saturday his team was playing in the semifinals, and one of the spectators was Déogratias Bugera, a man who worked as an architect in Goma but liked to spend his weekends in his native region. "When the match was over, Bugera said he would like to sponsor our team. He gave us sugarcane, bonbons, and cookies. If we won the finals that next week, he said, he would pay for everything: all the equipment, the jerseys, even new soccer shoes." Ruffin could hardly believe his ears: new soccer shoes! "That next week, there he was. We really wanted to win, and we beat the opponents by 2–0. We were all allowed to go along in his Daihatsu to pick up the new uniforms. It was one of those pickups with nets over it. There were thirteen of us. The oldest was sixteen, the others were fourteen or fifteen. My roommate Roderick went too." But the boys' exultation quickly changed to confusion.

> We drove off in the direction of Bukavu, but we didn't stop there. We went on, all the way to the Rwandan border. We crossed at the

bridge over the Ruzizi. There weren't even any customs formalities, no guards, no immigration service, nothing. We drove on until we got to an airfield. Wait here, Déogratias said, and he left. We didn't know exactly where we were, we were just schoolboys. It was already five thirty in the evening and it was getting dark. We were afraid the headmaster at the boarding school would punish us, and we started crying. At seven o'clock a big truck came by and we had [to] climb in. The drive took five hours. "What's the headmaster going to say?" we asked each other. That was our biggest worry. Finally, we arrived at the military training camp at Gabiro. We didn't get soccer shoes, but they gave us rubber boots, not leather boots like the ones at home. There were a lot of children at that camp, all of them kidnapped from Goma and Uvira. There were also a few Bunyamulenge, but they were there as volunteers. They cut off our hair right away. It was one o'clock in the morning, and as a sort of hazing we had to crawl through the mud. You have to rid yourselves of Mobutu, they screamed, you are the new liberators of your country.[39]

Young Ruffin's testimony is very important, not only because it describes the fate of what was then a relatively new phenomenon, the child soldier under duress, but also because it shows how Rwanda was preparing to invade Zaïre. The Tutsi regime that came to power in Kigali after the genocide was extremely wary of those 1.5 million Hutu refugees in Zaïre. Contrary to international directives, they were not a few dozen kilometers from the border, but bunched up almost right against it. In those camps, the recently routed Hutu regime was busy regrouping. They had money and weapons and were determined to retake Rwanda. Just as the Tutsis in exile had awaited their chance in Uganda from 1962 to 1994, the Hutus in eastern Zaïre were now waiting their turn. Most of the refugees, some 85 to 90 percent, however, did not belong to the national army in exile, had not taken part in the genocide, and had never belonged to Interahamwe.[40] They were innocent civilians who simply wanted to go home again, but who feared for a genocidal countercampaign.

An invasion was being planned in the refugee camps. The international community was aware of the problem, but did not seem inclined to do much. After the debacle in Somalia, America had no desire to once again see the corpses of GIs dragged through the dust. Belgium did not feel like losing another ten paratroopers. And UN Secretary Boutros Boutros Ghali could not round up enough support to deploy an international force. Any international action in Zaïre, in any case, would be seen as support for Mobutu and no one wanted to go that far. And so Kagame decided to take things into his own hands: his Rwandan Patriotic Front, by then redubbed the Rwandan Patriotic Army, the new government forces, would just have to neutralize those camps itself. His old friend Museveni, president of Uganda, promised his support.

Raiding the camps, however, meant raiding a sovereign country, an act of aggression against a foreign power. Kagame therefore went looking for a stalking horse for his initiative, and found one among the frustrated Tutsis in Zaïre. Humiliated for years by the "real" Zaïrians, they were now being bowled over as well by 1.5 million Rwandan Hutus. Déogratias Bugera, the soccer fan who had kidnapped Ruffin and his teammates, was a Tutsi from North Kivu and the leader of the Alliance Démocratique des Peuples (ADP). Along the same lines, there was also Anselme Masasu Nindaga, a Tutsi from South Kivu who was politically active in Bukavu and led the Mouvement Révolutionnaire pour la Libération du Zaïre (MRLZ). But there were also older nationalists, like André Kisase Ngandu, a Tetela who harked back to the Lumumbist tradition. And then there was Laurent-Desiré Kabila, no Tutsi either but a Luba from Katanga, the man who had kept the area between Fizi and Baraka out of Mobutu's hands ever since 1964 and made such a lamentable impression on Che Guevara. And if Kabila's "rebellion" had been an utter mess in 1964, by 1996 it was showing little improvement. Kabila lived semipermanently in Tanzania and earned a living by smuggling a little gold, doing a little gunrunning, and organizing the occasional kidnapping: the mixed farming of African crime, in other words.

In October 1996, at Kagame's instigation, these four men would set up the Alliance des Forces Démocratiques pour la Libération (AFDL). Kabila was their spokesman and, as the eldest of the four, received the honorable term of address *Mzee*, Swahili for *elder*. Bugera was the alliance's deputy, Kisase its military commander.

Ruffin Luliba saw it all "live":

> During our training, we were introduced to the men who later set up the AFDL. We already knew Bugera. But Kisase Ngandu, Masasu, and

Mzee came to visit too. Mzee even gave us two cows, which meant we had our first good meal in a long time! Normally we ate beans and corn from our mess tin. There were two battalions in the camp, and our training lasted six months. Three months of physical training for the battlefield and espionage, two months of ideological training to impress upon us the war's objective. And one month of concrete preparations. The first part was especially hard. Some of us died. Roderick, my roommate at the minor seminary, succumbed to dysentery. We buried him in a blanket, there weren't any coffins. At the end of our training we were given our real uniforms, and they told us again that we were "the new liberators" of our country.

It was clear from the start that Rwanda hoped not only to neutralize the camps, but to actually push on all the way to the capital, two thousand kilometers (about 1,250 miles) to the west. Kagame demanded that Mobutu step down, because he had given the *génocidaires* shelter and protection. Minuscule Rwanda would force Zaïre, the giant of Central Africa, to its knees, and the AFDL had to make it look like a domestic uprising. This was to be Kagame's third regime change in a Central African country: after Uganda and Rwanda, it was now Zaïre's turn.

At the head of the invasionary force was the fresh-faced but dauntless Rwandan officer James Kabarebe, one of Kagame's most trusted men. He was only twenty-seven: a boy with a baby face, but also with great charisma and a flexible conscience. The invading army was already known for youthfulness, because it made use for the first time of *kadogos* (child soldiers) from Zaïre. They were recognizable by their baggy uniforms and particularly by their black rubber boots, the true earmark of Rwandan involvement. Kalashnikovs seemed too big to fit in their hands, but the way they clutched the characteristic curved clip showed more venom than reluctance.

Ruffin remembered the first phase:

James Kabarebe said: I need ten *kadogos* from Bukavu, ten from Uvira and ten from Goma. I reported to him and we had disguise ourselves as street children and go in and spy. James told me: "I'm entrusting this mission to you. Go and watch the FAZ [Forces Armées Zaïroises, Mobutu's government forces]. See what kind of guns they have. See whether they have reinforcements." He gave me a Motorola to keep in touch with him. I crossed the border wearing rags and went to look at their camp in Bukavu. When I got there, the soldiers were busy plundering. One of them shouted to me to come and help him carry the booty! I hid the Motorola. It was complete chaos. Shots were being fired. Then I went back to Rwanda to tell James what I had seen. I didn't look up my family while I was in Bukavu. When you're in the army, you forget your family. The army was my family.

The FAZ, plundering? Kabarebe was delighted to hear that. Zaïre, he decided, was now a complete shambles. And indeed: in early October, when the deputy governor of South Kivu announced that he would start the next day with the ethnic cleansing of the Banyamulenge, the threatened group went on a rampage. That was the starting shot for the hostilities. Rwanda commenced the invasion. A few days later, the AFDL came to the fore as rebel movement. Uvira was taken on October 28, 1996, Bukavu two days later. One of the first casualties was Christophe Munzihirwa, the archbishop who had sharply criticized the Rwandan machinations. Ruffin and his buddies fought in the front lines. "There were Rwandans, Ugandans, and even Eritreans with us. We smoked big six-inch joints, that gave us the courage to be patriots." Mobutu's soldiers turned and ran right away, but the heaviest resistance came from the Mai-mai, the popular militias that hated all things Rwandan.

My first fight was with the Mai-mai who were guarding the offices of the RTNC, the public broadcaster. I had a short Kalashnikov. That took some getting used to. I'm left-handed, so I kept burning myself; the cartridges ejected on the right side of the gun, right against my stomach. A Mai-mai came running up to me with his red kerchief and his fetishes. He didn't have any ammunition. I put a bullet through his head. I was terribly upset. I had never killed anyone before and I felt terrible. Let me go back to the third section, I begged the officers. I didn't want to fight up in the first section anymore. You have no choice, they told me, and they gave me a hundred lashes.

After the fall of Uvira and Bukavu, Goma followed on October 31, 1996. Within only a few days the AFDL had taken the three most important cities in eastern Zaïre, and by no accident the three cities with the biggest refugee camps. The AFDL wanted to press on to Kinshasa as quickly as possible, but neutralizing the camps was crucial for the Rwandans. Ruffin clearly felt the tensions within the mixed invasionary forces: "Whenever we got to a refugee camp, the Rwandan Tutsis would do all the work. Hundreds, thousands of dead people . . . Fathers, mothers, women . . . The Hutus are serpents, they said. At the Kashusha camp, close to Bukavu, I went into a tent where they had just killed a grandmother and a pregnant woman. The only one alive was a child. A toddler. I was supposed to kill him, but I couldn't. He petted my gun. I let him go, sent him along with a few Hutus who were running away."

Around Goma, where the five largest refugee camps were located, the killing was particularly remorseless. Rwanda riddled the hovels with mortar and machine-gun fire, causing many of the Hutus there to flee in a panic back to their homeland. Within a few days, almost four hundred thousand refugees, a huge sea of humanity, fled east across the border.[41] In Rwanda a new traffic sign was posted: *"Ralentir: refugiés"* (Slow: refugees crossing).[42] But a great many Hutus, especially the more militarized among them, moved farther west into the jungles. By the time the United Nations had put together an intervention team to protect the refugees, the camps were empty. The ensuing struggle between the Rwandan Hutus and Tutsis, the sequel to the genocide, was to take place on Zaïrian soil.

At the age of fourteen, Ruffin learned about the horrors of war from close by. His battalion moved south, by way of Uvira, along Lake Tanganyika to Katanga. At Bendera, a town in north Katanga, he experienced the grimmest fighting. He and his comrades were pinned down by heavy shelling. "A firefight is like a drum set. Bombs and bazookas sound like the tom-toms and the floor tom. The bursts from our Kalashnikovs are like ruffles on a snare drum. The bass drum is represented by an 80-mm mortar. And the cymbals, those are our shrieks, because we were always screaming. We made ghostly sounds to drive the enemy mad, some in a low voice, others shrilling. We shouted their names and said that we would find them." War, madness, hysteria. Soccer, but then without the ball. Only the screaming. And with guns.

The screaming didn't help. Ruffin and three other soldiers were taken prisoner by Rwandan Hutus. He was terrified. "The Hutus were notorious for killing people with the machete, the way they had during the genocide.

They cut off your arms or pounded in your skull, so they could see your brains. That was typical of them." He was the youngest of the four prisoners, and that proved to be his salvation. One by one, the others had to lay their arms on the chopping block.

The Hutus had new machetes, they glistened: they were like mirrors. My friend looked the other way when they raised the machete. He screamed. I saw his hand, his hand that was still moving, even though it was already lying on the ground. Then they made him suffer terribly. They kept on chopping, they ran their blades through his body, until he was dead. And then they did the same to the second one, and the third. My friends were slaughtered one by one and I watched. When it was my turn, the commander told me that his name was Mungara and that he had been President Habyarimana's bodyguard before the president was murdered. He was going to spare me and started writing a letter in Kinyarwanda. "Here, take this to Kabarebe." They tore off my clothes and sent me away in my underwear. I came down out of the hills and returned to our position. That was the hardest moment of my life, I'll never forget it. When I finally got there, I handed the letter to James Kabarebe. He read it and said: *"Dieu le veut.* It is God's will. Mungara has murdered the whole family, but I'm going to keep you as my bodyguard."

Ruffin, a Zaïrian boy who until recently had known nothing about politics and found the offside rule already difficult enough to comprehend, had almost been killed in a conflict between Rwanda Hutus and Tutsis. "I didn't have to go into battle anymore. James loved me, he let me carry his duffle bag. 'Ruffin, bring me my bag!' he would shout to me. In the days that followed, I saw him studying the map of Congo. He had never been here before either. Kabarebe had no real education, but he was very logical and calm, he could analyze and listen well. He had lost his family and he said to me: 'You have to love your country, *kadogo.*'" And that was how Ruffin, the boy who liked soccer and wanted to become a priest, became bodyguard to the de facto commander of the invading force that would dethrone Mobutu.

THE AFDL USED A PINCER MOVEMENT to conquer Zaïre. Ruffin was with the southerly arm that moved toward Lubumbashi; the northern pin-

cer headed for Kinshasa, the city on the river. After three decades of dictatorship, many tens of thousands of civilians were now overrun by war as
well. A virtual exodus began. Many of the inhabitants of Kivu tried to get
out, but the last planes were chock full and anyone with a jeep had to hand it
over to the plundering soldiers of the FAZ. Thousands of them then decided
to travel to Kisangani on foot. It was a seven-hundred-kilometer (435-mile)
journey through the jungle, the first part of which went by way of Kahuzi-
Biega, a mountainous natural park where in better times tourists had come
to see the gorillas. Dr. Soki, a Bukavu physician, walked away after his house
was destroyed by a mortar shell.[43] Sekombi Katondolo, an artist from Goma,
left town with a few friends in search of safety.[44] Émilie Efinda, a relatively
prosperous female pharmacist from Bukavu, started the journey in high
heels.[45] It was an arduous trip through the jungle, at the height of the rainy
season. People took shelter beneath the leaves, slept on the ground, fought
off the ants, and ate rotten fruit. Hygiene was at a minimum. Stockings,
handkerchiefs, and rags were used as sanitary napkins.[46] The paths through
the interior were muddy; at many places there was no road left. Where the
bridges had been washed away, people waded through rivers. Trucks could
pass only here and there, but the drivers demanded exorbitant sums to carry
the sick, exhausted, and starving a little farther. The column of refugees was
huge and heterogeneous: plundering FAZ soldiers, panicky civilians, terrified Rwandan Hutus running for their lives, drugged child soldiers, hardened military men from Rwanda and Uganda. The only ones moving in the
opposite direction were the Mai-mai, off to combat the foreign elements. In
ragtag groups they moved eastward, with no central chain of command.

Deeper in the interior, the pursuit of the Hutus led to grave human
rights violations. As soon as the AFDL came in, villagers noted, the Rwandans would ask where the refugees were, then take off to massacre them.[47]
This led to massive carnage. The situation was particularly gruesome at
Tingi-Tingi, only a stone's throw from Kisangani, where a group of some
135,000 Hutu refugees had gathered. Many of them were in a pathetic
state. Cholera had thinned their ranks and their children were dying in
great numbers. When the AFDL approached from the east in late February 1997, the survivors ran into the jungle to hide. The Rwandan Tutsis
then misused the international aid organizations to help regroup the refugees in a number of makeshift camps. As soon as a new crowd of Hutus
was gathered, aid workers and journalists were barred from the area "for

safety reasons" and the ethnic cleansing could begin with impunity. It was
not only Hutu soldiers or Interahamwe who were murdered, but also malnourished children, women, old people, the wounded, and the dying. The
killing sometimes took place at gunpoint, but much more often with the
machete and the hammer. Ammunition was expensive and heavy to carry
through the jungle.

The international community was denied access to the area and the
true extent of the atrocities became clear only later. Eyewitness accounts
from perpetrators are rare. "Yes, I was at Tingi-Tingi," said Lieutenant Papy
Bulaya, a former soldier in the AFDL. Only after many bottles of beer was
he able to talk about it.

> Listen, our objective was Kisangani, and Tingi-Tingi was in the way.
> So we had to neutralize it. I was a *kadogo,* only fifteen, our commander
> was Rwandan, General Ruvusha. He's a colonel in the Rwandan army
> now, but he was terrible. Laurent Nkunda was there too. Drive out the
> enemy, those were our orders. Our Tutsi commander told us: They're
> *génocidaires,* they have to die. They would call out: *Kadogo,* kill this
> person. And we had to obey, otherwise we were executed on the spot.
> We had to keep going all the time. A lot of Rwandans were killed there
> back then. Afterward their bodies were doused with gasoline and
> burned, or buried. The supply trucks moved along behind us: food for
> us and gasoline for the "mopping-up," to "clean the slate." When I think
> back on it, it hurts so much. I regret it, but we were loyal to the AFDL.[48]

The emergency camps at Tingi-Tingi had provided shelter for eighty-
five thousand people; after the cleansing they were empty, deserted, desolate. Tens of thousands of Hutus were massacred. A group of forty-five
thousand fled west, to Équateur, where they were intercepted at Boende
and Mbandaka and murdered en masse. Eyewitnesses even saw soldiers
kill babies by crushing their skulls with a boot heel or dashing their heads
against a wall.[49] A few Hutus were able to escape and made it to Congo-
Brazzaville, some even as far as Gabon. By then they had covered more than
two thousand kilometers (about 1,250 miles) on foot, straight across Zaïre,
under conditions more miserable than anything Stanley had endured. All in
all there were only a few thousand survivors, a tiny fraction of their original

numbers. During the invading army's advance, an estimated two to three
hundred thousand Hutu refugees were murdered.[50]

THE WAR LASTED SEVEN MONTHS and was, in essence, a steady offensive
westward to Kinshasa. Real battles were waged at some places, like Bunia
and Watsa, but almost everywhere else the AFDL simply rolled on through.
Kindu fell on February 28, 1997, Kisangani on March 15, and Mbuji-Mayi on
April 4. The conquest of Kisangani was of particular strategic and symbolic
importance, because the city lay on the river, the Central African thorough-
fare to Kinshasa. Prime Minister Kengo wa Dondo had vowed that that city
would never be taken, but there you had it: the rebels overran it without a
hitch. The characteristic images of the AFDL's advance showed two long
rows of child soldiers in black rubber boots, moving down both sides of the
road in silence on their way to a town or city. They were foot soldiers in
the most literal sense: children moving on foot. By the time they arrived,
Mobutu's army had already fled, often after a bit of plundering. In Kikwit
the inhabitants paid the government soldiers to leave without sacking the
town.[51] Once they were gone, the local citizens welcomed their liberators
from the east with banners and singing. The democratic opposition was
pleased with the military liberation. "The UDPS welcomes the AFDL,"
some banners read.[52] The young soldiers who came from so far away and
who marched through the streets in such great earnestness were admired
for their courage and patriotism.[53] Everywhere they came, new recruits
signed up. The Katangan Tigers, whose invasion of Shaba had failed in 1978,
joined as well. The AFDL was engaged in a truly triumphal procession.

During grand rallies, Kabila spoke to the newly liberated crowds. For
the first time the masses saw the man about whom they had heard so much
on the radio. He usually went dressed in black and wore a cowboy hat on
his huge, bald dome. Kabila was a robust figure, a man with meat on his
bones who laughed broadly and, with one hand in his pocket, exuded an air
of ease, even nonchalance. In a firm voice he told overblown stories about
his liberation army, spoke of the need for popular militias and urged parents
to donate a child to the cause. His charisma was undeniable. Compared to
the grumpy old man in Gbadolite, he was a breath of fresh air. He exuded
power, but also conviviality. Everything was going to be different now.
Rwanda vehemently denied all involvement, but many inhabitants of the

interior still suspected that Kabila's cakewalk had been no purely domestic affair. But all was allowed, as long as it meant being rid of the *vieux léopard* (old leopard). "A drowning man will clutch at any piece of driftwood he can find, even at a snake if need be," the people in Kikwit told each other.[54]

KABILA'S AFDL WON THE SUPPORT not only of the people, who were tired of Mobutu, and that of Rwanda and Uganda, but also of the United States. Since the end of the genocide, Kagame's Tutsi regime—thanks to its carefully cultivated role of victim—had gained credit with the American authorities. Embarrassed by a genocide they had not succeeded in preventing, new partner countries like the United States, the United Kingdom, and the Netherlands began providing Kigali with lavish support. With Bill Clinton, furthermore, a president had been elected who wanted to force a definitive break with his predecessors' old, cynical Zaïre policies.[55] He believed in new African leaders, men like Mandela and Museveni—heads of state who in no way resembled the Mobutus, Bokassas, and Idi Amins of yore, he thought—a new generation: might Kabila perhaps be one of those? Although there was no internationally orchestrated approach, the Rwandan army in any event met with no obstruction in carrying out its plans. Just as the French had continued to support the Hutu regime, despite the rumors of genocide, so too did various American services provide logistical and material support for the invading army's offensive, despite the rumors of massacres.[56] The old-fangled cynicism that the Clinton administration wanted to do away with made way for a new-fangled cynicism: humanitarian in its intentions, highly naive in its analyses and therefore disastrous in its consequences. There was no long-term vision. The confusion was great, the policy off the cuff. The backing for Rwanda and the rebels would unleash years of misery. Kabila must have found it rather amusing: thirty years after being assisted by Che Guevara, he was now suddenly receiving support from the Satan of Imperialism itself.

Mobutu, though, had lost his allies. France briefly tried to help him with a detachment of soldiers, but without any particularly great enthusiasm. He then hoped to turn the tide with a few European mercenaries, but that was no more successful than it had been in 1964. The only ones who showed up were Bosnian Serbs who had fought in the Yugoslav wars, but they were no match for Kabila's troops.

Throughout the AFDL offensive, Mobutu spent most of his time in

MAP 8: THE FIRST CONGO WAR:
KABILA'S ADVANCE (OCTOBER 1996–MAY 1997)

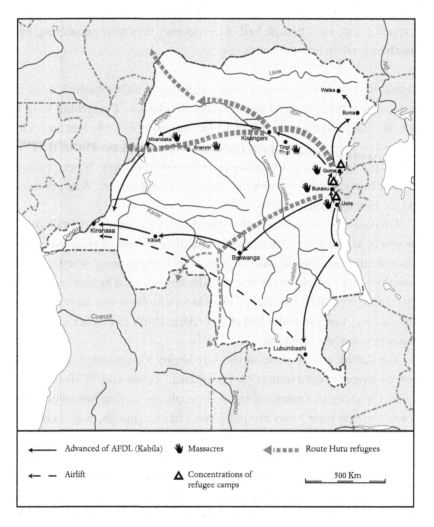

| ←—— Advanced of AFDL (Kabila) | ✋ Massacres | ◀▪▪▪▪ Route Hutu refugees |
| ←— — Airlift | △ Concentrations of refugee camps | 500 Km |

Europe, where he was operated on for prostate cancer (giving rise to a new name for the new batch of worthless Zairian banknotes: *les prostates*). He resided in Lausanne and at his villa in Cap-Martin. When he finally did return to Kinshasa it was as a deathly ill man who could barely walk. Nevertheless, he was welcomed by a enormous crowd of cheering compatriots. The chief had come back! He was going to save the country! Everything was going to be all right! But it didn't turn out that way. In the capital the

bickering between Mobutu and Tshisekedi went on unabated, as though no
massive invasionary force was approaching. They continued to squabble as
before over who was allowed to be the prime minister and who was allowed
to appoint him, even though half of the country they were squabbling over
had already fallen into the hands of others.

YOUNG RUFFIN, MEANWHILE, was on his way to Lubumbashi. He and his
crew carried their guns and bazookas on their backs. "Everything went by
foot. We followed the railroad tracks for long stretches. My feet hurt a lot.
We used to pour water into our boots, that eased the pain a little, it let you
walk easily again. But it also made your feet sweat terribly. When you took
off your boots, your feet stank like a three-day-old corpse!" Soldiers' tricks
and the humor of the trenches.

On April 9, 1997, Lubumbashi—the country's economic capital—fell to
the rebels. *Mzee* Kabila settled in and immediately received visits from international mining companies like De Beers and Tenke Mining, who knew that
from then on he would be the one to do business with. The first concession-
ary mining contracts were signed even before Mobutu was ousted.[57] It was
already clear that the scales had tipped. After thirty-two years of dictatorship, a new age was dawning.

For Ruffin, a new phase in the war began. Commander James Kabarebe no longer needed him as his bodyguard. "James said: 'This is the end
for us. I'm going to Kisangani, but you people are staying here with *Mzee*.'
It was the first time I was around *Mzee*. His son, Joseph, was there too."
Father and son stayed in Lubumbashi while the Rwandan Kabarebe led the
fighting elsewhere. The victory was within arm's reach, and that allowed a
certain amount of relaxation. Ruffin had fond memories of those days with
the president-to-be. "With *Mzee,* the good life started. I'm your father, he
said, but never forget your natural parents. He asked where I came from.
Bukavu, I said, I was kidnapped by Bugera. Ha, he said, then there's no more
playing priest for you! He liked to tease us. One day we plundered the some
storehouses that belonged to the FAZ. I dressed up in a government soldier's
uniform, with leather boots and everything. Is that you, *kadogo*? *Mzee* asked.
Yes, I said, it's me. We stole the enemy's supplies. You did? He laughed. He
shook my hand and said: very good, stay with me."

With that pat on the head, Ruffin's incredible youth took another unexpected turn: now he was one of Kabila's bodyguards. Within a year he had

been transformed from a naive, soccer-playing boy into a worldly-wise young man who stayed on his toes and experienced history live, as it happened. The price he paid for that was fear and the loss of innocence, but each phase brought with it new forms of appreciation. "Kabila liked me. He entrusted his money to me. Ten thousand dollars! He often used to eat with us, right out of his mess tin. Afterward we would arm wrestle and he would be the referee. It was a sport we'd taken part in a lot out in the *maquis*. We never went to nightclubs or brothels: the only lives I knew were those of the seminarian and the soldier. We lived in Hotel Karavia, the best hotel in Lubumbashi. *Mzee* had room 114. The diamond hunters would make appointments to come and see him. He gave me a Motorola."

In that same hotel room, Kabila regularly received phone calls from his chief of staff, Kabarebe, who was approaching the capital on seven-league boots. Coming down the Congo River, he had seen the two capital cities on their opposite shores and had to ask some local fishermen which one was Kinshasa; otherwise he might have accidentally liberated Brazzaville.[58] Kinshasa was about to fall, Kabila heard in his hotel room. He had never dreamed that things would go so quickly. Two weeks earlier he had flown to Congo-Brazzaville for direct negotiations with Mobutu. Nelson Mandela had called them both to meet on neutral ground, aboard a South African ship in the harbor of Pointe-Noire, but those nocturnal talks had led nowhere. Mobutu was unwilling to budge and Kabila saw no reason why he should add any water to the wine; after all, he had the upper hand. No, Kinshasa would be freed by force of arms and Ruffin would be there to see it happen.

"*Mzee* told us: 'Get going, all of you! And lots of luck! We'll see each other again in Kinshasa!' And we said: 'At your service!'" Kabila, that much is clear, was only the rebellion's front man: it was Kabarebe who did all the work. And the *kadogos* of course. Ruffin: "I was on the first plane that landed at Kin, a private plane from Scibe-Air. I'd never flown before. Our people had already taken the airport. Jeeps took us to the borough of Limete, we went on foot from there."

The lack of a peace agreement, of course, entailed major risks. Everyone was afraid that a violent confrontation was coming in Kinshasa. Mobutu had just appointed General Marc Mahele as his new chief of staff, a dyed-in-the-wool soldier who had earned his stripes during the Shaba wars and relentlessly crushed the plundering in 1991 and 1993. Mahele was, without a doubt, the most capable officer in the Zaïrian army at that moment. His integrity made him popular with the people, but he was feared for his toughness. Now it was up to him to defend Kinshasa against the rebel advance. On Friday, May 16, 1997, however, Mobutu fled at the crack of dawn to his palace in Gbadolite. The capital ran the risk of all-out anarchy; the next twenty-four hours would prove decisive. In Kinshasa, a city of millions, what everyone feared was a total free-for-all. The Kinois were more afraid of their own soldiers than of the rebels and shuddered at the thought of a new and devastating wave of plundering. General Mahele saw the hopelessness of the situation and decided not to offer up a megalopolis to the madness of one old man on the lam. To spare the civilian population, therefore, he contacted the AFDL and went late that evening to the camp at Tshatshi where Mobutu's last supporters had entrenched themselves. Among them was the president's eldest son, nicknamed "Saddam Hussein" for his legendary cruelty. Mahele tried to convince them to forgo all plundering; they, in turn, saw him as a turncoat officer. He was murdered in the early hours of Saturday morning.

A few hours later Ruffin walked in his black rubber boots down Avenue Lumumba in Limete. The arrival of the AFDL resulted in a festive frenzy. In the distance you could still hear the roar of heavy artillery, but he and his companions no longer had to fight. "We received an incredible welcome. The men shouted *"Libérateurs! Libérateurs!"*; the women spread their *pagnes* on the ground for us to walk over. The people gave us water. They spoke Lingala, we couldn't understand them. We were looking for the home of Prime Minister Kengo wa Dondo, and the people showed us the way. We didn't know our way around the city. We had orders to take the offices of the RTNC and Mobutu's Palais de Marbre."

In one of the houses they searched, Ruffin pocketed a solid gold ashtray. It was May 17, 1997, and within hours the AFDL held all key positions in the city. The Beach, the Hotel Intercontinental, the Memling . . . Some government soldiers began plundering, but the majority slipped into people's homes and begged for civilian clothing: to walk around in uniform now was to sign one's own death sentence. Highly placed women who owed their management positions to Mobutu hastily burned their *pagnes* printed with the MPR logo or the Great Steersman's portrait.[59] As accounts were settled, some two hundred people were killed in isolated incidents; very few compared to the way things might have been. In Lubumbashi, Kabila received a call from Kabarebe. "Kinshasa has fallen!" Kabila shrieked with pleasure and rolled laughing over the carpet of his hotel room.[60] He was coming, right away.

And once again, Ruffin was there: "That day I went back to the airport to meet *Mzee*. 'See, I was telling the truth!' he shouted to me. He held a press conference. I'm in all the pictures and film footage with him, along with Joseph and Masasu, another of the AFDL founders."

At that press conference, Kabila pronounced himself the new head of state of a new country, the Democratic Republic of the Congo. That "democratic" was a bit strange, for no one had voted for him and the nonviolent opposition of Tshesekedi and his followers had been completely ignored. The only thing Kabila adopted from the Sovereign National Conference was the idea to change Zaïre's name back to Congo. The struggle for a civil society carried out by people like Régine Mutjima had been passed in the fast lane by the military conquest in which Ruffin had taken part. She was now forty-two, he was fourteen. When Kabila was sworn in as president a few days later, on May 29, 1997, that did not take place in the houses of parliament where the conference had met, but just down the road, in the big new soccer stadium. His lieges, the heads of state of Rwanda and Uganda, were there, as were those of Angola and Zambia. But the impressive stadium was not jammed with cheering Kinois. In a city of millions, at least a third of the seats remained empty. The words of Kabila's inauguration rolled from the loudspeakers and echoed against the half-empty, concrete grandstands.

But Kabila kept a tight hold on the reins. By way of Togo, Mobutu fled to Morocco and went into permanent exile. Aware that the end was near, he had had the bones of his mother and a few other loved ones exhumed so he could take them with him. Barely four months later, surrounded by a few friends and family and the bones of his ancestors, beaten and bitter, he breathed his last.

IT WAS A DAY LIKE ANY OTHER and the waters of Lake Kivu rippled imperturbably. For Ruffin Luliba, however, it was an emotional day. When Kabila went back to Bukavu for the first time, Ruffin went with him. He hadn't seen his parents in years. "It was five o'clock in the evening and I walked back to my parents' home. I saw my mother outside, pounding *pundu,* and I fired three times in the air. She was startled and ran inside, and my father ran after her. Then I shouted: Papa, it's me! My mother came outside and wept. I had left as a seminarian and came back as a soldier. They had already held a wake for me a long time ago. Everyone cried, even my brother." For the family, it was as though Ruffin had returned from the dead. It was a fond reunion. But Ruffin also visited the mother of his roommate, Roderick, the boy who had been kidnapped along with him and who had died of dysentery after a few days in Rwanda. "I told Roderick's mother the sad news. I was staying then with *Mzee,* at the Hotel Résidence. He told me to bring my parents along. When I introduced them, the first thing he did was hand my father two thousand dollars. He said: 'Please accept my apologies, but I'm going to take him with me again. Your son is a patriot.'"

CHAPTER 12

COMPASSION, WHAT IS THAT?

The Great War of Africa

1997–2002

A NEW REGIME, A NEW SOUND. THE INHABITANTS OF KINSHASA must have thought their ears were playing tricks on them. The post-Mobutu era began with a low, metallic tone that rose to a high, shrill note and then back down again, before rising again, and again. It was a noise that cut through everything, splitting the traffic in two and echoing in the alleyways. Children stopped their soccer games and covered their ears. They grimaced and looked around for the red truck. Up and down, up and down went the hellish blare of the siren. For the first time in decades, Kinshasa, a city of millions with its endless slums, its tattered electrical networks, exposed cables and hundreds of thousands of little coal fires, had an indispensable "priority" vehicle: a fire truck.[1]

And that was only the beginning. Laurent-Désiré Kabila indeed seemed to be bringing about changes. The garbage that lay in huge, steaming piles all around the *cité* was picked up for the first time in years. The sewers were cleaned. The hallways of the ministries smelled of bleach. Even the airport at Ndjili, the world's most chaotic terminal with its tangle of passengers, customs men, immigration officials, policemen, soldiers and "protocols" who pushed and shoved to gain control of your passport and baggage receipts, even that anthill was gradually become well-ordered. Soldiers and policemen received their pay; they didn't get much, but at least they got it regularly. For the first time in decades, teachers and civil servants could start saving up again for a bicycle. The towering four-digit inflation receded to two digits, partly as a result of the strong dollar. Additional banknotes were no longer being printed, which made cash scarcer again and therefore more valuable. During the first half of 1998, inflation amounted to only 5 percent.[2] In June 1998 the *nouveau zaïre* was replaced by a new currency: the *franc congolais*. One Congolese franc equaled one hundred thousand new zaïres, which in turn equaled fourteen million old zaïres. The currency was stable, at least at first, and quickly became accepted all over the country. The bills did not bear the likeness of Kabila, but of neutral objects like a Chokwe mask or the Inga Dam. When the new currency was introduced, all the greats of Congolese music—from ancient Wendo Kolosoyi to Papa Wemba to the young star J. B. Mpiana—sang its praises, like a sort of Band-Aid for a banknote.[3]

But appearances were deceiving; the enthusiasm for Kabila quickly dwindled. As euphorically as he had been received, just as quickly did the people grow tired of him again. Making friends is an art, but Kabila mastered the even rarer art of rapidly turning friends into archenemies. Not just some of them, which could have been a sign of cunning—no, all of them, which was more a sign of ineptitude. It started with the democratic opposition from the Mobutu era. The many thousands of citizens who had courageously struggled against the dictatorship gave Kabila, at the very least, the benefit of the doubt. Many hoped that the resolutions of the Soveriegn National Conference would now truly be given the force of law and that Kabila would keep the promises Mobutu had broken. But that was the last thing Kabila intended to do. For him, his conquest was the start of a new story. After all, what did he—the perennial *maquisard*—have to do with the five-year-old blather of a hall full of starry-eyed idealists? The constitution, the parliament, the government, and the electoral committee of the transitional years all landed in the trashcan.[4] Union pour la Démocratie et le Progrès Social (UDPS) supporters ended up in prison and were beaten.[5] Only two months after the "liberation" of Kinshasa, Étienne Tshisekediwas arrested. He was interrogated, placed under house arrest, and then disappeared into exile in his native region. One of Kabila's ministers said: "We gave him seeds and a little tractor, so that he can start a farm."[6]

No, rather than instate full-blown democracy, the new government reverted to an extremely authoritarian regime in which everything revolved around the person of Kabila himself. The multiparty system was abolished; only his Alliance des Forces Démocratique pour la Libération (AFDL) was allowed to continue, even though it was merely an alliance of convenience set up at Rwanda's instigation a few days after the invasion of Zaïre. At first Kabila had merely been its spokesman, but he neutralized his three cofounders one by one. André Kisase Ngandu, the only one with military power, he had had murdered during the war itself. After being sworn in as president, he made sure Anselme Masasu was sentenced to twenty years' imprisonment, while Déogratias Bugera, Ruffin's kidnapper, was promoted out of the picture. The military alliance was transformed into a national party, but without much flesh on its bones. Congo became the AFDL, but the AFDL was, in actual practice, Kabila himself. The people were allowed to organize themselves politically only in the form of Comités du Pouvoir Populaire (people's power committees). No one knew exactly what that was supposed to mean, but it smacked of badly digested *maquis* Marxism. A new constitution came into effect on May 28, 1997, and essentially placed all power in the hands of the president. From then on, Kabila stood at the head of the legislative, executive, and judicial branches, and of the army, the administration, and the diplomatic corps. When it came to cabinet ministers, he preferred to surround himself with fellow Katangans or with former political exiles. Opponents who had for years been counting on receiving a political mandate saw strangers walk away with them. Just for the fun of it, Kabila granted ministerial posts to the by-then grown-up daughters of Joseph Kasavubu and Patrice Lumumba—a historical reference that lent him an air of legitimacy, but was in fact a farce.

"I graduated in 1994 from Lubumbashi University with a degree in international affairs," Bertin Punga, a leader of the later anti-Kabila protest movement, told me. "I was politically engaged and I had been against Mobutu. At the time of the campus killings in 1990, I saw three dead bodies. I was from Kasai and I remembered how the governor had chased us out of Katanga. So when the AFDL came along, I joined up. Before that, politics had been a matter of caste, but after that revolution it seemed open to everyone. I have a university education, I told myself, I should really go into politics. But when I got to Kinshasa I saw that the jobs were being handed out to poorly educated people from Katanga, while I, with my university diploma, was demoted to a much lower *diplomé d'État* [state-certified graduate]. When I saw how many Katangan ministers there were, I knew that Kabila was just another Mobutu. No, he was even worse, when you think that

Mobutu spread his abuses out over a period of thirty-two years. There were summary executions, the multiparty system was abolished, the single-party state made a comeback. That business with the Comités du Pouvoir Populaire, that was really just a repeat of the MPR as far as I was concerned."[7]

During his first year in office, Kabila seemed to be aiming for a strong, authoritarian, and extremely personalized state, but in practice that state remained quite feeble. There was no real policy, no vision, no government apparatus. Even the army was a joke. Mobutu's Forces Armées Zaïroises (FAZ) was disbanded and replaced by the Forces Armées Congolaises (FAC). It sounded official, but was in fact a hodgepodge of former FAZ soldiers, former Katangan Tigers, *kadogos*, Banyamulenge, and Rwandan Tutsis. The chief of staff was the Rwandan James Kabarebe. Kabila oversaw his country the way he had once overseen his rebel territory: laxly, very laxly. The only thing about which he was conscientious was maintaining control over the channels of information. It was no mistake, therefore, that his adviser for all things communications-related was once again Dominique Sakombi Inongo, the propagandist-turned-prophet. Kabila must have learned that from Mobutu: a strong regime needs to keep the media in an iron grip. The radio journalist Zizi Kabongo found out about that himself one night at 2 A.M., when the army came pounding on his door.

"Kabila had a very cool relationship with the public broadcasting organization," Kabongo told me.

He saw the entire staff as a clutch of Mobutists. One evening we rebroadcast one of his meetings. Kabila didn't sleep much and he heard the broadcast. Ever since Mobutu we'd had no money for equipment, so we always had to erase our tapes and use them over again. But this one tape was badly erased. After the recording of Kabila's meeting, there was a section of tape that still contained the tail end of a report on Mobutu. The technician on duty fell asleep, but at the end the listeners heard *papa Maréchal*'s voice again. *"Oyé! Oyé! Papa ndeko. Our friend!"* you heard the people shouting. Mobutu has returned, the listeners thought. That same night the army rounded up all the journalists to throw them into prison. They knocked on my door at two o'clock. In the prison, I ended up among men who had been sentenced to death and revolutionaries. The situation was quite grim. Kabila was out to eliminate all his enemies.

Zizi, whose shins bore the scars of resistance to Mobutu, stood accused of Mobutism. More than 160 journalists were imprisoned between May 1997 and January 2001.[8] "The next day we were all brought to the presidential palace. Kabila himself gave us a terrible scolding for our act of rebellion. For punishment, we were all obliged to study Marxism. But when it was over we finally got the new tapes we'd been waiting for for years."[9]

The democratic opposition and the UDPS had been stiff-armed, the AFDL resigned to the scrapheap, the press snarled at and then silenced. What other bridges were left for the new leader to blow up? Those connecting the country to its foreign allies, of course. Within no time Kabila blew his credit with the United Nations by first refusing, then obstructing, an investigation of the mass extermination of Hutu refugees. Foreign teams of experts were systematically boycotted. Kabila was faced with a choice: he could either place the blame on Rwanda (which was where it belonged) and thereby admit that his victory was not due to his own rebellion, an admission that would destroy his popularity at home, or he could take the blame himself, which would earn him an international reputation as a brutal mass murderer. Domestic interest and international interests were at a standoff. It would have been a high-wire act even for a seasoned politician, and Kabila was no seasoned politician. Diplomacy was mumbo jumbo to him; boorishness was his strong suit. He entered the international arena like a suspicious rebel rather than a senior statesman. Within no time he had accused France of neocolonialism and America of a lack of diplomatic courtesy, and had called Belgium a terrorist state.[10] All three of those countries had put up with a great deal from Mobutu in his day, but ludicrous statements like this were something new. This was no longer the voice of a sly fox, but of a clodhopper. Other African heads of state, too, soon became familiar with their new colleague. In 1997 Nelson Mandela waited for him for hours at the peace talks in Congo-Brazzaville; the affront threw the always-genteel statesman into a rare paroxysm of rage. Egyptian president Hosni Mubarak was already waiting for him at the airport in Cairo, with an honor guard and the red carpet, when Kabila called to cancel the appointment because he felt "a bit tired." Tanzanian president Benjamin Mkapa actually welcomed him to his residence, but in complete defiance of diplomatic protocol Kabila cut the visit short and flew back to Kinshasa.[11] President Yoweri Museveni of Uganda and Vice President Paul Kagame of Rwanda would also become acquainted with their protégé's unmannered ways. They had hoped to control their chaotic neighbor by installing a pawn of their own, but Kabila turned out to be an unguided missile.

And then something very important happened: Kabila turned his back on Rwanda and Uganda. He had little choice in the matter. All over the country there was growing protest against the foreign interference. Rwanda in particular became the whipping boy. Every Tutsi was seen as Rwandan and every Rwandan as an occupier. Things even reached a point where anyone with a pointed nose or high forehead was seen almost immediately as an infiltrator. People in Kinshasa were extremely annoyed by the highly visible presence of Tutsis in the armed forces, often in high-ranking positions. These were officers who spoke neither French nor Lingala, but English, Swahili, and Kinyarwanda. These new military leaders frequently behaved like arrogant victors and saw no problem in reinstating the *chicotte,* the strop of hippopotamus hide that summoned up so many bad memories of colonial days. Women who wore jeans or a miniskirt in public, which had been allowed since 1990, received a public lashing. Taxi drivers committing a traffic violation did too. The number of lashes was not limited to twenty-five, as officially established in colonial times, but was determined by age: a fifty-year-old received fifty lashes. It became a widely accepted idea that overpopulated Rwanda was longing for raw materials and lebensraum, and therefore had its eye on Kivu, where so many Tutsis already lived. People believed that Rwanda was out to establish a Grande République des Volcans (great volcanic republic), a new state consisting of Rwanda and Kivu. It did not help any when a group of prominent Rwandans publicly called for a "second Berlin Conference" to reconsider the borders established in 1885.[12] Some Congolese felt that their huge country had already been annexed by the dwarf state of Rwanda.[13] A deep, deep hatred arose between the two countries, reminiscent of the relations in more distant times between China and Japan or Ireland and England. Many Rwandans considered Congo to be a country of lazy, chaotic bunglers who cared more about music, dancing, and food than about work, infrastructure, and public order. Many Congolese saw Rwanda as a cold, authoritarian country where plastic bags were banned for reasons of public cleanliness and motorcycle helmets were mandatory, a country of arrogant, pretentious parvenus who looked down on them in contempt. Many interpreted the differences between the countries in terms of an ancient cultural conflict between "Bantus" and "Nilotes," even though those were highly problematic concepts from colonial anthropology. As long as Kabila's court was filled with those hateful foreigners, he could forget about his authority being recognized: the president knew that was how the people felt. So there he was at the head of a vast country, in a city that was new to him, with a population he neither knew nor understood. Little by little, the cheers died out. "We need to give our liberators back their liberty," people on the street said scornfully.[14]

And that was precisely what Kabila did. In a nighttime broadcast on July 26, 1998, more than a year after his glorious entry into Kinshasa, he announced that Rwandan and other foreign soldiers were to leave the national territory. This time it was not a matter of a badly erased tape. The Congolese people were thanked "for tolerating and giving shelter to the Rwandan troops."[15] That communiqué sealed for good the break with Kigali and Kampala. In the days that followed, hundreds of soldiers left Kinshasa. Chief of staff James Kabarebe, the man who had taken Congo in Kabila's name, was thanked for services rendered. He returned to Rwanda in a fury. A new escalation was now inevitable. And indeed, less than one week later he invaded Congo again.

THE WAR THAT LASTED FROM OCTOBER 1996 TO MAY 1997 and brought about Mobutu's fall is known by many names: the Banyamulenge uprising, the war of liberation, the AFDL offensive. These days it is more commonly referred to as the First Congo War. On August 2, 1998, the Second Congo War broke out. Rwandan troops crossed the border again, Kabarebe again led the invasion, the objective was once again regime change in Kinshasa. This time, however, conflict would not last seven months but five years, until June 2003. Officially, that is, for unofficially the war simmered on, at least until the moment that I write this, in spring 2010.

The Second Congo War was an extremely complex conflict in which, at a certain point, no fewer than nine African countries and some thirty local militias took part. It was a showdown on an African scale, with Congo as the central theater of war. The promptness with which a number of states, from Namibia in the south to Libya in the north, chose sides (for or against Kabila) was reminiscent of the formation of the ententes in Europe on the eve of World War I. Because of its continental scope, it is sometimes referred to as the First African World War, but that is an unfortunate term that skims too lightly over the ponderous impact that the World Wars I and II had on Africa. The term *Great African War* is therefore more useful, even though the hotbed of the conflict was limited mostly to Congo, and the local militias were active for a longer period than any foreign national troops. In terms of casualties, this Great African War or Second Congo War developed into the deadliest conflict since World War II. Since 1998 at least three million and perhaps as many as five million people have been killed in hostilities in Congo alone, more than in the media-saturated conflicts in Bosnia, Iraq, and Afghanistan put together. And their numbers continue to rise. In 2007, an estimated forty-five thousand casualties were still being reported as a result of the indirect consequences of that forgotten war. Most of those were civilians. They did not die in the course of fighting, but as a result of malnutrition, dysentery, malaria, and pneumonia: afflictions that could not be treated because of the war. One must note, however, that many of those maladies were not being treated before the war either. Congo already had an above-average mortality rate and the conflict did nothing to ameliorate that. In 2007 that rate was still 60 percent higher than in all the rest of sub-Saharan Africa.[16] Average Congolese life expectancy at birth was fifty-three.

The Second Congo War disappeared from the international media reports because it was considered incomprehensible and obscure. And indeed, there were no two clearly delineated camps; even more, there was no clear division of roles into villain and underdog. After the Cold War, Western journalists increasingly came to apply a moral frame of reference in reporting on armed conflicts: in Yugoslavia, the Serbs were the major culprits; in Rwanda, the Tutsis were the innocent victims. In both cases that led to disastrous misrepresentations and policy measures. In Congo it was not particularly easy to find a "good" side. Anyone viewing the conflict from close up knew that all those involved had their own skeletons in the closet. The grievances often seemed justified and the methods chosen often problematic. None of the parties seemed able to step back from the fray, either literally or figuratively, in order to consider the legitimacy of the other's perspective and search for common ground. For a grindingly poor country with a young, uneducated population that had known only Mobutu's dark despotism, that was definitely too much to ask. The children of a dictatorship are rarely model democrats. The Second Congo War became a conflict

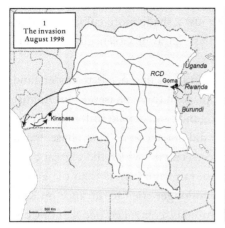

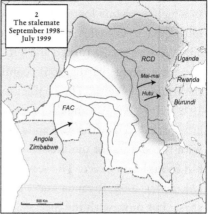

Rwanda, backed by Uganda and Burundi, invades Congo. The cities in the east are taken immediately, an air link to the far west of the country is intended to hasten the taking of Kinshasa. The invasion is made out to be the work of a domestic rebel movement: the RCD.

Kabila's foreign allies (most notably Angola and Zimbabwe) put an end to the rebels' advance. The front stabilizes. In the east of the country, the rebels are still engaged by the Mai-mai, and the Rwandan Hutu militias supported by Kinshasa. Uganda sets up a second rebel movement: the MLC. The Lusaka Peace Agreement proves ineffective.

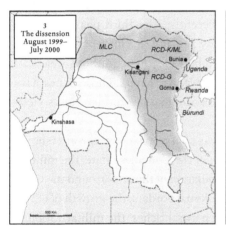

With Kinshasa beyond reach, attention is turned to the available booty. But the dividing of it leads to dissension. The rebel movement falls apart into a pro-Rwandan and a pro-Ugandan schism: the RCD-G (for Goma) and the RCD-K (for Kisangani), respectively. Rwanda tries to take Kisangani, a major diamond center, away from Uganda. After an initial confrontation in August 1999, the RCD-K flees to Bunia and becomes the RCD-ML. In May and June of 2000, Rwanda takes Kisangani.

In the north the rebellion crumbles completely. Pro-Ugandan rebels no longer fight against Kinshasa or pro-Rwandan rebels, but simply among themselves. New, smaller armies come along. In Ituri the snarl of interests can no longer be disentangled. In the end, the motif is that of plundering, even in Rwandan-controlled territory. The 2002 peace agreement pacifies a large part of the area. The MLC and RCD-G are allowed to put forward a vice president, but in Ituri and Kivu the conflict simmers on for years.

in which everyone found everyone else just a shade more culpable, so that hitting back was allowable and an endless spiral of violence could ensue. The Western media turned and left.

Yet a simple cartographic comic strip is all one needs to understand the course of events. The conflict took place in three phases. From August 1998 to July 1999 Rwanda, along with Uganda and a makeshift native rebel army, tried to overthrow Kabila. They did not succeed. That phase ended with the signing of the Lusaka Peace Agreement, which did a great deal, but brought no peace. The second phase ran from July 1999 to the end of December 2002. Rwanda and Uganda no longer tried to advance on Kinshasa, but now, with the help of local militias, controlled one-half of Congo's territory, allowing them to help themselves on a massive scale to the raw materials present there. Now that booty had taken precedence over power, schisms arose within the rebellion and there were violent confrontations in Kisangani. This turbulent phase ended with the Pretoria peace agreement in December 2002, which was to enter into effect in June 2003. The Rwandans and Ugandans withdrew to their own countries and the United Nations increased its presence. That put an official end to the war; unofficially, however, things went differently. The third phase began in 2003 and, in Kivu, is still going on today. During this long period the war has been limited to the extreme eastern part of Congo, in those areas that border directly on Uganda (Ituri) and Rwanda (Kivu). Those zones have been subjected to bouts of extreme violence, massive human rights violations, and incredible human suffering.

In each of its phases the conflict was characterized by the aftershocks of the Rwandan genocide, the weakness of the Congolese state, the military vitality of the new Rwanda, the overpopulation of the area around the Great Lakes, the permeability of the former colonial borders, the growth of ethnic tension due to poverty, the presence of natural riches, the militarization of the informal economy, the world demand for mineral raw materials, the local availability of arms, the impotence of the United Nations, and so on and so forth.

On June 25, 2007, in the Rwandan capital of Kigali, I had breakfast at the celebrated Hotel des Milles Collines, the place of refuge during the genocide that served as inspiration for the film *Hotel Rwanda*. It was still an exorbitantly expensive multistar hotel. I did not spend the night there, but arrived that morning for an interview with Simba Regis, an introverted

Rwandan war veteran only a few years older than I. At the buffet we used tongs to pick out croissants glistening with butter. The waitress brought us wonderfully fresh fruit juice. Simba Regis was born in 1967 and his life story reflects the history of the Rwandan Tutsis in a nutshell. In 1959, when the Hutu uprisings began, his parents fled to Burundi. He was born there, but throughout his childhood and youth he was constantly reminded that not Burundi, but Rwanda was his homeland. He sympathized with the struggle of the Tutsis in exile and went to southern Uganda in 1990 to join up with Kagame's army, the Rwandan Patriotic Front. He took part in the invasions of Rwanda, he was among the first to reach Kigali, and he escaped the genocide of 1994 by the skin of his teeth. "Six-year-old children lay there wasting away, young mothers were slaughtered by the Interahamwe. It was maddening. When you've seen that, you have to put up a fight." And so he was there in 1996 when Rwanda first invaded Congo to neutralize the Hutu threat. And in 1998, during the second Rwandan invasion, he was once again in the front lines; this time too—in addition to dethroning Kabila—the elimination of the remaining Hutu militias was a major objective. Thousands of Rwandan Hutus were still hiding in the forests of eastern Congo and, more than ever after the AFDL massacres, were out for vengeance.

The fighting began on August 2. Rwanda received backing from Uganda and Burundi, who were also worried about the rumbling on their western borders and knew of the mineral riches of the eastern Congo. Goma and Bukavu fell immediately. Two weeks later, rumor had it that the conquests had been the work of a Congolese rebel movement, the Rassemblement Congolais pour la Démocratie (RCD). Ernest Wamba dia Wamba, a former history professor, was pushed forward as its leader. But the RCD was as much a phantom construction as the AFDL had been in 1996. As he slowly picked apart his croissant, Simba Regis spoke of that in no uncertain terms. "We trained those rebels. Rwanda was simply better organized. The Congolese wore Rwandan uniforms and boots. They were under our command. We were their godfathers."

Simba Regis fought on Congolese territory for four years, from 1998 to 2002, the full length of the official war. He was in Katanga and in Kasai. On occasion the Rwandans fought against the Interahamwe and the Mai-mai, who were supported by Kabila, but usually nothing happened at all. "On faisait la vie," he said, "we made a living," by which he seemed to suggest that exploitation of the mineral resources was more important than waging war. Katanga was still brimming over with raw materials, Kasai was still extremely rich in diamonds. The fight against the organized Hutus he referred to as "just and noble," but he was sick and tired of war as a way of life. "I'm finished. I've been at war ever since 1990. The ones who make the decisions about the war are never the ones fighting, but I lost my brothers and my friends. There were eleven of us, all friends from Bujumbura; we came from the same neighborhood and went to the same elementary and secondary schools. Of those eleven, two are still alive. Me and someone who lives in Canada." The patio outside the breakfast room looks out over Kigali. The city glistens in the morning light. "When I've been drinking beer, I have nightmares. I see houses being blown up. I see my friends crying because they've lost an arm or a leg. And I'm always powerless, I can't do anything to help. Then I wake up with a start. I can still taste the war. I've had a bad life, really. I want to go to Europe, because in five or ten years' time things are going to explode here again."[17]

James Kabarebe thought the job would be over quickly. In 1996 it had taken his forces seven months to get to Kinshasa: this time he could do better. His plan was as risky as it was audacious. At Goma airport he hijacked a few planes, filled them with RCD soldiers and forced the pilots to fly to the west, to the military base at Kitona on the Atlantic Ocean. From there it was only four hundred kilometers (250 miles) to Kinshasa. His air link seemed to work: on August 5 he took Kitona and succeeded in convincing the soldiers present—most of them demotivated former FAZ soldiers being "reintegrated" into the new army—to help him fight against Kabila. On August 9 they took the crucial port town of Matadi, on August 11 the Inga hydroelectric plant. Kabarebe now had his finger on Kinshasa's switch and could cut off the capital's power supply. Night after night, he plunged the hungry megalopolis into darkness. Anti-Tutsi sentiment flared up in the working-class neighborhoods. A few hundred Tutsis or people with Tutsi features were lynched by the crowds in a horrible fashion. As in the South African townships, a car tire was hung around their neck, filled with gasoline, and then ignited.

All indications were that Kinshasa would soon fall. Kabila's army was no match for Kabarebe's troops. Still, things took a different turn. Kabila was saved in the nick of time by foreign troops: on August 19, 1998, four hundred Zimbabwean soldiers entered Congo; on August 22, the Angolan army began the liberation of Bas-Congo. Angola's role was particularly decisive. During the First Congo War it had remained neutral: no one in Luanda mourned the imminent departure of Mobutu, whose support for the right-wing UNITA rebels had caused so much suffering. During the Second Congo War, however, the cards were reshuffled. There was a distinct possibility that, in order to bring down Kabila, Rwanda would this time support UNITA. That could not be allowed to happen. Zimbabwe, on the other hand, had interests in Katangan mining operations and therefore more economic motives. In addition, a sort of ideological brotherhood existed between presidents Robert Mugabe, José Eduardo Dos Santos, and Kabila; all three had flirted with what is referred to in Africa so exquisitely as le marxisme tropicalisé (tropicalized Marxism). Fidel Castro had supported Angola for years, just as he had Kabila with the visit from Che Guevara. During his time as President Kabila's bodyguard, Ruffin Luliba had noticed those close ties. "Mzee liked revolutionaries. Men like Mugabe and Castro, he thought they were wonderful. His personal physician was a Cuban. I went with him to Cuba a few times. There were four of us kadogos, and we went to see Castro; I even shook his hand. We had dinner with him in Havana."[18] It was probably Castro who urged President Dos Santos of Angola to send his army into Congo.[19]

Kabila's coalition grew. After Zimbabwe and Angola, Namibia joined in as well. Northern allies were found in Sudan, Chad, and Libya, each of which had its own reasons for preventing Kabila's fall. Sudan offered its services because of a perennial conflict with Uganda over its support for rebels in southern Sudan. Libya provided a few planes in order to break out of its international isolation. Chad sent two thousand soldiers as a gesture of solidarity with Sudan and Libya. In the end, Kabila had a seven-nation army at his disposal: in addition to his own forces, there were troops from three countries to the north and three to the south. This was the coalition that stood up to the three countries from the east and operated behind the blind of the RCD: Rwanda, Uganda, and Burundi, with Rwanda as leading player and America's undisputed favorite. Once again, Congo's central location played a crucial role in the course of history. The armies now facing off were large: Kabila's coalition had approximately eighty-five thousand troops, the rebels some fifty-five thousand.[20] This impressive military presence led to a complete stalemate. Western Congo was soon back in Kabila's hands, but the east was still held by the RCD. There was no real front, only clearly defined zones, often divided by a very broad stretch of no-man's-land. Kabila's authority extended only as far as Bas-Congo, Bandundu, western Kasai, and a large part of Katanga; Kigali and Kampala controlled northern Katanga, North and South Kivu, Maniema, and Orientale province. When Chad withdrew from Équateur in 1998, that part of the country fell into rebel hands as well. The occupying force this time, however, was not the RCD but a new rebel army supported exclusively by Uganda: the Mouvement pour la Libération du Congo (MLC). Its commander was Jean-Pierre Bemba, son of the wealthiest businessman of the Mobutu era. His troops consisted largely of veterans of the Division Spéciale Présidentielle (DSP), Mobutu's remorseless private army.[21]

Rwanda's second invasion was intended to be a repeat of that in 1996, but this was not the case. Due in part to the sentiments of the local population, the situation in Congo had become intractable. If the AFDL had once been welcomed as a liberation army, the RCD was seen right away as an invasionary force. In a city like Goma, Kabila was still extremely popular. When Wamba dia Wamba tried to recruit local boys for his RCD, Jeanine Mukanirwa mobilized her influential association of rural women. She had helped found the women's movement in Kivu at the end of the Mobutu era. "There were five thousand of us women. Wamba dia Wamba came by to try and win us over for his rebellion. He called it a war of rectification, but we knew that Rwanda was behind it. We said: 'In 1996 you people took away our children to fight. Now you come to take the rest of them to fight against their own brothers. Your war has no justification!' Yes, we women were courageous back then."[22]

Wamba's RCD was hated outright. Even the Banyamulenge hesitated about whether to go along with Rwanda this time; compared to two years earlier, their enthusiasm had dwindled considerably.[23] People from Goma told me about how their entire municipal administration had passed in Rwandan hands. The tax office, the immigration service, the intelligence service . . . There was no fighting when the city was captured, but once the new authorities were installed there began an endless series of kidnappings and disappearances.[24] Intellectuals, journalists, civil-society activists, and church leaders met with intimidation and arrest. Hundreds of dissidents and counterinsurgents from the interior lost their lives.[25]

In the garden of the Caritas guesthouse in Goma, on the shore of lovely Lake Kivu, I interviewed someone who referred to himself as "Muhindu."

He walked with a limp and had a huge scar on his right arm. He had served for five years as truck driver for an RCD commander. He chose his words carefully. "A lot of young people were kidnapped back then. I always had three soldiers with me when I drove to a house. All the able-bodied boys and men were picked up and tossed into the back of the truck. The door was closed. Then I would drive to Kinyogote, close to Mugunga, at the edge of the lake. There was a garage there, where they used to keep speed-boats. That was the prison. We threw them in there. After a few days they were killed. With ropes. I would take a motorboat out onto Lake Kivu. You have to tie big rocks to the bodies." The waves on the lake splashed against the shore, but he wasn't hearing them. It was quite cold there in the eastern highlands, but he was wearing only a T-shirt. He took a sip of beer and went on. "If you had a problem with someone, you went to see a friend in the RCD. You gave him some money and he made sure your enemy was killed. I took about sixteen people in my truck every day, and I drove for the RCD for five years. Sometimes there were a hundred of them in that garage. They died from the cold and the wind. The waves crashed in too."[26]

The RCD operated quite brazenly in the towns. Those were under their control, but the countryside was not. That was the territory of the Interahamwe and other Hutu forces, who received support from Kinshasa. It was an improbable reversal of history: in 1996 Kabila had led a rebellion that carried out massacres among the Hutu refugees; two years later he was giving those same refugees weapons to fight against Rwanda. . . . In Congo, nothing was what it seemed. Alliances came and went, depending on the situation. Ideological conformity? Political affinity? Those were of no importance. The only thing that mattered was military (and later, pecu-niary) opportunism. Your enemy's enemies were your friends; they were the ones you hooked up with. In eastern Congo that logic kept the struggle between Rwandan Hutus and Rwandan Tutsis going for a long time. The echoes of the genocide did not die out.

The Mai-mai, too, received supplies from Kinshasa. Kabila no longer had troops in the east, but by way of the Mai-mai he could still keep the RCD from gaining total control over the interior. And so he farmed out the war there to two subcontractors who were extremely strange bedfellows: the Interahamwe and the Mai-mai. The one group consisted of Rwandan Hutus who had committed the original genocide, the other of Congolese hypernationalists who swore by their superstitions. In Bukavu, in June of 2007, I had dinner in deepest secrecy with four Mai-mai. Leery as they were of the city, they appeared only long after darkness had fallen, at the anonymous private home of a mutual friend. The atmosphere was edgy at first. Their "colonel," a man in his thirties with bloodshot eyes, loudly related endless stories about the history of the Mai-mai, heroic tales that spoke of both rage and combativeness but went on for so long that his cohorts fell asleep. Later in the evening, though, once they had woken up, they talked at length about the war and their rituals. After a while, with the consumption of food and beer, they even showed me their magic leather armbands, their *grigris*. They rolled up their pants legs to point to where they had been struck by bullets that didn't kill them ("And this is where it came back out!"). They invited me to feel their upper arms where, now that you mentioned it, there really was a bullet still under the skin ("No doctor, nothing like that: I just put a plant on it"). At our next meeting, they promised me, they would apply all their immunization rituals to one of their comrades, then shoot him. I would see that the bullets slid off his chest like water. Or wait, they had an even better idea. Because I was clearly so fond of Congo, that made me a potential Mai-mai too, they felt. They would apply all their rituals to me, that's right, that was more like it, and one of them would shoot me. Wouldn't that be an unforgettable experience!?[27]

The Mai-mai were never too particular about your ethnic origins, as long as you loved Congo with a passion. Yves van Winden knew all about that. He was a Belgian who had run a small airline in Congo for years, a sport pilot who had made his hobby his profession and who had made Congo his fatherland. During the war he acted as contact person between Kabila and the Mai-mai. We met at a nightclub in Goma. The other patrons included Russian pilots who had a sideline in gold smuggling and some seedy types in military uniforms I couldn't quite identify. A few young prostitutes sat around the pool table, sipping their cola through straws. "They called me the 'white Mai-mai,'" Yves van Winden said, "I brought them weapons from Kabila. I carried out more than four hundred solo flights, five or six hours each. That's a very long flight. Most of the time I flew my Cessna, but sometimes a DC-3 or a little Antonov 26. I took six hundred kilos [1,320 pounds] of cargo on each flight. Probably about twenty-thousand Kalashnikovs, three to five hundred bazookas, two hundred 60-caliber mortars, twenty 90-caliber mortars, and ten 120s. And also two SAM-7 missiles, anti-aircraft."

Why would anyone take 240 metric tons (about 265 U.S. tons) of weapons to rebel territory? "Patriotism. Arming the Mai-mai was what stopped the RCD. The government still owes me a lot for all those flight hours. One time my Cessna was shot at during takeoff, the bullet flew right past my seat. I wasn't touched. The Mai-mai weren't surprised at all. After all, they had baptized my plane!"[28]

The map of Congo was frozen in place: to the west and south there was Kabila with his Angolan and Zimbabwean allies; to the north Bemba with his Ugandan-supported MLC; and to the east Wamba dia Wamba with his Rwandan-supported RCD, which fought against the Kinshasa-backed Interahamwe and Mai-mai. Peace talks had already started in early 1999, but it was not until July of that year, under pressure from France and the United States, that an agreement was reached in the Zambian capital of Lusaka. The foreign armies promised to withdraw their troops, the United Nations would send five hundred observers by way of a peacekeeping force, and Congo was to initiate a national dialogue about the arrangements for a postwar transitional period. Yet another transition. Ever since Mobutu had allowed the start of a democratization process in 1990, the country had lived in a permanent state of provisionality.

BUT THE WAR WAS NOT OVER. After Lusaka it merely entered a new phase, a messy, dirty phase. All wars are dirty, but when the political motive makes way for a pecuniary one, things go completely sour. And that is precisely what happened. The RCD no longer aimed for Kinshasa, but ensconced itself in a state of rebellion and noted that there was good money to be made in eastern Congo. Westerners have become used to seeing wars as exorbitantly expensive, money-guzzling enterprises that are disastrous to the economy. But in Central Africa, exactly the opposite was true: fighting a war was relatively cheap, especially in light of the magnificent profits to be made from raw materials. And this was no high-tech war. The oversupply of light, secondhand firearms, often from the post-Communist regimes of Eastern Europe, pushed prices down, and (child) soldiers who were allowed to plunder their own salaries cost nothing at all. They kept the population cowering, while the ore was there for the taking. War, in other words, became a worthwhile economic alternative. Why would one want to call a halt to such a lucrative business? Under pressure from the people themselves? But that's what the guns were for, right? And what if a part of that impoverished population profited from the mineral wealth as well?

When I first met Dr. Soki, he was sitting alone in a Greek cafeteria in Kisangani, eating an omelet. It was a blisteringly hot day, but the air-conditioning kept things bearable inside. I had heard about him and we soon entered into conversation. He was originally from Bukavu and had been one of the many Congolese who fled to Kisangani in 1996, at the time of the first Rwandan invasion. A mortar had destroyed his house. He and his family trudged through the jungle for three weeks. But a few years later the war would reach his new place of residence too.

The most important event during the second phase of the war, Dr. Soki would find out soon enough, was the breaking of ties between Rwanda and Uganda. Now that profit had taken precedence over victory, the friend-ship between Kagame and Museveni hit the rocks. They no longer fought together to gain Kinshasa, but against each other to seize Kisangani. The rebels had already taken Dr. Soki's town in 1998. Kisangani was the main regional trading center for diamonds. All over the city there were *comptoirs du diamant* (exchange offices), often Lebanese-operated, where prospectors and couriers from the interior came to cash in their stones. Uganda ruled the roost at first, but Rwanda was also interested in the proceeds and decided to dislodge its northern neighbor. On at least three occasions, gun battles broke out in the streets of Congo's third largest city. Even today the inhabit-ants of Kisangani speak of "the one-day war"(August 1999), the "three-day war"(May 2000), and the "six-day war" (June 2000). The latter conflict was particularly violent, Dr. Soki recalled. Officially, the city was supposed to be demilitarized. Jeeps were already leaving, but both parties were afraid that the other would rush in to fill the vacuum.[29] The troop strength of the UN peacekeeping force, the Mission de l'Organisation des Nations Unies au Congo (MONUC) had increased significantly, but not enough to keep things on an even keel. The Ugandans were camped to the north of the city, close to the Tshopo River and on the grounds of the Sotexki textile plant. The Rwandans were in the south, along the Congo River. It is unclear who provoked whom, but the planned withdrawal escalated instead into a full-blown firefight with heavy artillery. Within the next six days more than a thousand shells flew over the residential neighborhoods of the city with the most modernistic architecture in all of Congo.[30] The people moved into their basements and had nothing to eat for days. At night the sky was filled with roaring, falling stars. There was neither water nor electricity. The people drank stagnant water from puddles and cisterns and endured a war that was not their own.[31] Uganda and Rwanda were fighting over crippled-but-wealthy Congo, the way a jackal and a hyena might tug at the same carcass.

A makeshift cemetery was set up behind the public hospital. During the final, six-day war alone more than four hundred civilians were killed. There were countless wounded and innumerable houses were destroyed. "The war broke out on a Monday morning at ten o'clock, I was talking to a client about some building plans," said Utshudi, an engineer, who was not at home when one of the first shells landed on his house. "We lived at Deuxième Avenue 11, in the borough of Tshopo. When I returned, not a house was standing. It was a wasteland. Just bodies everywhere. They lay there for six days. We had to run away. The soldiers even shot at the people digging graves. When the war was over we went back to collect the corpses. We put them in bags and buried them in the cemetery behind the hospital. At one go I lost my wife, my younger sister, my sister-in-law, and my four children: seven family members. These days I pray to God to let me forget."[32]

The break between Rwanda and Uganda ran parallel to a schism within the RCD: the rebel movement fell apart into a pro-Rwandan faction (the RCD-G, for Goma, led by Émile Ilunga and later largely by Azarias Ruberwa) and a pro-Ugandan faction (the RCD-K, for Kisangani, led by Wamba dia Wamba and later by Antipas Mbusa Nyamwisi, also referred to as the RCD-ML, for Mouvement de Libération, or RCD-K/ML).[33] Congo was rich not only in raw materials, but also in abbreviations. Dr. Soki viewed it all with a certain distance: "I didn't think about politics much. We knew nothing about the motives behind the war." When the six-day conflict began, the international aid organizations withdrew their personnel. Dr. Soki remained behind in a besieged city of half a million inhabitants. Like the physician in Albert Camus's La Peste, he simply tried to sustain human dignity in an inhumane world.

For six days I worked alone at the Kisangani public hospital. There were three nurses and fifteen interns. An American surgeon from the Red Cross arrived only later. The people slept on the floor, on mats they'd woven themselves. It took a long time for blankets and medicine to get there. We worked from seven in the morning till eight at night. We treated two thousand people, people who had been shot in the stomach, in the chest, in a limb, or even in the head, people whose stomachs had been torn open by shrapnel. We removed blood from their lungs, shrapnel from their bladders. We performed amputations. It was real wartime surgery, but we had almost no trouble with infections. At first, though, there wasn't enough diesel fuel to run the generators. We had to sterilize our instruments over a coal fire. And then a shell actually hit the hospital. One of the two operating rooms was destroyed and our five-thousand-liter reservoir burst and all the water ran out. That caused a lot of panic among the patients and personnel. We weren't safe, not even there.

Dr. Soki spoke calmly about that week-long inferno. There were no heroics; his tone was more that of resignation and sorrow. "We treated soldiers too. Four Ugandan soldiers came in, their abdomens torn apart, their intestines hanging out. We were able to save them. We treated everyone, we didn't discriminate. When Rwandan soldiers came in, we put them in another room. I just kept on, because of the suffering I'd known myself. I had traveled seven hundred kilometers [about 430 miles] on foot, I had seen children and grown-ups die along the way. Apparently I had the courage needed to give myself to others." These days he eats his omelet alone. He doesn't like to talk much. "We had a birth that week too. A lot of women went into labor too soon, because of the shock. We performed a Caesarian. I held the child in my arms. May God grant him life, I thought."[34]

AFTER TAKING PART IN THE TINGI-TINGI MASSACRE outside the city in 1997, Lieutenant Papy Bulaya finished his career with the AFDL in Kisangani itself. He married there, put away his guns, and went to live with his wife's family in the brousse, where he farmed a plot of land. At last he was living the life of the average Congolese in peacetime: the farming life. But then Wamba dia Wamba came to Kisangani in 1999 because of the schism in the RCD. "He said: 'You people want to fight for your country, but the Rwandans are trying to occupy us. Look at what's happened in Goma!'" Farmer Papy figured he had turned enough soil already, and once again became Lieutenant Papy. He received three months of training, this time from a Ugandan colonel. Wamba had most definitely switched camps.

In August 1999 he was there when Rwanda shelled Kisangani and took it for the first time. When Wamba moved his headquarters to Bunia, he went east too. Now he wanted to sign up with Roger Lumbala, who had started his own rebel army, the RCD-N (for National, although the *L* for *local* would have been more fitting) in Bafwasende, the heart of the diamond region. Lumbala originally came from the RCD-G, had flirted with the RCD–K/ML, set up the RCD-N, and, finally, was in league with Bemba's MLC.[35] The rebel movement was falling apart, especially on the Ugandan side, and Papy was tossed to and fro. At first he wanted to go to Lumbala, then changed his mind; then he wanted to go back to Wamba, but Wamba had been replaced by Mbusa; what he really wanted was to go back to his family, which was in Beni, but that was too far away, so he stayed with Mbusa anyway. Loyalty was, above all, a matter of opportunism. In the end he wandered about for years with a handful of troops through the jungle of what was known in better days as the Parc National de l'Okapi, with its eighteen-thousand square kilometers (seven thousand square miles) one of Congo's biggest nature reserves, a world heritage site since 1996, and inhabited as a rule only by Mbuti Pygmies.

There were seven men in their group; Papy was the squadron leader. Deep in the jungle they arrived at the little town of Bomili, where they had a glorious view of the confluence of the Ituri and one of its tributaries. The place was ruled by a man named Mamadou, a Malinese poacher who had adopted the airs of a village chieftain. It was reminiscent of the way Msiri, the Afro-Arab slave trader from the east coast, had had himself crowned king of the Lunda in 1856. As official political authority faded, there was room for new structures from the inside out: foreign traders could pursue their affairs with impunity and, with a bit of force, achieve real political power. In the year 2000, the Congolese interior was as wild and woolly as it had been in the mid-nineteenth century. Even the merchandise was the same. "Mamadou had a house full of ivory. I saw fifteen tusks there, all of them almost two meters (6.5 feet) long. He had four hunters: a man named Pascal and three Pygmies. There were also okapi hides and a rhinoceros horn. Mamadou took everything we had, even the little chains around our necks. He beat us for three hours. Then he said: 'Carry that ivory for me, otherwise I'll kill you.'" A command that could have come straight from the nineteenth century. Papy and his men walked seven kilometers (about 4.3 miles) with the tusks on their shoulders, in precisely the same region where the *Arabisés* had once done their raiding. When it grew dark, they built three little lean-tos for the night. They had no intention of continuing to play porter. "One hour later, Mamadou arrived. He had been following us and he opened fire. One of us was killed, so we killed three of his hunters, including Pascal. We ran away and buried the ivory. I still have to go back and get it someday."

Listening to Papy was like rereading Joseph Conrad's *Heart of Darkness*, an immersion in a gloomy, dark-green world full of lethargic violence. A world of shady characters as cruel as they were bleak and drunken. "Mamadou was in cahoots with the king of imbeciles, Ramses. That was the number-two man in Bemba's MLC. There was a lot of rivalry with Mbusa's RCD-ML." A sultry world full of misty thinking. The pro-Ugandan rebels were no longer fighting against Kinshasa, not even against Rwanda, but simply against each other. "The MLC wanted to expand to the east. They attacked Isiro, and later Beni and Butembo too. Ramses was their commander. At Mambasa his men captured Pygmies and ate them." A feverish world with bizarre rituals and gruesome incidents. The Pygmies' family members were even forced to consume parts of their murdered relatives' bodies. The hearts of newborn babies were cut out and eaten. . . . [36] A clammy world with water dripping from the leaves, and the distant cries of animals. Papy sneered, snorted. His somber words dripped with contempt. "One day I lost track of my friend, my comrade. At first we couldn't find him. Then we saw him at a bend in the road. Ramses had got hold of him. His head was impaled on a stick. His penis was tied to the stick a little farther down."

A world of fearful sweat and the smell of bodies. Two million civilians took off running to nowhere. Deep in the jungle, villagers were so absolutely cut off from the outside world that they could no longer find clothes to replace their rags. *Les nudistes*, they were called. Naked, they walked through the forest in search of food, as though it were 1870 all over again—but this time they were ashamed.[37]

In colonial times, the area around Bomili was known for its wealth of smaller gold mines. The lodes were not as rich as those in the more easterly Kilo-Moto, but still worth exploiting. Papy became involved in gold mining and proved a good deal more successful at that than in the ivory trade. Soldiers became entrepreneurs, killers became traders. "I ran thirty-five gold mines around Nia-Nia. That was my sector. No one paid me and my men, but each mine had its own CEO." Although those were mostly teenagers in torn T-shirts, the term CEO (PDG in Papy's French, *president-directeur-*

général) still showed that the economy of plunder had attained a certain degree of formalization. "I called all the CEOs together and held a speech: 'You people have to start making contributions, otherwise the soldiers will start helping themselves and then you'll have people leaning on you. Everyone has to contribute: *l'effort de guerre* [the war effort]. Every month I want to have five grams of gold from each of you.' There was some discussion, and finally we agreed to three grams. There were five thousand *creuseurs* working at some of the mines, but the CEOs only got a little bit of what came out of them."

Industrial mining was a thing of the long distant past. The machines used during the colonial period had been idle and rusting for decades. The work was done now by *creuseurs*, young men and children who scraped away the sediment with a hoe or pick. It resembled the earliest days of Katangan mining, a century earlier, except this time no one was on a payroll; they were all independent businessmen who paid taxes to a superior in the form of a portion of the proceeds. "I went around to all the mines to collect the tax. I had to use that to feed my men, but also to satisfy my superiors. I sold the gold to brigade or battalion commanders. I also demanded a few square meters of the mine for my own use. I had digs everywhere, and about a dozen *creuseurs* who panned sand for me in the river. That brought in about five hundred grams a month, *bon*, if I was lucky."

As medium-sized fish, Papy occupied a place somewhere halfway up the pyramid of the wartime economy. The artisanal mining activities formed a long chain: from *creuseur* to mine manager (CEO) to the ranking officers, and then on the *comptoirs* in the urban centers or even directly to Uganda, where it was sold in turn to international gold buyers. Salim Saleh, President Museveni's brother, was a key figure in such bulk transactions. At the big gold mines of Kilo-Moto, however, all these middlemen were skipped over. There the Ugandan army had direct control over the pits. Mine workers had to do their digging with no safety equipment and without pay, without shoes, and often without tools, at gunpoint. Accidents were common. When a tunnel collapsed in 1999, at least one hundred miners were killed.[38] In 1999 and 2000 Ugandan gold exports rose to between $90 and $95 million annually. Rwanda at the time was exporting $29 million worth of gold each year. This is a great deal, especially when one realizes that neither country has any significant domestic gold deposits.[39]

The same thing went for other mineral raw materials. Before the war started, Uganda exported less than two hundred thousand dollars worth of diamonds; by 1999 that had multiplied fourteen times over, for a total of $1.8 million.[40] Rwanda, a country without diamonds, exported perhaps as much as $40 million in such stones each year.[41] That immediately explains why control over Kisangani was so important. But there was more involved than just precious metals and gems. From Congo, Rwanda also raked in tin, a much more workaday ore used around the world to manufacture food packaging. Between 1998 and 2004 that country produced some 2,200 metric tons (2,420 U.S. tons) of cassiterite (tin ore) itself, but exported 6,800 metric tons (nearly 7,480 U.S. tons), more than three times that amount. The difference came from the mines in Kivu.[42] The area around the Great Lakes resembled a sort of African Schengen Area, a unified market where goods could cross borders freely. Tropical hardwood, coffee, and tea disappeared eastward as well. Congo became a "self-service country."[43] The scramble for Africa was now being organized by the Africans themselves.

And then there was coltan. An unseemly substance that resembles black gravel and is heavy as lead, the ore was mined from muddy deposits. But suddenly the whole world was clamoring for it. It was to become Rwanda's major economic asset in Congo. What rubber had been in 1900, coltan was in 2000: a raw material, available locally in huge quantities (Congo contained an estimated 80-some percent of the world supply), that was suddenly in acute demand around the world. Cell phones were the pneumatic tires of the new century. Coltan comprises columbium (niobium) and tantalum, two elements that are adjacent in the periodic table. While niobium is used in the production of stainless steel for, among other things, body piercings, tantalum is a metal with an extremely high melting point (almost 3,000°C), which renders it extremely well-suited for superconductors in the aerospace industry and capacitators in electronic equipment. Tear open any cell phone, MP3 player, DVD player, laptop, or gaming console and inside you will find a little green labyrinth with all manner of obscure little elements. The drop-shaped, brightly colored beads are capacitators. Break them open and you will be holding a bit of Congo in your hand.

The year 2000 witnessed a veritable coltan rush. Nokia and Ericsson were hoping to bring to market a new generation of cell phones, while Sony was poised to launch its PlayStation 2 (the company actually had to postpone the introduction due to a dip in the supply of coltan).[44] Within less than a year the price rose by 1,000 percent, from thirty to three hundred dollars a metric pound (1.1 U.S. pound). With the exception of one Australian quarry, eastern Congo was the only place on earth where it was mined. Down Under it served as a welcome source of state revenue, but in Congo it was more a curse than a blessing. A feeble state with great wealth beneath its soil, that is asking for trouble. All the coltan mines were controlled by Rwanda; in 1999 and 2000 Kigali exported a mind-boggling $240 million worth of coltan annually. Most of that was sheer profit. Rwanda had to pay the traders and rebels in Congo, of course, but that was peanuts compared to what coltan brought in. The profits made from the war were three times higher than the losses.[45] The occasional crate of Kalashnikovs, therefore, was a minor write-off.

But Rwanda and Uganda were not the ones who profited most from the pillage of eastern Congo's raw materials. In an increasingly globalized economy, governments were only intermediaries in a mass of complex, international, and rapidly mutating trading networks. Kagame and Museveni were not at the end of any supply line; it was the multinational mining companies, shady fly-by-nights, notorious but highly evasive arms dealers, and crooked businessmen in Switzerland, Russia, Kazakhstan, Belgium, the Netherlands, and Germany who made a killing by selling Congo's stolen raw materials. They all operated in an extremely free marketplace. In political terms Congo was a disaster area, but in economic terms it was a paradise—at least for some. Failed nation-states are the success stories of runaway, global neoliberalism.

But Papy had other things on his mind. One day he decided to try his luck again at ivory. With a little help from a few Pygmies, it shouldn't be too hard. "The village chieftain gave me his permission. It took us four days to locate a track, and we followed it for a week. When we finally saw the elephant, he had only one tusk. Later we found a whole herd. I shot one of them, a female. That evening we ate the trunk. It was good."

Of the some six thousand elephants in the Okapi Reserve where Papy wandered, more than half were killed for ivory or meat. Poaching became big business in Congo. In Kahuzi-Biega Park, almost 50 percent of the 130 mountain gorillas, already a very rare species, disappeared as well. Virunga Park had more than twenty thousand hippos; only thirteen hundred of them survived the war.[46] Given a mobile population that consumed between 1.1 and 1.7 million metric tons (between nearly 1.21 to 1.87 million U.S. tons) of bushmeat each year and burned 72 million cubic meters of wood, nature suffered greatly under the war.[47] Industrial forestry came to a halt but, with the supply of electricity cut off, all of Congo went back to cooking on wood fires, consuming one cubic meter (1.3 cubic yards) per person annually. Bushmeat usually came from monkeys and antelopes. For sale on every market one saw smoked, almost charred monkeys with eyes seared shut and mouths wide open. During my first trip to Congo in 2003 I even saw elephant meat being sold on a Kinshasa street market.

Papy's poaching career, however, was short-lived. "The next day we went back to get the tusks. We found a little one, standing beside its dead mother. I shot it too. Compassion, what is that? When I got closer I saw that it had only two weensy little tusks. *Bon,* gold was more my sort of thing anyway."[48]

THE SECOND PHASE OF THE WAR lasted so long because so many profited from it; not just the big multinationals far away, not just the slick traders in their climate-controlled suites, not just the military leaders in the neighboring countries, but everyone at every level of the pyramid. After the miserable Mobutu years, the common people were finally getting the chance to earn a little money of their own. That became clearest of all during the coltan rush. Farmers in Kivus north and south abandoned their bedraggled fields, children left school by the bunches, even teachers turned their backs on their classrooms. "We know that digging coltan can't solve our daily problems," a few *creuseurs* said, "but here we earn a lot more than we used to." The risks they took were all part of the game. Men in particular had a chance to regain their financial autonomy. The informal economy of the 1980s had provided women with new opportunities, but artisanal mining during the war was strictly a man's world. "Digging for coltan is very profitable," two *mamans* told me, "but the husbands are the only ones who gain from it. As soon as they have some money they go off and find another woman in Goma; they even buy a house for her, and that while our own children barely get along and can't even go to school."[49]

The war had not begun with profit in mind, but now that so many were turning one it simply went on.[50] Commerce and war held each other in a stranglehold: in addition to the militarization of the economy, there was also the commercialization of violence. Soldiers like Papy offered their services anywhere the money was good. The informal economy of the past had now become a military economy: it was still about the large-scale smuggling of Congo's riches, but with a Kalashnikov added to the equation. Extreme violence became extremely common; ethnic hatred looked suspiciously like commercial rivalry.

"They took Kasore, a Lendu in his thirties, away from his family and attacked him with knives and hammers," an eyewitness reported in gold-laden Mongbwalu in Orientale province. There the Hemas and Lendus, the Ituri district's two major population groups, had fought for control over the pits. The mines, which once generated an ethnic melting pot, now were sources of dissension. Uganda's policies fanned the fires of racial hatred.[51] The Hemas, the eyewitness said, "killed Kasore and his son (of around twenty) with knives. They cut the son's throat and tore open his chest. They cut the tendons at his heels, crushed his head and removed his intestines." After some attacks, the assailants told those left to listen that now they knew who was the boss. The flipside of globalization was tribalization: the international raw-materials robbery was accompanied by the revival of old rituals or the creation of new ones. A *féticheur* forced one Heman man to undergo a bizarre test: "He had two eggs. They tied me up, I was scared to death. He rolled the eggs over the ground at my feet. I was told that if the eggs rolled away from me I would be considered innocent. But if they rolled toward me, then I was a Hema and therefore guilty. I was lucky, the eggs rolled away. But Jean, who was with me, was not as fortunate. The eggs rolled the wrong way and they told him to run for it. While he was running away, the Lendu shot arrows at him. He fell. They cut him to pieces with their machetes, right before my eyes. Then they ate him."[52]

In addition to profits, the war was also about new forms of morality. The reports from human rights organization rarely include eyewitness accounts from combatants. But in Kasenyi, a fishing village on Lake Albert, I succeeded after some difficulty in getting a number of them to talk. There is little truth to the prevailing idea that all child soldiers were kidnapped. Many went into the army voluntarily. "Our village was attacked twice. My grandfather, sister, and brother were killed. I was twelve and I joined up. Of my own free will. Our massacre was the reply to their massacres. I stayed with the UPC [the major Hema militia] for three years." This young Hema, who insisted on remaining anonymous, was now a veteran: "We were trained by Rwandan mercenaries. Bosco Ntaganda was our general. He fought with Joseph Kony too. I was at the bloodbath at Mahagi. We took mothers, fathers, children. I was told to kill and I killed. Killing women and children, that was hard for me. Fortunately I had a rifle; I was afraid to kill people with a machete. The soldiers took girls to marry. I had to watch as they raped them. Bosco said: 'When you're a soldier, women are free. Everything is free.'"[53]

In a country where the educational system had been destroyed, where there were no jobs, where dowries were unaffordable and the average life expectancy was only forty-two, the war provided not only profits but also a sense of purpose. Children with no future suddenly had an ideal and an identity.[54] "My brothers are fishermen now, they work on the lake in their canoes," another young man told me. "In the war they were with the PUSIC [another Hema militia]. They were twelve and fourteen in 2002. When they came back from the war, they spoke laughingly of their pillaging and raping. The war was a joke, perhaps a joke that brought death along with it, but still a joke."[55] Veterans among themselves bantered stories back and forth, like students after a drunken night on the town. The fighting had been a bacchanal of blood and beer, a Dionysian ritual of running, grabbing, and biting, a blowout with roasted goat meat, smooth female flesh, screaming voices, gun smoke, female flesh that grew moist anyway, there you go, a rush, a curse, a carnival, a temporary upending of all values, a conscious transgression, a forbidden pleasure smothered in a sauce of fear, goose flesh, and humor, lots of humor. A gruesome feast of life's brittleness.

At a certain point, as I sat drinking beer on the waterfront with Muhindu, the man who had dumped bodies into Lake Kivu, he said something disconcerting. "A soldier is like a dog. If you open the gate, he causes damage. In the morning, before we were sent out, our leader would say: 'Go out and do something foolish.' We ransacked houses. We took cell phones, money, and gold necklaces from people. We raped. If you give someone permission to kill, after all, what difference does a rape make?"

I WAS SITTING IN A HALF-DARKENED OFFICE IN GOMA. You could hear the noise from the street. There were no banners of international NGOs in front of the door, no logos, no air-conditioning. This was the anonymous, discreet workplace of La Synergie des Femmes, the city's only shelter for Congolese women, run by the women themselves. Across the wooden table from me is Masika Katsua, a forty-one-year-old Nande woman. She used to live in the interior. The Nande were successful traders in places like Beni and Butembo, and that had created a lot of bad blood. "It was in 2000. We were at our own home. My husband imported goods from Dubai. The soldiers came in. They were Tutsis. They spoke Rwandan. They sacked everything and wanted to kill my husband. 'I've already given you everything,' he told them, 'so why do you want to kill me?' But they said: 'We kill big traders with the knife, not with a gun.' They had machetes. They started hacking at his arm. 'We have to chop hard,' they said, 'the Nande are strong.' Then they butchered him, like in a slaughterhouse. They took out his intestines and his heart."

While she spoke, she never looked up. She scratched incessantly with the plastic cap from a ballpoint pen in the grain of the wooden table.

> I had pick up all the pieces. They held a gun to my head. I wept. All the pieces of my husband's body. I had to gather them together. They cut me with a knife, that's how I got this scar. I have another one on my thigh. I had to lie down on his body parts, to sleep on them. I did that, there was blood everywhere. I wept and they started raping me. There were twelve of them. And then my two daughters in the next room. I lost consciousness and ended up in the hospital. After six months I still wasn't healed. I was still bleeding and gave off terrible smells. My daughters were pregnant. A boy and girl were born, but my daughters refused to have them. I took those children under my wing. When I came back, it turned out that my in-laws had sold everything, the house, the land, everything. They said it was my fault that my husband was dead. I had no sons and therefore no right to stay. The family turned me away. When my grandchildren ask me now about that scar, I can't tell them. It was their fathers who did it.

In 2006 Masika was once again beaten and raped, this time by Nkunda's men. They had come looking for her because, after fleeing on her own to the interior, she had started organizing classes for other rape victims. New victims came to her each day, girls who didn't dare to press charges. "I want to kill Nkunda. God forgive me. If I die for doing that, at least I will have done something that gives me release. I'm still alone. The men don't want me anymore, and I hate all men. I want to help other women. My home is open to them. I pray a lot. I hope for nothing. I try to forget. But when I think back . . . on how my husband and I lived together . . . all that sadness."[56]

And the waters of Lake Kivu slap against the docks. The top of the Nyiragongo volcano vanishes amid the clouds. Jeeps with tinted windows drive slowly over the roundabout. Two boys are pushing a big wooden bicycle through the mud. The bike groans beneath a man-sized bag of colorful flip-flops. And inside a half-darkened office a woman rubs the cap of a ballpoint pen slowly back and forth over the wood, as though trying to scratch something out.

CHAPTER 13

LA BIÈRE ET LA PRIÈRE
(SUDS AND SANCTITY)

New Players in a Wasted Land

2002–2006

ASK ALMOST ANY CONGOLESE WHERE HE WOULD REALLY LIKE to live, and there's a good chance that he will say *"na Poto,"* in Europe. *Poto* in Lingala comes from Portugal, the first European country with which Central Africa became acquainted. In more concrete terms, *Poto* means Brussels or Paris, because the rest of Europe—with the possible exception of London—doesn't count. Jamais Kolonga, the first Congolese man to dance with a white woman in the 1950s, proudly told me that all eight of his children now live in Europe. *Poto* means success. Ask the same Congolese where he definitely does not want to go, and you will be sure to hear *"na Makala."* *Makala* means charcoal, but it is also an outlying district of Kinshasa where charcoal was once made, and where today Kinshasa's Centre Pénitentiaire et de Rééducation is located, the central prison. In the popular imagination, the word *makala* stands for everything the Congolese fear and hate. Ever since the days of Mobutu, that horrible word summons up images of starvation, torture and murder. *Makala* is where the state shows its fangs, a gloomy, pitch-black place dripping with blood and death. Taxi drivers often refuse to take you there.

"And you must absolutely, absolutely not lose this little blue slip," the guard says to me before opening the gates. In a chaotic entrance hall where everyone shouts that he is responsible for *la sécurité*, I am frisked a number of times and have to turn in my cell phone and my money. In exchange for my phone I am handed a crumpled piece of cardboard with a number on it; I had taken out the SIM card beforehand. My cash—twenty dollars, I brought no more than that on purpose—disappeared into a drawer. An official tore off a scrap of paper and wrote on it that I, Monsieur David, had turned in twenty dollars. But even more important than these two hat-check stubs, it seems, is the little slip of blue paper that I hadn't asked for. Smaller than a cigarette paper, it nevertheless turns out to be essential to my future well-being. "When you come out later on, you have to hand this in. If you don't have it, we can't let you through. Then you'll have to wait for evening roll call, to see if everyone is still here." My questioning look receives a reply. "We have to make sure you didn't give it to a prisoner who's taken off, you see." And what if a prisoner happens to be missing? "Then you're the prime suspect." And what if I really don't have it anymore? "Then you stay here." Welcome to Makala.

The bolt is slid aside. I cross a patch of dry lawn and arrive at a building with turnstiles, where I nod to a few apathetic-looking guards. "Pavilion 1?" I ask as casually as possible, as though it were something I did every week, pay a visit to death row. One of them raises his chin slowly to indicate a door. I find myself in a narrow corridor between two high concrete walls. The realm of the guards ends here, this is where the realm of the criminals begins. The guards haven't been paid for years, which means they're on something like permanent strike. They still show up, but they don't do diddly-squat. They drape themselves listlessly over their plastic lawn chairs and fiddle with their broken walkie-talkies. The warden has therefore farmed out the upholding of intramural discipline to the prisoners themselves—with all that obviously entails. The sky is a blue strip far above. In the corridor, hundreds of eyes stare at me. Raucous noise. No one is wearing a prison uniform. Basketball jerseys. Tank tops. Muscular bodies. Shaven heads. Makala was originally built for fifteen hundred prisoners; today it holds six thousand.

Standing still is a sign of weakness, so I worm my way through a row of young men who ask for, no, who demand, money and cigarettes. A little farther along I arrive at the notorious pavilion. The bright daylight is suddenly cut off. The long, gloomy corridor with cells on both sides is plunged in darkness. A few of the cell doors are open; there is laundry hanging out to dry. Hubbub. Here and there in the darkness I see the faces of prisoners sitting around little coal fires. It reminds me of a Russian Orthodox church just before midnight liturgy, but these are not icons illuminated by flickering candlelight. These are condemned men, preparing their meal on primitive stoves, for there is no official prison grub distributed in Makala. If your family doesn't bring you something, you eat grass or gravel.

"I've been here for eight years," Antoine Vumilia tells me in his bare-bones cell. I look around: the entire cell seems to measure about seven by three-and-a-half feet, narrower than a good-sized twin bed in Europe. "I share this cell with two other people." He has me sit down on his cot. There are a few books on the nightstand: Celine's *Voyage to the End of the Night*, *A Hundred Years of Solitude* by Gabriel García Márquez, books by Abdourahman Waberi, Zadie Smith, Colette Braeckman. . . . It is a good thing for him that he has them, the books. "The most hardened prisoners are lord and master around here. The warden's office lets them do whatever they want. They run the drug trade, money changing, the trade in mobile-phone vouchers." And then, in a whisper: "Last year they 'executed' three prisoners." With a gun? "No, they just kicked them to death."

IT WAS JANUARY 16, 2001. Antoine Vumilia was employed by the Conseil National de Sécurité, Laurent-Désiré Kabila's intelligence service. His office was right beside the Palais de Marbre, the residence of the head of state. Only a single wall separated him from the presidential quarters. A little after noon he was startled by an infernal racket. "I heard shots," Vumilia told me there on death row, "three of them. And a couple of minutes later, eight or ten more."

On the far side of that wall, Kabila was meeting with an adviser when a *kadogo* came up to him. *Mzee*'s guard still consisted of faithful child soldiers from Kivu. Although Ruffin had been demobilized by UNICEF a year earlier—he was seventeen and had to go back to school in short pants, amid city kids of twelve who didn't even know how to take apart an AK-47—Rashidi, one of his former comrades in arms, was still in service. It was Rashidi who walked up to the president now. It looked like he meant to whisper something in his ear, but then he drew an automatic pistol and fired three times. One of the bullets went through the back of the president's colossal head. Kabila died immediately, forty years less a day after Lumumba was murdered. A few minutes later, young Rashidi was riddled by bullets fired by a colonel in the palace.

Vumilia had heard that gun battle. One week later he was arrested on suspicion of conspiring in the murder. As a security officer, Vumilia had written a report in which he warned of irritation among the child soldiers from Kivu. The *kadogos* were Kabila's most loyal followers, but it seemed as though they too were starting to feel passed over. Vumilia himself came from Kivu; he knew what was going on there, but because he was personally acquainted with the ones involved he had decided not to tell everything he knew. "I was in a dilemma: I had to protect the regime, but these were my friends. They were extremely dissatisfied. What do you expect? Masasu had been murdered in November 2000." Young Anselme Masasu Nindaga was their hero: a street fighter like themselves, a man of verve and daring, one of the founding fathers of the AFDL.[1] After Kinshasa was taken in May 1997, however, Kabila shoved him aside and had him thrown in prison. When he was released in fall 2000, he dreamt aloud of Kivu's secession and won a large following. Soon after that, he was shot. In the violent protests that followed among the child soldiers in Kinshasa, dozens of people were killed. The love affair with *Mzee* was over for good. Kabila had now even blown up the bridge connecting him to the ones he called "my children." Bitterly, the children began plotting. Revenge, blood, murder. Vumilia tried to talk them out of it: "These were really young boys. All they wanted was to show that they were fed up. I told them it was pure suicide, that there was no future in." But he was arrested along with them and refused to testify against them in a trial that was no trial at all. "They wanted me to testify against people I knew, people I ate with every day in prison."

Besides, wasn't it possible that Kabila had been murdered for very different reasons?[2] Could one really be sure that the plot came from Kivu? What if Angola was involved? Couldn't it perhaps all have been about diamonds? There were rumors that Kabila, who owed so much to Angola, had entered into cahoots with the hated UNITA rebels who controlled northern Angola with its wealth of diamonds. Hadn't there been Lebanese men who acted as mediators between Kabila and UNITA? And weren't eleven Lebanese diamond dealers murdered in Kinshasa right after Kabila's assassination? Yes, that was all true. But it was all so vague, so shadowy. No one could get to the heart of the matter, especially not Vumilia. "I tried to get the boys off the hook, but that made people conclude that I must be one of them." Along with thirty others, Vumilia was sentenced to death with no chance of appeal. International human rights organizations called the trial a miscarriage of justice.[3]

For the thousandth time, Vumilia's eyes traced the walls of his cell. "I've

been here for eight years already. It's unspeakable, it's incredibly hypocriti-
cal. The leaders of the regime know the truth, but all they wanted was to
keep the public quiet by quickly giving them a scapegoat."[14]

KABILA'S DEATH WAS A TURNING POINT in the Second Congo War. His
son, Joseph Kabila, was quickly appointed to replace him. With his wavering
voice and extreme youth (he was only twenty-nine), he at first cut a rather
feeble figure. The Congolese barely knew him; the West figured he was a
marionette. But less than one month later, he met his Rwandan counterpart
and archenemy Paul Kagame in New York and delivered a number of striking
speeches. He spoke of peace, national unity, and the role of the international
community. Could this mean that a new era was dawning? Yes, it could.
After a number of United Nations reports had clearly shown how Rwanda
and Uganda were pillaging the country's raw materials, Kagame and Yoweri
Museveni could no longer claim that they were in Congo only for national
security reasons. This resulted in a long series of peace talks in Gabarone
(August 2001), Sun City (April 2002), Pretoria (July 2002), Luanda (September
2002), Gbadolite (December 2002), and again in Pretoria (December 2002).
At this final meeting, thanks to brilliant negotiations by Senegalese UN
negotiator Moustapha Niasse and considerable pressure from South Africa
and the African Union, the crucial agreement to put a complete stop to the
war was signed at 3 A.M. on December 17, 2002: the *Accord Global et Inclusif.*
Rwanda and Uganda had already agreed to a withdrawal, but this time the
agreement also applied to their domestic militias. The signatories included
the government in Kinshasa, a few representatives of Congolese civil society, Tshisekedi's UDPS, Jean-Pierre Bemba's MLC, Azarias Ruberwa's RCD-
G, Antipas Mbusa Nyamwisi's RCD-ML, Roger Lumbala's RCD-N, and the
Mai-mai. The term *inclusif* was apt enough. In fact, the agreement was so
all-inclusive that war criminals, to keep the peace, were not prosecuted but
promoted to the office of vice president.

The accord allowed for a two-year transitional period, during which
power was to be distributed according to the formula "1 + 4": beside President Kabila there were to be four vice presidents, two from among the rebels (Bemba and Ruberwa), one from Kabila's entourage (Abdoulaye Yerodia)
and one from the peaceful opposition (surprisingly enough, not Étienne
Tshisekedi, who had been carrying out a nonviolent struggle for the last ten
years, but Arthur Z'Ahidi Ngoma). Within that two-year period all existing

militias were to be combined into a new national army and preparations were to be made for democratic elections. The term could be extended by two six-month periods. In anticipation of the long-awaited popular vote, an interim parliament and cabinet were installed.

The agreement was absolutely historic. Now, after years of despair, there was a major chance of achieving peace and reconstruction. The new Congo therefore received the international community's concerted support: the troop strength of MONUC, the UN peacekeeping force, was raised to 8,700 blue helmets and rose in subsequent years to 16,700, making it the biggest UN operation in history (and, with an annual budget of around $1 billion, also the most expensive).[5] Led by the always-optimistic American William Swing, the soldiers were to safeguard the ceasefire and supervise disarmament. "*Ça va swing!*" a popular song on Congolese radio in those days, parodied his pronounced Anglo-Saxon accent. The new regime was assisted in policy matters by the the Comité International d'Accompagnement à la Transition (CIAT), a unique instance of bilateral and multilateral diplomacy in which the ambassadors of the five permanent members of the UN Security Council, along with those from Belgium, Canada, Angola, Gabon, Zambia, and South Africa, and representatives from the African Union, the European Union, and the MONUC, actually helped to run the country. The CIAT was no external advisory body, but a formal transitional institution.[6] "We were, in fact, a supervisory committee," said Johan Swinnen, Belgium's ambassador in Kinshasa at the time. "We had no legislative power, but we served to provide momentum and to stimulate. We supplied expertise. We didn't want to be busybodies, but partners. Still, there were frictions between the CIAT and the 1 + 4. When the process was over, we issued a few highly critical and outspoken communiqués. We had abuse heaped on us. After that, they didn't like us anymore."[7] There was talk of "monitored sovereignty"; the country, however, was at least partly in de facto receivership. The MONUC and the CIAT were more than just the training wheels of the new Congo.[8]

And that support was badly needed. The new leaders did not do a particularly good job; they emulated the abuses of Mobutism with a zeal that would have startled Mobutu himself. While crucial dossiers dealing with military reform and the electoral process awaited action, one of the first laws to pass through parliament stipulated . . . higher wages for the members of parliament. The fixed salary of six hundred dollars a month (already generous in a country where a university professor earned thirty) was doubled to twelve hundred. The senators, as political elders, even jacked up their pay to fifteen hundred dollars a month.[9] In 2005 the members of parliament as a whole (620 souls) treated themselves to a respectable vehicle: each representative was given a brand-new SUV valued at twenty-two thousand dollars—the terrible condition of Kinshasa's roads, after all, called for solid coachwork.[10] That those same roads could have been repaired for that money seemed hardly relevant. Rather than an opportunity for a lasting reconstruction of society, political mandates were still the fast lane to individual financial gain. There were no incentives for good governance, not as long as corruption was so rewarding, both financially and socially: it was considered praiseworthy. "You mustn't forget that our politicians are the children of poor people," a Congolese school principal told me once.[11] While corruption in the West is viewed as unjustifiable, in Congo it is seen as extremely justified: it is the person who misses out on a perfect opportunity to feed his family who is acting in a completely unjustifiable fashion.[12]

The cabinet ministers and vice presidents were not about to miss out. All of them felt that they had a right to "special treatment" for their "logistical requirements." In everyday language that meant: a villa and a large automobile. The four vice presidents even received villas with three bathrooms, in addition to a Mercedes limousine, a luxury passenger car, and two escort cars. The hope that the *quinquevirate* of president plus vice presidents would serve to maintain an ethical balance soon proved quite naive. The gentlemen gave each other plenty of room and shared only one concern: making sure the transition lasted as long as possible. In 2004, they all exceeded their budget by more than 100 percent, Bemba by even 600 percent.[13] The 2005 budget awarded the head of state a sum eight times that reserved for health care in Congo as a whole and sixteen times the country's agricultural budget. Politics was war by other means. The state-owned Gécamines enterprise still had all the resources needed to breathe new life into the national economy, but the president's circles signed a series of dubious contracts with often extremely shady foreign businesses. Those contracts established joint ventures that allowed foreign cowboys free rein within certain operational arms of the mining giant. They were allowed to exploit and export at will, while the Congolese state received little or nothing in return—and the well-filled envelopes went on changing hands under the table.[14] Once again, a tiny elite was being given the keys to the kingdom. Clientelism was in ruddy good health. "$1 + 4 = 0$" was the equation popular painters sometimes added to their satirical canvases.

The army fared no better. Officially, all of the militias were to be fused into a new, 120,000-man army.[15] And yes, quite a few former rebels suddenly received a national army uniform, while many of their commanders received a high rank (always good bait for pulling in warlords), but in actual practice this *brassage* (intermingling) barely scratched the surface. Behind the facade, nothing changed. Soldiers who have been each other's enemies for five years do not embrace each other that quickly. By 2006 only three of the planned eighteen brigades had actually fused.[16] What's more, the *brassage* had rendered the Congolese army top-heavy: after the promotions of all those former rebels, there were now almost twice as many officers and noncoms as foot soldiers.[17] Within the Congolese armed forces, commanding was considered more pleasant than obeying—no, not commanding, but *commandeering*. The extensive officers' corps busied itself with the mass misappropriation of funds. The salaries of the rank and file vanished into the pockets of colonels and generals who did not hesitate to roundly exaggerate the numbers of their enlisted men in order to receive more funding. The underpaid and badly trained soldiers themselves were neither motivated nor disciplined, and behaved accordingly. The new government army, the Forces Armées de la République Démocratique du Congo (FARDC), should have been the cornerstone of the resurrected state, but instead became as much a whited sepulcher as the FAC of Kabila *père*, Mobutu's FAZ, or even Lumumba and Tshombe's ANC. Jokingly, people sometimes twisted the acronym FARDC to make of it: *phare décès* (dead beacon). Since gaining independence, Congo has never had at its disposal an army comparable in efficiency and discipline to the former Force Publique. For that reason, that army has never been able fulfill the primary function of statehood, that of maintaining the monopoly on violence.

Does it come as any surprise then that the war was never completely over? As long as the security forces remained a sham, the MONUC stood alone. But one cannot hold together a territory half the size of Europe with only seventeen thousand men; even the biggest UN mission in history was no more than a drop in the ocean. In Iraq, six times smaller than Congo, there were stationed at that moment some 150,000 American troops, and even they were unable to contain the violence. The blue helmets' presence had a calmative effect is some areas, but elsewhere impotence was their portion.

EASTERN CONGO remained in turmoil, even after the *Accord Global et Inclusif*. There the conflict entered its third phase. Its theater of operations was much smaller, but the human suffering was still great. In essence, the violence was now concentrated in two areas: Ituri and the two Kivu provinces. By no coincidence whatsoever, both areas were rich in ore and bordered on Uganda and Rwanda, respectively.

In Ituri, the conflict flared up precisely because of the peace agreement. When the Ugandan army withdrew definitively from Bunia on May 6, 2003, Lendu militias pounced on the city's center and killed dozens of Hemas. A few days later the Hemas rolled in, in turn, and killed off dozens of Lendus. The conflict resembled a miniature version of the 1994 genocide. The cattle-breeding Hema felt affinity with the Tutsis: an ethnic minority that formed the social elite. The Lendu were farmers who compared themselves to the Hutus: numerous, but at the bottom of the social ladder. It was, in fact, the ancient conflict between herders and farmers concerning access to the land, the conflict between the pastures and fields and the crops that were eaten by cows.[18] But this time that Cain-and-Abel conflict was stirred up by overpopulation, and put to good use by a Uganda greedy for gold.[19] The ethnic tension in the region rose to such heights that devoutly Catholic women from both sides told me: "Even we, *les mamans*, took up arms. We felt threatened." Or: "We were accomplices. We carried weapons in our baskets and water jugs."[20] The ethnic violence in Ituri was no atavism, no primitive reflex, but the logical result of the scarcity of land in a wartime economy in the service of globalization—and, in that sense, a foreshadowing of what is in store for an overpopulated planet. Congo does not lag behind the course of history, but runs out in front.

Hundreds had already been killed in the provincial town of Bunia within one week in May 2003, but the entire region was entangled in a bloody war of inextricable complexity. And in terms of that complexity, in the Ituri district, the Second Congo War reached its absolute nadir. A dozen militias were active, loosely organized little armies of children in flip-flops toting guns, led by dodgy young men in their twenties and thirties who often operated under an assumed name and switched alliances back and forth with other warlords. With its countless mergers, schisms, joint ventures, and takeovers, this new brand of armed conflict looked more like the business world than war as we know it. In the offices of the MONUC, dispirited officials hung charts on the wall tracing the organization of the various militias: it made them only more dispirited. Every month a new militia came along, or the orderly chart had to be rearranged—more columns, more arrows, more acronyms, more additions to the rogues' gallery of photos beside them—until the chaos finally jibed with that in the field and lost all explanative value. But there was one constant factor: sooner or later, all parties received weapons from and were trained by Uganda.[21] That was less an indication of a conscious divide-and-conquer policy in Kampala, however, than of the internal rivalry within the Ugandan army; each general had his own militia in Congo, a militia he could abandon or resuscitate as the situation required. Even more arrows, even more connections, for on the Ugandan side too there was no solid ground. Rwanda backed the occasional militia as well. No, the war was not over yet. It had become a small but stubborn snarl, a form of self-perpetuating armed banditry.

Exactly one year later, in May 2004, Kivu was the scene of extreme outbursts of violence. The major rift there remained that between Hutus and Tutsis; there too overpopulation played a role, but particularly the overpopulation in Rwanda itself. Ten years after the genocide, Rwandan Hutus still could not return to their overfull fatherland, where partisan prosecution awaited them. "Kabila doesn't chase them away and Kagame won't have them," was the pithy way Belgian diplomat Johan Swinnen summarized the situation.[22] Their long-lasting exile was still causing unrest, prompting Rwanda's continued support for Congolese Tutsis willing to deal with the Hutus. In May 2004, therefore, Laurent Nkunda's men, along with those of Mutebusi, moved killing and plundering through the streets of Bukavu. They raped dozens of women and girls, usually as a gang. Some of the girls were no older than three.[23] Nkunda was a Tutsi from North Kivu and a welcome guest in Kigali. From 1990 on he had fought alongside Kagame; in 1996 he had advanced along with the AFDL. He had held a top position in the RCD-G in 1998 and terrorized the population of Kigali with an iron fist in 2002. Because of his leading role in the massacres, he was leery of accepting a position in the new government army. And so Nkunda became Kigali's new golden boy, taking Bukavu in his own, characteristic fashion. For a time, the fragile peace process seemed to ground to a halt. Was this the start of a third war?

Amid general outrage, the UN blue helmets (most of them Uruguayan) in both Bunia and Bukavu stood by and watched powerlessly, to say nothing of faintheartedly. But, thanks to a number of historical firsts, calm soon descended again in Ituri. For the first time in history the European Union carried out a joint military action, with something like a European army. With UN approval, primarily French commandoes pacified the city of Bunia during Operation Artémis. International arrest warrants were issued for the most important warlords. Three of them were detained and sent to The Hague, including Thomas Lubanga, the head of the biggest Hema militia. In 2010 he was the first defendant to be tried by the new International Court of Justice. In that regard, too, Congo is at the vanguard rather than the rearguard of history.

In Kivu the transition seemed about to founder and make way for a new war when 160 refugees—most of them Congolese Tutsis—were brutally murdered at the Gatumba camp in Burundi, in reaction to Nkunda's violence. Rwanda once again sent troops to Congo to protect the befriended Tutsis. For a while everything seemed to be starting all over again, but the United Nations, South Africa, and the CIAT did everything in their power to ease the pressure.

During the third phase of the war, the conflict gradually reverted to what it had been at the start: a clash between Rwanda and Congo concerning the treatment of Hutu exiles in Kivu. Kagame still hoped to neutralize them, fearing as he did that they were plotting a coup in Rwanda. Just as his own regime had been molded during banishment in Uganda, he now believed that a Hutu takeover was being plotted in eastern Congo. And he had no intention of letting that happen: Rwanda was full and firmly in Tutsi hands. The chronic conflict has now lasted for more than fifteen years. The suffering in the area around the Great Lakes can be traced back to that fateful day in spring 1994 when the French government decided to allow the Hutu regime to escape to eastern Congo, weapons and all.

Today, small but powerful Rwanda—still in great favor among the donor countries—displays all the earmarks of a blossoming military dictatorship, while neighboring Congo remains huge, sluggish and weak, and unable to deal with the problems of the day. It is as though a lone professional soldier, Rwanda, were living in a rigorously simple one-room flat in a chaotic apartment building inhabited by an enormously loud and dysfunctional family that has fights, neglects to pay its bills, and sometimes even forgets to turn off the gas stove. On more than one occasion the soldier takes his gun down off the wall and storms into the neighbor's kitchen, where he causes more damage than necessary. Rather than simply turning off the gas, he shatters all the cups and saucers, shoots holes in the kitchen ceiling, and marches out again with a boiled ham under his arm. The result is more noise and more fighting. A neighbors' quarrel: that, in effect, is what is going on in Central Africa today, a neighbors' quarrel in which one party roundly curses the other. Not without reason, by the way, for Kigali is every bit as culpable as Kinshasa. The conclusion, however, remains bitter: as long as the crucial transition simply refused to take place in Kinshasa, the Second Congo War in the east simply would not stop.

THE CLINKING OF BEER BOTTLES—hundreds, thousands of beer bottles, big brown bottles jostling for position on the conveyor belt—drowned out the other factory noises. It sounded like a carillon, a high, insistent tinkling that rose above the hiss of the rinsing machine, the clack of the labeler, the rattle of the conveyor, and the sigh of the pressure hoses—like the sound of chimes ringing out above the bustle of a busy city. The nervous, cheerful tinkling rolled through the noisy factory hall and mingled with the smell of malt and alcohol. It was 2002 in Kinshasa and Bralima, and the Primus brewery had just opened two ultramodern, fully automatic packaging lines that could process seventy-two thousand bottles an hour. The war was barely over, but the brewing industry was bursting at the seams. Bralima (from Brasserie et Limonaderie de Léopoldville) started as a small colonial brewery in 1923, but has been owned by Heineken since 1987. In the war's wake, the Dutch conglomerate had every intention of gaining control of and expanding the beer market in perennially thirsty Congo. A million and a half hectoliters (about 39,625,807.85 gallons) of suds went out the factory gates in 2002; by 2008 that was almost 3 million. That spectacular doubling of production, though, was still a far cry from the record that had been set in 1974, the magic year of the boxing match, when Bralima had produced 5.5 million hectoliters (over 200,000 gallons). But the future looked bright.[24] In Kinshasa alone, Bralima once again had fifty thousand retail outlets and bars.

The way the politicians were dragging their feet did not, in any case, much faze the multinationals. The start of a new period of peace held the promise of new markets, which had to be conquered with all due speed. The same applied, a fortiori, to mobile telephone systems. Vodacom, the South African telephony operator, had started laying the first cables while the ethnic violence was still in full swing. During the worst firefights, the workers would simply stop digging and take cover for a few hours.[25] Whence all the hurry? In a country where the telephone infrastructure had been in ruins for decades, there was an enormous demand for cell phones. The MONUC troops alone accounted for thousands of subscriptions. And the Congolese rank and file, too, began dreaming of owning a GSM. When I was in Kinshasa for the first time in 2003, a Congolese cell phone number had only seven digits; by 2006 it was up to ten. Mobile telephony is to Africa what movable print was to Europe: a true revolution that has profoundly redefined the structure of society.[26]

A weak state like Congo left plenty of room for new international players. During the Cold War it had been foreign nations (France, Belgium, and the United States) that helped determined Zaïre's fate, but now it was increasingly foreign private partners, such as companies, churches, and NGOs. Since the end of the last war, large parts of Congo were run by international charities that took over government tasks. The reason Kabila could grant himself a budget eight times that for national health care was because he knew the money for that care would come from abroad anyway. The same went for education and agriculture: the favorite domains for international donors. The assistance granted by many hundreds of NGOs was often impressive, but not devoid of consequences. The corruption endemic within the Congolese civil service apparatus prompted many NGOs to remain "nongovernmental" in the host country as well and to work only with regional and local partners.[27] Understandable, but hardly conducive to restoring the bond of trust between government and people. In addition, the influx of foreign funding created something like "aid addiction": the Congolese began doubting their ability to manage for themselves. Monsieur Riza Labwe, a friendly but hardworking man who ran a modest hotel in Bandundu, was kicking against the pricks of such passivity when he said to me: "All these NGOs here make us too dependent. Someday an NGO is going to come along and start telling us how to brush our teeth."[28] Nowhere was this "NGO-ization" more obvious than in Goma, blasted to pieces by the war and overrun with lava since 2002. In December 2008, while crossing town at evening rush hour on the back of a scooter—the public transport of the common man—I looked around at the traffic we were jauntily zipping past:

all jeeps, all belonging to NGOs, all of them with a logo decal on the door
or a pennant on the antenna. For Justine Masika, the founder of La Synergie
des Femmes, this was a source of great irritation.

> There are more than two hundred organizations for women's rights
> in Goma alone right now. There are a lot of fake NGOs among them,
> local organizations that line their own pockets with foreign money at
> the expense of women who are ill. Everyone just comes here and starts
> something up. The money from the donor countries goes by way of the
> United Nations, but they take a substantial commission, up to 20 or 30
> percent. There is a true UN Mafia! I don't work with them anymore.
> The UN Food Program, UNICEF . . . they come here with enormous
> budgets, but 60 percent of it goes to logistics, without any results being
> booked first. All these foreigners apparently have to receive "danger
> pay," all these offices have to be air-conditioned, they have to be well-
> furnished and guarded. And a terrible amount of money goes to public
> relations. They want to have a high profile, even here. But the women
> it's all about are in danger and require discretion.[29]

Tough language, and Masika is not just anybody: in 2005 she was one
of the thousand women jointly nominated for the Nobel Peace Prize, and in
2009 she received the prestigious Human Rights Defenders Tulip from the
Dutch government and the Pax Christi International Peace Award.

COMPARED WITH THE HUMANITARIAN PRETENSIONS of development
aid, trade and industry were at least straightforward about their financial
priorities. Profits were the thing. That there is nothing dishonorable about
this requires no explanation for Dolf van den Brink. After taking a degree in
philosophy and business administration, this irrepressibly dynamic young
Dutchman became commercial director of Heineken in Kinshasa: the number two man at Bralima. In that position, he was one of those responsible
for the exceptional growth seen in recent years. "When I came here in 2005,
Primus had a 30 percent market share in Kinshasa, while Skol, the brand sold
by our competitor Bracongo, had 70 percent. Now the situation has been
reversed: we have 70 percent and Skol only 30."[30] He showed me a slide from
a PowerPoint presentation. The line on the chart showed a rising curve.
Written above it, in hip management lingo: *On a gagné beaucoup de batailles,*

mais pas encore la guerre! We've won lots of battles, but we still haven't won the war! In Bralima's conference room there hung a plaque, to remind the staff of their first obligation: *Esprit de combat!* (fighting spirit!)—as though this country had not just emerged from a hideous conflict.

And a war it was. The main reason Bracongo had done so well at first was that it had Werrason, while Bralima had to make do with J. B. Mpiana. Werrason and Mpiana were wildly popular pop musicians who had taken part in both breweries' promotional campaigns. In 2005 Werrason was clearly more successful and it was unthinkable that any of his fans would ever order a Primus. At a time when politicians were not elected, people had no jobs and three-quarters of the urban population was under twenty-five, pop musicians wielded immense power.

The rivalry between J. B. Mpiana and Werrason was legendary. Each generation in Congolese pop music had known its own clash: between Franco and Kabasele in the 1950s, between Franco and Tabu Ley in the 1960s, between Papa Wemba and Koffi Olomide in the 1980s. In the late 1990s, though, the mood became even grimmer. In 1981, Mpiana and Werrason had joined forces in a band with the megalomaniacal name Wenge Musica 4x4 Tout Terrain Bon Chic Bon Genre. That was asking for a row. It was a legendary lineup that bestowed upon the world in general and Kinshasa in particular the *ndombolo,* the most popular dance style of the nineties and of the new century, a choral dance in which the men bent down low and seemed to be boxing while the women rolled their buttocks in truly spectacular fashion. The *ndombolo* was provocative, obscene, hilarious, and, as often goes with trendy dance styles, also kind of fun. Onstage Werrason and Mpiana showed off their Telecels, the first generation of cell phones, which were the size of a rubber overshoe. At that point those devices were reserved almost exclusively for top military officials and cabinet ministers, but now fans shoved beer bottles far down into their back pockets to create the impression that they, too, owned such a voluminous example of cutting-edge technology. Wenge Musica was *the* sensation of the nineties. When Kabila took Kinshasa, people danced the *ndombolo.* But Wenge Musica went the way of all Congolese pop groups or political parties once they gain a bit of success: it fell apart. Werrason and Mpiana were at each other's throats, the fans were splintered, and even today they will talk to you of the power struggle with a passion and precision one rarely hears when they speak of the war.

People speak without irony of *la guerre des albums, la guerre des salles,* and *la guerre des stades.* Initially, Werrason was the contender who displayed exceptional militance: his first CD was called *Force d'Intervention Rapide* (rapid intervention force). With song titles like *Attentat* (Attack), *État d'Urgence* (State of Emergency), *Ultimatum, Couvre-feu* (Curfew), and *Cessez-le-Feu* (Cease-fire), it was clear that military jargon was already trickling into pop culture.[31] Each album brought with it a new dance and a new fashion. Fans waited to buy clothes until the new CD was released. But in 1999, when Mpiana became the first of his generation to fill the legendary Parisian concert halls Zénith and Olympia, Werrason took revenge by playing to twenty-thousand fans at the Bercy sports palace, and then moving on to the Zénith and Olympia. In France, *bien sûr*; from now on, Congolese history would also be acted out in the haunts of its expatriates.

Around the metro stations Chateau d'Eau in Paris, Porte de Namur in Brussels, and Seven Sisters in London, Congolese entertainment districts arose, complete with hairdressers, music shops, and greengrocers selling manioc and smoked grubs. The country's misery had caused tens of thousands to emigrate. In Kinshasa, Werrason and Mpiana tried to score points off each other during concerts at the Stade des Martyrs, where the audiences sometimes numbered more than one hundred thousand. In 2005 they held a showdown *fara-fara* (face-to-face), with podiums set up at opposite ends of the field. This *concert du siècle* was intended to be a war of attrition, to decide who was strongest. The bands began playing at 10 P.M. and went on all night. When police marshals tried to pull the plug the next morning, street children formed a living shield around the electrical generator. At one in the afternoon, the army put an end to the event with the use of tear gas. Despite the more than two hundred thousand spectators the battle remained undecided and Werrason's star continued to rise.[32]

Roi de la forêt they called him, king of the jungle. His bodyguards were the *manzaka na nkoy*, the leopard's angels.[33] With his deadpan expression, his bombastic sunglasses, and razor-sharp goatee he became the living epitome of Congolese "cool." Born on Christmas Day 1965, he seemed destined for greatness. The UNESCO appointed him its peace ambassador. The pope received him in Rome, and on CNN Jamaican superstar Shaggy called him "the greatest living African artist."[34] But to Kinshasa's thousands of street children—little boys thrown out of their homes on suspicion of witchcraft, children who ran away voluntarily, AIDS orphans who camped out permanently in the sand in front of Werrason's rehearsal rooms, all those who called themselves *sheges*, after Schengen, for they lived in an extremely free market place—to all those young bodies in worn-out rags he remained *Igwe*, the high priest. For him they were prepared to die.

And then, in July 2005, the news came in: Werrason was switching from Bracongo to Bralima! It struck like a bombshell. Werrason had been in Europe for months. On a Bralima expense account? To keep from having to serve out his contract with Bracongo? Everyone was speculating about it because music in Congo is more important than soccer in Italy. How much must they have paid him? To this day, the price of that transfer is the best-kept secret in Kinshasa. Dolf van den Brink knows how much it was; he is the one who arranged the deal. "I'm afraid I can't tell you that," he laughs from across the desk when I ask him. "Believe me, pop music costs us many hundreds of thousands of dollars each year. It accounts for two-thirds of our marketing budget. We've invested in a concert stage that cost three hundred thousand dollars, the biggest one in the country. We have trucks, genera-tors, and stewards. We have an agency with thirty employees, so we can stage free concerts in the city. Once a year the artists write a Primus song for us. We pay for the studio and for the CD and video clip. That alone costs us between 100,000 and 150,000 dollars. We pass out four thousand free CDs and nine thousand cassette tapes through the bars in the *cité*. Everywhere, people dance to those Primus tracks."

He seemed slightly dazzled by it himself. Given, he had written his the-sis under the supervision of Dutch sociologist Dick Pels on the subject of "the aestheticization of business," but he had never dreamed that he would become the patron of an African international superstar. "For me, there's a symbiosis between music and the brewery. Werrason has three orchestras; more than a hundred people depend on him for a living. He can't get by from the sales of CDs and cassettes alone. Concerts are prohibitively expensive. So sponsoring is crucial for him, in addition to VIP concerts and perfor-mances in Europe. And we handle that too. When he goes to le Zénith, we pay for fifty airline tickets; if we didn't, he would leave."

Werrason, an informal survey shows, has a reputation for being very difficult. A peace ambassador, true enough, but above all a pain in the ass. His sponsors are expected to import dozens of cars for him and his entou-rage and get them past customs. Appointments are of no importance. The rare person granted an interview catches at best a glimpse of him, and then futilely waits for hours for him to return, as this writer found out one icy cold December day in Paris. Dolf van den Brink sighs. He rummages around on his desk and shows me a scrap of paper. "Sylvia Mampata just came by, that's his wife. She's going to be giving a party soon and she's asking us for fifty garden chairs, thirty crate loads of beer, and fifty thousand dollars. That's how it goes the whole time. Do you get what I'm saying?"

Of course we do, because Dolf has just explained his PowerPoint graph. "Look, you can see it very clearly here," he says contentedly. "Werrason came to us in July 2005. Within two months our market share rose by 6 percent: from 32 to 38. And that upward trend has held. Today we're at 70." Bralima became one of the Heineken concern's fastest-growing subsidiaries. In 2009 it even had a market share of 75 percent: figures that European managers can only dream of. Van den Brink was rewarded with a transfer to America where, at the age of thirty-six, he became CEO of Heinken USA.

Bralima had indeed done its homework before Werrason's historical switchover. A few days after his return from Europe—tens of thousands of young people accompanied him from the airport to Samba Playa, his rehearsal hall—he gave a Primus concert in his home town of Kikwit under the title *Changement de fréquence* (change of frequency). He had never performed there before. It was the biggest pop concert ever. *Changement de fréquence,* those were the riverboats full of sound equipment and lighting, electrical generators, and fifty thousand crates of Primus that Bralima had sent from Kinshasa months earlier. *Changement de fréquence,* that was the huge field close to the airport where the podium was erected and to which tens of thousands of people came on foot, from everywhere, sometimes even 120 kilometers away. *Changement de fréquence,* that was Werrason who arrived on the day of the concert in an Air Tropic Fokker, with traditional chieftains and village chiefs, and who kissed the ground when he had landed. *Changement de fréquence,* that was the King of the Jungle who was welcomed like a head of state, perched atop a Bralima truck. *Changement de fréquence,* that was his twenty-man orchestra, knocking the first notes through the sound system a few hours after sundown. The phenomenally tight rhythms of Kakol, the crystalline guitar solos of Flamme Kapaya, the effortless falsetto of Héritier, the burlesque raps of Roi David. The latter was the successor of the unforgettable "Bill Clinton," the *animateur* who had gone solo and was now under contract to Kerrygold, where he composed tunes for powdered milk. *Changement de fréquence,* that was finally seeing in real life all those names you had been hearing for years. Seeing the improbably supple buns of Cuisse de Poulet roll as she danced the *ndombolo* on stage next to Bête Sauvage and Linda la Japonaise. My God, what a party! *Changement de fréquence,* finally, that was Werrason, coming on stage after midnight, looking out implacably over that sea of ecstatic humanity (how many were there? 300,000 according to the most sober estimates, 700,000 according to the fans), singing three songs and then passing out medicine to the widows and the sick, a gesture the government could learn from, with all its messing-about and infighting! *Changement de fréquence,* that was the heavyweight bout with Muhammad Ali revisited, the difference being that this time it was not the president footing the bill, but a limited-liability conglomerate from Amsterdam. That too was a change of frequency.[35]

"The crowd in Kikwit was enormous," Flamme Kapaya told me one morning in Kisangani. It was the start of a sultry day and we were sitting in the overgrown garden of a house beside the river. For ten years Kapaya had been Werrason's star guitarist and artistic director. Ask any young person in Congo who is the greatest living guitar player and they will inevitably reply: Flamme Kapaya. "We had to warm up the audience, tell them how fantastic Werrason was. We had to play and dance to get him to come up. But he sang for maybe fifteen minutes in total and he raked in all the money. We didn't even get paid. The whole switch from Bracongo to Bralima didn't change a thing for us. He took everything, we got nothing! Werrason became filthy rich and bought a house close to Brussels. He was like an heir to Mobutu." And since profits were more important than reeducation, Bralima kept the system going, because the Heineken shareholders wanted to see pleasing charts and graphs. There is a fundamental similarity with the foreign interference of yesteryear: just as America gnashed its teeth but kept Mobutu in the saddle, so Heineken learned to live with Werrason's whims, for otherwise he would switch to the competition. Loyalty came at the expense of integrity. Kapaya is still angry about that: "I composed the songs, I arranged them, but he had the songs registered in France under his own name. *Arrangeur—compositeur: Werrason,* that's what it says on the CD. I'm only mentioned as the guitar player." Kapaya was the musical brain behind *Kibuisa Mpimpa,* which is generally considered Werrason's best album, one those in the know refer to as "culturally and musically revolutionary."[36] "I handled the recordings in Europe, I mixed the album, but when it was finished I didn't even get a copy! Werrason even stole my five author's copies."

It seems hard to believe, but while I was spending those three hours in a cold Paris studio, amid female groupies in big, flashy winter coats, waiting fruitlessly for my interview, Werrason was nowhere to be found. Kakol and Héritier, the drummer and singer, did all the work. They instructed the background singers, operated the panel, and made the tough musical decisions. "We were so naive," Kapaya sighed. "He wanted musicians who weren't on to him. If you were, he didn't want you around anymore. Music is the passion of all young people, but he misused that. What it really boils down to is people's exploitation of other people. That's why I left. I don't want young people to go down that same road. They need to know their rights." With his fingers he did a little drum roll on the edge of his chair, looked at the river and then said: "Werrason is a businessman and a politician. Lots of the women who danced on stage with him stayed in Europe. People paid him to be allowed to go along to Europe as a member of the band."[37] And Bralima just kept paying for dozens of tickets when Werrason flew to Paris with his "band." His colleague Papa Wemba was sentenced in Paris to a few months in prison for similar practices. Frontier running, the French court ruled.

COMPANIES ARE NEVER NEUTRAL PLAYERS, particularly not in defective nation-states. With a promotional budget many times that of the local ministries of education or information, they reach more people than the government does. Kinshasa today is infested with billboards for multinationals like Nestlé, DHL, Vodacom, and Coca-Cola. The concrete walls around factories, stadiums, and army barracks are daubed with commercial slogans. Television stations spew more publicity than programming. The Primus songs performed by Bralima's artists are seen on a number of channels all year round. They often last ten minutes or more. The dividing line between advertising and entertainment is fading. Kinshasa dances to promotional tracks.

The message is hammered in at other spots as well. Mobile-phone operator Tigo, a multinational active in sixteen countries and with its head offices in Luxembourg, was generous enough in 2006 to perk up the national airport's dilapidated arrival hall; all big companies have their charity programs (scholarships, hospitals, teaching materials: anything as long as it's visible). For the first time in decades the drab walls at Ndjili received a new coat of paint, but anyone coming off the plane into the hall today might think he has ended up in the Tigo stand at a trade fair rather than in a public building. The walls are festooned with dozens of the GSM operator's banners and plaques, and there is no other advertising. And in the midst of that whirlwind of glitter the traveler stands, passport in hand, cursing the sluggish state.

Concerns such as Bralima and Tigo do, of course, pay taxes, more than they would like, for in a corrupt country new taxes are invented each week. But if things seem to be getting out of hand, they threaten with the ultimate sanction: closure. And that would mean not only unemployment for all their personnel, who are paid very reasonably, and poverty for those small-scale sellers of beer or phone vouchers, but above all a stop to the fiscal revenues for all those officials. No hungry tax inspector relishes that thought. Multinationals are the country's biggest taxpayers. Governments therefore have a tendency to listen to them.

Back at the Berlin Conference of 1885, it was decided that the Congo Free State was to be open to international trade. Competition between market and state still exists today, in fact more than ever. In those days the focus was solely on the purchase of raw materials, today it's about the selling of products as well—even in a desperately poor country, there is a great deal of money to be made with the trade in little commodities like phone vouchers, bottles of soda pop, or bags of powdered milk. To win the souls of all those dispossessed, foreign companies colonize the public spaces of the destroyed country with a temerity only thinly disguised by the bright smile of slick marketing.

In October 2008, for the period of one week, I became a minor celebrity in Kinshasa almost without lifting a finger. Strangers came up to me on the street, saying they recognized me from my pictures and were surprised that, despite my status, I had no car of my own. Dolf van den Brink had called me a few days earlier. "We're organizing a concert with Werrason in the cité. Feel like going along?" The performance was to take place in Bumbu, one of Kinshasa's poorest neighborhoods. As we drove there in convoy, he explained things to me. "Bracongo has started playing dirty. They're running spots saying that Bumbu has 'fallen,' and that Primus is no longer market leader there. That's patently untrue, but we've been forced into the defensive. Now we're going to demonstrate the opposite with something big. Not a commercial, not a campaign, but a free Werrason concert! It's the first time he's ever played in Bumbu. I'm expecting a big crowd."[38] The air-conditioned SUV swerved around potholes. Dolf told me that Primus had gone through different phases in its marketing. At first the baseline had been *Pelisa ngwasuma*, freely translated as: get the groove started. The emphasis on ambiance went down well in a war-torn country. Then they changed the color of the label to match Congo's national colors: blue, yellow, and red. Now that the war was over, Primus had to manifest itself as the national beer, bar none. The state was rotten to the core, but national pride was apparently still intact. Bralima took skillful advantage of that. Meanwhile they had arrived at a new baseline: *Primus, Toujours leader*—Primus, Still the Leader—because the point was to make the newly won market leadership seem unshakeable. The desire for dominance was an important issue among the people, Dolf believed; they needed to know who is "the strongest." He was going to Bumbu to make that clear.

Interesting, I thought; GSM operator Vodacom hammered on precisely the same themes: national feeling and leadership. *Un réseau, une nation* had been their baseline in Congo for years: one network, one nation. Now they are presenting themselves as *Leader dans le Monde Cellulaire*, leader of the world of cell phones. Their Congolese website states that "our best is better than the best of all the rest. Losing is not an option. We are one team, and competition is our sport." Which company is the most Congolese? And who is the leader? Weren't those also the central themes in the electoral battle that was rolled out between Kabila and Bemba? In July 2006 the elections were ready to take place and the two favorites for the presidency were at each other's throats like pop stars. Bemba, still more warlord than states-man, accused Kabila of being a quasi-Rwandan without the necessary *con-golité*—a bizarre claim when one realizes that Bemba himself is one-quarter European. As president, Kabila tried to rise above the tumult by saying that "he who carries eggs doesn't bicker"—a statement that would haunt him for months. The reference was to the street children who went from bar to bar, balancing a box of hard-boiled eggs on their heads to sell as snacks. But after that comment, all of Kinshasa thought the president was a mean bas-tard. The brusque accusations back and forth resembled the rivalry between Werrason and Mpiana, or between Bralima and Bracongo. In the struggle for the country's highest office, the notion of leadership was linked directly to national identity. Commercial and political slogans were cross-pollinating back and forth.

As we pulled into Bumbu, Dolf peered out the window. The working-class neighborhood was dark, but the bars and sidewalk cafés were packed. Contentedly, he noted that about 80 percent of the bottles on the tables were Primus. A little farther along we saw Bracongo trucks parked along the streets: the competitor was bound to be passing out thousands of bottles of Skol during and after the concert. Dolf even wondered aloud whether Bracongo might not have hired a few youth gangs to stir things up. Bralima, in any case, had brought along its own security. And that was no unnecessary precaution; the young people of Bumbu—in fact the only generation present—had turned out in force. The closer we came to the concert grounds (the band was already playing, we could hear them from far away), the more young people began recklessly clinging to one of the cars at the back of the convoy. It was an SUV with tinted windows, painted in the Primus colors. They seemed convinced that Werrason was inside it. After we had been forced to a halt for a moment amid an ecstatic crowd, we were able to take a detour to the rear of the podium. The cars parked facing out: all the better to drive away quickly if things got out of hand. We climbed out and walked to the podium, shaking a few hands along the way. In the half-light backstage, with the basses thumping so hard you felt it in your midriff, I didn't recognize him right away. He looked much more normal than I remembered from the pictures, more timid too. "Monsieur Werrason," I said, *"bon concert."* "Mmm," he replied. And there it was, my shortest interview ever.

We climbed onto the podium. A row of dancers, behind them a row of musicians, all wearing Primus T-shirts. A wall of sound. I waved to Kakol, the drummer. Behind him, the back wall of the stage was covered by a huge banner: *Primus, Toujours Leader!* I shielded my eyes with my hand to look out at the audience. The podium had been set up at a broad intersection. In all three directions: hundreds of meters of people all crushed together. I tried to count one section, in order to extrapolate. Thirty thousand people? Forty thousand? Someone handed me a bottle of Primus. Cameramen filmed the two white men on the podium. And then, then the seemingly shy man with the goatee came up the steel steps to the left of the podium. Slowly, almost listlessly, he stepped up to the spotlights. He peered out into the restless darkness. Thousands of arms were raised, fists crossed at the wrists. *Igwe! Igwe!* was the deafening sound.

After the show, Dolf van den Brink was delighted. Not only had Werrason pulled little teenage girls up onto the stage to dance the *ndombolo* for him, but on two occasion, between songs, he had held a bottle of Primus aloft to tell the crowd that Bumbu was still in the hands of Bralima. Invaluable brand promotion. The show had cost ten thousand dollars. That was peanuts. The footage would be broadcast on TV nonstop in the next few days. Bralima paid thirty or forty thousand dollars a month to Antenne A, one of Kinshasa's biggest broadcasters, which used that in turn to pay its personnel. Bralima, in fact, owned the station.

"But I know you," a number of Kinois said to me a few days later when I sat down beside them in the backseat of a dilapidated taxi. "You're the white guy who was up on stage at the Bumbu concert. Don't you have a car?" It says something about Bralima's clout. In a city of eight million, where I happened to be staying, I was suddenly more famous than in the city of one million where I had been living for the last ten years.

I USUALLY BUY MY MOBILE PHONE VOUCHERS from Beko, on the shady Avenue des Batetela, one of the few truly pleasant streets in Kinshasa. Beko, who is in his early twenties and holds a degree in education, sits beneath a parasol from six in the morning until eight at night, selling prepaid cards for Tigo, Vodacom, CelTel, and CCT. Every day. On Sundays, however, only from eleven o'clock on, for he attends mass first. That is his only diversion. The tree-lined footpath along the Avenue des Batetela has a little street market. Sitting beside him is a female money changer, beside her an old woman who fries little fish that, for reasons I still don't understand, are referred to as Thomsons. A little farther along is a boy who sells pocket diaries, ballpoint pens and shoelaces, beside a young woman deep-frying beignets over a charcoal fire. A beignet is the only thing many people eat here in the course of a day. Tasty and filling.

On a good day Beko has a turnover of one hundred dollars, but less than eight dollars of that go to him. For every five-dollar voucher he sells, four dollars and sixty cents goes to the GSM operator, sometimes even as much as four seventy-five. "And it's only the good customers who buy a five-dollar recharge," he clarifies. All right, an eight-dollar profit, on the best of days. But Beko lives far away from the Avenue des Batetela, very far away. He is one of the 1.6 million people who commute to the city center each day in exhaust-belching, packed VW buses.[39] His ride costs him hours of his time and one dollar and fifty cents. If he wants to eat something during the day, even if it's only a chunk of manioc loaf with a little slice of fish, that easily costs him another dollar and a half. When he gets home he gives one dollar to the aunt he lives with, because his parents are dead. He is the sole bread-winner for his brothers and sisters. Of those eight dollars, he has already gone through more than half. And he is still not finished.

While we are talking, a loudmouth comes by and begins shouting at him and the other vendors. Without protest, Beko hands him two hundred Congolese francs. A little farther along is a man in a police uniform. "Offi-cially, we're not allowed to be here. He's supposed to give us a fine, but he never does. Instead, he sics that man on us. If we give him two hundred francs, he leaves us alone. The only thing is, he comes by three or four times a day. If we don't pay, he takes our wares. This way it only costs me a dollar or a dollar fifty."[40] Call it extortion or a form of ultradirect taxation, as long as the government doesn't pay the policeman's wages it won't stop. Which is not to say that a police uniform is no longer a highly valuable asset. It guar-antees its wearer a regular income, not from on high, but from the bottom up. No wonder that a trade has arisen in positions with the constabulary. Rumor has it that one can purchase a job with the police for a lot of money, the way one might take over a business.

Seven days a week, a little later on Sundays, the best years of Beko's life are going by. Tigo has introduced another new service, he sees. For a pittance customers can receive a daily text message that, the company claims, will "brighten up your day." Using Tigo Bible, you are sent a Bible verse each day; Tigo Foi provides religious counseling; Tigo Amour gives advice on your love life; Tigo Riche tells you how to make money. If you want cheering up, you can get it. The company offers no service with news flashes.

Beko laughs uneasily when I ask about his dream. "To become an ambassador," he says guardedly. Politics fascinates him. At the newsstand he *rents* a paper each day: for a few cents he is allowed to read it for half an hour. Buying one is out of the question; a newspaper costs one dollar and they are a rarity in Kinshasa. The few that do exist have a circulation of no more than fifteen hundred, microscopic in a city of eight million. Outside the capital there is no printed news to be had. And the contents of the papers that do exist are generally meager. *Le Potentiel* and *Le Soft* do their best, but the rest are dominated by gossip and partisanship. Journalists accept pay from the ministers they write about.[41] The layout is miserable, the quality of the printing depressing. But each day Beko hands his copy back neatly at the newsstand. Will his dream ever come true? He was twenty-two when I first met him in May 2007. "In Congo, people usually don't live past forty-five," he said with a wan smile, *"c'est comme ça."* In that same year, Tigo grossed a profit of $1.65 billion.[42]

BEKO IS AN EXCEPTION. More than half of all Kinois consider themselves poorly informed, women even more than men. The only ones with a sense of being up to date are the men older than fifty with a university diploma, the last generation to receive a decent education.[43] Yet there is no lack of media in Congo. Radio remains the most popular by far, television does particularly well in the cities, the Internet is bloodcurdlingly slow everywhere. At home, no one is on line. Surfing and drafting your résumé are things you do at Internet cafés, the so-called *cybers*—at least when the electricity hasn't gone out.

The national broadcaster has been breathing its last for as long as anyone can remember, but in 2002 the MONUC, in cooperation with the Swiss NGO Fondation Hirondelle, set up Radio Okapi, a station with national coverage and editorial desks in ten cities. For years it has been the only national medium in Congo. Foreign and local journalists there press on courageously, day after day. Okapi reporters are among the best (and the best paid) in their profession. The daily news broadcasts are extremely worthwhile, but the annual $10 million price tag makes one wonder what will happen in the long run. Who is going to pay for that, once the United Nations leaves?

Television is everywhere in the big cities. Congolese men watch more than two and half hours a day, the women often more than three.[44] During the 1 + 4 period, the medium experienced a remarkable boom. In February 2003 there were some twenty-five stations in Kinshasa alone; by July 2006, the month of the first round of presidential elections, there were thirty-seven.[45] The vast majority of those were local broadcasters. One can begin a television station in Congo for less than twenty-five thousand dollars. Any self-respecting politician, entrepreneur or clergyman has his own station these days. Zapping past those channels is an educational experience, but not necessarily by reason of their content. Tropicana, Mirador, and Raga are commercial stations showing mostly music clips, interspersed with commercials, insofar as there is any difference. Digital Congo is President Kabila's station, run by his twin sister and rivaled at that time by Canal

Congo and Vice President Bemba's Canal Kin. With the means at their disposal, Antenne A and the RTNC try to remain informative. Ratelki belongs to the Kimbanguists; Amen TV and Radio TV Puissance represent more recent Christian movements. More than half the channels belong to the Pentecostal churches.[46] Pausing at RTVA, it is good to know that the station belongs to Pastor Léonard Bahuti, the man who admonishes his (largely female) viewers to swear off jewelry, nail polish and hair attachments. RTAE belongs to "Général" Sony Kafuta "Rockman," the devout leader of l'Armée de l'Eternel. RTMV is in the hands of his archrival, "Archbishop" Fernando Kutino, founder of l'Armée de la Victoire, who has been in prison for years. All these religious broadcasters switch back and forth between sermons and soap operas. The dramatic installments deal with moral issues concerning life and survival in present-day Kinshasa (poverty, adultery, witchcraft, fertility, success) and emphasize that only charismatic Christianity can offer redemption amid the chaos of the day. In 2005 I was present when one such soap opera was recorded. What was striking was not so much the modest means (one camera, one lamp, one microphone) or the production-line approach (shooting today, editing tomorrow, broadcasting the next day), but the extreme youth of the actors. Young people in their twenties were doing their best to grant meaning to their lives and those of the viewers by means of fanatical religious discourse. The oddest station one comes across while zapping is NTV. There one watches as Pastor Denis Lessie, the owner, holds up his hands and invites you to place yours on the TV screen, touching his, because the Lord moves in ways that include optical fiber and airwaves. Hear the crackling of the Almighty, feel the hair rise on the back of your neck at the touch. Recently he asked his believers, by way of benediction, to sprinkle water on the picture tube or plasma screen.

I LEAFED THROUGH THE WELL-THUMBED GUESTBOOK of the little hotel in the interior. There hadn't been many foreign guests before me. In fact, only one: Andrew Snyder from Florida. His handwriting was clear and firm. Occupation? Pastor. Reason for visit? Crusade. *Ah, bon.* The American evangelists' crusades had apparently reached the provincial towns as well. It made me wonder how Fernando Kutino was getting along these days.

Kutino was a case unto himself. In Kinshasa in the early nineties he had seen the arrival of the first generation of American evangelists, a new kind of missionary who brought a charismatic variation of Christianity: Pentecostalism. Mobutu was so incensed over the power of the Catholics who had organized the March of Hope that he allowed other clerics to come and spread God's word. Divide and conquer; that went for souls too. Fernando Kutino, still an unremarkable boy at the time, heard about Jimmy Swaggart, the American TV evangelist who had achieved world fame in the West with his weepy confession of sexual infidelity. In Kinshasa Swaggart became known for his rousing services that brought many thousands into a state of ecstasy. But the German evangelist Reinhard Bonnke came to town as well, as did the Dutchman John Maasbach, married men in neat suits who bore witness to their faith with lively shows and impeccable coiffures. They had not been sent out by a central ecclesiastical authority but operated on their own initiative, often assisted by their families. These "reborn Christians" hooked up with the local prayer groups that gathered weekly to lift up their hearts unto the Lord outside the regular Sunday services. It did not take long before native men of the cloth arose as well and Fernando Kutino was a key figure among them.

Kutino put on a tie, called himself "Reverend" and delivered a message that ran quite counter to the traditional churches and rituals. It was the starting shot for the Congolese *églises du réveil*, the churches of the awakening, the revival, the new beginning. The curious were drawn in by the emphasis placed on charismatic worship, in which "healing" and "salvation" could be obtained during moments of intense religious rapture. With its rituals of trance that the believers experienced as the presence of the Holy Spirit, Pentecostalism was a variation of Christianity that closely matched the spiritual cosmos of African ancestor worship. Praying aloud, casting out demons, speaking in tongues: it reminded one of Simon Kimbangu's rise in 1921. Then too, fervent faith had been a remedy against witchcraft. Then too, people had begged for instant healing.

But Kutino added another layer, that of *la prospérité*. Redemption was not only spiritual, but also material by nature. During the bitter crisis years of the 1990s, this was the message people wanted to hear. What good did it do the poor, in spirit or otherwise, to be blessed when their children were dying of starvation? When your measly banknote turned out to be worth only half as much as it had when you got up that morning? No, not poverty but riches were the evidence of contact with the exalted. And to demonstrate his piety, Kutino decked himself out richly. A man of God, after all, could hardly appear in rags before his big boss? Seated on a bombastic throne he called on his followers each week to give gladly to his church. Ostentatious donorship became evidence of devotion and virtue. Kutino accepted the luxury automobiles and intergalactic GSMs with good grace. "I love money," he told a French journalist, "it helps you to live well."[47] Revolting? Yes, but no different from the forces in medieval Europe that had seen to the building of cathedrals while the members of their religious orders walked around in gold brocade and filigree. Postmaterialism is a luxury only the wealthy can afford. The pauper looks up to the pimp. Just as Papa Wemba had brought a spark of hope to youth culture with la Sape, so did Kutino introduce a notion of prosperity via the detour of faith. Kutino himself, with his gold watches and crocodile-leather shoes, was nothing but a sapeur. He embodied success, strength and welfare.[48] He was the Werrason of the liturgy. In December 2000 he brought a crowd of more than 100,000 believers to the heights of rapture in the Stade des Martyrs. His services were adorned with live pop music and offered plenty of opportunity for singing and dancing. "Sing, sing, sing, dance, dance, dance for the King of Kings," a religious pop artist told his audience, "because if you people don't do it here, it must be because you do it elsewhere, in the world of darkness."[49] Kinshasa had become the devil's city; only God granted mercy and Kutino was his treasurer.

During the 2002–2006 transition, the églises du réveil experienced enormous growth, particularly in the cities. Kutino's example inspired imitators everywhere. Under lean-tos, on city buses, and at busy intersections self-proclaimed pastors preached the word with verve. In Kinshasa one began finding stores that sold only lecterns, wooden or glass pulpits from behind which one could spread the good word. A new prophet arose every weekend. By 2005 there were an estimated three thousand Pentecostal churches in Kinshasa.[50] Most of them were quite modest, a few were massive. "Full Gospel" filled the stadiums for marathon sessions lasting three days or more. Preachers from Nigeria and the United States came by with fiery confessions of faith. The songs of praise and actions de grace (thanksgiving) came raining down. An ad on the front page of Le Potentiel promised a "festival of miraculous healing" in the huge Stade des Martyrs, with Reverend Dr. Jaerock Lee, a South Korean: "The dead are raised, the dumb speak, the blind see, and the deaf hear. All manner of incurable illnesses, including AIDS, cancer, and leukemia, can be healed. With tangible proofs that God is alive: you yourself can be where the miracles happen. Free admission."[51]

The churches tried to outdo each other with bellicose names like

l'Armée de l'Eternel, l'Armée de la Victoire, Combat Spirituel and la Cha-
pelle des Vainquers. It made one think of the warlike titles of the pop albums
and the struggle for leadership in commerce and politics. Normally believers were faithful to a single church, but now the turnover was large. There
arose something like serial monotheism. "If your God is dead, then try
mine," was the slogan of Pastor Kiziamina-Kibila, as though speaking of
detergent. Many people shopped among the various churches. After Joseph
Ratzinger was elected pope and adopted the name Benedict XVI, Koffi Olomide adopted a new stage name: Benoît XVI. When the Catholic Church
reprimanded him for that, he simply changed it to Benoît XVII.[52]

But this was more than mere rivalry. In fact, it was about the struggle
between good and evil, between Christ and Satan, between the true faith
and sorcery. The *églises du réveil* held a simple, binary worldview that helped
people place the contradictions in their lives within a framework. Adversity
could be blamed on evil spirits in a world of shadows; good fortune was a
gift from God. At l'Armée de l'Eternel, young women paid ten, twenty or
fifty dollars to have the preacher, Général Sony Kafuta "Rockman," perform
the laying on of hands and so help them to find a husband, become pregnant,
or get a visa for Europe. Wasn't that brazen moneygrubbing at the expense
of desperate people? "We want schools to be built too," that church's spokesman explained to me. "We feel that people need to work to earn money and
not just pray. We organize free AIDS tests and teach young parents how
to raise their children."[53] For a hardworking orphan like Beko, the church
provided a social safety net. Religion rushed in where government failed.
Some pastors were able to establish peace between rival youth gangs, something the police never tried.[54] They took "witch's children" off the street and
tried to "cure" them.[55] Like the multinationals, they filled the vacuum that
had arisen when the state withdrew. Desperate people found a cozy shelter
amid the warmth of fellow believers. Shops were rechristened La Grâce, Le
Christ, Le Tout-Puissant, cybercafés became "Jesus.com," exchange offices
"God Is My Bank." A new generation of first names even came into fashion:
children were now called Touvidi (from *Tout vient de Dieu*, everything comes
from God), Plamedi (*Plan Merveilleux de Dieu*, God's marvelous plan), Emoro
(*Éternel Mon Rocher*, the Eternal my rock), and the unlikely Merdi (from *Mer-
veille Divine*, divine marvel, which had to be explained to me as well).[56]

On November 2, 2008, I attended a Sunday morning service of Parole de
Dieu in Yolo-Sud, a poor neighborhood in the capital. More than a thousand

people were there in the courtyard, packed in close together beneath a zinc roof. They sang, they danced, they shook homemade rattles. It was then that I understood something of the success of these churches: the atmosphere was incredible. No collection was held. Anyone wishing to make a donation could do so at the church entrance. Wearing sneakers, the prophet Dominique Khonde Mpolo sat on the podium. Simplicity was his motto. Not every pastor is a money-grubber. During his extremely lengthy sermon he railed against "Jésus Business" and suggested it be replaced by "Jésus Verité." "All these other churches that promise people money . . . We have no need of luxury, we don't even eat meat. No one here wears a suit. We need to work for our country rather than for our own self-image." He himself specialized in resurrections. He claimed to have raised four people from the dead already. The first one had been the hardest, but now he had a magic potion. All you had to do was brush it onto the dead person's lips.[57]

Abbé José Mpundu, the Catholic workingman's priest who had helped to organize the March of Hope, considered it a disturbing development. "These new churches only rock people to sleep. They do nothing to liberate. They promise an easy kind of happiness in the form of 'miracles,' but they call no one to account. *Nzambi akosala*, the people say, God will take care of it. Let me be perfectly frank: those churches are a blessing for the regime. They make things easy for the politicians. That's why the regime supports them so generously. Sony Kafuta, the one who calls himself 'Rockman,' is quite close to Kabila and his mother; he is their spiritual leader."[58]

Sucking up to the powers that be, that was at least one thing of which Kutino could not be accused. In the course of his career, he had run up successively against Mobutu, Kabila *père* and Kabila *fils*. While the Kabilas had handed "Rockman" an appointment as head chaplain of the national army, Kutino had started *Sauvons le Congo*, the "Save Congo" campaign. Guests on his TV channel delivered straightforward diatribes against the 1 + 4. That made his organization one of the few critical voices heard from Pentecostal circles. Sharp criticism was leveled at what he called the *anti-valeurs* (the un-dignitaries). The tenor was extremely anti-Rwandan. Following insinuations that Joseph Kabila was letting himself be led by the Rwandan lobby—or, even worse, was himself a Rwandan Tutsi—the station was shut down and the "bishop" fled to Europe. He returned only in 2006.

But that was not the end. In May 2006, six weeks before the elections, Kutino—known by now as the Archbishop—landed at Kinshasa and held a huge rally in the Stade Tata Raphaël. He wore a scarlet bishop's robe and hung out of a jeep, waving to great crowds of supporters. He continued to rant against "foreign" influences and accused Kabila of a lack of *congolité*. Sowing doubt about the president's origins (his mother was not his real mother, he was Rwandan, etc.) became an approved tactic for the opposition. Not that there was proof of any such thing.[59] The service was broadcast without interruption on the channel belonging to Kabila's major rival, Jean-Pierre Bemba. As soon as it was over, Kutino was handcuffed and led away, and Bemba could go to visit him in Makala. One month later Kutino was sentenced to twenty years' hard labor, including ten years' probation. He was found guilty of the illegal possession of firearms, conspiracy, and attempted murder, but it was clearly a settling of accounts. International human rights organizations condemned the extremely shaky judicial process.[60] On the website of *Sauvons le Congo* is a dramatic film clip of the prophet's "last words," filmed on the closing day of the trial. In it Kutino speaks more hesitantly than ever. There is nothing left of his legendary *flux de bouche*. The footage is mixed with the bloodiest scenes from Mel Gibson's *The Passion of the Christ*. This prophet, too, was crucified: that is the message. But in the courtroom he still wears a neat, tailored suit and pocket handkerchief. A martyr in a bespoke suit, that perhaps is the entire ambivalence of *les églises du réveil*.

AND SO A COUNTRY that was no country at all dragged itself toward its first free elections in forty-one years, as agreed in the *Accord Global et Inclusif*. Companies and churches—*la bière et la prière*—had co-opted the public spaces, befogging and gladdening the people's minds. In the run-up to the proverbial "high day of democracy," set for July 30, 2006, after much dragging of feet, the population consisted more of consumers and obsequious believers than alert citizens. In colonial days, the super-alliance of church, state, and industry—the notorious colonial *trinitas*—had seen to it that the population remained servile and obedient. Something similar was going on now as well. The state, it is true, was much weaker, but still fond of snuggling up to the two remaining pillars. The "post-colonial trinity" consisted of a corrupt political caste that entered an alliance with newfangled religions and pop stars raised on high by the business world. President Kabila, who had not distinguished himself during the transition by any excessive amount of dash, made full use of these alternative power blocs.

As early as April 2002, during his concert at the Zénith, Werrason had called on the people to support Kabila because of his "efforts for peace."[61] An invaluable piece of promotion, for Kabila enjoyed little support in Kinshasa's working-class neighborhoods, where Bemba was *toujours leader*. At the signing of the Sun City peace agreement in 2003, Werrason, *ambassadeur de la paix* after all, gave a concert for the delegates.[62] In 2004, when Nkunda took Bukavu and the the people turned against the UN blue helmets, he was even called in to quiet things down. The pop artist made great by a multinational was now charged with keeping the masses in line.

On January 25, 2005, Kabila invited all the greats of Congolese music to the presidential palace for a glass of champagne. Werrason and J. B. Mpiana were there, as were Papa Wemba and Koffi Olomide and a few other archrivals. The president was able once again to play the great conciliator, the one who had brought peace not only to the eastern hill country but also to the bars of Kinshasa. A photograph taken at that party was seen around the world. It was an exact copy of the snapshot Jamais Kolonga had shown me, when he and Franco and Kabasele stood with raised glasses beside Mobutu. In Congo close ties had always existed between politics and music. Hadn't Kabasele gone along to the round-table conference in Brussels, when he composed his "Indépendance Cha-Cha"? Hadn't Franco been closely involved in Mobutu's policy of *authenticité*? Hadn't Papa Wemba sung along when Kabila's new currency was introduced? Yes, they had all done that.

But now things went a step further. In the 1990s it had become fashionable for private persons to pay artists to use their name in a song's lyrics. For a fistful of dollars, Mpiana, Werrason, and their colleagues were willing to do a little name-dropping. With Mpiana, the results looked something like this: "Love, love, what's that get us, *Ruphin Makengo?* / They start with love and that's where it ends, *Jean Ngendu*. / Is it just a matter of pride, or what, *Lidi Ebondja?*" With Werrason, it sounded like: "You should have told me before, *Hugues Kashala*. / You're wasting my time, all my friends are married, *Chibebi Kangala*. / Even my little sisters. / *Claudine Kinua*, she's mad."[63] The phenomenon was referred to as *kobwaka libanga*, tossing pebbles, to draw attention. It has since become a regular feature of Congolese pop music. The second half of the song, the *sebene*, is the instrumental part in which guitar solos move the dancers to a climax, swept along by the *animateur* who belts out a whole list of names. Politicians and prominent figures not only pay journalists for an article, but also pay pop stars for a mention. If you go out to *Le 144* on Louizalaan in Brussels, the chicest Congolese disco in town, you will even hear the DJ screaming to be heard above the music, telling the audience whose birthday it is and how many bottles of champagne they have ordered. In Kinshasa things occasionally got out of hand. "Treize ans" by Werrason contains no fewer than 110 names, "Lauréats" by Mpiana actually mentions two hundred.[64] This was no longer simply the paying of tribute; this was serial product placement. Artistic autonomy? Of no importance, on the contrary. The real deadbeat was the one who couldn't refer to the rich or powerful. That was proof of social isolation, and therefore deadly for an artist who wanted to be the leader. Werrason's opportunistic collaboration with Kabila and his entourage was so obvious, as was Mpiana's sympathy for Bemba by the way, that the Haute Autorité des Médias, the Congolese FCC, felt obliged in the weeks before the elections to ban broadcasts of their all-too-sectarian pop songs. Before that, however, they were aired nonstop. By that time, however, the popular singer Tabu Ley, a friend of Kabila father and son, had already been appointed vice governor of the city of Kinshasa, and Tshala Muana, one of the few female pop stars, had a hit with: "Vote, vote for Kabila / Everybody vote for Kabila / We're all going to vote for Kabila, our boss / He's the only right leader for Congo."

The Pentecostal churches, too, with the exception of Kutino's, hopped on the presidential bandwagon. "All authority comes from God," the believers heard on Sunday morning, "so pray for your leaders." And as if that weren't explicit enough, the prophet of the moment would gladly add: "Let everyone who loves Jesus and Kabila stand up and clap."[65] The army chaplain Sony Kafuta became so caught up in his Kabila mania, both at his temple and on TV, that the communications authorities had to reprimand him for hate mongering.[66] The Catholic Church watched it all from a distance and shook its head. This was a far cry from the critical role it had played in the struggle against Mobutu.[67]

IT WAS JULY 27, 2006, three days before the big day and Kinshasa was buzzing with electoral fever. That the elections were coming at all was due to the international pressure exerted by the CIAT, but above all to the brilliant work done by the Commission Électorale Indépendante (CEI), led by the inspirational priest Abbé Malu Malu. The preparations were truly impressive. Congo had become a country without an infrastructure. It was impossible to cross the country by car. Even the major urban centers were no longer connected. Congo was more an archipelago than a *pays-continent,* an archipelago whose islands could be reached only by plane, helicopter, or canoe. No one knew how many people lived there, no one kept track of the births, no one had an ID. The last forms of personal identification were the MPR membership cards from the Mobutu era. But on June 15, 2005, the CEI succeeded in registering twenty-five million voters, an overwhelming success. The outline of a new constitution was established by referendum on December 19, 2005. The new electoral law was ready on February 21, 2006. The campaign could begin. Tshisekedi, the opposition's historical leader, boycotted the process from the start and fell victim to his own obstinacy. Vice President Ruberwa, who was still seen as Rwanda's puppet, did not stand a ghost of a chance. After Operation Artémis in Bunia, the European Union launched a second military initiative: EUFOR, a 1,400-man European intervention force to keep the peace in Kinshasa; African elections, after all, tend to result more in rows than in democracy.

On July 27, Jean-Pierre Bemba, the man from Équateur, the fellow whose troops had practiced cannibalism, made his triumphal entry into Kinshasa. He was received with open arms: he was the *mwana ya mboka,* the country's son, the true Congolese. More than a million people accompanied him on the classic route from airport to city center, the same twenty-kilometer (twelve-mile) route taken, amid loud cheering, by Baudouin, Mobutu, Tshisekedi, and Werrason. Bemba was going to speak to his followers at the Stade Tata Raphaël, the arena bound up with so many historic moments in Congolese history, ranging from the riots in 1959 by way of the heavyweight bout of 1974 to Kutino's sermons in 2006. A group of drunken boys had brought a dog along with them, which they decked out with a campaign sweater bearing Kabila's likeness. Always good for a laugh. The animal was completely unnerved and barked at its own tail. Others carried a gigantic portrait of Mobutu, that other strongman from Équateur, for by now a generation had arisen that only knew about Mobutism by word of mouth. Even the old green MPR flag waved over the stadium. Bemba represented the promise of a restored state and powerful leadership. Like Mobutu, he could easily deliver a ninety-minute speech without resorting to notes. His sturdy frame and direct language made him much more popular in a Kinshasa gone wild than timid Kabila with his fractured Lingala and his French, which still bore a trace of an English accent. To many Congolese, Kabila seemed a youthful pawn of the international community (he was only thirty-four, Bemba was forty-three), not someone who could infuse the country with new pride.

And then something telling happened. After Bemba's rally, a frenzied crowd of young people moved through the city, attacking the major pillars of Kabila's campaign. The postcolonial trinity linking President Kabila with the evangelist Sony Kafuta and singer-cum-beer promoter Werrason were the targets. The young Bemba supporters wreaked havoc at the Haute Autorité des Médias, who they suspected of partisanship in favor of the incumbent president. Then they moved on to the temple of Kabila adept Sony Kafuta, a little further way. They wrecked the huge meeting space of his Armée de l'Eternel, leaving that "army of the everlasting" in a state that looked more like the smoking ruins of the present. From there they marched on a few hundred meters to Samba Playa, Werrason's rehearsal and concert hall. And that former place of pilgrimage for so many young, poor Kinois was also turned upside down in a flash by a furious crowd of those same young, poor Kinois. They felt betrayed by Werrason's transparent support for Kabila.[68] That month, Bralima lost 3 percent of its market share. The alliance between suds, sanctity, and the system might have been out to keep the people ignorant, but that did not mean the young voters were going to take it lying down. These were their elections.

CHAPTER 14

THE RECESS

Hope and Despair in a Newborn Democracy

2006–2010

T HE LIGHT WAS STILL FRAGILE AT SIX IN THE MORNING. PAS-
cal Rukengwa had to get used to the silence in his native village.
What a difference with Kinshasa! Bushumba was thirty-five kilometers
(about twenty-two miles) from Bukavu; this was his home region, this was
where he came from, and even though he had lived for years in the frenetic
capital, this was where he would vote. For the first time. Pascal was forty-
two. The last time there had been free elections in his country, he was one
year old. "I'm voting for life," he said, "to be able to live. This act is a new
beginning."[1]

He saw how busy it was already. At this early hour there were rows of
people waiting in front of the polling place. Some of them had spent the
night in front of the door.[2] This was not a Sunday like any other. *Mamans* had
put on their best *pagnes*. Some men wore ties and spit-shined shoes. Teenagers were showing off their mirrored sunglasses. Young women had had new
extensions braided into their hair. Patiently they stood in line, holding their
orange polling cards.

Pascal Rukengwa had no time for pride or sentimentality, but in a way
this was his day too. This was what he had been pushing for for years. He
was one of the twenty-one members of the CEI, the national electoral committee that had organized the enormously complex voting process. "All hope
was focused on us, but we had to learn everything as we went along. Sometimes I felt like a stranger in the jungle, where an animal could come along

and tear you apart [at] any moment. Wasn't all this hope out of proportion
to our capacities? At some places people had never even seen a computer."
The United States and the European Union provided massive financial support. At almost half a billion dollars, the largest part of it from Europe, these
were to be the biggest and most expensive elections ever organized by the
international community.[3]

Pascal looked around. Around the country, fifty thousand polls were
opening their doors at this very same moment. Forty thousand observers
from home and abroad were keeping an eye on things.[4] During the last few
months, a quarter of a million polling officials had crossed the country, providing the people with information.[5] The ballot boxes had been taken by
helicopter, truck, and motorbike to the remotest corners of Congo. At some
spots in the jungle, they even went by canoe or were carried by porters.

But today it was happening. Sixteen million people descended on the
polling places, refugees even left their plastic huts. Pascal's background
was in South Kivu's *société civile,* the congeries of nongovernmental organizations. "Free elections, that was the most urgent desire of the Sovereign National Conference. For the people it was transformed into a magic
moment, but for me it was a day full of stress. A pregnant woman waiting
in line fainted, and the closest hospital was ten kilometers [about six miles]
away. A child became unwell and then died. I kept driving back and forth.
That day I didn't have a single moment to myself. But honestly, I had no idea
that the people attached so much importance to choosing their leaders."

Despite a few minor incidents, the voting took place with great dignity.
In the polling places—often nothing more than a larger hut—the voters
were handed the necessary forms. The ballot card for the presidential election listed thirty-two names. Joseph Kabila was among them, of course,
alongside Jean-Pierre Bemba and Azarias Ruberwa, the rebel leaders who
had been made vice presidents. Antoine Gizenga was on it as well, the man
who had once served as deputy prime minister under Patrice Lumumba.
And Nzanga Mobutu, the son of. In addition one had Pierre Pay Pay, the
former director of the national bank and Oscar Kashala, a physician who
had come back to Congo from America. The ballot card for the parliament
was a good deal more complicated. There were ten thousand candidates for
five hundred seats in the house, divided over more than two hundred and
fifty political parties. The form consisted of six large sheets of paper with a
passport photo of each candidate beside the name; one-third of the country,

after all, was illiterate. Little old ladies asked officials to cross off "Monsieur Sept" (Mr. Seven) for them. That was Kabila, whose Parti du Peuple pour la Reconstruction et la Démocratie (PPRD) had been placed seventh on the ballot list.

When the polls closed, the counting began. Wherever possible that took place on the spot, to avoid ballot-box fraud, although that meant things were not always simple. "We had no electricity," Pascal Rukengwa said, "and the flashlights they had given us didn't work. There was no money to buy candles, but the people went off looking for them. We handled things ourselves. At some of the polling places the people actually slept beside the ballot boxes, to make sure nothing went wrong."

The image of brave citizens counting votes by candlelight in a hut, often after having eaten nothing all day, is intensely moving. The image of fatigued men and women sleeping with their arms around a sealed ballot box, as though it were a shrine or a child, can leave no one unmoved. The elections' biggest winner was the common Congolese man and woman.[6] Before sunup the next day, many of the counts had already been phoned or texted through to the computing centers. The miracle had taken place.

Pascal Rukengwa flew back to Kinshasa. On August 20, 2006, three weeks after the elections, the definitive results were announced. None of the many thousands of observers had reported any large-scale fraud, and the surprising outcome seemed to confirm that: no one had an absolute majority. Kabila took 45 percent of the votes, Bemba 20 percent. Coming in third with 13 percent was old Gizenga, a man who had not campaigned but relied on his historical aura. Pascal: "The electoral returns led to enormous frustration: Bemba knew he hadn't won and Kabila realized that he was not a shoo-in. There was a lot of shooting in the city. The Bemba supporters focused all their rage on Kabila and on us. They suspected the CEI of partisanship, while we were actually amazed that Bemba had raked in so many votes! We had to meet in a basement. I wasn't sure I would be alive at the end of the day. The Mission de l'Organisation des Nations Unies au Congo (MONUC) tanks took us to the public broadcasting station to announce the official results. I sat on the floor, between the soldiers' legs. It was an old tank, it had trouble starting. They actually make a sound like a huge diesel generator, did you know that?"

The election results showed that a striking fault line ran through the electorate. Kabila had won in the east of the country. In provinces such as

North Kivu, South Kivu, Maniema and Katanga, he achieved Stalinist scores of more than 90 percent (of up to 98.3 percent in Maniema and Katanga). Hardly remarkable, though, when one realizes that he himself came from the east and was seen there as *l'artisan de la paix* (the crafter of peace), the man who had stopped the war. Bemba triumphed in those western provinces not directly affected by the war (Bas-Congo, Kinshasa, Bandundu) and his own Équateur. The fault line corresponded roughly with the border between Lingala-speaking and Swahili-speaking Congo. For a time, people feared for a macroethnic conflict.

The day after the results were announced, Kabila's guards fired on Bemba's residence in Kinshasa, by their own account after being provoked by his bodyguards. What they didn't know was that Bemba was at home at that moment, in a meeting with virtually all the major Comité International d'Accompagnement à la Transition (CIAT) ambassadors. The shooting went on for hours; Bemba's private helicopter was destroyed. The skirmish came to an end after an intervention by the MONUC and EUFOR, the European peacekeeping force.

Calm was restored, however, and the second round of the presidential elections was held in a largely peaceful atmosphere on October 29. As is usual in such second rounds, the number-one candidate made a deal with number three. Kabila promised to appoint Gizenga prime minister in exchange for his following. He also won the support of number four, Nzanga Mobutu, who was given the post of minister of agriculture. Mobutu Jr.'s conversion to the presidential fold was portentous; he came from Équateur, which was Bemba's province as well. Kabila's cartel, the Alliance pour la Majorité Présidentielle (AMP), now housed both his own PPRD and Gizenga and Mobutu's parties. Truth *can* be stranger than fiction: that the son of *mzee* Kabila was now cutting deals with the son of Marshal Mobutu probably caused more than one ancestor to turn over in his grave. It was as though the children of Churchill and Hitler had banded together to form a political party.

Kabila took 58 percent of the votes, Bemba 42. On December 6, 2006, two days past his thirty-fifth birthday and newly married, Kabila was inaugurated as the Congo's first democratically elected president since Joseph Kasavubu. With that, the Third Republic finally became a fact. Mobutu had heralded the end of the Second Republic in April 1990, but the transition to a new political system had taken more than sixteen years, sixteen years of hunger, poverty, war, and death, sixteen years of despair and hopelessness.

WERE THINGS REALLY GOING TO CHANGE? From day one, many in Kinshasa were skeptical. Kabila considered himself the candidate of the Western world. Although the elections had proceeded correctly by all accounts, the people of Kinshasa, the Kinois, remembered all too well how Louis Michel, then European commissioner of development cooperation and a former Belgian minister of foreign affairs who was extremely active in Central Africa, how this "big Loulou" with his cigar and his pats on the back and his guffaws, how this man, who for many Congolese had been the very epitome of the always-elusive "international community," had said on TV in a less guarded moment that Kabila represented "the hope for Congo."

Abbé José Mpundu, the highly acute cleric who had helped organized the March of Hope, was quite scornful. "From 1990 to 1995 I fought for elections that would not be like the charade we got this time. It was a parody, orchestrated by the international politico-financial Mafia! I wanted to vote for Tshisekedi, but he had relegated himself to the sidelines, so I just voted for Bemba. They let us play a bit part. It was one big, worthless Mafia gambit. For a lot of money, the international community bought itself the president it preferred; we would have been better off passing the hat around to finance the elections and building our own ballot boxes. At least then they would have been our own."

Abbé Mpundu's extremely critical comments were not all the exception in the capital. Electoral commissioner Pascal Rukengwa came from the east, where the people had voted en masse for Kabila. On December 6 he was present at the inauguration, but he was not impressed by what he saw. Of course, there were lots of important guests, lots of heads of state. And yes, Tshala Muana had sung beautifully. But everything seemed to be organized so amateurishly. "There weren't enough seats. People had to stand outside in the sun for hours. I was invited to the dinner, but it was a complete chaos. The room was full of people who hadn't been invited, so I couldn't get in. It wasn't organized very well, it wasn't very professional." Those were only details, of course, but Pascal found the day's substance fairly dubious too. Western observers were pleased with the president's speech. Hadn't he spoken powerfully about "the five building sites," *les cinq chantiers,* of national reconstruction? Wasn't that a reference to infrastructure, water and electricity, education, employment, and health care? And hadn't he literally said that "the recess was now over"?

Pascal wasn't so sure: "I had no faith in it. That story about the five building sites, I thought that was rather childish. If a government doesn't address those essential tasks, then what is it there for? He wasn't on the campaign trail anymore, was he? As far as I was concerned, the recess was still going on. It was the same hesitant, immobile man. But looking back on it now, I can only conclude that I was being too positive about the whole thing."[7]

How to describe Congo on the eve of the Third Republic? Statistics, percentages, and figures are not enough. The world reveals itself in crumbs and grains of sand. How to describe this vast area?

By saying that it was a fertile country where many ate only once every two days? That countless of its people suffered from hemorrhoids due to an imbalanced diet of manioc? That people who had no money to buy hemorrhoid salve, if it was there to be bought at all, simply treated themselves with cheap, imported toothpaste? Yes, good friends of mine have told me that. Cuts were treated with brake fluid, burns with vaginal fluid. They shined their shoes with a free condom; the lubricant made the leather glisten. Women who wanted fatter buttocks, they said, inserted a bouillon cube in their vagina. Others gave themselves enemas of beef extract.

How to describe a country? A country that was not a state but still had more than half a million civil servants, older men and women who didn't enter retirement because that didn't exist and so just kept going to the office, amid file cabinets brimming over with moldy, termite-ridden files, hoping for a smattering of wages and dreaming of a bit of good governance.[8] In a patient hand they filled out endless forms and treated the civil service hierarchy with great awe; just because a state exists only on paper does not mean it is unreal, on the contrary. In Bunia, a letter had to make its way across seventeen different desks before receiving a reply.[9] In Boma I met a city librarian without a library.

How to describe a country? Through the jungle of Équateur there walked a man, leading a pig. He was on his way to his village along the Congo River. When he got there he would wait until a boat came by, which happened about once a month. If such a boat came by—they were more like floating villages with a market square, courthouse, and menagerie on board—he would paddle out to it to sell his hefty pink yearling to the crew or to the passengers who hung shouting over the railing. But the river was still far away, two hundred and fifty kilometers (155 miles) from here. He walked through the jungle in solitude, for three whole weeks, sometimes toting his pig, sometimes letting it walk at the end of its rope. At night he slept beside the animal. The river was far away, so terribly far away. And all he had on his feet were flip-flops.[10]

THE TASK KABILA SET OUT TO PERFORM was anything but simple. Manfully, he dictated: "There will be punctuality, and discipline. I will take up matters again with determination and regain 100 percent control of the situation."[11] The new constitution, in any event, provided a well-thought-out system of checks and balances. Congo was neither presidential nor parliamentarian, but something in between (the head of state did appoint the prime minister, but parliament could take legal steps against them both in the event of high treason). Congo was neither unitarian nor federalist, but something in between (the provinces were smaller but received more powers and funding). Congo had a new parliament and senate (the former chosen directly, the latter through the provincial councils). And a constitutional court was set up, with extensive powers to settle conflicts between the prime minister and the president. This complex construction was intended to keep too much power from collecting in the hands of one and the same branch of government.

For the parliament and cabinet, there was little danger of that. The house of representatives was highly fragmented: its five hundred members represented no fewer than seventy parties, plus an additional sixty-four one-man parties. The two largest ones, Kabila's and Bemba's, held only 175 seats, but even they were internally fragmented. The cabinet was an obese monstrosity with some sixty ministers, not because there was so much to arrange, but because there were so many to mollify. (Later the cabinet team would shrink to forty-five posts, still twice as many as in Lumumba's cabinet in 1960.) At first, eighty-one-year-old Prime Minister Gizenga received great acclaim, but soon his status proved more antiquarian than alive. One of his ministers was dubbed with the unusual title of *minister près le Premier ministre*. A minister in proximity to the prime minister? In actual practice, the good man was charged with keeping the prime minister awake during meetings.

In late January 2007, after less than two months, a clear indication was seen of the new political culture. The provincial councils had to elect their provincial governors, and the results, to put it mildly, ran contrary to expectations. The PPRD, Kabila's party, won eight of the nine provinces, even in places where it had not made a dent in the parliamentary elections—Équateur was the only province with a governor owing allegiance to Bemba. The country had been sprinkled generously with bribes: afterward, candidates who failed to be elected even went so far as to publicly demand a refund on their boodle.[12] Provincial council members eventually admitted to taking bribes. In Bas-Congo this fraud created so much bad blood that rioting broke out. Few inhabitants felt like having one of Kabila's flunkies at the head of their glorious province. Bundu-dia-Kongo, a religious-political movement of ethnic bent that had championed the rights of the Bakongo even in Mobutu's day, called for public protest. The movement dreamed of restoring to its former glory the historic Kongo Empire, which had once reached from Angola to Congo-Brazzaville. During demonstrations at Moanda, Boma, and Matadi, there was massive rioting: ten policemen were killed and the army opened fire on the demonstrators. The final toll: 134 fatalities.

In March 2007 Kabila once again chose the path of violence. As vice president during the 1 + 4 regime, Bemba had had a right to a private militia. Now that he was a mere senator, however, he refused to disband his corps of guards. There was no way, of course, that he could continue to avail himself of a little army of five hundred freebooters. But after the attack on his house in August, he—not without reason—feared for his safety. With his Garde Républicaine, after all, Kabila still had a private army of no less than fifteen thousand troops, an elite corps he had assembled during the transitional period. On March 21, on Boulevard du 30 Juin, the city's busiest arterial, Kabila's men opened fire. For three days Kinshasa was in a state of paralysis. Office buildings and embassies had been hit by mortars. Bodies lay scattered about on rotundas. A storage tank for fuel exploded in flames. More than three hundred, perhaps as many as five hundred, were killed. Afterward the presidential services arrested and tortured an additional 125 persons, most of them from Équateur. Dozens were murdered.[13] Bemba himself, despite the international warrant out for his arrest, fled to Portugal. He figured that his senatorial immunity would protect him. But in May 2008 he was arrested in Brussels and handed over to the International Criminal Court in The Hague, where at the time of this writing he is still awaiting trial.

"There shall be discipline," Kabila had said. His violent actions in August, January, and March, however, did not bode well. It reminded many of the way Mobutu, immediately after his coup, had four cabinet ministers hanged to underscore his authority. Kabila's Garde République also reminded people of the former DSP and his intelligence services of those maintained by Mobutu. But was there any substance to that? Perhaps the truth was all much more tragic, much more banal. In all three cases the incidents had followed upon skirmishes that got out of hand and ended inadvertently in a bloodbath. Kabila could hardly admit it of course, but what these incidents were, in fact, was proof that he had no control over his troops, not even his own elite troops, rather than any demonstration of active malice on his part. Mobutu had wanted to show that he was a powerful figure with powerful principles; Kabila had to hide the fact that he was a weak figure surrounded by feeble institutions.

But there was no hiding: soon enough, Kinshasa was buzzing with rumors that Kabila was hooked on cocaine—no, that he spent the whole day playing with his Nintendo—no, that he had been shot and wounded and that was why he showed his face so rarely. People went looking for the wildest explanations for the inefficacy they saw. "Après les élections = avant les elections" (after the elections = before the elections) they mumbled, a sarcastic nod to the situation at independence in 1960. His popularity plummeted, even in the east of the country. Only rarely did one see him smile; rarely did he appear in public. Only now and then did he appear on television: seated at his desk like a stony sphinx, he would read aloud an announcement.

But still, at the start of the Third Republic, a new élan could be noted here and there. The huge, unwieldy parliament voted on fifteen laws during the first ten months of its existence, called sixteen ministers to account, set up eight investigative committees, and discussed a budget. An investigation was started into reported corruption and unlawful mining contracts.[14] The new spirit became even more tangible in Lubumbashi, where the public spaces were given an impressive face-lift. The holes in the asphalt were repaired, playgrounds and school were renovated, sixteen hundred garbage containers were placed around the city, and a public sanitation service was set up.[15] When I arrived there in June 2007, workers were busy checking and repairing streetlights and pruning trees along the city center's long, straight streets.

This was, however, invariably the work of a few energetic individuals.

That parliament seemed to be getting down to brass tacks was thanks to its dynamic speaker, Vital Kamerhe, one of the president's confidants who understood the art of concisely paraphrasing interminable debates and leading the way to a decision. Katanga once again showed initiative thanks to Moïse Katumbi, a flamboyant businessman who combined cunning with populism and remained unconditionally loyal to *le grand chef* in Kinshasa. Kabila needed such dynamic figures to convince the people that his five workplaces were coming along well, but at the same time he made sure their popularity did not exceed his own. New elections, after all, were to be held in 2011. When the widely popular parliamentary speaker Kamerhe openly criticized Kabila's military operations in the east of the country in January 2009, he was forced to resign and the regime lost one of its more intelligent players. Since that time, the Katanagan governor Katumbi has—by his standards—been strikingly quiet. His voluntaristic approach had initially served to illustrate the disadvantages of highly personalized government. In June 2007 I saw that the general hospital at Lubumbashi had recently installed two brand-new coolers in its morgue and acquired a truck for picking up corpses. *Don de Moïse* (donated by Moïse) was painted in huge block letters on the sides of both. Generous, to be sure. But the hospital itself, the country's second largest, had not had a drop of running water for the last four years.[16] Patients going to the toilet had to first wade through a four-centimeter (1.6-inch) thick layer of dreck and piss. I saw it with my own eyes.

The elections cost a ghastly amount of money and generated great expectations, but before long people began concluding that the results had been quite meager. In keeping with age-old custom, the members of parliament bumped up their own monthly salaries—to forty-five hundred dollars in 2007 and then to six thousand dollars in 2008—and bestowed upon themselves and their secretary chairperson shiny new Nissan Patrols: it was one of the few agenda points that met with a little dissent.[17] "I don't get it," a Kinois told me once. "During the campaign all those candidates looked us straight in the eye, but the first thing they do after being elected is to have themselves driven around in a four-wheel-drive with tinted windows so they don't have to see us anymore." As a result, important dossiers like military reform, governmental decentralization, and legal reforms remained untouched, with all the consequences one might imagine.

At the hospital in Lubumbashi I was introduced to Luc, a handsome young man. He was confined to a wheelchair. Nine months earlier he had been arrested one night while trying to steal a roll of electrical wire. In the absence of formal jurisdiction, Congo is marked everywhere by kangaroo courts. The crowd avenged itself on Luc by dousing his hands and feet with gasoline. He watched himself burn. His left foot, his right foot, his left hand. A few months later he went to the toilet and saw his right hand fall off. All he has left now is a thumb. He cannot operate his own wheelchair. And the judicial system is still a farce.

With the cabinet, things went no differently. In October 2008 Kabila replaced soporific Prime Minister Gizenga with Adolphe Muzito, who had previously been his minister of the treasury: a well-behaved, harmless man who has in the meantime done nothing in particular to distinguish himself, beyond accumulating some suspicions of corruption. With a few celebrated exceptions, most of the cabinet ministers proved equally indisposed to action. And why should they act? To display initiative was to risk falling from grace and losing a lucrative post (as happened in late February 2010, when Kabila once again reshuffled his cabinet team and invited twenty new notables to join him at the banquet table). Besides, real policy was established elsewhere, within the president's own closest circles. In the Third Republic, true power did not rest with the country's democratic institutions, but with a handful of the president's most trusted advisers, including his mother and his twin sister. The real policymakers are often those like Augustin Katumba Mwanke, who owes more to his years of loyal service to Kabila than to any charisma or competence. Concerning military affairs, for example, the most powerful man since 2009 has been John Numbi. He is not the defense minister or chief of staff of the national army, but inspector general of police and long one of the president's favorites. He has had almost no military training.

Bright spots? Yes, a few. Until the international banking crisis in 2008, Congo's currency had remained relatively stable: one dollar equaled some five hundred Congolese francs at that time, only to fall afterward to nine hundred to a dollar. Year after year the budget grew, but by 2010 it still added up to only $4.9 billion, comparable to the annual means of a medium-sized European city or half that of New York's Columbia University in the course of a single academic year. Hardly enough to finance the reconstruction of a gigantic country in ruins. What's more, half that sum is coughed up by international donors: a quarter of it goes to repaying the nation's debts. GNP rose by a few percentage points annually, due largely to mining activities, but that too is marked by total dependence on foreign capital.[18] In 2009 per capita GNP was $200, clearly higher than the $80 seen in 2000, but still a far cry from the $450 of 1960. To arrive at a par with the current level of neighboring Congo-Brazzaville ($4,250 annually, thanks to oil revenues), the population will have to wait until 2040, an internal document from the prime minister's office said in February 2010. And that will only be achieved if one can assume consistent real growth of 13 percent annually and an unchanged population growth of 3 percent.[19]

In macroeconomic terms, therefore, slight progress is being seen, but such trends say nothing about the life of the common people. The Human Development Index, calculated by the United Nations each year for every country in the world, provides a much better view of citizens' welfare than does per capita GNP; the index takes into account such things as the degree of literacy, education, health care, and life expectancy. In 2006 Congo found itself on the tenth lowest spot; in 2009 only five countries had a lower index score. Not a particularly propitious trend.[20]

Each year *Foreign Policy* magazine, along with the Fund for Peace, publishes the Failed States Index, a list of the world's sixty most defective states. In 2009 Congo was number five, worse than Iraq, and two places up from 2007.[21] After a period of slight improvement, Congo seems once again about to descend into chaos and mismanagement. The Doing Business Index for 2010 places Congo in 182nd place in a field of 183 countries: only the Central African Republic scored worse. Anyone hoping to start a business in Congo must count on reserving 149 workdays for administrative purposes. Obtaining a building permit easily takes 322 days. On an average, one pays taxes thirty times a year. The tax on profits equals almost 60 percent—money that never ends up with the common Congolese citizen.[22]

What that common Congolese citizen does end up with is disease. The infant mortality rate is one of the world's highest: 161 out of every 1,000 children do not live to the age of five. One out of every three children under the age of five is underweight. Life expectancy at birth is forty-six years. Almost 30 percent of the population is illiterate, 50 percent of the children do not attend elementary school, 54 percent of the population has no access to clean drinking water.[23]

SO WHY DON'T THE PEOPLE RISE UP? Within an eighteen-month period, a government investigation revealed in 2007, some $1.3 billion disappeared into the pockets of three national financial institutions and six state-owned businesses.²⁴ A dizzying sum, yet it produced no public outcry. Of the sixty mining contracts with international concerns such as Anvil Mining, De Beers, BHP Billington, AngloGold Kilo, and Tenke Fungureme Mining scrutinized by parliament under Kamerhe's leadership, not one was shown to be sound.²⁵ State-owned Gécamines in 2008 contributed only $92 million to the state treasury, rather than the $450 million the government was owed.²⁶ The diamond mines of Bakwanga and the gold mines of Kilo-Moto provided almost no revenues. But public outrage? Belligerence? Fury? Given, there were a few incidental strikes by civil servants and teachers, but the common Congolese makes the best of a bad job and is almost ashamed of the hope he once cherished in the run-up to the elections. "Ca va un peu," it goes, it goes, he will answer when you ask how things are going.

In November 2008 I talked about this with Alesh, a twenty-three-year-old rapper from Kisangani and one of the most promising figures in Congolese hiphop. Rap is a relatively new genre in Congo, but for Alesh it is a way to break through the lethargy. In his song "Bana Kin" he points an accusing finger at the deadening music scene in Kinshasa: "Your music is rich and shows us the tradition / but ethically speaking offers no contradiction." Figures like Werrason and J. B. Mpiana had not roused the country from its slumber, even if their commercial ditties may have had some artistic value. Alesh viewed religion in equally nuanced terms: "I've got nothing against praying / But for them it was a mosquito net / That kept them tangled, deep in debt / like a spider's web." To talk to Alesh was to talk to a new, self-aware generation freed of colonial or postcolonial inferiority complexes. "We have to dare to criticize ourselves; too many dreams die because of a lack of hope," he told me. In 2008 he recorded "L'élu," a merciless song in which he reminds the elected representatives of their promises: "You add to the dissension, with all your condescension / you have all these pretentions, but the people long for your detention."²⁷

WERE THE ELECTIONS, THEN, nothing but a show, after all? A difficult question. For millions of citizens they were of undeniably great symbolic importance. The zeal with which people voted and counted showed that this was more than a pipe dream on the part of the international community. But the elections were more meaningful before and during the actual polling than they were afterward. The ritual was at least as important as the result. It was, after all, an illusion to hope that proper elections would immediately lead to a proper democracy. The West has been experimenting with forms of democratic administration for the last two and a half millennia, but it has been less than a century since it has started putting its faith in universal suffrage through free elections. How then could the West expect that particular method to magically transform a deep-rooted culture of corruption and clientelism into a democratic constitutional state in accordance with the Scandinavian model? And then in a region that, during its precolonial, colonial, and postcolonial eras, had known almost nothing but forms of autocratic rule? How naive would one have to be to suppose that it would all land on its feet after that initial electoral impulse? Democracy must be the objective—it is, after all, the least bad of all forms of government—but in Congo very little emphasis was laid on the vitally necessary steps along the road to a democratic system, or on the pace at which those steps were to be taken. In 1955 Jef Van Bilsen had predicted that thirty years would be needed for the switch from a colony to a sovereign state, but today the situation is in many ways worse than it was then. Free elections should not be the kickoff to a process of national democratization, but the crowning glory to that process—or at least one of the final steps. Peace, security, and education should go before, as well as local elections that can stimulate the formation of a grassroots culture of political accountability. In principle, the local elections should come first, but Kabila disdained and ignored them.

Western political experts often suffer from electoral fundamentalism, in the same way macroeconomists from the IMF and the World Bank not so long ago suffered collectively from market fundamentalism: they believe that meeting the formal requirements of a system is enough to let a thousand flowers bloom in even the most barren desert. Nobel Prize laureate Joseph Stiglitz, however, has clearly showed that "sequencing" and "pacing" are essential to the introduction of a market economy.[28] One does not start cultivating the desert by first sowing the best of seed. The same goes for introducing a democracy.

With its enthusiastic attempt to install democracy once and for all in Congo by means of the formal electoral procedure, the international community has above all sidelined itself. Democracy was the aim, obscurantism the result. For, as the democratically elected president of a once-again sovereign country, Kabila was not about to tolerate any more foreign busybodies—after four years of patronization by the CIAT, he had had enough of that. To put it cynically: America and Europe paid enormous sums to gag themselves diplomatically in Congo. One can, of course, flour-ish promises of loans and attach to them preconditions for good gover-nance (the new buzzword at the IMF and World Bank in particular, but the European Union is all-too-willing to hop on the same bandwagon), but why would an African head of state respond to such advances when China offers much more money and is far less cantankerous about what's done with it?

Some political scientists claim, by way of hypothesis, that three or four elections are needed to make things go in the right direction. That one must not despair too soon. That it is normal for a country to sputter a bit after a cold start. Repeated elections can, indeed, generate a pattern of change in which responsibility is taken; leaders may ultimately feel called upon to consider governing well. But it can just as easily become a hol-low ritual and one that provides autocratic regimes with a thin veneer of legitimacy. It is much too soon to decide whether such elections actually promote democracy in Congo. It should be noted, however, that in Sep-tember 2009 Kabila—with an eye to the elections in 2011 and 2016—set up a commission to determine whether the term of presidential office should not be extended from five years to seven and whether the constitutional limit of two mandates should not be scrapped, making him permanently eligible for reelection.[29] It should also be noted that in 2009 a number of human rights activists were arrested for their critical stances.[30] One of the president's bosom buddies (who didn't know that I knew he was a bosom buddy) once told me casually during a lunch shortly before the elections: "As president, Mandela was much too Western; Mugabe and Mobutu, those were real African leaders."

LATE NOVEMBER 2008. I was having a meal with two brothers, both young actor-directors, at a little Indian restaurant across the street from the MONUC base in Goma. We were sitting outside under the awning, waiting patiently for our food, when my phone rang. Tomorrow's trip would have to be canceled, I heard, the driver had run into problems, his battery was dead or his tank was empty, no, no, it was all very compli-cated, there was no way I could help, he was very sorry and wished me a good evening.

"Ça va?" Sekombi, the older brother, asked when I snapped down the cover of my cell phone.

"No," I said, "I was all set to meet with Nkunda tomorrow and now I hear that it's not going to happen."

I had arranged for a jeep, a driver, fuel, and a guide familiar with the rebel territory. That morning I had purchased my press accreditation at the Ministry of Communication and Media for a measly $250—the most expensive sheet of paper I'd ever bought—I'd had passport photos taken, I had gone by the State Security offices. I had told the MONUC officer in charge about my plans. And, most importantly: I had called the number-two man at Laurent Nkunda's civil staff. It had not been easy to reach him in rebel territory, where there was almost no cell phone coverage, but the appointment had been made: tomorrow morning at nine he would meet me at the old mission post.

"You want us to drive you?" Sekombi interrupted my lament.

Sekombi and Katya, his younger and more taciturn brother, were solid folk. To run a cultural center for young artists in bullet-riddled, lava-ridden Goma, one had to be made of stern stuff. Their eldest brother, Petna, had set up the center. One month earlier, with rebel leader Nkunda at the city gates and Kabila's Forces Armées de la République Démocratique du Congo (FARDC) looting the town, the Katondolo brothers' cultural center had gone on imperturbably with its idiosyncratic film festival. But to venture into the theater of war with two actors? In that old, beat-up jeep of theirs?

"But do you two have the papers you need to get through?"

To find Nkunda we would have to pass three roadblocks manned by the FARDC, a few kilometers of no-man's land, and then three roadblocks guarded by Nkunda's Congrès National pour la Défense du Peuple (CNDP). The rebel barriers would be no problem, I had been assured. Nkunda had his troops under control. But the national army roadblocks could prove to be a nightmare. Passports and press credentials could not always stand up to their frustration.

"No," Sekombi said, "but we've got our hair."

Excuse me? I almost choked on my *poulet tikka masala,* which had finally appeared on the table after a two-hour wait. I looked at their wispy hairdos. With plenty of goodwill, one could see them as the start of something like dreadlocks.

"We're Rasta's. Everybody loves us. *Nous sommes cool.* They'll let us through."

IT WAS ALREADY LIGHT when we left the city just after six. We had filled the tank and bought a few packs of cigarettes. "Always comes in handy," said Sekombi, a nonsmoker, as he took a bite of his cookie. The jeep bounced over the dirt road. Its steering wheel was on the right: almost all cars in eastern Congo come from the neighboring countries, which are former British colonies. The silhouette of the two-thousand-meter-high (6,500-foot-high) Nyiragongo volcano with its eternal plume of smoke rose up in the distance. Sekombi was waxing lyrical. "That volcano is our mother, our sister, and our mistress, all in one. When I see that wisp of smoke I'm always reminded of a huge breast that keeps giving milk. Once you've drank of it, you always come back." But sometimes that breast produced a milk black as night: in 2002 the volcano had buried half of Goma beneath a flow of lava. The second floor of some houses became the ground floor that day. The city had asphalted itself in a whirling intoxication. Goma, the black city in a rust-brown landscape, is the only place in Congo where the roads don't have potholes, but bumps.

A little farther north we came past the first refugee camps, the same camps occupied by Rwandan Hutus back in 1994. Now they provided shelter for the quarter of a million civilians who had fled from Nkunda. A festival campground without the festival, a sorry jumble of canvas and cardboard. In North Kivu someone is always on the run.

Eight kilometers (five miles) later we arrived at the first roadblock. A thin rope with a branch dangling from it had been tied between two oil drums; half a dozen soldiers were hanging about listlessly. Sekombi rolled down the window. "Ya, man!" he laughed to the men in khaki. His brother Katya was sitting quietly in the backseat, but he was now wearing the trademark of the true Rastafarian: a thick knitted cap. "Rastaman!" the soldiers cheered, "wo-woow!" They joked, they shot the breeze, they accepted cigarettes from us and wished us a nice day. "Peace and love!" With those words, Sekombi put an end to the border formalities. Peace and love! To soldiers! During a war! But they untied the rope and waved to us as we pulled away. The same scene was repeated at all the other roadblocks. I had never realized that embryonal dreadlocks and nicotine were all you needed to get to Central Africa's most feared warlord.

After the brutal taking of Bukavu in 2004, Nkunda had kept his head down for a time. As a trained psychologist he became the pastor of a Pentecostal church in Kivu.[31] He only entered the public eye again in 2006. Immediately after the results of the parliamentary elections were announced, he set up the CNDP, the Congrès National pour la Défense du Peuple.[32] The names of Congolese rebel movements are, often enough, gratuitous abbreviations, but Nkunda's brainchild took the cake: it was not a "congrès" at all, but a militia, it was not "national" but regional, and what was meant by "the defense of the people," well, you could ask around at the refugee camps about that. Yet still, that last part of the name was probably the most accurate, as long at least as you read it as the defense of "a people," one particular population group, the group that had been mocked and pestered for the last twenty years and to which Nkunda himself belonged: the Congolese Tutsis. Had a colonial ethnographer in the 1920s wished to photograph an archetypal Tutsi, he would undoubtedly have dragged Nkunda in front of the camera. With his tall, bony frame, his high forehead, and pointy nose, he embodies all the clichés about the Tutsi male. He and Kagame could have been brothers.

The CNDP arose when it became clear that the elections would produce little or nothing for the Tutsis. Vice president Ruberwa's Rassemblement Congolais pour la Démocratie (RCD), which was supposed to defend their interests, turned up empty-handed: no ministerial posts, no provincial governor, not even a provincial council member: nothing more than fifteen seats in parliament.[33]

On November 25, 2006, just before Kabila was sworn in, Nkunda bared his teeth and overran Sake, a town thirty kilometers (about nineteen miles) from the provincial capital of Goma. The hilly, volcanic area north of Goma, along the border with Uganda and Rwanda, became his stomping grounds. And although the movement was not exclusively Tutsi, it received Rwanda's support from the start. Nkunda's CNDP fit in the same category as Kabila's Alliance des Forces Démocratiques pour la Libération (AFDL) and Wamba dia Wamba's RCD, the only difference being that this was no Rwandan initiative operating under Congolese flag, but a Congolese initiative with Rwandan backing. His main enemies were the Hutu refugees in eastern Congo, now organized in the Forces Démocratiques de Libération du Rwanda (FDLR)—yet another questionable name, for there was not much democracy to be found and that liberation of Rwanda was a relative thing: many of them had Congolese spouses, farmed plots in Kivu, supervised a few little mines, and ensured themselves, raping and plundering as they went, of a steady income, so why go to war against Kagame's powerful army?

The struggle between Tutsis and Hutus in Congo, therefore, now continued under the aegis of the CNDP and the FDLR. The motives were both ethnic and economic.[34] On neither side did the troop strength exceed ten thousand men, but the brutality with which those troops were applied was indescribable. Civilian suffering became the norm, gang rape a right. As they had during the Second Congo War, the Hutus received support from Kinshasa—FARDC and FDLR officers sometimes worked mining sites together—and once again the Mai-mai joined in too. Sexual violence was a weapon wielded by both sides. Lawlessness reigned supreme. Even civilians began raping on a massive scale, not as a weapon this time, but just for the fun of it.

The years 2007 and 2008 were marked by repeated attempts to stop the violence. January 2007: Nkunda agrees to let his CNDP warriors be absorbed into the government army, but rather than any far-reaching *brassage* (intermingling), he receives a much more superficial *mixage*. His rebel army is not disbanded and spread over barracks far away, but is allowed to merge on the spot. The result is predictable: the FARDC does not swallow up the CNDP, but the CNDP the FARDC. Nkunda becomes a general in the national army and is able to get on with his rebellion. "FARDC?" the joke goes. "Forces Armée Rwandaises Déployées au Congo [Rwandan troops deployed in Congo]!" December 2007: The fate of the Hutu refugees is discussed during peace talks in Nairobi. January 2008: after lengthy negotiations in Goma, the Amani process is launched. Abbé Malu Malu, the former chairman of the electoral committee, succeeds in convincing all the militias to sign a provisional peace agreement.

But it doesn't work. In May 2008 I fly in a MONUC helicopter from Goma to Masisi, where Malu Malu, in the presence of Belgian Minister of Foreign Affairs Karel De Gucht, will announce the armistice. Thousands and thousands of people have gathered in Masisi. There is singing, dancing, and drumming. It is exceptionally moving. Peace, yes, the people have been waiting for peace for a long time. But two young Hutus tell me: "It's going well now, all we need is one more genocide, a little one, to wipe out Nkunda's men."[35] The hatred remains endemic. In late October 2008, while

Sekombi and his brother were screening art house movies, Nkunda moved on Goma.

THE JEEP RATTLES ITS WAY through the demilitarized zone, which largely coincides with the Virunga National Park. This is, quite literally, no-man's-land: there is not a soul in sight in this dark green landscape, which is of a beauty so raw that it leaves you speechless. Volcanoes, forests, silence, mist.

The CNDP roadblocks are a piece of cake: they don't even want our cigarettes. As we penetrate further into rebel territory, we see more people out on the road. Women carrying yellow water jugs on their back, men leading reddish-brown cattle, boys with wooden bicycles loaded with sugarcane, bananas, or charcoal. After kilometers of bumping along through jungle and plantations with tall banana plants we finally arrive at the ruined Jomba mission post. Hundreds of children crowd around the jeep with its cargo of two Rasta's and a white man. They run their hands over the coachwork and race off hysterically when Sekombi honks the horn. The man I had agreed to meet comes walking up: René Abandi, a lawyer in jeans and a denim shirt, not yet in his forties, with a friendly face and a quiet voice. Could this really be the CNDP's number-two man? He has friends in Antwerp, he tells me, and he worked on his doctorate at the university of Urbino. But when Nkunda started his offensive, he became the first member of his civil staff. Abandi is a Congolese Tutsi. From spokesman he has been promoted to something like the movement's minister of foreign affairs, for the rebel territory has its own government. He suggests we drive on to a nearby village, where Nkunda is going to speak to the people.

The road turns muddy. We cross a stream lined with huge papyrus plants, then wind uphill to Rwanguba, a hilltop aerie. The view is breathtaking. We are able to see a dozen kilometers in all directions: hills, volcanoes, emerald green valleys, a wisp of smoke through the trees, a distant lightning bolt. It looks like a nineteenth-century panoramic painting, an idyllic fresco of nature with, in the foreground, in 3-D, the turmoil of war. Hundreds of people are packed together in front of the central building on the hill. CNDP soldiers frisk us, then let us through. We wade through a tractable crowd to the front. There, sitting beneath a lean-to, are all the rebel movement notables and officers. Bosco Ntaganda is there, the army chief of staff sought by The Hague for crimes against humanity. In the middle, in full uniform, is Laurent Nkunda himself. He is toying with a

black walking stick, the silver handle of which has the form of an eagle.
His remarkably long fingers never stop caressing the head of the cane. The
chairman's eyes are set back so far that his head resembles a skull. From
beneath his military cap I can see the veins twisting along his temples. He
stands up to welcome us and makes sure we get a seat. During these weeks,
Nkunda is at the summit of his fame. His rebel territory is almost half the
size of Rwanda, the international press is writing about him, he considers himself unbeatable. Children holding spears come and dance for him,
young girls prance about in the grass. In Rwanguba he will demonstrate
his authority; he is the new chief. When the war dances are over he stands
up and walks slowly toward the crowd. He starts talking and never stops.
He waves his eagle-head cane sternly, sternly points his bony index finger.
Then he cracks a joke. Charm and terror in one. He praises the villagers for
not having run away. "You are real people, you have stayed. Good. Work in
your fields, go about your business. Don't judge me by my face, but by my
actions." When he is finished, he walks back to his seat calmly and you can
hear the grass rustle around his boots.

That afternoon Nkunda meets with his civil and military staff in a house
on the hillside that was once built by a Protestant mission. I wait for hours
in the garden with Sekombi and Katya. There is cola and beer. A group of
about twenty child soldiers keeps watch, their bazookas and Kalashnikovs
at ready. They will not be lured into conversation, but they do want to know
what that heavy object is in my pocket. Obediently, I show them my two
cell phones. At that moment, thirty kilometers (about nineteen miles) to the
north, their comrades are waging a bitter fight against the Mai-mai. They
are extremely tense.

The meeting takes a long time. Nkunda has granted an audience to
local traders who want to pay fewer taxes. The rebel territory is not rich in
mines: the CNDP receives its funds through the sale of cattle, coffee, and
charcoal and the taxation of traders and truck drivers. Sekombi and Katya
grow nervous. It is already three in the afternoon and it looks like rain.
They want to be back in Goma before dark, for safety's sake. I hesitate,
reconsider, then let them go. A bit later I see their white jeep wind down the
hill and disappear into the greenery. I am going to spend the night with a
gang that, two weeks earlier in nearby Kiwanja, was involved in the slaughter of 150 civilians.[36]

Major Antoine raises a half-liter bottle of beer to his lips and wants to talk to me about history. Is it true that the Egyptians mistreated the Jews so badly, the way the Bible claims? Have the Egyptians ever apologized for that? Why did the Belgians cut off the hands of their Congolese subjects? Was that in order to get more coffee? ("Rubber," an eavesdropper whispers, "coffee, that's only around here.") Why is the price of every raw material determined in Belgium? Why are there only three Frenchmen playing on the French national soccer team? Is that because of globalization? But then why does the International Criminal Court prosecute only Africans? The most absurd questions are punctuated by shrewd remarks. There is one thing about which he wants to be perfectly clear: "The CNDP is Congolese through and through, no matter what anyone says. That fellow in Kinshasa is a worthless do-nothing who is selling the country to the Chinese. You can tell by his soldiers. When we fight against them, it never takes more than half an hour. After that they run away. But if it goes on for hours, then we know for sure that we're fighting against the FDLR, even if they're wearing the uniforms of the government army that supports them. They just keep on fighting. They're like wounded animals, you know. For them, it's either winning or nothing at all."[37]

It is pitch black outside now and I haven't eaten anything since six o'clock that morning. Headache. Chills. We're high in the mountains here. Finally, around ten o'clock, I'm allowed to go inside. First everyone has to eat: goat meat with rice, prepared by a few Tutsi women. A group of about eight officers and traders sit down at the tables, which are arranged in a U. Nkunda sits alone at his own little table, like an umlaut over the U. Behind him is a bodyguard with a machine gun and a receiver plugged into his ear. No one speaks. When the chairman says something, everyone pretends to be interested. When he tells a joke, they laugh a little too loudly. He is finished before all the rest. While the guests continue eating uneasily, he cleans his teeth slowly with a toothpick and looks at the others around the table, one by one. His teeth are bared in a horrible grimace. One of his eyelids droops badly. Occasionally he relaxes his features and swallows a leftover morsel.

"Come, let's talk," he says. He leads me to a dormitory at the back of the building. His bodyguard and René Abandi follow. We sit down on three low stools between bunk beds and mosquito netting. The teenage boy with the loaded rifle remains standing and never takes his eyes off me. Nkunda starts in right away. He doesn't talk, he whispers. He speaks beseechingly and looks at me wide-eyed the whole time, as though he had to drive out a demon in me: "There are so many fault lines in this country, between the east that voted for Kabila and the west that wanted Bemba, between Mobutu's former FAZ and the *kadogos*, between the Hema and the Lundu, between Tutsis and Hutus. Congo needs to go through a process of national reconciliation."

I can hardly believe my ears. Is he, the ruthless tyrant, suddenly going to play the great conciliator? Is he trying to use this talk to cuddle up to the West, or what? Rational discourse in order to hold off a robust intervention force? In any case, he makes masterful use of the international disillusionment with Kabila. "I know Kabila. He is incapable of debate. He destroyed both Bemba and Bundu-dia-Kongo. This country has the right to be liberated. This country has never been independent. This country should finally be able to profit from all its opportunities, otherwise the Congolese people will turn against Kabila the way they turned against Mobutu."

At the height of his fame, he has clearly tweaked his ambitions. He is no longer concerned solely with protecting the Tutsis, or even with the fate of the Banyarwanda, but with nothing short of liberating all of Congo. "There will not be a Tutsi territory in Congo. The CNDP is not a Tutsi rebel army, because Tutsis make up only 10 to 15 percent of our movement. We are a Congolese rebellion. The West condemned the genocide, but not its perpetrators. They are still here. And it's unacceptable, after all, that foreign troops are operating within our country's borders and are even armed by our government! Normal countries do not tolerate illegal aliens, but here we give them guns!"

Nkunda, liberator of the nation: it takes some getting used to. He, in any case, seems ready for the role: "I set up the CNDP as a sort of core for the national army of the future." Well, do tell. "It was sort of an experiment: I wanted to prove that with very little funding you could set up a disciplined army that would not pillage the surroundings." Excuse me? "The violation of human rights is something you rarely see within the CNDP. We have a clear code of conduct. My soldiers don't receive salaries. They receive rice, beans, and corn—that is their salary. But we've showed them that there is a future. They live for that dream." With all due respect, I object, but your army is hated throughout the rest of Congo. "That's because the only voice that is heard is that of the MONUC. They claim that we are rapists and that we massacre people. They claim that we are an armed Rwandan division. But those days are past! It was not a happy time, back when Rwanda and Uganda were here." But you were there yourself, weren't you? You led the Rwandan troops in Kisangani! "That's right. I defended Kisangani. That made me the most popular officer in the city."

Wait a minute, it occurs to me, he's still hated there, even today! Under his reign of terror in 2002, dozens of young people from the poor neighborhoods were murdered. At the bridge over the Tshopo, two hundred policemen and soldiers were slaughtered and thrown into the river. They were tied up and had a gag stuffed in their mouths. Some of them were shot and killed or decapitated, others had their neck broken or were bayoneted. Their stomachs were cut open so that they wouldn't float to the surface after a couple of days. Nkunda was there. He supervised the operation, with Rwandan support.[38] And now he's trying to claim that foreign interference is a bad thing?

"When Germany threatened England, didn't Churchill call on his people to resist? They applauded him for that. So why should we have to accept that the FDLR occupies this place, like the Germans did back then?" Churchill was chosen by the people, general, you were not. "In times of war, that doesn't matter. Hitler was an elected official, and look what happened. De Gaulle wasn't chosen by the people, but he freed France, didn't he?" For a moment, I'm dumbstruck. Is he trying to claim a comparison with the most important French statesman of the twentieth century? "Yes, I am the General De Gaulle of Congo!"[39]

DAZED BY THIS ADVANCED COURSE IN RHETORIC, I squeeze into a jeep with René and seven others. Crammed into the baggage compartment is a child soldier with a Kalashnikov. It is almost midnight. We drive east through the wet, dripping hills and hope that we don't run into a Mai-mai patrol. I'm scared and confused. What I don't know is that, at that moment in New York, the finishing touches are being put to a UN report demonstrating Rwandan participation in the CNDP. What I don't know is that Human Rights Watch is preparing a report on Nkunda's atrocities.[40] I have arrived at the point where history is still warm, fresh, and elusive. I can't see the big picture. No one can see the big picture.

All I know is that I would rather talk to normal people than with politicians, that I learn more from anecdotes than from rhetoric. All I know is that I once sat in Grâce Nirahabimana's plastic hut at the Mugunga refugee camp, block 48, number 34; it was too small for me to stand up. Grâce was a beautiful twenty-three-year-old woman with two children, Fabrice and David. Her two brothers, twelve and sixteen, had been taken by Nkunda, her two sisters died of dysentery, she was raped by three soldiers. She left everything behind. Her sisters died in the camp—too little to eat, no toilets—and were buried among the banana plants. It was cold, sitting on her bed. A harsh wind was blowing across the lunar landscape of lava and rattling the plastic walls of her little hut. "I don't feel safe at all," she sobbed. "I'm afraid, very afraid. Afraid of Laurent Nkunda."[41]

After what seems like an endless drive, the jeep stops at an old colonial-style house. "This is the Ugandan border," René tells me, "this house once belonged to the chief customs official. There, where those trees are, that's where Uganda starts." The place is called Bunagana, it is a safe place to spend the night. But to René's surprise, the house turns out to be full of child soldiers, at least twenty of them. They are sleeping in the armchairs, on the floor, in the kitchen. There is neither water nor electricity, but a bed is arranged quickly enough.

I get up early the next morning. Shirtless, I go out onto the patio to read through my notes. A thirteen-year-old boy tells me that his rifle is called a Chechen. Around eight I walk into the village with René to get some break-fast. He did not sleep well. "Gastritis," he sighs, "I worry too much, it's just the way I am. Nkunda suffers from it too, alongside his asthma. The war is not good. It's the worst thing there is, but it's all we can do."

We arrive at a very normal-looking house. It turns out to be the civil headquarters of the CNDP. There I meet all the dignitaries I saw yesterday as well. Nkunda's sister is there too: two peas in a pod. The courtyard is an open-air garage. There, a half-dozen Humvees captured from the FARDC are being fixed up for the rebels' war effort. For the first time in weeks I eat cheese, Kivu cheese, a Tutsi specialty. The rebel leaders discuss the news of the day. Desmond Tutu and Romeo Dallaire, former UN commander in Rwanda, have just called for a large-scale intervention force to be sent to North Kivu. "What a bunch of hooey," René snorts. "Now that they don't have any more political arguments, they bring in the moral heavyweights. Humanitarianism is being misused to cover up military domination." The others chime in in agreement. "We'll all be going to the International Court anyway," he jokes, "so we might as well rape and murder, otherwise we'll be there for no good reason."

That intervention force never arrived. The European Union had no desire to act on UN Secretary General Ban Ki-moon's appeal, and the African Union, the South African Development Community, and Angola didn't feel much like coming to Kabila's aid. That morning in Bunagana I concluded that Nkunda might very well go on ruling his territory for a long time to come. The Uganda border was officially sealed, but I saw a truck full of flour come into Congo. Who's going to stop him anyway? I thought. Congo has no army, the MONUC won't intervene, a more extensive intervention force is not going to arrive, and besides, he has food and he collects taxes. Maybe he'll keep up this bush war for as long as Laurent-Désiré Kabila did.

But I was wrong. One month later, in January 2009, what no one dreamed would happen actually did: the Congolese and Rwandan armies, those sworn enemies, joined forces and arrested Nkunda. A totally unexpected move, but they had little choice: Kagame had lost a great deal of international credit after the appearance of that UN report detailing his support for the CNDP, and Joseph Kabila was cutting a bad figure with his worthless army that no one wanted to help. Strange bedfellows in pursuit of a common cause, they even tried to put the FDLR out of commission. They only succeeded in part, but Nkunda ended up in custody in Rwanda and has been awaiting trial in Congo ever since. The CNDP was handed over to war criminal Bosco Ntaganda and simply "fused" once again with the government army.

The joint Congolese-Rwandan operation became known under the name Umoja wetu (in the first half of 2009) and was followed by the operations Kimia II (2009) and Amani Leo (2010), proactive campaigns against the FDLR by the national army (in fact by former CNDP forces, led by the scoundrel Ntaganda) in cooperation with the MONUC, which resulted in more civilian suffering than glory.[42] By 2010 the FDLR numbered six thousand men, no more than a homeopathic residue of 1994's 1.5 million refugees. Less than three hundred of them were suspected of crimes of genocide.

If Rwanda is overmilitarized, then Congo remains undermilitarized. The country's armed forces are still more an apparition than any real force to be reckoned with. And that is clear to see. The FARDC is unable to stop Ugandan rebel leader Joseph Kony's Lord's Resistance Army from sowing unrest in the northeast, let alone effectively defend more than seven thousand kilometers (over 4,300 miles) of national borders. And that at a moment

when geopolitical tensions are rising: with Uganda concerning oil in Lake
Albert, with Rwanda concerning methane gas in Lake Kivu, and above
all with Angola concerning oil concessions in the Atlantic—sometimes
resulting in skirmishes. The army cannot even keep order within Congo
itself. In the course of a quarrel over fishing rights in a few ponds in Dongo
(Équateur) in November 2009, at least a hundred people were killed and
ninety thousand fled. The will to change seems minimal.[43] With an army,
Kabila could assert himself more; without an army, however, he need fear
no putsch.[44]

AND LIFE FLOWS ON LIKE THE RIVER. On the other side of the country, in Nsioni, people are walking up and down the red, dusty main street
of the village. I watch them from the terrace where the music has been
turned up deafeningly loud for me and two other customers. Think away
the cell phones and there is little difference between today and the 1980s.
The same soft drink bottles, the same cars driving around, the same
rickety market stalls selling slices of dried fish. The only thing that has
changed is the size of the slices; today they are little more than cubes. But
across the street it looks as though a UFO has landed. Towering above
the dull barracks and the faded housefronts is a pristine white building
that gleams on all sides. Neatly parked before the door are four brand-
new motorcycles with sparkling chrome. Their seats are still covered with
protective plastic. Beside them are ten men's bikes, bound together, their
handlebars turned parallel to the frames and still packed in cardboard.
The glistening rod brakes are a glory to behold. From inside the building
comes the flickering blue glow of a plasma screen. Hanging above the door
is a sign that explains a lot: CHINA AMITIÉ COMPANY. In Nsioni, the first
Chinese traders have touched down.

I go in and say hello to a wary-looking Asian couple who speak not a
word of French or English, but whose merchandise speaks for itself: a *horror
vacui* of flashy sports shoes piled up to the ceiling next to TVs and clocks
and racks full of perfume. The China Amitié Company creates the same
impression of luxury and comfort for the people of Nsioni as the supermar-
kets did in the farming villages of Europe in the 1950s. What a contrast to
those sorry market stalls where you went to buy razor blades or candles one
at a time! What luxury when you compare those perfumes to the homemade bars of soap you've been scrubbing yourself with all your life! What

easy comfort when you realize that you no longer need to go to Boma or Kinshasa to buy such products! And affordable as well!

The shopkeepers even sell paintings in gaudy frames, showing mountain landscapes and alpine pastures. Asiatic merchants coming to the African interior to sell European landscapes: this, I believe, is what they call globalization. The world as marketplace. It reminds one of the ingenious graffiti sprayed on the old railroad trestle at Matadi, less than a hundred kilometers (sixty-two miles) from here. That bridge from the 1890s, when Nkasi's father and Chinese workers built the rail connection to Kinshasa, today serves as canvas for an act of vandalism that brilliantly summarizes this third millennium: www.com.[45]

From the late 1990s, more and more Chinese began coming to Africa. They arrived not only to sell their wares, but much more frequently to buy raw materials. The formidable explosion of the Chinese economy, as a result of Deng Xiaoping's controlled experiment with capitalism along the country's coastlines, enormously boosted the demand for mineral riches. In 1993, for the first time, China imported more oil than it exported.[46] The first African countries with which it established intensive ties, therefore, were the oil states of Nigeria, Angola, and Sudan. Zambia and Gabon entered the picture later as well, because of their copper and iron ore. As "geological scandal," and despite the war and the grim business climate, Congo also caught China's eye. In Katanga Chinese adventurers soon caught scent of a golden opportunity and swooped down on the wreckage of the once-prosperous mines. In 2003 Gécamines, at the insistence of the IMF and the World Bank, had fired eleven thousand superfluous miners.[47] They received severance pay, but most had spent it all on cars and TVs. Many of them then were *creuseurs*, artisanal miners. Like in Kivu, they were prepared to use limited means to scratch away at the old mines and fill sacks of ore that they then sold to Monsieur Chang or Monsieur Wei.

In February 2006 I had the chance to visit the mine at Ruashi. There, hundreds of *creuseurs* were digging for heterogenite, an ore that contains both copper and cobalt. I saw children clamber down into poorly shored-up wells of up to twelve meters (thirty-nine feet) deep. I saw a five-year-old boy covered in dust, wearing a "Plop the Gnome" T-shirt. If they were lucky, they received five dollars a sack. Sometimes a group of friends would bring up as many as ten sacks in the course of a day. It was strenuous, dangerous work, they said, but they could live from it. What a contrast with my visit later that day to the enormous spic-and-span Luiswishi cobalt mine owned by Belgian businessman Georges Forrest, where I saw no more than a few dozen Congolese at work. They all wore safety helmets and operated excavators with hubcaps bigger than a human.

The Chinese buyers were private entrepreneurs and received no support from the Chinese government. Some of them began their own makeshift foundries, in order to export ore in more concentrated form. Their Congolese day laborers worked under ghastly conditions. They were badly paid, they breathed in noxious fumes, and they had no work clothes, let alone any collective labor agreement. Take Jean, for example. He went to work for Jia Xing, one of the larger copper-processing companies with a depot at Kolwezi and a foundry in Lubumbashi. The concern employed two hundred people and Jean received a permanent contract: he was an experienced smelter. A day worker, therefore, could sometimes work his way onto the payroll—although the contracts were often written only in Chinese. Jean's shifts lasted twelve to thirteen hours a day with an ultrabrief lunch break, seven days a week. The company worked with a day shift and a night shift. There was no protective clothing, Jean's tools were worn-out, the heat from the blast furnace was unbearable. Jean earned $120 a month, plus a $100 bonus if he ran the blast furnace: with that, Jia Xing was the best-paying Chinese employer in Katanga.

One morning he and twelve colleagues arrived a few minutes too late for work: they had been held up by a traffic accident. For punishment they were locked in a container, where they sat from seven in the morning until five in the afternoon. At the end of the day they were all fired. There were, after all, plenty of others waiting to take their places. And so Jean went to work as a *creuseur*. He sold his sacks of ore to his former boss, but the spots where artisanal mining was allowed were limited. Perhaps he should hook up with the teams that trespassed in the middle of the night on the big mining companies' concessions? It was dangerous, working in the dark. Some workers drowned or suffocated on the job, others were shot by the security guards. He could always go to work at the Emmanuel Depot in Kolwezi, but the workers there always got drunk during their lunch break because they had to process radioactive ore without gloves or a mask.[48]

Katanga became home to a rough capitalism reminiscent of that in the 1920s but the 2008 banking crisis caused forty of these private firms to pack up and leave. The copper price fell from almost $9,000 to $3,600 a metric ton (about $9,900 to $3,960 a U.S. ton) and the province imposed stricter conditions. Tens of thousands of artisanal diggers were left without work. Suddenly, Katanga was looking more like it did in the 1930s.[49]

Chinese state-owned companies started moving in; not hit-and-run fortune hunters but mammoth concerns with virtually unlimited funding. The road from Kinshasa to Matadi was rebuilt, as was the road from Lubumbashi to the Zambian border, raced along now by trucks filled with ore. CCT, a Chinese telecom company, became one of the country's major cell-phone operators. And yet another Chinese company began laying a 5,600-kilometer-long (3,500-mile-long) glass-fiber cable to open up Congo to the digital revolution.[50] Relations between Mobutu and Mao had been hearty even as early as the 1970s; in those days, the focus was on cultivating ideological comradeship (the single-party state, the *abacost,* and the parades in Congo were the result—no mean feat for a pro-American country), but now it was about business. Congo became one of China's newest trading partners. In 2006, President Hu Jintao organized a crucial Sino-African summit in Beijing attended by no less than forty-eight African heads of state. During that meeting, $2 billion in contracts were signed and China promised up to $5 billion in loans and a doubling of its aid efforts by 2009, while purging the countries' outstanding debts and lifting a whole slew of import duties on African products. With an eye to trade relations, Chinese dignitaries visited almost every country in Africa. Beijing stuck rigorously to its policy of noninterference in domestic affairs and championed the principle of fraternal South-South cooperation, as opposed to paternalistic North-South meddling. It all sounded lovely, but the gist was also that China apparently had no objections to doing business with unsavory characters like Robert Mugabe in Zimbabwe and Omar al-Bashir in Sudan. The new China was businesslike, efficient, and pragmatic. All it asked in return from its new trading partner was that it state the opinion—once each year, during the UN General Assembly meeting—that Taiwan actually belonged to mainland China.

In September 2007 Minister of Infrastructure, Public Works and Reconstruction Pierre Lumbi announced that Congo had closed a megadeal with China. The country would set up a joint venture under Congolese law with three Chinese state-owned enterprises (a bank, a road-building company, and a general contractor). Through Gécamines, Congo would maintain a 32 percent share in the enterprise; the remaining 68 percent would be Chinese.

The joint venture would be allowed to excavate 10 million metric tons (11 million U.S. tons) of copper and six hundred thousand metric tons (660,000 U.S. tons) of cobalt in Katanga—gigantic volumes, when one realizes that only eight million metric tons (about 8.8 million U.S. tons) of copper were mined during the entire colonial period and that the country's total reserves were estimated at 70 million metric tons (77 million U.S. tons).[51] In return, the new partnership would invest $3 billion in restoring the country's mining infrastructure, and $6 billion in the construction of paved roads (34,000 kilometers, or 21,000 miles), unpaved roads (2,738 kilometers, or 1,700 miles), railroads (3,215 kilometers, 2,000 miles), houses (5,000), polyclinics (145), hospitals (31), hydroelectric plants (2) and universities (2). Investments, all in all, of $9 billion. And because the joint venture had no revenues as of yet, the People's Republic of China would advance the funding for these major works: the venture would simply pay it back over time. Kabila was elated: "For the first time in our history, the Congolese people will see the usefulness of all its copper, its nickel, and its cobalt!"[52]

It was, indeed, an impressive agreement. Only seven pages long, shorter than a normal rental contract, it was the most important document concerning Congo since the ten-year plan of 1949. Congo would become a construction site unlike anything seen there since the 1950s. In the Western press the deal was often depicted as a "loan" from China, while in fact it was a tradeoff: ore for infrastructure. An exchange of that sort did not imply a return to a precolonial economy, but was a handy way to skirt around corruption: a hospital, after all, is not easy to slip into one's pocket. But it was very much a tradeoff, with a crucial clause attached. Should the deposits fail to produce the quantity of ore hoped for, Congo would be obliged to meet the terms of contract by other means.

As soon as the announcement was made, the West began screaming bloody murder. Neocolonialism! A new scramble for Africa! Rapacity disguised as win-win gobbledygook! To some, the contract seemed a twenty-first-century variation on the agreements Stanley had asked the village chieftains to sign. The Congolese had let themselves be hosed! It hadn't even been discussed in parliament! It wouldn't generate any jobs! Rumor had it that the Chinese were flying in their prisoners to do the work! Et cetera, et cetera.

Some of these reservations were justified, but others were pure panic; panic in the face of this complex, up-and-coming world order, a world in which China was rapidly acquiring superpower status. It reminded one of the skittishness at the time of the Berlin Conference or at the start of the Cold War. Congo has been drawing the attention of foreign powers for a century and a half, and that has often led to tensions—between European and Arab traders around 1870, between European nation-states after that, between America and Russia during the Cold War, and now between China and the West. Every time a newcomer claims a position on the Central African chessboard, it results at first in suspicion and nervousness.

But was it true, had the Congolese government been taken for a ride? It is hard to say. Inherently, there is no objective standard in trade by barter other than the mutual satisfaction of the trading partners. China was pleased to gain access to raw materials; Kabila was pleased with the promised reconstruction of his country. In any case, the contract had not been forced down his throat, but followed upon two months of vigorous negotiations in Beijing.[53] Attempts to nevertheless quantify the fairness of the deal are doomed to fail as well. Whether 10 million metric tons (11 million U.S. tons) of copper for $9 billion in investments is a fair deal depends, after all, on the international price of copper. In light of the pronounced fluctuations on world markets in recent years, it may amount to $14 billion, but it could amount to $80 billion. Yet one thing is clear: China is not out to plunder the Katangan substrate in the short term, for the simple reason that China's economic policy is characterized by gradualness and planning. Beijing had absolutely nothing to win by depleting and destabilizing Africa. On the contrary. The view of China as a quack physician offering a deathly ill patient a family pack of vitamin C in exchange for, say, a kidney and a lung, does not apply. China has started on a long, structural presence in Africa that will change the face of the world in the century to come.

How democratic that presence will turn out to be is, of course, still very much the question. The Sino-Congolese contract was negotiated behind closed doors, without consulting parliament. And even though the Congolese parliament has, by now, had the opportunity to comment on it, its say in the matter has remained very limited. What's more, the generous trade relations that China maintains with Zimbabwe and Sudan demonstrate that for Beijing human rights are no sacred criterion; no more, after all, than in China itself. For China, commercial interests currently take precedence over humanitarian ones. Although a permanent member of the UN Security Council—which gives it a great deal of power—China is, for example, too dependent on high-grade Sudanese oil to take exception there to al-Bashir's regime. That sounds opportunistic, but it is no more or less opportunistic than the way France, Belgium, and the United States kept Mobutu in the saddle in the 1980s. Among Western regimes, respect for human rights dates only from the 1990s. And even then . . .

The most stubborn opponents to the Sino-Congolese contract were the international financial institutions. The IMF and the World Bank were not pleased with the clause that stipulated that, should there be too little copper or cobalt in the ground, Congo would have to meet its obligations by different means. By putting up such collateral, Congo ran the risk of accumulating even greater debts . . . and it already had such a staggeringly huge pile. There was something to that. The country still drags along behind it the debts acquired during the Mobutu era, and by 2010 the deferred payments and interest accrued on them totaled an astronomical $13 billion. That equaled one-quarter of the country's spending each year, more than 90 percent of its GNP, 150 percent of all exports, and more than 500 percent of government revenues (not including foreign aid).[54] The horse trading with China now added a potentially huge slab of debt on top of all that.

What the IMF and the World Bank did not say, however, was that they were in a position to do something about that burden. Year after year they continued to insist that it be repaid, even though Erwin Blumenthal had roundly stated in the 1980s that it would never happen. The unfairness of weighing down a newly elected government with the twenty- or thirty-year-old squandermania of a former dictator dawned only gradually on the Bretton Woods institutions. It was, to be sure, a huge sum of money and it would set a bad example to erase outstanding liabilities all too readily, but $13 billion effectively crippled all attempts at reconstruction. It was as though the new inhabitants of a tenement apartment were being charged for the exorbitant phone bills of the former tenants, who had hung on the line all the time. Rigobert Minani, a Congolese intellectual, once rightly claimed that the international financial institutions were "holding the national economy hostage."[55]

The reason the IMF clung to its demand that the debts be repaid was that the obligation was the only thing still providing the rich Western countries with a toehold in Congo. The IMF is international by name, yet it awards votes according to the financial contributions from member countries. That means that the United States and the European Union, as the major contributors, control almost half the votes; China, home to one-quarter of the world population, has only a 4 percent vote.[56] Diplomatically speaking, the West had little voice in Congo's affairs once the elections were over; the IMF, however, whose president is by statute always a European, acted as the ultimate big stick in posing conditions regarding anticorruption measures, fiscal matters, and monetary and economic policies. The debt might be allowed to dwindle, but not to disappear completely.

As part of a large-scale aid program for "heavily indebted countries," the IMF stated its willingness to forgo claims on $9 billion out of the total of $13 billion if Congo complied with a series of strict conditions. Those conditions included a revision of the contract with China. At first Kabila was unwilling but in early 2009 the government found itself so strapped for cash—due to the war against Nkunda and the low price of copper as a result of the world economic crisis—that it had hardly enough funds to finance two or three days of imports. Scattered across the bottom of the state coffers was only a measly $30 million. The IMF and the World Bank reacted with lightning speed and a donation of $300 million. Since then, the authorities in Kinshasa realize that it is prudent to continue the dialogue with those institutions and that China is not the country's sole source of redemption. It is better for them to make sure their bread is buttered on both sides.

In December 2009, after months of renewed negotiations, a deal was struck: the collateral clause was scrapped and in return China would lower its investments from $9 billion to $6 billion. The IMF promptly coughed up $150 million and announced that Congo was now much closer to remitting its debts: of the original $13 billion, the country now had to pay back "only" $4 billion.

Meanwhile, India too is poised to enter into a business partnership with Congo, a cooperative arrangement on which the IMF is sure to keep a close eye.[57]

BEHIND THE HIGH WHITE WALL I could see the colossal machines for mixing asphalt: on October 17, 2008, I drove around the perimeters of the Chinese Railway Engineering Company (CREC) in Kinshasa's outlying riverine district of Kinsuka. The CREC is one of the Chinese state-owned companies in the consortium with Congo and one of the biggest construction concerns in Asia, with one hundred thousand workers on its payroll. Kabila had given the company a huge terrain close to the riverside quarries and two other concessions in the city. Rumor had it that the CREC fired Congolese workers if they refused to obey orders, even when those orders were given in Mandarin. Their monthly wages of $150 are paid out at an extremely low rate of exchange, which means they actually take home only $70.[58]

But there was no way I would be allowed in, I soon discovered, let alone carry out interviews on the grounds. All I got to see were those high white walls around the concession, hundreds of meters long. Driving around them to the rear, I saw that the concession bordered on a working-class neighborhood. There was only a sandy path leading to it. As I climbed toward the houses, a little boy of about four came running up to me. He looked at me, pointed his finger and said, loud and clear, because children like to name the things they know: "Chinois!"

A generation is growing up in Kinshasa today for whom a European is more exotic than a Chinese. In Congo there are once again children, just as there were in the late nineteenth century, who have never seen a white man in real life. One finds them even in the working-class neighborhoods of Kinshasa. On any number of occasions I have noticed toddlers running away and shrieking at my monstrous appearance as I walked through their alleyways.

Congolese adults, however, waver between East and West. Europe and America are still admired for their know-how, but many people wonder why they see so little of that, while the Chinese carry out one ambitious project after the other. They have the impression that the West is no longer interested. Still, the election of Barack Obama brought new hope. Old Nkasi couldn't believe it when I spoke to him that first time, the day after America had elected its new president. At six in the morning after his historic acceptance speech, young people gathered at the busy Kintambo Magasin rotunda in Kinshasa and cheered: "He's one of us! He's one of us! He's a Mutetela!" Because the president's surname begins with an O, people thought he was a member of the tribe belonging to the Batetela, where names like Omasombo, Okito, and Olenga are common. But even those better informed about his lineage were convinced that a new chapter in African American relations had begun. And indeed, Hilary Clinton came to Goma, the first American secretary of state to visit the country since 1997. That she visited Congo and not Rwanda, which after all borders on Goma, made people hope that America would alter its uncritical pro-Rwandan policy. A special U.S. envoy was appointed for the Great Lakes region, and during his Nobel Prize acceptance speech in December 2009 Obama explicitly referred to the sexual violence in Congo. Yet in actual practice, the new American government has as yet developed no coherent vision for Central Africa.[59]

So then what about the Chinese? During my conversations I noticed that the Congolese often speak in ambivalent terms when it comes to the Chinese presence. Their view is a mixture of admiration and suspicion, a paradox that often expresses itself in mild mockery. In their social dealings, they see the Chinese as aloof, stiff, and uncongenial. They don't smile much, many people feel; they don't mix with us; thirty of them occupy the same house and they forget to live! The language barrier and the huge cultural differences, of course, do little to promote contact. Those who work for a Chinese boss adopt a subservient attitude, but laugh about him (not her, for there are no women) a little behind his back—an attitude no different from that with which European men were received a century ago. That does nothing to detract, however, from many people's admiration for the speed with which the contracting companies go about their work. "Bachinois batongaka kaka na butu," a humorous popular song says: the Chinese always do their building at night, and when you wake up in the morning another floor has been added.

It took a while for the actual work to get rolling, but people were impressed when the CREC—less than a year after the banking crisis—began renewing the sewers and the surface of Boulevard du 30 Juin in the center of Kinshasa, even if all the trees did have to be cut down and the arterial reduced to a four-lane road where many fatal accidents occur. The Congolese realize all too well that Kabila has farmed out his celebrated *cinq chantiers* to the Chinese in order to mask his own immobility, just as he has farmed out the war to the Rwandans and the MONUC. After all, he needs to have something to show before the 2011 elections arrive. *Cinq chantiers?* More like *Cheng Chan Che!* Whenever young people see a Chinese on the street or a Congolese woman wearing an Asiatic blouse, they will roar: "Cheng Chan Che!"

IF THERE IS ONE PLACE IN CONGO where the awe for China becomes almost tangible, it is along the walk in front of the Chinese embassy in Kinshasa. Three mornings a week, long rows of Congolese crowd together here in the hope of obtaining a visa. Some of them arrive as early as five in the morning to be sure of a spot. Others pay one of the boys hanging around to save their place in line. Early one morning I myself once stood in line there for three hours. Most of the applicants turned out to be young women, applying for a Chinese visa not to settle there permanently, but to buy things: after all, if the Chinese come here to buy up our ore, we might just as well go there to buy their products straight from the wholesaler. What the China Amitié Company did, they could do too.

It was an exhausting morning, but a fascinating one as well. The Chinese embassy is located right across the street from MONUC headquarters. The long row of applicants paid little heed to the white tank, in which a Pakistani blue helmet with an impressive handlebar mustache was guarding the entrance to the compound. He stood bravely at his machine gun, behind a wall of sandbags and thick rolls of barbed wire that the street children used as their laundry line. These women, however, had literally turned their backs on the United Nations and were putting their hope in the new savior, the People's Republic of China.

As I stood in line, I started up a conversation with Dadine and Rosemonde. Dadine was an unemployed, twenty-seven-year-old actress. She had heard about other women who went to Guangzhou, the big industrial town in southern China that in Cantonese is simply called "Canton." In 2007 she tried her luck for the first time and spent a week there buying trousers, shoes, wigs, and body stockings. Back then it had been easy to get a visa, but after the Beijing Olympic Games in 2008 the procedure had become much more stringent. She had left with only her handbag and come back with sixty-four kilos (140 pounds) of baggage. The sandals she had bought there for three dollars a pair she was able to sell in Kinshasa for nine, sometimes even fifteen dollars. She didn't have a shop of her own. She simply went by the homes of friends or to the student hostels in the city. "My customers are able to buy original articles that are a lot cheaper than they're used to, and suddenly I'm earning some money. It's been good for my morale, I've become independent. A hundred dollars is no longer such a big deal for me. I still don't have a husband, but there are definitely a lot more candidates these days."[60]

Rosemonde, an impish twenty-six-year-old, cherishes even greater ambitions. She and her younger sister have been going to China since 2006, to Guangzhou too. Their parents have died, she has a child. None of the people waiting out on the pavement went to Shanghai, Hong Kong, or Beijing; Guangzhou was the place to be. "I buy plates and glasses for restaurants, I buy ice makers, plasma screens, and computers. The trick is to find things that other people don't already import, so you can demand a higher price. Every time I go I fill a sea container with merchandise, all on my own. The container comes in by ship to Boma, Matadi, or Pointe Noire. A shipment like that costs twelve thousand dollars, which is a lot of money, but in two years' time I've earned fifty thousand dollars, and so has my sister. We've both been able to buy our own homes." Young women with the means to own their own property in Kinshasa: that is an absolute novelty. Just as women found new opportunities in the informal economy of the 1980s, today the globalized variation on that economy is offering new prospects as well.

The Congolese market is being flooded with inexpensive Chinese goods. That has actually put an end to the local textile industry, one of the country's last remaining process manufacturing industries. A *wax chinois* (a dyed fabric from China), the women tell me, can't compare to the legendary *wax hollandaise* from Vlisco with which they once made their best clothes. "But what do you expect? A *wax hollandaise* cost $120, and a *wax chinois* only $5." Because the clothing, televisions, and generators that are "made in China" have a strikingly short product life, the Lingala language now has a new adjective: *nguanzu*. It comes from *Guangzhou* and means "not particularly durable," or "unreliable." Meanwhile, a woman who cheats on her husband is now also said to be *nguanzu*.

Rosemonde wore a jumper printed with the words *Dior, j'adore*; no, what it said was *Dior, j'ddore*—after all, with so many ideograms of his own one can hardly expect a Chinese factory worker to master the Roman alphabet as well. On the streets of Kinshasa, women who frequent China as often as they do clearly go dressed differently too. More flamboyant, more extravagant, almost like pop stars. They stand out in a crowd. A young woman in a miniskirt or white boots is almost certainly a Guangzhou trader. "Elles sont 'guangzhouifiées,'" they've become Guangzhou-ified, people say. But Rosemonde has adopted the real hallmark of the new Congolese female. She rolls up the sleeve of her *Dior, j'ddore* jumper to show me her bare shoulder. There, hard to see against her dark skin, is the pride of this third millennium: a tattoo. "They're so good at it over there. You should really go and see for yourself."[61]

CHAPTER 15

WWW.COM

AHIGHWAY IN THE MIDDLE OF THE NIGHT, BUT THAT'S NOT
the way it feels. Even after midnight, the cabs weave an invisible
web as they glide from one lane to the other in search of the fastest way
through. Compared to Kinshasa, however, the traffic is quiet as a graveyard.
Not much honking. No rumbling DAF trucks of prehistoric vintage, driving
at a snail's pace and discharging a cloud of diesel fumes thick as marsh gas.
No battered VW vans with thirty passengers or more on wooden benches,
the last row dangling their legs out the back. And absolutely no holes the
size of a volcanic crater in the asphalt. The green and white cab skims over
a busy eight-lane highway through endlessly expanding suburbs, past drab
residential housing blocks. Closer to the center of town we cross highway
overpasses suspended between office buildings and apartment complexes.
Sometimes there is a highway above us and one below. A vertical loom.
And below, much further below, we see little food stands with lanterns and
bright-red neon signs. Guangzhou.

I'm sharing the taxi with three Congolese; we're on our way in from
the airport. We left Kinshasa a day ago. Kenya Airways brought us first to
Nairobi, where we waited for seven hours, and then, after a stop in Bangkok, to Guangzhou, seven time zones to the east. The other flight path goes
by way of Dubai. From its hub at Addis Ababa, Ethiopian Airlines also uses
that route. In the last few years both airlines have started offering almost a
dozen weekly flights between the African continent and southern China,

flights that leave with an empty hold and return full to the brim. "Why would we take clothes with us? We can buy them there, can't we?"

The first time out, Dadine had been a bit wary. "Right after takeoff I went to the toilet. I hid all my money and my passport beneath my clothes, because I'd heard you had to watch out for the Nigerians. They drug you with something and then they take everything you have. I had fifteen hundred dollars with me that time: the big traders go with as much as twenty thousand dollars in their pocket. You have to stay on your toes."

The cab driver is in no danger. Plastic bars have been installed behind his and the passenger seat. We, prisoners in the backseat, are kept entertained. The seats are comfortable, and at the base of the plastic bars is a built-in TV screen showing cartoons and commercials for skin cream. The volume is turned down low. One of the Congolese is up in front, talking to the driver about the fare. They've been at it for twenty minutes already. He, Georges, speaks fluent Cantonese. After a few years in Guangzhou he has the language down pat. I knew that almost all Congolese are multilingual and learn new languages easily, even when they are older, but that a person could learn Chinese without going to school was more than I could imagine. Georges didn't think it was anything special. One young African woman had taught herself the language within three months.

The taxi takes us to the area close to the Tianxiu Building, in the north of town, right beside busy Huanshi Dong Lu and Guangzhou's big inner ring, a neighborhood of dilapidated high-rises, TV towers, switching yards, and messy urbanization. In recent years, a real African neighborhood has arisen here. This section of the city is home to about a hundred thousand Africans, most of them here only very temporarily. This is where Georges has his cargo office, alongside hundreds of others. His sector is "air and ocean freight, full and groupage container," as his impressive business card says. In the days that follow I notice that all Africans here have equally impressive cards, flashy cardboard rectangles printed in English, French, and Chinese and showing six different mobile numbers, in China and in Africa. The streets around the Tianxiu feature a host of hotels offering, for twenty dollars a night, extremely comfortable double rooms. Those hotels are full of Africans. I will spend the next ten days in the New Donfranc Hotel and not see a single Westerner.

The taxi drops us off at a pedestrian way crowded with men and women, Chinese and Africans. After checking in, I go out to explore the neighborhood; the shops, as it turns out, are open round the clock and sell shoes, suitcases, T-shirts, mobile phones, and lingerie. The streets are lined with farmers wearing reed hats and peddling piles of fruit completely unfamiliar to me: apples smaller than cherries, still on the branch, and grapefruits bigger than soccer balls, which are peeled patiently and skillfully; beside wooden handcarts bearing tanks of butane, men in sleeveless T-shirts are wokking like mad, the sweat pouring from their faces; they mix up noodles, pak-choi cabbage, and oyster sauce, sway the pan to and fro, fill little Styrofoam containers. Suddenly a shrill whistle rings out: the police are coming and they all race off with their handcarts, the gas fires still burning vigorously—the blue flames flicker like torches, the oil hisses hysterically, soy sauce flies in all directions—and within a matter of seconds they have disappeared into a darkened alleyway amid the garbage cans and fleeing rats, leaving the customer alone on the shopping street, bewildered and supperless. I buy a kilo of mandarin oranges from an old farmer; he weighs them on a bamboo scale that he holds up in front of his piercing eyes; I pay in a currency that is strange to me, not even knowing if they think in kilograms around here, and nod by way of thanks, wondering whether that's actually the appropriate gesture. The little man with the weathered face, in any event, smiles, baring two rotten teeth. My hotel is not a separate building, but part of a labyrinthine mall where hundreds of boutiques sell the same gold necklaces, imitation Nokias and soccer shirts, Barça jerseys, Chelsea jerseys, and Dutch national team jerseys reading: Ruud van Nistelrooy, number 9. I locate the elevator doors that lead to the hotel, but when I climb out on the sixth floor, I find myself not in the corridor with rooms on each side, but in a darkened space completely unfamiliar to me. The situation has something dreamlike about it; the strains of stringed music cut through the darkness, two *koi* carp swim slowly in a softly lit aquarium, and, as I stand there with the bag of mandarin oranges in my hand, gradually coming to terms with having taken the wrong elevator, an extremely charming young lady comes up and asks if I am here for the "very special massage."

Later, when I finally reach my hotel room, I see that the display containing city maps also features a decree noting that "according to the Regulations of the People's Republic of China on Administrative Penalties for Public Security whoring legally forbidden is" because, as it turns out, "recently some aliens suffered stealing or robbery during whoring." What a thing to have happen to you, pity the poor alien; nevertheless, the same display contains three packages of condoms, two packaged pairs of panties ("Antisepsis & Healthy"), and four packets of the unfamiliar South Pole ("Liexin Resispance [sic] the Germ Liquid"); as that description clarifies very little indeed, I read on the back that the product is made from natural Chinese herbs and effectively kills 99.9 percent of all bacteria "for male and female privates itch and other social disease."

Night has come, but it doesn't feel that way; the jetlag and the deluge of impressions keep me awake for hours; sleepless, I zap past thirty-six channels full of screaming samurai and businessmen in the throes of debate and become stranded at last in the middle of a game show in which candidates in colorful outfits have to negotiate a perilous obstacle course; very few of them succeed, most end up ignominiously in a tank full of water, to the vast merriment of the audience and the host, who seize the chance to laugh at them mercilessly. It is 4 A.M., I miss Kinshasa and slide open the curtains; on the other side of the courtyard, in a smoky room two floors down, four bare-chested men are playing mahjong under dingy neon lighting—a gambling hall, an opium den, who's to say. Their voices are just out of range, but every once in a while I see them rise to their feet and shout at each other furiously.

JULES BITULU SAW IT ALL CHANGE. I met him in his office on the tenth floor of the Taole building, in the hectic business district of Dashatou. "In 1993 I was the only African here. Along with a Chinese partner, I started a company in Shunde, not far from here; two years later we moved to the city center. To the Chinese I was a creature from outer space, an attraction. There was no racism back then, more like curiosity. Wherever I went they pulled up a chair for me right away. Now there are about two or three thousand Congolese living here. Most of them come from Kinshasa, Lubumbashi, Goma, and Bukavu. Five hundred of them have no visa and live here illegally. Some of them get into trouble with drugs, but there are also lots of Nigerians here who carry Congolese passports."

Guangzhou is the capital of Guangdong Province, an area five hundred kilometers (310 miles) in diameter with about one hundred million inhabitants, almost twice the entire population of Congo. It was here, back in the late 1970s, long before Shanghai, that Deng Xiaoping relaxed the reins of the planned economy for the first time. It was, after all, his home region. The great distance between the coast and Beijing made it a safe laboratory for an experiment in liberalization. What's more, Guangdong is located right across from the even-freer Hong Kong and Macau, and so could enter competition with them. Thirty years later, it is the manufacturing center of the world. The province is the leading global producer of air-conditioning units, microwave ovens, computers, telecom systems, and LED lighting. Guangdong is the third largest exporter of textiles and makes 30 percent of all our planet's shoes. The factories of Shenzhen export toys to all corners of the globe, and until recently produced two-thirds of all the world's artificial Christmas trees—not bad for an officially atheistic region. This tightly circumscribed area accounts for 12 percent of the Chinese economy and more than one-quarter of the country's total exports. That astounding success was due in part to a system of highly subsidized raw materials, but the financial crisis of 2008 gave the region a major buffeting—the Chinese state-owned banks remained solvent, but foreign customers disappeared. Hundreds of thousands of workers lost their jobs. Today an attempt is being made to transform a serial production economy, based solely on export, into an innovative, knowledge-based industrial center that can also serve a fast-growing local market. And that seems to be working: in the crisis year 2008, telecom giant Huawei closed contacts worth more than $23 billion, an increase of 46 percent.

With its propitious location in the delta of the Pearl River, Guangzhou has always been a spot for international trade. It constituted the point of departure for the maritime silk route and established contacts with Christianity and Islam early on. The city still has a lovely mosque, dating back to perhaps as early as the seventh century, the century when Islam arose, and a Catholic cathedral of much more recent date. Persians, Arabs, Portuguese, and Dutchmen found their way here. Little surprise, therefore, that today as well it is the hub for new foreign trade relations, this time with Africa.

Bitulu came to China on a scholarship in 1988. He was part of a group of seventeen Zairians selected to attend university in Beijing in the context of friendly ties between the two nations. The first year was taken up by a mandatory linguistics course, followed by four years of computer science. Today his Mandarin Chinese is better than that of most Cantonese (according to Beijing, Cantonese is not a language but a variation on Mandarin, the standard language), and he draws Chinese characters at a pace equaled by few foreigners. An African who can write in Chinese, that takes some getting used to.

One day, during his linguistics year, he saw the word *democracy* written on the wall of the administration building.

I wondered: what's going on here? I had noticed some unrest, but I didn't understand. There was nothing about it on TV. Our professors warned us not to go to Tiananmen, but I took the bus there and saw that the square was packed with students. There were no classes being held, everything had been suspended. At the university I saw two coffins; occupied or vacant, I couldn't be sure. Back at my apartment I saw, from the ninth floor, the American students being picked up by a minivan from the embassy. The students from the former French colonies, like Gabon, were also leaving. We, the Zaïrians, were the last to go, then our consul picked us up too. On the way to the embassy we saw burned-out army trucks along the street. There was a massacre going on. The Japanese students told us later that it was very well organized, with trucks to pick up the bodies and cleaning crews. We slept on the floor of the embassy for nine days. It was cold and there was no food.

As a recently qualified computer-science engineer in a land full of recently qualified computer-science engineers, Bitulu did not find a job right away, but he possessed another talent: music. During his student days he had already led a band manned by the few Congolese students present in Beijing, and now he joined a traveling Chinese orchestra. "We spent six months going from village to village in the interior. I visited the provinces of Guangxi, Hunan, Yunnan, Huizhou, and Sichuan. At first I only played guitar, later I sang in Chinese too. For the audiences, that was a real attraction. But I didn't feel comfortable. I never saw another Congolese and we didn't eat well, only that Chinese food." His experiences remind one of the fate of the Congolese who had built their huts at the Tervuren expo a century earlier. Then too, a black person was more circus attraction than human being.

But later, after I had already started my business, I kept playing. At the weekend I played in a reggae band in Hong Kong, at the Africa Bar.

Later on I sang Chinese songs in bars and restaurants, sometimes three sets a day, sometimes six days a week. I earned good money. I sang in Mandarin, Cantonese and English. *Un Congolais, c'est bizarre.* I played in big hotels. Lots of karaoke, too. My work took me all the way to the Mongolian border. In 2000, when I had a gig in Beijing, I met six Congolese students who were here without work and without a visa. I took them back with me to Guangzhou. That was the start of the Congolese community here. They worked in the discos. That phenomenon was becoming immensely popular. After that everyone left the music scene and started their own businesses.

From sideshow attraction to migration pioneer. Bitulu had been more or less the Peter Stuyvesant of Congo, I realize. He turns out to be a gifted storyteller and very well informed. During our talk, I fill ten pages with notes. He tells me how it had all started in Guangzhou in 2000, when a group of West Africans, Senegalese, and Malians, arrived within a few months of each other. They stayed at a Muslim hotel close to Tianxiu. He tells me how easy it was to obtain a visa back then, even for six months, even for a year. What a difference compared to the situation today, where you're lucky to receive a visa for just two weeks, he sighs, where people go underground once their visa runs out and risk prison sentences of one to six months. "The situation is becoming unbearable, even for people with an official visa or residence permit, like me. The flights keep getting more expensive, the price of merchandise has gone up, transport is pricy, the Congolese customs costs are sky-high, and the market in Kinshasa has become saturated."

He's not homesick for Congo. "I've really become permeated with Chinese culture. The Congolese should organize themselves better, the way the Chinese do. They should start working collectively, but they don't want to do that, even though that would help them to negotiate much better prices. It's like a virus. The contract that Congo signed with China, that was badly negotiated too. No one in the Congolese delegation spoke Chinese. Now China is going to build a few roads, quickly, which no one will keep up." What he says is what many Congolese in China are thinking: this deal, the biggest in their country's history, was a rush job; the country has been sold downriver for a little pocket money. "I'm perfectly willing to admit that I'm ashamed to be Congolese. Since independence, Congo has never been a real country. Nothing there works. All the people think about are their own wallets. In twenty years' time, I've seen China develop." In the villages where Bitulu went to sing his songs in 1990, less than 5 percent of the families had a television in 1990; by late 2006 that was 90 percent.[1] "I've watched Vietnam grow. I've been to Dubai and I was amazed. It's a desert, right, but flowers grow there, they put tubes under the lawns. No, they've done a good job of developing their country. If God had put the Congolese out in the desert, would they have done the same? *Papa, c'est fini!* It's not the white people's fault, or Mobutu's fault, that things are going so badly at home; they're just the scapegoats—that's all over and done with. Look at the Chinese. They learn from Europe and they know that there's no magic involved, only hard work."[2]

THE DASHATOU BUSINESS DISTRICT is dedicated entirely to electronics. There are shopping malls just for digital cameras, next to shopping malls for laptops or LCD screens. After my meeting with Bitulu, I seize the chance to get lost there. That leads me to a windowless megastore where they sell only cell phones. Hundreds of boys and girls man the little stalls there; when they get hungry, they duck down behind the cash register and wolf down a paper cup of noodles. Participative observation being unequaled as ethnographic methodology, I inspect their wares. "Chinese copy!" they say frankly when I hold up something that looks like a perfect iPhone. "This one good copy. This one bad copy." That seems clear enough. "This one original." No, I'm not interested in an *owigina*. A real fake seems much more original to me, especially since these copy phones offer features the original does not, like room for two SIM cards, useful for the frequent traveler. In the brave new world, the line between real and fake fades. Fake is no flimsy replica, but technological avant-garde. And so I buy a few fake iPhones and imitation Ericssons for about fifty dollars apiece. Back in Kinshasa I will sell a couple of them to help pay for my ticket home. What I don't know at this point is that I should have bought thirty of them instead of five: within a single day, I will sell them for many times the price I paid.

At one of the stands I meet Enson, a young, hyperactive Chinese who, installing SIM cards and replacing batteries all the while, speaks to me in fluent Lingala. No, he's never been to Congo, he says, he works in this windowless space every day, but a lot of his customers are from Kinshasa. He doesn't speak French, though: without realizing it, however, half the technical jargon he uses consists of French words. "Ozana besoin sim mibale?" (Do you need one with a double SIM card?) "Ay, papa, accessoires mpo modèle oyo eza te." (Sir, this model has no accessories.)

THE TIANXIU BUILDING is a multilevel shopping mall with bright neon lights and a cacophony of Muzak. The narrow corridors consist of tiny glass shops where extremely extroverted merchants ply their wares. Many of these shops are outlets for factories elsewhere in Guangdong. Along a twenty-meter (sixty-five-foot) length of corridor I see industrial batik textiles, flip-flops, sneakers, boots, dress suits, jogging suits, T-shirts, g-strings, jewelry, cell-phone chargers, cell phones, electric fans, antimosquito coils, chainsaws, generators, motor scooters, and drum sets. As soon as the customer comes in, the salesperson jumps to his feet. Your wish is their command. Lengths of cloth are unfolded and put away again. Suits are lifted down off the rack with a stick and held up to see if they fit. The verbal communication is a disaster, but the adding machine always saves the day. This results in absolute gems of pantomime. The merchant types in his asking price in renminbi, as the yuan is called these days, and holds up the little screen for the customer to see. The African converts this to dollars, frowns, says: "No, no, no!" and types in a price that is half that. The Chinese smiles but looks pained, shakes his head and punches in a number that hurts him less. Upon which the African rests his arms in despondency on the glass counter and casts a bored look out the window. After a dramatic pause full of inconsolable longing and deep indignation, he keys in a new number and turns the calculator around for the merchant to see. This goes on for a time, back and forth, until the African makes as if to leave for another shop and new sources of possible friendship are found after all.

One shop sells only g-strings, including a fantastic model printed with the Angolan flag. The strings and triangles bear the national red and black colors; the Communist logo—a gear, a machete, and a star against the jubilant yellow of dawn—is printed at vulva level. When I ask cautiously what it might cost, it turns out they are sold only by the thousand. "Thousand," the woman says, "not one," as she types in a one and three zeros on her calculator.

Dadine is hesitating over a few pairs of jeans. The price, seven dollars, appeals to her; in Kin she can get thirty-five for them, but jeans are so heavy in her baggage. That means she can't take as many with her, and the way customs is at home . . . She's going to think about it a bit first. "At home it's a war zone. You come back from China, exhausted, and the customs people at the airport pounce on you while you're still waiting for your bags. They demand thirty dollars a bag to let you through, sometimes even up to a hundred, but often enough they just open your baggage with a pen or a key and take a shirt or a pair of pants, right before your eyes."

The sisters Fatima and Fina, rare Congolese Muslims, are in a fix. I met them on the plane and a few days later I see them sitting on a bench, recovering from their bout of shopping. They had been planning to fill a sea container with cans of tomato puree, they explain to me, a twenty-foot container, not forty, those are too expensive, but at the factory they were told that the order could not be ready before December. That means the cans would arrive in Kinshasa no sooner than February, too late for the year-end parties they had been counting on. Maybe they should try their luck with nutmeg? But then again, the price of nutmeg rose from $7,200 to $8,200 a metric ton (2,200 pounds) between January and October 2008, and a container easily holds twelve metric tons (over thirteen U.S. tons). And then the transport! It costs $5,600 to have a twenty-foot container shipped to Matadi, $10,000 for a forty-foot container. Plus you have the import duties, and Congolese customs are the most expensive in the world: up to $15,000 for a small container, $20,000 for a big one. They explain it all to me. The official rates are, as always, negotiable, but a lot of people these days prefer to have their cargo shipped to Pointe Noire in Congo-Brazzaville. Maybe that's what they should do? The container would be brought by truck to Brazzaville and then their hundreds of bags of nutmeg would be loaded onto the ferry to Kinshasa, where the cargo handling is traditionally done by people in wheelchairs, because they don't have to pay as much for the crossing. Crippled porters, I've seen that with my own eyes. The handicapped people I spoke to considered it an acquired right to accept pay for loading their wheelchair with sacks, piled so high they couldn't see over them, and then roll onto the boat as a passenger.

Lina is, without a doubt, the most successful young businesswoman I've met. Within four days she has had two large sea containers filled with building materials: tiles, doors, air-conditioning units, glazed earthenware, sinks and toilets, and lighting fixtures. In Kinshasa these days you find Aomeikang brand toilets, Meijiale brand sinks, Hefei Chenmeng brand fire alarms, and, yes, even Wij Mei brand toilet paper. Lina's first container is already sealed; now she is looking around for a couple of plasma screens to go in the second one. When she's done with that, she's going to have some clothes made for herself. She brought along a few photographs from an African magazine, it's up to the Chinese to do the copying. The only thing is: she has this nasty pain in her stomach. Her niece came along with her this time; the younger woman wonders whether it might be a good idea for them to undergo fertility treatment in China. Why, after all, buy only goods when there are also services to be had? But Lina will become acquainted with the Chinese medical system sooner than she thought. When I see her again a few days later, she tells me she went to a clinic. The nasty pain in her stomach was an inflamed appendix. "Normally, I would go to South Africa for an operation," she says, "but this time I'm going to have it done in China. They say Chinese medicine is good."

THE AFRICAN GROUP MIGRATION TO GUANGZHOU is becoming a factor of growing significance. More and more people are arriving all the time and becoming a deeper part of the country itself. Some of the migrants share a home as though they were family: while everyone is out buying goods, one of them stays home to prepare the most African meal possible with the available ingredients. Others eat with chopsticks as though they have never done otherwise. One Congolese man had started a café and dance hall, Chez Edo, which every African I spoke to said was the most fun place in the whole megalopolis, but the government closed it down because he didn't have the right papers. Others have started barber shops or design clothes. Homosexuals, who have a bitterly hard time of it in Africa, have discovered new possibilities in China and have no plans to return home. I met a young Congolese gay man who had been disowned by his family in Kinshasa, but had started a relationship with a Nigerian in China. For him, China was not the land of repression, but of freedom.

One of the big merchants, Monsieur Fule, is informally recognized as "chairman of the Congolese community in Guangzhou." Neither the function nor the organization itself are official, but the role he plays is rather like that of consul. Anyone arriving in town goes by to talk to him. When I meet him, he is sitting at a desk covered in women's shoes. "I've been here for nine years and I have a residence permit," he says confidently. Fule was one of the needy students who Jules Bitulu convinced to go with him from Beijing to Guangzhou. "But for foreigners without a visa, the Chinese have a prison. The golden years are over. Commerce here has become slippery ground, but in Congo it's even much worse. Our country is destitute and things aren't getting any better. Everything is dirty, but thanks to China everyone is now at least dressed properly." He is fairly positive about the big contract between the two countries. "It may sound a bit vague," he says, "but people have been stealing ore from Congo for years already. Now at least there are billions of dollars being paid for it. Congo is still flat on its rear end," he concludes from behind his wall of ladies' footwear, "but we'll go back someday anyway. The Congolese migrants in Europe don't care about their country; their social life takes place there, but those of us here in China realize that commerce alone is not enough to satisfy us. Someday we'll go back."[3]

One Sunday morning I enter office number 3105, on the thirty-first floor of the Tianxiu Building, high above the shops. It is a sparsely furnished space with a worn-out carpet, but a Congolese merchant has set up his own church here under the ambitious-sounding name Église Internationale pour la Réconciliation. Prayer meetings are held three times a week; on Sundays there are two services of three hours each. As I enter, I noticed that, in this particular diaspora, God has lost a little of his sparkle. He matches the interior. There are only eight worshipers, including the Chinese keyboard player. During a lengthy meditation on a Bible verse, the preacher says: "God's word is like the rain. It only rises back up to Heaven after watering the earth, so that we know . . ." "WHAT SUCCESS IS!" the congregation answers in unison. This game of call and response is nothing new to them. "In all our . . ." "PROJECTS!" "So that they all may . . ." "SUCCEED!"

Then the congregation stands to pray. Their eyes closed and arms raised, they talk out loud, beseeching the Lord loudly for strength and commercial insight. The pastor also asks them to pray for *notre frère* David, who is here today for the first time. During the singing afterward the Africans dance limberly, while the Chinese organist simply shifts his weight back and forth from one foot to the other. "It's not easy for them," the evangelist tells me afterward, "they don't know much at all. They don't even know who Abraham is. If you have to explain all that first . . ."

THAT AFTERNOON I pass by the home of Patou Lelo, a trader who sends a hundred to a hundred and fifty containers to Africa each month. He took his

MBA at Wuhan and now lives in a modest apartment on the ground floor
of a housing block where the sun rarely enters. His daughter, who is almost
two, is playing on the carpet. She has African features, but Chinese eyes. Her
skin has a warm, ochre tint to it.

> When I first got here a lot of people asked whether they could touch my
> skin. They thought I was Chinese, but that I had stayed out in the sun too
> long and would soon turn white again. When I walked down the street
> with my girlfriend, a lot of people thought she was my interpreter, or
> even a prostitute. We've been married for two and half years. Her mother
> was dead set against it. "It's either him or us!" she said, but my wife's
> stepfather didn't make a fuss. "Listen, he's a calm and serious man," he
> said. In Congo, it was the same way: my father didn't give a damn, but
> my mother was very upset. She didn't accept my wife until after our
> child was born. In China, the family is as sacred as it is in Congo; it's not
> like in Europe, where the couple is the most important thing. Here the
> grandparents are very important, we care for them. The couple with one
> child and the grandparents, that's the nuclear family here.

Atop a chest of drawers are some photographs of Lelo's wedding. They
show him and his wife in traditional Chinese, Japanese, and Western outfits.
A radiant couple. His nephew and his brother flew in from Congo for the
wedding; the entire Congolese community in Guangzhou was there. Still,
things here are not always easy, he admits.

> It's a totally different culture, and diametrically opposed to our Congo-
> lese one. The Chinese are hypernationalistic. My wife will automati-
> cally start defending someone, simply because they're Chinese. She's
> atheistic too. Not many Chinese are religious, or maybe they're Bud-
> dhists, but that's *une petite religion*. Here they burn their dead. That's
> hard for us to take. When a Congolese person dies, the community
> gets together to raise money to have the body flown back. In economic
> terms, the Chinese are highly developed, but they're morally back-
> ward. That spitting on the floor in big restaurants . . . Although I have
> to admit that Chinese women are much more open than the men, my
> wife certainly is.

He knows he's lucky; racism is rapidly becoming more common in Guangzhou. More and more taxi drivers refuse to take a Congolese fare. They no longer call them *hçi rén* (blacks), but *hçi gǔi* (black devils). The streets around Tianxiu are known as the neighborhood of the black devils or chocolate city. If an African woman touches the vegetables at the market, the sellers will sometimes throw them away.

"But the blacks themselves are partly to blame. They don't integrate, they don't adapt. The drugs gangs of Nigerians and people from Sierra Leone give us a bad name, while a lot of Congolese people here work very hard." Harder than in Congo, Lelo insists. "Look, people who are a hundred percent honest don't exist in Congo. They're always out to make some easy money fast. They don't understand the principle of investment, because the family always takes all the money. There's no room for reinvestment. But here there's more distance between the businessperson and the family, you understand?"

Everyone in his own family has emigrated—his brother lives in Spain, his sister in France, another sister in Manhattan; his old mother was the only one who stayed behind in Kinshasa. Many Congolese go abroad to escape suffocating family ties. The oft-praised African solidarity has something touching about it in times of crisis, but in times of reconstruction it generates an infernal logic that makes long-term projects impossible: the little bit of money that *is* available is immediately distributed to feed many hungry mouths. Reinvestment and planning are not highly valued. In China, things are much easier. There are no uncles and nephews to accuse you of sorcery when you refuse to share the little bit of money you've earned; witchcraft in Congo is the ultimate argument for enforcing solidarity.

"No one here ever talks about witchcraft," Lelo says, visibly relieved to be rid of that higher metaphysics. In Congo, many people have turned to the Pentecostal churches to protect themselves from witchcraft, but this morning I witnessed how little need there is of that in China. "Fake pastors and false shepherds only proliferate in Congo because of the poverty, but here work is more important than religion."[4]

THAT EVENING I stop in at the office run by Georges, the man who picked me up from the airport. Even on Sunday, he is hard at work. "We have to work while we're still young," he says, "because someday we'll be old." His transport company's motto is *Vous server, c'est notre devoir* (serving you is our duty) and that is definitely no empty slogan. Two employees, César and Timothée, drag huge cardboard boxes around and lug them up onto a scale, where they can barely even read the display. Georges is on the phone constantly. Can that container be sealed yet? How many tons can still go in? When does the truck leave? Has someone already gone to the airport? Wait a minute, David, how many kilos of baggage allowance have you still got? What, forty kilos! But what have you been doing for the last few days? Didn't you buy anything at all? Only five cell phones and two suits? Forty kilos, are you sure? Do you want to sell them? Fourteen dollars a kilo, okay?

And while I am literally selling thin air, at the back of the little office, two Chinese staff members, Iso and Jodo, are filling out forms. Iso, a young woman with a delicate-looking pair of reading glasses, flips through a dictionary; she's trying to learn English and French. Working for a Congolese trader is a good way to earn some money and to brush up on your languages. On the wall is a DHL poster and a world map with China in the middle: Europe and America have become outlying areas, Asia and Africa constitute the new center. European-American relations may have been the most important intercontinental contacts of the twentieth century, but Sino-African relations will be those of the twenty-first.

A printed sentence in Lingala is hanging on the wall: "svp Ndeko awa ezali esika ya mosala" (Dear friend, this is a place of work). "I printed that out and hung it up there," Georges says, "because otherwise the Congolese come in here and want to chew the fat all day." The Congolese in Guangzhou are incredibly industrious. One of the traders I called for an interview told me: "Today I'm much too busy, but tomorrow I have forty minutes for you. Will that be enough?" Vastly different from Congo, where almost everyone is available all the time, and where most people are disappointed when you make moves to leave after only four hours.

When the two cargo personnel are done with their weighing and stacking, they suggest we go out for a beer. Right next door is a snack bar with a few chairs out on the pavement. Darkness has already fallen, but in Guangzhou nighttime is a relative notion. We sit on the sidewalk and watch the girls from the massage parlor across the street. They wear white robes and a red ribbon draped over their shoulder. They are experts in the traditional techniques of Chinese massage, and they are trying to draw in customers. For a real massage, César explains, not "the very special one."

César is in a class all his own. His eyes are bloodshot and his voice vac-
illates between mirth and blues. In Congo he was a police commander for
years; "Commandant César" was what he still liked to be called. He served
under Mobutu, Kabila *père,* and Kabila *fils.*

> You still had the tough training back then. I once spent two days stand-
> ing in a pool of water, up to my chest. Dirty, filthy water, if you fell
> into it you were dead. Or four days' guard duty, on your feet the whole
> time, without sleeping, no problem. But in 2002 I'd had enough. My
> whole family had taken off, all six of them. My parents were the only
> ones who stayed behind, with my sister to take care of them. I went to
> Thailand and from Thailand I tried to get to Germany. A friend of mine
> who was already in Germany sent me his passport by DHL. But when I
> got to the German border, the immigration people saw that something
> was wrong. I was thrown into jail for a month, then put on a flight back
> to Thailand. From there I traveled around to all the countries: Singa-
> pore, Vietnam, Malaysia, Hong Kong, Korea, the Philippines . . . I had
> to move every month to keep my passport valid. That's how I ended up
> in China, but my visa has already expired. They could come and pick
> me up any time.

He puts down his glass and shouts to the proprietress in Cantonese that
he wants another beer. The alleyway is drab. On the ground beside our plastic chairs is a fat rat that doesn't move but keeps chewing on something the
whole time. "I met a beautiful woman here, a woman with long black hair.
She came from western China. She didn't look Chinese at all, more Indian
or Russian, I don't know." Uighur, probably, but I don't interrupt him. Timothée is plucking at the label on his beer bottle. César starts in on his second
draft. "It was all going wonderfully. We ran a phone shop together and we
did good business. She wanted to have children, but I already have eight in
Kinshasa. Then she started trying to corner me. She demanded that I sever
all ties with my friends and family. It had to be just her and me. *Mais je suis
un africain!*" He shouts it out, but the rat still doesn't budge. "I felt so imprisoned, I was ready to kill myself. But she was such a beautiful woman, everyone looked at me when I went out with her. The phone shop was doing well.
Then, after hesitating for a long time, I broke up with her; it was really hard.
She kept the shop, but she blackmailed me: if I ever went into the phone busi-

ness again, she would report me. So here I am. No job, no visa; all I can do is a few odd jobs for Georges."

The rat is gone, and Timothée suggests we go out dancing. Have I ever been to Kama? It's really something special. In the taxi on the way over he fills me in on the disco's history. "The owner of Kama is Chinese, he's married to an Arab woman. But the DJ is Nigerian." When we enter I see that the disco is housed in a pitch-black temple, there are both Asians and Africans walking around. The DJ plays Chinese techno, Asian beat, and, of course—what else, it is after all the country's major export product after copper ore—Congolese rumba. We find a table and order our beers. Commander César is starting to perk up. He swings along with Magic System's "Bouger bouger" (Move, move), the catchiest number to come out of Africa in this third millennium. No Western music is played; pop and rock are irrelevant genres from the far corners of an old world. A band is setting up to play. The DJ makes way for a female Cape Verdean reggae vocalist; her backup band comes from the island of Mauritius. The go-go girls are three Philippine singers whose latex boots are four times as long as their skirts.

There are all kinds of side rooms one can rent for a karaoke party. César doesn't understand why you would want to go off and croon on your own when you could watch such an, um, interesting performance. Cutting back and forth between the tables is a gorgeous Chinese girl selling flowers; she even has a teddy bear for sale that's almost as big as she is. That, too, is puzzling to César. "Les chinois," he sighs.

After a couple of hours we head off for a beer at an outdoor café in the African neighborhood. There we're able to hear each other again, although our ears are still buzzing. The traffic leaves red and yellow trails as it zooms past, neon signs scream for our attention, ladies of easy virtue float up and float away again. Timothée, who hasn't said much all evening, begins to loosen up as well. "I've started discovering all kinds of new flavors," he laughs. "Russian, Chinese, Thai, Tanzanian, Rwandan . . . whoa, no, no Rwandans, I hate them! But the most expensive women are the Africans; there aren't a lot of them. For an African girl with a nice butt you easily pay two hundred RMB for one go. That's thirty dollars!"

"Or four hundred RMB!" César chimes in. "For one go!"

"I always pay 150 RMB for two goes, sometimes only a hundred. The Chinese girls only get thirty RMB, less than five dollars. What do you expect? They've got nothing, and they do nothing!"

"In Bangkok I saw some weird things," César laughs. "Boys who turned into girls, no kidding! Their . . . their . . . how do you say that? Their thingamabob was cut off, their penis, yeah, that's it. And then they had a hole drilled in it. *Vraiment!*" Once again he shakes his head at this remarkable Asia where a twist of fate has brought him. And to think that he had been planning to live in Germany. He turns to watch the girls as they walk by and smiles at them. His eyes are red, his face is weathered, but something vulnerable comes over him. Is it the alcohol? Is it lovesickness? The nostalgia of the exile? "I don't want women anymore. On very, very rare occasions I'll take a girl home with me, but almost never. I usually just go into the bathroom at night and take some douche gel. I rub it on myself and that's how I relax."[5]

IT IS NO LONGER THE SOUND of the slit drum that spread the news from village to village, no longer the dull thump of the tom-tom, no longer the crack of the whip, not the pealing of the mission bells, not the thunder of the train or the rattling of the drill in the mineshaft, no, it is no longer the ticking of the telegraph, the crackle of the radio or the cheering of the people that sounds the nation's heartbeat today. It is not in the stamping of manioc in the mortar, not in the slap of water against the canoe's hull. The heart of this country is not in the rattle of weapons in the jungle, not in the table pounding against the wall while a woman screams that she never wanted this, no.

It is night, but that is not the way it feels.

The new Congo reverberates to a different tone, the new Congo sings in the arrival hall of an airport thrumming with noise. It is the sound of tape, brown rolls of tape around packages and boxes, tape that screams as it is unrolled and grunts as it is torn, *grrrreeeeee . . . clunk,* tape that scrapes and shrieks and bawls, tape, meters and meters of tape in the airport arrival hall, a quiet wailing around the baggage trolleys, as in an incubator. Everywhere people are swaddling their things in brown plastic. And once the goods are packed, they are inscribed in magic marker with name and district and street.

That shrilling sound is no complaint, but the cry of new life.

I HAD NOTICED THEM ALREADY during boarding: two women with bleached-blond hair, no, with bleached-blond, bobbed wigs. They were chattering happily, slapping each other on the back, laying their heads on each other's shoulders and giggling wildly. Their suitcases and bags were in the hold, their names scribbled on the tape. They were both wearing the same brand-new outfit, pants and a blouse with a floral motif. With the labels still attached. Just wait till the people saw them in Kinshasa! When you have something new, you flaunt it. The men didn't cut the label off the sleeve of their new suit, did they? Children didn't take the plastic wrapping off the brakes of their new bicycles, did they? Well then!

The atmosphere on board was festive. The two bleached-blondes had put on headphones, they were watching a cartoon and commenting loudly on what they saw. We were flying back to Central Africa. This was only the second direct flight between Guangzhou and Nairobi, the first had left two days ago. No stopover in Bangkok or Dubai, just straight across the Indian Ocean in one shot: it felt like a historic happening.

In Nairobi I saw two young Dutch tourists with sunburned faces sprinting for their gate. They were wearing short pants and sandals and carrying a big wooden giraffe, a souvenir wrapped in local newsprint. I didn't know what exactly, but something about the scene irritated me. In the last few days I had felt as though I were being granted a glimpse into the third millennium, but now I was being tossed back rudely into the last century, the century when Europeans bought wooden giraffes in Africa. My reasoning wasn't completely lucid, but I was too tired to worry about being consistent.

During the last stretch of the journey we flew straight across Congo. The bleached-blond women lay sleeping, their mouths open. Through the little window I saw the huge, moss-green broccoli of the equatorial forest, crisscrossed on occasion by a brown river glistening in the sun. That Congo's natural riches have helped lend hue to the world's economy is a familiar enough story. From billiard ball and rubber band by way of bullet casing and atomic bomb to the cell phone. But that purely utilitarian jingle seems to me too limited and too cliché, as though Congo, this wondrously beautiful country, were only the world's storehouse, as though—with the exception of its raw materials—it had not contributed much to world history. As though its subterranean layers were important to all mankind, but its own history merely a domestic matter, richly permeated with dreams and shadows. While I, in my conversations and reading had so often seen the exact opposite. In the early twentieth century the rubber policies gave rise to one of history's first major humanitarian campaigns. During both world wars Congolese soldiers contributed to crucial victories on the African continent. In the 1960s it was in Congo that the Cold War in Africa began, and that the first large-scale UN operation was held. The point is not whether those were achievements on the part of the Congolese themselves: the point is that Congolese history has helped to determine and form the history of the world. The wars of 1998 and 2003 prompted the biggest and most costly peacekeeping mission ever, as well as the first joint military effort by the European Union; their conclusion produced a unique combination of multilateral and bilateral diplomacy for the purpose of minutely monitoring agreed policies. The 2006 elections were the most complex ever organized by the international community. The International Criminal Court is currently establishing invaluable jurisprudence with the prosecution of its first defendants—three men from Congo. Clearly, the history of Congo has on any number of occasions played a crucially important role in the tentative definition of an international world order. The contract with China, accordingly, is a major milestone in a restless world in motion.

They walked out in front of me, across the tarmac, on their way to the yellow airport terminal. A few planes were parked haphazardly here and there. The jet engines of one of them ripped the world in two like a giant buzz saw. In the midst of that extraterrestrial roar hung the odor of burned kerosene, mixed with the odor of smoldering plastic from the nearby slums. The air sizzled in the heat and it was not yet noon. I had been too tired to approach them, too tired from traveling and from my attempts to understand. But I saw them walking, still in high spirits, clearly proud of the journey they had just made. I saw the blond hair of their wigs bounce with every step they took. I saw how the wind tugged at a few strands of it. And while they stepped briskly across the crumbled tarmac on the way to their homecoming, I saw the labels on their sleeves flap and spin in the morning air, frisky and playful, as though they had something to celebrate.

ACKNOWLEDGMENTS

The idea for this book arose one evening in November 2003 at Café Greenwich in Brussels. I was sitting alone at a table having a drink. In the years that went before I had traveled a great deal in Southern Africa and written about that; now I was about to visit Congo for the first time. In preparation for my trip I had just visited a few bookstores in Brussels, without really finding what I was looking for. Maybe I should write it myself, I realized then, for apparently I belong to that genus of writers who happen to write the books they themselves would like to read. At the time there was no way I could have known that, with that playful brain wave, I was embarking on a project that would take years and result in countless unforgettable encounters. But even at an early stage I decided to surround myself with a few of those whose judgment I value highly: Geert Beulens, Jozef Deleu, Luc Huyse, and Ivo Kuy. According to good Central African tradition I referred to them as "my uncles": I could call on them whenever necessary, and enjoyed their confidence even before I had proven myself worthy of it. The sense of their silent involvement meant more to me than they realized.

It was clear from the very start that this book, sweeping gesture that it is, would be easier to write if I were not associated with any university. The freedom of authorship was dearer to me than the certainty of academic tenure. For the funding I decided to stick to the rule applied by Amnesty International, that is, to accept no money directly from any government institution: only in that way could I preserve my independence. It was therefore a true stroke of luck to be able to receive the support of five institutions, all of which work with autonomous—and often even anonymous—assessment committees. I am sincerely indebted to the Flemish Literature Fund, the Dutch Foundation for Literature, the Pascal Decroos Fund for Investigative Journalism, the Foundation for Special Journalistic Projects, and the Netherlands Institute for Advanced Study for the means they placed at my disposal. During two of my ten trips to Congo I traveled with the press corps, within the context of a Belgian ministerial visit. During my longer stays I regularly made domestic flights aboard planes used by the UN peacekeeping forces. That was the full extent of my embeddedness. I received no money from any government ministry, was not sponsored by any company, and accepted no lodgings from any NGO. Those who offered me such favors I teasingly told that, were I to accept, I would be placing them at far too great a risk.

Independence is the greatest good, but that is not to say that I played the knight errant. I was greatly nourished by the ideas of many others, first of all the numerous informants whose names have been mentioned in the previous chapters. They are the life's blood of this book. As time went by, I even became friends with a few of them. But help also came from a great many people behind the scenes. A number of eminent Congo experts were extremely generous with information from the start. Lieve Joris provided me with books and contacts with a generosity rarely seen in this day and age. Walter Zinzen, Filip De Boeck, and Benoît Standaert were inexhaustible sources of erudition and friendship. Guy Poppe, Katelijne Hermans, Ine Roox, Peter Verlinden, Koen Vidal, Maarten Rabaey, and John Vandaele were more than willing to share with me their views on Congo. A number of people who knew I was working on this book drew my attention to interesting source material. My special thanks in this regard goes out to Colette Braeckman, Raf Custers, Roger Huisman, Piet Joostens, Luc Leysen, Alphonse Muambi, Sophie de Schaepdrijver, Mark Schaevers, Vincent Stuer, Margot Vanderstraeten, Pascal Verbeken, Paule Verbruggen, and Honoré Vinck.

In Kinshasa I profited greatly from my conversations with Zizi Kabongo, Annie Matiti, Noël Mayamba, consul Benoît Standaert, and Johan en Mieke Swinnen, the Belgian ambassador and his wife at that time. Chauffeur Didier Catu, Colonel Frank Werbrouck, ambassador Geoffroy de Liedekerke, and Brother Luc Vansina helped in various ways to solve logistical problems. In Kisangani I was assisted by Pionus Katuala, Faustin Linyekula, and Virginie Dupray. In Bunia it was a privilege to become acquainted with radio journalist Jean-Paul Basila. Sekombi Katondolo, Chrispin Mvano ya Bauma, Cléon Mufingizi, and Carine Tchoma came to my assistance during my time in Goma. In Bukavu I was the guest of Adolphine Ngoy and her family. In Lubumbashi I spoke at length with Jules Bizimana, Father Jo De Neckere, and Paul Kaboba. In Rwanda I was guided expertly by Gady Byabagabo. In Nkamba, the sacred city of the Kimbanguists, I learned a great deal from the young journalist Tétys Danaé Samba. In Nsioni I had the great good luck to listen to the accounts of Dr. Jacques Courtejoie and his friends Roger Zimuangu and Clément Nzungu. In Boma I met the remarkable town archivist Placide Munanga, who told me about his city's history. In Kikwit I spent hours talking with the headmaster of the local school, Rufin Kibari Nsanga, whose desk was literally piled high with books and documents. It was a great joy to make his acquaintance. His knowledge of history was stunning and exceeded only by his curiosity and glowing hospitality.

During the time I spent with the MONUC (Mission de l'Organisation des Nations Unies au Congo), during the 2008 Nkunda offensive, I had fascinating encounters with William Elachi, Sylvie van den Wildenberg, and Bernard Kalume. In China I learned a great deal from my conversations with Belgian consul Frank Felix, Flemish economic envoy Koen De Ridder, Congolese journalist Jaffar Mulassa, and the African entrepreneurs Georges Ndjeka, Dadine Musitu, and Lina Garcia Mendes.

During my travels I regularly met and engaged in edifying and enjoyable conversations with journalists or researchers. I think in particular of Caty Clement, Samuel Turpin, Greg Mthembu-Salter, Kipulu Samba, Hery Mambo, Delphine Schrank, and Kristien Geenen. Most of the time I traveled alone, but it was a wonderful experience to venture out on a few occasions with intrepid travelers like Jan Goossens, Carl De Keyzer, and Stephan Vanfleteren. I met Kris Berwouts, director of EurAc, the European network of NGOs in Central Africa, during a flight from Kinshasa to Bukavu. Even without the aborted plane crash upon our landing at Bukavu, we would have become friends. But as we stepped unharmed from the aircraft and made our way through the tall grass, the torrential rain and the red mud, away from a plane that might explode at any moment, we both realized that we had been very lucky indeed and that from then on we would be bound not only by our love for Congo but also for life itself.

During the actual writing phase, I was regularly able to turn to historians Jean-Luc Vellut, Daniel Vangroenweghe, Zana Aziza Etambala, Guy Vanthemsche, and Vincent Viaene, to anthropologists Filip De Boeck, Peter Geschiere, Klaas de Jonge, David

Garbin, and Anne Mélice, to art historians Roger Pierre Turine and Sabine Cornelis, to archaeologist Els Cornelissen, economist Frans Buelens, and filmmaker Valérie Kanza. Walter and Alice Lumbeeck and Frans and Marja Vleeschouwers, friends of my father from the 1960s, helped me to understand the Belgian perspective regarding the Katangan secession, while Michel and Edith Lechat and Jean Cordy served as very special informants concerning the colonial era. Many people who I never had the pleasure of meeting in person were nevertheless willing to answer my e-mails and phone calls. Reverend Martin M'Caw, Robert Lay, Julian Lock, and Betty Layton helped with information concerning the very first generation of Protestant missionaries. Aldwin Roes, Fien Danniau, Nancy Hunt, Myriam Mertens, Bob White, Bodomo Adams, and Bram Libotte sent me unpublished manuscripts, while Dominiek Dendooven, Didier Mumengi, Steven Spittaels, and Didier Verbruggen helped me with specific information or documents that I was looking for. Bogumil Jewsiewicki, Tom De Herdt, Stefaan Marysse, and Erik Kennes also answered a few thorny technical questions. Odette Kudjabo told me on the phone about her grandfather, who had fought in World War I, Michel Drachoussoff talked about his father, who had kept such a fascinating wartime diary in 1940–45, and Dorothée Longeni Katende related stories about her grandfather, Disasi Makulo, who she had unfortunately never met.

When the manuscript was finished, I asked a few experts to read through it. Vincent Viaene, Guy Vanthemsche, and Filip Reyntjens examined the chapters dealing with the Free State, Belgian Congo, and independent Congo, respectively, while Frans Buelens double-checked the passages on economics. I am deeply indebted to all of them for their careful comment.

In the world of Dutch-language letters it is uncommon to thank one's editor ("Only doing my job," is the slightly embarrassed comment one usually hears from them), but that rule does not apply to Wil Hansen, for the simple reason that he was much more than an editor, but *un honnête homme* of the rarest and most noble sort, and a great pleasure to work with.

I wrote *Congo: A History* at my studio in Kuregem, the oft-cited "problem neighborhood" in the Brussels district of Anderlecht, although I must add right away that I was bothered more by the noise of the police helicopters circling the neighborhood as part of the city's zero-tolerance policy than by anything in the neighborhood itself, where I have worked with pleasure for more than four years. I could never have dreamed of finding in Europe a better place to write about Congo: my studio has a view of the street where every day dozens of second-hand cars are bought and sold before being shipped to Central Africa. The street corners are adorned with posters for concerts by Werrason or services by faith healers. From the outside this neighborhood seems so poorly integrated into Belgian society, I am sometimes told, but from here it seems more like Belgium is poorly integrated into the world. Kuregem is a lesson in globalization, but also in empathy and involvement.

The best classroom for such lessons is probably the Royal Flemish Theater in Brussels. My research on Congo began more or less simultaneously with the start of that theater's artistic Congo project, a long-term exchange program between Congolese and Belgian artists. I was involved in the start of it; I taught several workshops for Congolese authors in Kinshasa and Goma, and meanwhile worked on my theater monologue *Mission*, which had its premier at that same theater. The wonderful work of people like Jan Goossens and Paul Kerstens convinced me that broad social debate is often carried on with greater urgency at such sanctuaries for critical thought than at many universities or in the increasingly commercialized media. It was in that context that I met a number of my dearest Congolese friends. I think in particular of the writers Bibish Mumbu and Vincent Lombume, of dramaturgists Papy Mbwiti and Jovial and Véronique Mbenga, of the actresses Starlette Mathata and Dadine Musitu, of filmmaker Djo Munga, choreographer Faustin Linyekula, visual artist Vitshois Mwilambwe, and sculptor Freddy Tsimba. Not only have they helped me to understand their country, but also to love it, for a country that brings forth such intelligent and courageous artists is far from being lost.

This book could also not have been written without the proximity of a number of very dear friends in Europe: each in their own way, Natalie Ariën, Geert Buelens, Emmy Deschuttere, Jan Goossens, Maaike Pereboom, Grażyna Plebanek, Stephan Vanfleteren, Francesca Vanthielen, and Peter Vermeersch supported me in the long process of writing this text. But above all I would like to thank Bernadette De Bouvere and Tomas Van Reybrouck, my mother and brother, for their untiring wisdom and warmth.

Brussels, April 2010

SOURCES

GENERAL

Congo: A History is the product of much listening and reading. In the notes to this volume, I have listed my sources as comprehensively as possible, yet a number of them deserve special notice. Either because I am so greatly indebted to them, or because they may help curious readers to perform their own research, or simply because I wish to overtly express my enthusiasm.

The first time I flew to Congo, the book I had with me was *The Congo from Leopold to Kabila: A People's History* by Georges Nzongola-Ntalaja (London, 2002), an excellent and animated introduction to the country; unfortunately, I left that first copy in the magazine compartment of the seat in front of me. But the margins of the copy that subsequently replaced it are also scored with pencil marks. The same goes for Isidore Ndaywel è Nziem's classic: *Histoire générale du Congo* (Paris, 1998). A great deal more academic than the former, it has nevertheless often impressed me with its comprehensiveness, its abundant exegeses, and its numerous maps. As I was writing this book, it always lay within arm's reach. Another useful reference work through which I thumbed regularly was the *Historical Dictionary of the Democratic Republic of the Congo* by Emizet Kisangani and F. Scott Bobb, the third edition of which appeared only recently (Lanham, 2010). Jean-Jacques Arthur Malu-Malu's *Le Congo Kinshasa* (Paris, 2002) constitutes a readable and personal overview, and one which deserves to be more widely known.

To get my bearings when it came to new periods and subjects, I started with the better-known reference works. Twenty years after their writing, the chapters dedicated to Central Africa in the seven-volume *Cambridge History of Africa* remain excellent. I read them alongside the entries in the eight-volume *Histoire générale de l'Afrique,* which were often written by African researchers. The recent *A Historical Companion to Postcolonial Literatures: Continental Europe and Its Empires* (Edinburgh, 2008) by Prem Poddar et al. helped me along with its thematic résumés and useful bibliographies.

A few older books are still very useful indeed, including *The River Congo* by Peter Forbath (New York, 1977) for the precolonial period and *Leopold to Lumumba* by George Martelli (London, 1962) for colonial times. Robert Cornevin wrote *Histoire du Congo (Léopoldville)* (Paris, 1963), a lucid but rather Eurocentric work whose lovely maps make up for a great deal. The collection of Jean Stengers's articles in *Congo: Mythes et réalités* (Paris, 1989) remain extremely important, particularly by reason of his analyses of the Free State.

Concerning the workings of the colonial economy, the reader of Dutch now has an excellent reference work: *Congo 1885–1960. Een financieel-economische geschiedenis* by Frans Buelens (Berchem, 2007). In addition to historical information dealing with colonial enterprises, it also provides a good overview of the background of colonial capitalism. For the social aspects of that capitalism, refer also to the classics by Pierre Joye and

Rosine Lewin, *Les trusts au Congo* (Brussels, 1961) and Michel Merlier, *Le Congo: De la colonisation belge à l'indépendance* (Paris, 1962). Specifically concerning the social aspects of Katangan mining, the works of Congolese historian Donatien Dibwe dia Mwembuie, *Histoire des conditions de vie des travailleurs de l'Union Minière du Haut-Katanga/Gécamines (1910–1999)* (Lubumbashi, 2001) and *Bana Shaba abandonnés par leur père: Structure de l'autorité et histoire sociale de la famille ouvrière au Katanga, 1910–1997* (Paris, 2001) make extensive use of oral sources.

For a long time, colonialism was seen as a form of one-way traffic between metropolis and colony, from Europe to Africa. In recent years that view has begun to change, and researchers have started looking at the repercussions of the colonial adventure on Europe. In his interesting book *Congo: De impact van de kolonie op België* (Tielt, Belgium, 2007), Guy Vanthemsche demonstrates convincingly that it was not only Belgium that formed Congo, but also vice versa. He focuses particularly on the Belgian economy and domestic and foreign policies. Along with Vincent Viaene and Bambi Ceuppens I helped to compile a reader that looks at the colonial impact on other parts of Belgian society, such as culture, religion and science, *Congo in België: Koloniale cultuur in de metropool* (Louvain, Belgium, 2009). In addition to this two-way traffic, attention is now being paid increasingly to the diversity of the colonial presence. Besides Belgians, after all, there were also Greeks, Portuguese, Scandinavians, and Italians active in the Belgian Congo. Works such as *Pionniers méconnus du Congo Belge* (Brussels, 2007) by Georges Antipas, about the Greek community in Congo, and *Moïse Levy, un rabbin au Congo (1937–1991)* (Brussels, 2000) by Milantia Bourla Errera, broaden the historical view.

Fascinating diachronic studies exist on various themes. Their cross-sectional perspective makes them worth noting here. Concerning education and science one has the work of Ruben Mantels, *Geleerd in de tropen: Leuven, Congo, en de wetenschap, 1885–1960* (Leuven/Louvain, Belgium, 2007), as well as that of Benoît Verhaegen, *L'enseignement universitaire au Zaïre: De Lovanium à l'Unaza, 1958–1978* (Paris, 1978). Kuvuande Mbote has written about architecture, as has Bruno De Meulder in *Een eeuw koloniale architectuur en stedenbouw in Kongo* (Antwerp, Belgium, 2000) and Johan Lagae in *Kongo zoals het is: Drie architectuurverhalen uit de Belgische kolonisatiegeschiedenis (1920–1960)* (Ghent, Belgium, 2002). For Congolese pop music (which is always more than just music), see Gary Stewart: *Rumba on the River* (London, 2000). Silvia Riva has written about Congolese literature in *Nouvelle histoire de la littérature du Congo-Kinshasa* (Paris, 2000). For film and visual culture, see Guido Convents's *Images et démocratie: Les Congolais face au cinéma et à l'audiovisuel* (Leuven/Louvain, Belgium, 2006). And for the visual arts and some truly marvelous illustrations, see Roger Pierre Turine's: *Les arts du Congo, d'hier à nos jours* (Brussels, 2007). Contemporary artists often provide the viewer with multiple layers of commentary on the history of their country. That certainly applies to the Congolese poets assembled in Antoine Tshitungu Kongolo's lovely anthology *Poète ton silence est crime* (Paris, 2002).

A few other books amazed, surprised and baffled me with their images: *Congo Belge en images* (Tielt, Belgium, 2010) by Carl De Keyzer and Johan Lagae derails all the existing clichés concerning the Congo Free State by means of its sublime selection from the Royal Museum for Central Africa in Tervuren's collection of photographic plates. Every bit as unsettling when it comes to present-day Congo is *Congo (Belge)* (Tielt, Belgium, 2009) by Carl De Keyzer, and *Congo Eza* (Roeselare, 2007) by Mirko Popovitch and Françoise De Moor, a collection of work by contemporary Congolese photographers. It is because I value photography highly as an autonomous form of discourse that the only illustrations found in my own book are maps.

INTRODUCTION

The broad geographical sketch contained in this introductory chapter was gleaned from a wide variety of sources on the Internet and from my own bookshelves. A useful source, replete with maps, is *Géopolitique du Congo (RDC)* by Marie-France Cros and François Misser (Brussels, 2006).

My own first attempt to write a "bottom-up history," based on interviews with those whose perspectives usually do not make it into the written sources, took place in a convalescent home in Brugge/Bruges, Belgium, in 2007. There I spoke to elderly people who had never themselves been to Congo concerning their memories of colonialism, about what they had thought at the time, and above all about what they had done (collecting silver paper, as it turned out, in addition to sewing and patching clothes for the missions, fishing for prizes at the mission-benefit carnival, and doing a great deal of praying for the "poor Congolese"). That study, and the methodological (im)possibilities presented by that combination of oral history and material culture studies, were expanded upon in the collection I edited with Vincent Viaene and Bambi Ceuppens: *Congo in België: Koloniale cultuur in de metropool*. But my analysis was in fact no more than the explicit formulation of the method I have been using for a long time in my earlier journalistic and literary work (e.g., the play *Mission*). And of my conviction that the most highly underestimated archives in Congo are the people themselves.

In addition to my background as archaeologist of pre-history, the importance I attach to the precolonial period is due to Eric Wolf's classic *Europe and the People without History* (Berkeley, CA, 1982). The earliest human population of Congo is virtually unknown, as Graham Connah has shown in *Forgotten Africa: An Introduction to Its Archaeology* (London, 2004). Even the more recent surveys serve only in part to fill in the blanks; see, among others, *African Archaeology: A Critical Introduction*, edited by Ann Brower (Oxford, 2005), and above all Lawrence Barham and Peter Mitchell's *The First Africans: African Archaeology from the Earliest Toolmakers to the Most Recent Foragers* (Cambridge, UK, 2008). I therefore based my snapshot of life some ninety thousand years ago on the excavations at Katanda performed by John E. Yellen: "Behavioral and Taphonomic Patterning at Katanda 9: A Middle Stone Age Site, Kivu Province, Zaïre," *Journal of Archaeological Science* (1996). For a good survey of the rise of modern human behavior in Africa, see Sally McBrearty and Alison S. Brooks, "The Revolution that Wasn't: A New Interpretation of the Origin of Modern Behavior," *Journal of Human Evolution 39* (2000). My snapshot of Pygmy life around 2500 BC makes grateful use of recent studies by Julio Mercader, "Foragers of the Congo: The Early Settlement of the Ituri Forest," in *Under the Canopy: The Archaeology of Tropical Rain Forests*, edited by J. Mercader (New Brunswick, NJ, 2003).

The period around the year AD 500 and the phenomenon of the Bantu migration became more familiar to me through reading Jan Vansina's impressive *Paths in the Rainforest: Toward a History of Political Tradition in Equatorial Africa* (Madison, WI, 1990), supplemented by the painstaking archaeological work of Hans-Peter Wotzka, *Studien zur Archäologie des zentral-afrikanische Regenwaldes: Die Keramik des inneren Zaïre-Beckens und ihre Stellung im Kontext der Bantu-Expansion* (Cologne, 1995). Concerning gongs and drum languages I turned to John Carrington, *La voix des tambours* (Kinshasa, 1974) and Olga Boone's *Les tambours du Congo-belge et du Ruanda-Urundi* (Tervuren, Belgium, 1951).

A better understanding of the rise of the first states I gained after reading Jan Vansina's unique ethno-historical work. My own far-too-summary sketch of the local kingdoms of the savanna was based on his classic *Les anciens royaumes de la savane: Les états des savanes méridionales de l'Afrique centrale des origines à l'occupation coloniale* (Léopold-ville, 1965) and his *How Societies Are Born: Governance in West Central Africa before 1600* (Charlottesville, VA, 2004). Concerning the Kongo Empire around the year AD 1560, I turned to Anne Hilton, *The Kingdom of Kongo* (Oxford, 1985), to David Northrup, *Africa's Discovery of Europe* (New York, 2002) and to Paul Serufuri Hakiza, *L'évangélisation de l'ancien royaume Kongo, 1491–1835* (Kinshasa, 2004).

For the section on 1780 and the impact of the Atlantic slave trade, I made extensive use of Robert W. Harms's masterful *River of Wealth, River of Sorrow: The Central Zaire Basin in the Era of the Slave and Ivory Trade, 1500–1891* (New Haven, CT, 1981).

CHAPTER 1

This chapter relies in part on a booklet that is not available outside Africa: Makulo Akambu's *La vie de Disasi Makulo, ancien esclave de Tippo Tip et catéchiste de Grenfell, par son fils Makulo Akambu* (Kinshasa, 1983). That book presents the life's story of old Disasi Makulo, as dictated to his son. It fell into my hands through a stroke of sheer luck.

Although an enormous amount has been written about the African explorers (see, among others, Christopher Hibbert, *Africa Explored: Europeans in the Dark Continent, 1769–1889* [London, 1982]), there is no truly integral overview of the period 1870–85. Tim Jeal's wonderful *Stanley: The Impossible Life of Africa's Greatest Explorer* (London, 2007), however, is more than an unusually richly documented and thoughtful biography: it paints the panorama of an entire age. Insight into the wild and wooly mid-nineteenth century I gained through Jan Vansina's, "L'Afrique centrale vers 1875," in *La conférence de géographie de 1876 [Bijdragen over de Aardrijkskundige Conferentie van 1876]* (Brussels, 1976), as well as through Jean-Luc Vellut's, "Le bassin du Congo et l'Angola," in *Histoire générale de l'Afrique*, vol. 6: *L'Afrique au XIXᵉ siècle jusque vers les années 1880* edited by J. F. Ade Ajayi (Paris, 1996) and David Northrup's "Slavery and Forced Labour in the Eastern Congo, 1850–1910," in *Slavery in the Great Lakes Region of East Africa*, edited by H. Médard and S. Doyle (Oxford, 2007). More about the Muslim slave trade can be found in Edward A. Alpers, *Ivory and Slaves in East Central Africa* (London, 1975), Abdul Sheriff's *Slaves, Spices, and Ivory in Zanzibar* (London, 1987), and Ronald Segal's *Islam's Black Slaves: The Other Black Diaspora* (New York, 2001). Regarding the life and work of the two most powerful Afro-Arab traders in Congo, see François Bontinck, *L'autobiographie de Hamed ben Mohammed el-Murjebi: Tippo Tip (ca. 1840–1905)* (Brussels, 1974) and Auguste Verbeken, *Msiri, roi du Garenganze: "L'homme rouge" du Katanga* (Brussels, 1956).

For the native reactions to the European explorers, see Frank McLynn, *Hearts of Darkness: The European Exploration of Africa* (London, 1992). Johannes Fabian turned the anthropological gaze 180 degrees with an impressive ethnography of the European explorers: *Out of Our Minds: Reason and Madness in the Exploration of Central Africa* (Berkeley, CA, 2000). I was able to document my findings on the first generation of missionaries with the help of E. M. Braekman's *Histoire du protestantisme au Congo* (Brussels, 1961) and Ruth Slade's *English-Speaking Missions in the Congo Independent State, 1878–1908* (Brussels, 1959).

A very great deal has been written about the division of Africa. Thomas Pakenham wrote the hefty *The Scramble for Africa, 1876–1912* (London, 1991), but H. L. Wesseling's crystal-clear and entertaining *Verdeel en heers: de deling van Afrika, 1880–1914* (Amsterdam, 1991) helped me the most in understanding the international context within which Leopold II maneuvered. Wesseling in turn made great use of Jean Stengers's still-indispensable *Congo, mythes et réalités: 100 ans d'histoire* (Paris, 1989). Stengers's article

"De uitbreiding van België: tussen droom en werkelijkheid" in *Nieuw licht op Leopold I en Leopold II: Het archief Goffinet*, edited by G. Janssens and J. Stengers (Brussels, 1997) provides an update based on unique archive materials. Belgium's Royal Academy for Foreign Studies published two important collections dealing with the events between 1876 and 1885: *Bijdragen over de Aardrijkskundige Conferentie van 1876* (Brussels, 1976) and *Bijdragen over de honderdste verjaring van de Onafhankelijke Kongostaat* (Brussels, 1988).

CHAPTER 2

For more than a decade, the discussion concerning the Congo Free State has been dominated by Adam Hochschild's *De geest van koning Leopold II en de plundering van de Congo* (Amsterdam, 1998). That book's achievement was to inform a broad public about the abuses in Congo and to make academic knowledge accessible and exciting. Unfortunately, however, it depended more upon a talent for generating dismay than on any shades of subtlety; Hochschild's perspective is often very black and white. In understanding the complexity of a person like Leopold, I profited more from studies by Jean Stengers cited earlier, but also from more recent studies placing him in the context of his day. In his thesis, *Koningen van de wereld: De aardrijkskundige beweging en de ontwikkeling van de koloniale doctrine van Leopold II* (Ghent, Belgium, 2008), Jan Vandersmissen has pointed out the impact of geographical science. Vincent Viaene has shed light on the "fever of empire" in Belgian high society and drew my attention to the king's national and social agenda in "King Leopold's Imperialism and the Origins of the Belgian Colonial Party, 1860–1905," *Journal of Modern History* 80 (2008): 741–90. Recently, Jean-Luc Vellut examined the African context of Leopold's colonialism in "Contextes africains du projet colonial de Léopold II" (unpublished lecture, Louvain-la-Neuve, March 2009). See also Viaene, Vellut, and Vandersmissen's contributions in *Leopold II: Schaamteloos genie?* edited by Vincent Dujardin, Valérie Rosoux, and Tanguy de Wilde d'Estmael (Tielt, Belgium, 2009). The world, however, is still waiting for a definitive biography of Leopold II.

I found a nuanced view of officials, traders, and soldiers in the Free State in L. H. Gann and Peter Duignan, *The Rulers of Belgian Congo, 1884–1914* (Princeton, NJ, 1979). The catalogue for the exhibition *Het geheugen van Congo: De koloniale tijd* (Tervuren, Belgium, 2005) does its best, in the able hands of its editor Jean-Luc Vellut, to avoid old and new clichés concerning the Free State. Two scholars have carried out true pioneering work in the extremely fragmentary archives dealing with the rubber policies: Daniel Vangroenweghe with his *Du sang sur les lianes [Rood rubber]* (Brussels, 1985) and *Voor rubber en ivoor* (Leuven/Louvain, Belgium, 2005) and Jules Marchal with his *E. D. Morel tegen Leopold II en de Kongostaat* (Berchem, 1985) and *De Kongostaat van Leopold II* (Antwerp, Belgium, 1989, published under his nom de plume, A. M. Delathuy).

The story of the Free State, however, involves more than just the atrocities produced by the rubber policies. A good overview can be found in Jean Stengers and Jan Vansina, "King Leopold's Congo, 1886–1908," in *The Cambridge History of Africa*, vol. 6: *From 1870 to 1905*, edited by R. Oliver and G. N. Sanderson (Cambridge, UK, 1985), 315–58. It was from them that I adopted the dividing line of pre-1890 and post-1890. Analyses concerning international diplomacy and border issues are dealt with in the classic standard works (Cornevin, Stengers, and Ndaywel). Concerning the pacification of the region and the rise of local resistance, I consulted Allen Isaacman and Jan Vansina, "Initiatives et résistances africaines en Afrique centrale de 1880 et 1914," in *Histoire générale de l'Afrique*, vol. 7: *L'Afrique sous domination coloniale*, edited by A. Adu Boahen (Paris, 1987), 191–216. Jean-Luc Vellut wrote a thoughtful analysis of the role of violence in the Free State, "La violence armée dans l'État Indépendant du Congo," *Cultures et développement* (1984).

This chapter also depicts the Africans' growing familiarity with Europeans and their lifestyle. For information about the Africans who went to Europe as part of the world exhibition, see Maarten Couttenier, *Congo tentoongesteld: Een geschiedenis van de Belgische antropologie en het museum van Tervuren (1882–1925)* (Leuven/Louvain, Belgium, 2005), and Maurits Wynants, *Van hertogen en Kongolezen: Tervuren en de koloniale tentoonstelling 1897* (Tervuren, Belgium, 1997). Concerning the development of the state administration at Boma, the CD-ROM by Johan Lagae, Thomas de Keyser and Jef Vervoort proved a real gold mine: *Boma 1880–1920: Koloniale hoofdstad of kosmopolitische handelspost* (Ghent, Belgium, 2006). For the passages concerning the encounter between colonials and Congolese women, I consulted the fascinating study by Amandine Lauro, *Coloniaux, ménagères, et prostituées au Congo belge (1885–1930)* (Loverval, Belgium, 2005).

A few key works dealing with the Protestant missionaries were already mentioned under Chapter 1 above. The distinctions between their working methods and those of the Catholic missions were taken from Ruth Slade, *King Leopold's Congo: Aspects of the Development of Race Relations in the Congo Independent State* (London, 1962). A huge amount has been published concerning the person of Grenfell, most of it hagiographic in nature. The most important work is the two-volume biography by Harry Johnston, *George Grenfell and the Congo* (London, 1908). Concerning the role of native catechists, see the thesis by Paul Serufuri Hakiza, "Les auxiliaires autochtones des missions protestantes au Congo, 1878–1960: Étude de cinq Sociétés missionaires" (Louvain-la-Neuve, Belgium, 1984). A critical approach to the relations between the Catholic Church and the state can be found in the works of A. M. Delathuy (a pseudonym of above-mentioned author Jules Marchal): *Jezuïeten in Kongo met zwaard en kruis* (Berchem, 1986) and his two-volume *Missie en staat in Oud-Kongo* (Berchem, Belgium, 1992 and 1994). Vincent Viaene unpublished paper, "Leopold II en de Heilige Stoel" (2009), taught me a great deal about the ties between the royal house and the Vatican.

The earliest days of the Force Publique are described with military precision and visible pride by Lieutenant-Commander F. Flament in *La Force Publique de sa naissance à 1914: Participation des militaires à l'histoire des premières années du Congo* (Brussels, 1952), which proved nonetheless useful despite its partisan nature considering its age and point of view. See Philippe Marechal's ambitious study *De "Arabische" campagne in het Maniema-gebied (1892–1894)* (Tervuren, Belgium, 1992). Veterans like Oscar Michaux and Joseph Meyers have recounted their experiences with the mutiny in *Au Congo: Carnet de campagne* (Namur, Belgium, 1913) and *Le prix d'un empire* (Brussels, 1964), respectively. The soldiers' uprising has been the subject of much documentation, including Marcel Storme, *La mutinerie militaire au Kasai en 1895* (Brussels, 1970), Auguste Verbeken, *La révolte des Batetela en 1895* (Brussels, 1958), and Pierre Salmon, *La révolte des Batetela de l'expédition du Haut-Ituri (1897)* (Brussels, 1977).

The building of the first railroad is discussed and illustrated extensively in Charles Blanchart et al., *Le rail au Congo belge, 1890–1920* (Brussels, 1993). And despite the passage of time, René J. Cornet's *La bataille du rail: La construction du chemin de fer de Matadi au Stanley Pool* (Brussels, 1947) remains highly readable. Concerning the financing of the railroad and the rest of the Free State, I consulted *Combien le Congo a-t-il couté à la Belgique* (Brussels, 1957) by Jean Stengers. As institutional and diplomatic historian, he also wrote the standard work on Leopold's transfer of the Free State to Belgium, *Belgique et Congo: L'élaboration de la Charte coloniale* (Brussels, 1963). That crucial transaction was recently subjected to new review by Vincent Viaene, who examined its cultural impact in "Reprise-remise: De Congolese identiteitscrisis van België rond 1908," in *De overname van België door Congo: Aspecten van de Congolese "aanwezigheid" in de Belgische samenleving,*

1908–1958, edited by V. Viaene, D. Van Reybrouck, and B. Ceuppens (Leuven/Louvain, Belgium, 2009), 43–62.

CHAPTER 3

The period 1908–21 is, without a doubt, the most sparsely documented in all of Congolese history. The literature about the early years of Belgian colonialism is as scanty as the literature on the Free State is prolific. Fortunately for me, I had a few recent and excellent studies at my disposal. Concerning the social consequences of the fight against sleeping sickness, Maryinez Lyons wrote the classic *The Colonial Disease: A Social History of Sleeping Sickness in Northern Zaire, 1900–1940* (Cambridge, UK, 1992). I took information about the pharmaceutical experiments from a paper by Myriam Mertens, "Chemical Compounds in the Congo: A Belgian Colony's Role in the Chemotherapeutic Knowledge Production during the 1920s," presented at the Third European Conference on African Studies, in Leipzig, Germany, on June 5, 2009.

For the development of colonial anthropology, I refer the reader to the above-mentioned work by Maarten Couttenier (see Chapter 2). With specific regard to the compilation of the *Collection des Monographies ethnographiques,* see the doctoral dissertation by Fien Danniau, " 'Il s'agit d'un peuple': Het antropologisch onderzoek van het Bureau international d'ethnographie (1905–1913)" (University of Ghent, 2005). Concerning the broader context of colonial science, see Mark Poncolet *L'invention des sciences coloniales belges* (Paris, 2008).

This chapter takes a look at the rise of tribalism in early colonial Congo. The information about education at the Catholic missions and ideological representation of "tribes" in textbooks and school songs I drew from Marc Depaepe, Jan Briffaerts, Pierre Kita Kyankenge Masandi, and Honoré Vinck's *Manuels et chansons scolaires au Congo Belge* (Leuven/Louvain, Belgium, 2003). Honoré Vincks's online publication *Colonial Schoolbooks (Belgian Congo): Anthology* proved a true gold mine (www.abbol.com). A great deal has, of course, been written from a Catholic perspective about the first African priest, Stefano Kaoze. The most interesting study, however, is that by Allen F. Roberts, "History, Ethnicity, and Change in the 'Christian Kingdom' of Southeastern Zaire," in *The Creation of Tribalism in Southern Africa,* edited by Leroy Vail (Berkeley, CA, 1989). In that article, Roberts links the history of missions to Kaoze's own political ideals.

In the sections on industrialization, proto-urbanization, and proletarization, I was pleased to make use of the fascinating writings of André Yav. These sources can be consulted on line, including an integral English translation by Johannes Fabian: "Vocabulaire de la ville de Elisabethville," *Archives of Popular Swahili* 4 (2001): 29, http://www2.fmg.uva.nl/lpca/aps/vol4/vocabulaireshabaswahili.html.

There are a few excellent English-language studies of the social aspects of the earliest mining activities. Concerning the gold mines of Kilo-Moto, see David Northrup, *Beyond the Bend in the River: African Labor in Eastern Zaire, 1865–1940* (Athens, OH, 1988). For mining in Katanga, see John Higginson, *A Working Class in the Making: Belgian Colonial Labor Policy, Private Enterprise, and the African Mineworker, 1907–1951* (Madison, WI, 1989), and certainly also Charles Perrings, *Black Mineworkers in Central Africa: Industrial Strategies and the Evolution of an African Proletariat in the Copperbelt 1911–41* (London, 1979). For social conditions in Équateur, not including mining, see Samuel H. Nelson, *Colonialism in the Congo Basin, 1880–1940* (Athens, OH, 1994). With regard to the various forms taken by the recruitment of mineworkers, Aldwin Roes sent me his unpublished but extremely lucid lecture entitled "Thinking with and Beyond the State: The Sub- and Supranational Perspectives on the Exploitation of Congolese Natural Resources, 1885–1914," presented at the conference entitled The Quest for Natural Resources in Central Africa: The Case of the Mining Sector in DRC held in Tervuren, Belgium, on December 8–9, 2008. Bruno De Meulder also wrote the particularly interesting *De kampen van Kongo: Arbeid, kapitaal, en rasveredeling in de koloniale planning* (Amsterdam, 1996), concerning the housing of Katangan mineworkers. Labor conditions at William Lever's Huileries du Congo Belge have been described by the indefatigable Jules Marchal in *L'histoire du Congo 1910–1945*, vol. 3: *Travail forcé pour l'huile de palme de Lord Leverhulme* (Borgloon, Belgium 2001).

Concerning World War I, the reader may with pleasure consult Hew Strachan, *The First World War in Africa* (Oxford, 2004), and Edward Peace, *Tip and Run: The Untold Tragedy of the Great War in Africa* (London, 2007). For the administrative aspects of that period, see Guy Vanthemsche, *Le Congo belge pendant la Première Guerre mondiale: Les rapports du ministre des Colonies Jules Renkin au roi Albert Ier, 1914–1918* (Brussels, 2009). Concerning the armed struggle for Lake Tanganyika, Giles Foden wrote the successful *Mimi and Toutou Go Forth: The Bizarre Battle of Lake Tanganyika* (London, 2004). For more about the taking of Tabora, see Georges Delpierre, "Tabora 1916: De la symbolique d'une victoire," *Belgisch Tijdschrift voor Nieuwste Geschiedenis* (2002). I learned a great deal about the human side of the German East Africa campaign from Jan De Waele, "Voor vorst en vaderland: zwarte soldaten en dragers tijdens de Eerste Wereldoorlog in Congo," *Militaria Belgica* (2007–2008). The reader will learn more about the African presence on the European battlefields during World War I from the lovely exhibition catalogue assembled by Dominiek Dendooven and Piet Chielens, *Wereldoorlog I: Vijf continenten in Vlaanderen* (Tielt, Belgium, 2008), which includes an article about the ethnographic recordings made among prisoners of war in Berlin. Zana Aziza Etambala also deals with this subject in his *In het land van de Banoko: De geschiedenis van de Kongolese/Zaïrese aanwezigheid in België van 1885 tot heden* (Leuven/Louvain, Belgium, 1993). The most recent study is that by Jeannick Vangansbeke, "Afrikaanse verdedigers van het Belgisch grondgebied, 1914–1918," *Belgische Bijdragen tot de Militaire Geschiedenis* 4 (2006): 123–34. For more about Rwanda and Burundi under German and Belgian colonial rule, see Helmut Strizek, *Geschenkte Kolonien: Ruanda und Burundi unter deutscher Herrschaft* (Berlin, 2006), and Ingeborg Vijgen, *Tussen mandaat en kolonie: Rwanda, Burundi en het Belgische bestuur in opdracht van de Volkenbond (1916–1932)* (Leuven/Louvain, Belgium, 2005).

CHAPTER 4

That the period in Africa between the wars was anything but peaceful was recently illustrated in Jonathan Derrick's impressive overview *Africa's "Agitators": Militant Anti-Colonialism in Africa and the West, 1918–1939* (London, 2008), in which he, naturally, gives attention to the events in Congo. A great deal has been written about Simon Kimbangu, by historians, anthropologists, and also by his followers. In 1959 Jules Chomé rocked the colonial boat with his *La passion de Simon Kimbangu, 1921–1951* (Brussels, 1959). The best historical study is that by Susan Asch, *L'église du prophète Simon Kimbangu: De ses origines à son rôle actuel au Zaïre* (Paris, 1982). Jean-Luc Vellut recently wrote a meaty but extremely worthwhile introduction to the first volume of his source book *Simon Kimbangu, 1921: De la prédication à la déportation* (Brussels, 2005). The writings of Kimbangu's followers and sympathizers often include a historical perspective as well. The movement's former spiritual leader, Joseph Diangienda Kuntima, wrote his own extensive overview, *L'histoire du Kimbanguisme* (Châtenay-Malabry, France, 2007). One would also do well to consult the seminal work by Marie-Louise Martin, *Simon Kimbangu: Un prophète et son église* (Lausanne, Switzerland, 1981), and the much more recent work by Aurélien

Mokoko Grampiot, *Kimbanguisme et identité noire* (Paris, 2004). I also found a thorough-going study dealing with the deportations in Munayi Muntu-Monji, "La déportation et le séjour des Kimbanguistes dans le Kasaï-Lukenié (1921–1960)," *Zaïre-Afrique* (1977).

Information about other messianic movements can be found in Martial Sinda, *Le messianisme congolais et ses incidences politiques: Kimbanguisme—Matsouanisme—Autres mouvements* (Paris, 1972), in André Ryckmans's *Les mouvements prophétiques kongo en 1958* (Kinshasa, 1970), and in Jacques Gérard's *Les fondements syncrétiques du Kitawala* (Brussels, 1969). In addition, I had the good fortune to read the unpublished but well-documented typescript by Rufin Kibari, headmaster at Kikwit, "Mouvements 'anti-sorciers' dans les Provinces de Leopolville [sic] et du Kasaï, à l'époque coloniale" (1985). There is a broad contextual account of native Christianity and colonialism in Paul Raymaekers and Henri Desroche, *L'administration et le sacré (1921–1957)* (Brussels, 1983). Also, see Wyatt MacGaffey's classic *Religion and Society in Central Africa* (Chicago, 1986).

The most worthwhile study of capital punishment in the Belgian Congo is by Jean-Luc Vellut, "Une exécution publique à Elisabethville (20 septembre 1922): Notes sur la pratique de la peine capitale dans l'histoire coloniale du Congo," in, *Art pictural zaïrois*, edited by B. Jewsiewicki (Paris, 1992). More recent is Bert Govaerts's "De strop of de kogel? Over de toepassing van de doodstraf in Kongo en Ruanda-Urundi (1885–1962)," *Brood en Rozen* (2009).

Much ink has been dedicated to the revolt by the Pende (or Bapende), yet the extremely thorough study by Sikitele Gize remains unparalleled: "Les racines de la révolte Pende de 1931" *Études d'histoire africaine*, 1973. A more recent and detailed version of the facts was published by Louis-François Vanderstraeten, *La répression de la révolte des Pende du Kwango en 1931* (Brussels, 2001). After scraping away the inevitable layers of propaganda, a Russian study of the 1930s provides an extremely solid base for understanding the deeper issues: A. T. Nzula, I. I. Potekhin, and A. Z. Zusmanovich, *Forced Labour in Colonial Africa* (London, 1979). Nowhere else have I found a clearer link between the raising of the tax on labor and the process of proletarization.

The financial-economic history of the interbellum is clearly described in G. Vandewalle, *De conjuncturele evolutie in Kongo en Ruanda-Urundi van 1920 tot 1939 en van 1949 tot 1958* (Ghent, Belgium, 1966). For the social dimension, I once again turned to the works of Northrup, Nelson, Perrings, and Higginson mentioned in the previous chapter. The effects of industrialization on the natives' material culture and mentality has been described in very lively fashion in a study written in the 1930s by John Merle Davis, *Modern Industry and the African: An Inquiry into the Effect of the Copper Mines of Central Africa upon Native Society and the Work of the Christian Missions* (London, 1933). The social policies of Union Minière are discussed in a well-documented but company-partisan publication by René Brion and Jean-Louis Moreau, *De la mine à Mars: La genèse d'Umicore* [*Van mijnbouw tot Mars: De ontstaansgeschiedenis van Umicore*] (Tielt, Belgium, 2006). I recommend that it be read alongside two works of Bruce Fetter, *L'Union Minière du Haut-Katanga, 1920–1940: La naissance d'une sous-culture totalitaire* (Brussels, 1973) and *The Creation of Elisabethville* (Stanford, CA, 1976). The opening chapters of Johannes Fabian's *Jamaa: A Charismatic Movement in Katanga* (Evanston, IL, 1941) are also very lucid. An important source work on labor in the palm-oil sector is Jacques Vanderlinden, *Main d'œuvre, Église, capital, et administration dans le Congo des années trente* (Brussels, 2007).

To better understand the growth of urban culture, I turned to a collection edited by Jean-Luc Vellut, *Itinéraires croisés de la modernité: Congo belge (1920–1950)* (Tervuren, Belgium, 2000). In it I found fascinating chapters dealing with Scouting, soccer, media, the color bar, and daily life in the colonial city. Concerning the unique role of Tata

Raphaël de la Kéthulle I read, in addition to the chapter contributed by Bénédicte Van Peel to Vellut's collection, the article by Roland Renson and Christel Peeters, "Sport als missie: Raphaël de la Kéthulle de Ryhove (1890–1956)," in *Voor lichaam en geest: katholieken, lichamelijke opvoeding en sport in de 19de en 20ste eeuw*, edited by M. D'hoker, R. Renson, and J. Tolleneer (Leuven/Louvain, Belgium, 1994). More information about the Catholic Church's work among young people was found in publications including Karl Catteeuw, "Cardijn in Congo: De ontwikkeling en betekenis van de Katholieke Arbeidersjeugd in Belgisch-Congo," *Brood en Rozen* (1999). Sara Boel wrote an interesting doctoral dissertation about the regime's attempts to control the media and the arts: "Censuur in Belgisch Congo (1908–1960): Een onderzoek naar de controle op de pers, de film, en de muziek door de koloniale overheid" (Vrije Universiteit Brussels, 2005). Bruce Fetter has discussed local club life and the Catholic Church's attempts at recuperation in his classic article "African Associations in Elisabethville, 1910–1935: Their Origins and Development," *Études d'Histoire africaine* (1974). An older work by Georges Brausch, *Belgian Administration in the Congo* (London, 1961) remains worth reading due to its nuanced chapter dealing with the color bar. Benoît Verhaegen wrote an excellent article about the exaggerated fear of the Red Menace, "Communisme et anticommunisme au Congo (1920–1960)," *Brood en Rozen* (1999). Concerning body politics, the medicalization of Congolese society, and the local reactions to that in the Congolese interior, see the fascinating study by Nancy R. Hunt, *A Colonial Lexicon: Of Birth Ritual, Medicalization, and Mobility in the Congo* (Durham, NC, 1999).

In his above-mentioned *In het land van de Banoko* (Leuven/Louvain, Belgium, 1993), Zana Aziza Etambla dedicates a very enlightening chapter to Paul Panda Farnana and his Union Congolaise. See also François Bontinck, "Mfumu Paul Panda Farnana, 1888–1930: Premier (?) nationaliste congolais," in *La dépendance de l'Afrique et les moyens d'y remédier*, edited by V. Y. Mudimbe (Paris, 1980). In contemporary Congolese circles one notes renewed interest in this early champion of his countrymen's interests. Didier Mumengi honored him with *Panda Farnana, premier universitaire congolais, 1888–1930* (Paris, 2005). Antoine Tshitungu Kongolo examined his connections with the Belgian intellegentsia in: "Paul Panda Farnana (1888–1930), panafricaniste, nationaliste, intellectuel engagé: Une contribution à l'étude de sa pensée et de son action," *L'Africain* (2003).

CHAPTER 5

A clear overview of World War II in Africa and its impact on colonialism is found in Michael Crowder's article "The Second World War: Prelude to Decolonization in Africa," in *The Cambridge History of Africa*, vol. 8 (Cambridge, UK, 1984). A more recent survey of the situation in the Belgian Congo, unfortunately, remains lacking. The latest attempt at such a survey dates from the 1980s, from Belgium's Koninklijke Academie voor Overzeese Wetenschappen, *Bijdragen over Belgisch-Congo tijdens de Tweede Wereldoorlog* (Brussels, 1983); of greatest use to me were the articles by Léon de Saint-Moulin, Jean-Luc Vellut, Benoît Verhaegen, Gustaaf Hulstaert, Jonathan Helmreich, and Antoine Rubbens. The collection does not deal with the military aspects, however, which are covered by Emile Janssens, *Contribution à l'histoire militaire du Congo belge pendant la Seconde Guerre mondiale, 1940–45* (Brussels, 1982–1984). The Abyssinian campaign was documented by a few Belgian officers who took part in it, including R. Werbrouck, *La campagne des troupes coloniales belges en Abyssinie* (Léopoldville, 1945), and Philippe Brousmiche, *Bortaï: Faradje, Asosa, Gambela, Saio: Journal de campagne* (Doornik/Tournai, Belgium, 1987). Felix Denis placed online the diary and above all the fascinating photo album compiled by his father-in-law, Lieutenant Carlo Blomme at http://force-publique-1941.skynetblogs.be/. See also Christine Denuit-Somerhausen and Francis Balace, "Abyssinie 41: Du mirage à la victoire," in Jours de lutte, edited by F. Balace (Brussels, 1992).

Concerning the role of Katangan uranium in the development of the atomic bomb, see Jacques Vanderlinden, À propos de l'uranium congolais (Brussels, 1991), and Jonathan E. Helmreich, "The Uranium Negotiations of 1944," in Bijdragen over Belgisch-Congo tijdens de Tweede Wereldoorlog (Brussels, 1983). I also refer the reader to his Gathering Rare Ores: The Diplomacy of Uranium Acquisition, 1943–1954 (Princeton, NJ, 1986), and L'uranium, la Belgique, et les puissances by Pierre Buch and Jacques Vanderlinden (Brussels, 1995).

The social unrest in the mines has been lavishly documented in the above-mentioned book by Perrings, Black Mineworkers in Central Africa. I also consulted J.-L. Vellut, "Le Katanga industriel en 1944: Malaises et anxiétés dans la société coloniale," in Bijdragen over Belgisch-Congo tijdens de Tweede Wereldoorlog (Brussels, 1983). Especially useful was the study by Tshibangu Kabet Musas, "La situation sociale dans le ressort administratif de Likasi (ex-Territoire de Jadotville) pendant la Guerre 1940–1945," Études d'Histoire Africaine (1974), and that by Bogumil Jewsiewicki, Kilola Lema, and Jean-Luc Vellut, "Documents pour servir à l'histoire sociale du Zaïre: Grèves dans le Bas-Congo (Bas-Zaïre) en 1945," Études d'Histoire Africaine (1973). The clearest overview I found, however, was that by Bogumil Jewsiewicki, "La contestation sociale et la naissance du prolétariat au Zaïre au cours de la première moitié du XXe siècle," Revue Canadienne des Études Africaines (1976).

Vladimir Drachoussoff's fascinating wartime diary appeared in a modest print-run under his pseudonym Vladi Souchard, Jours de brousse: Congo 1940–1945 (Brussels, 1983). It was one of the most gripping documents that I had the pleasure to read while preparing this history. Governor General Pierre Ryckmans and Father Placide Tempels held nuanced views of colonial reality, see Ryckmans's Dominer pour servir (Brussels, 1948), and Tempels's Bantoe-filosofie (Antwerp, 1946). Also see Jacques Vanderlinden, Pierre Ryckmans, 1891–1959: Coloniser dans l'honneur (Brussels, 1994). Concerning this postwar period, see Nestor Delval's highly readable essay, "Schuld in Kongo?" (Leuven/Louvain, Belgium, 1966).

With regard to the postwar years, Anton Rubbens's work, Dettes de guerre (Elisabethville, 1945), comes highly recommended. It comprises a number of critical articles that appeared in the newspaper L'Essor du Congo. In addition, the reports drawn up by the Commission Permanente pour la Protection des Indigènes are required reading; besides containing useful social information, they illustrate quite characteristically the colonial paradigm; see L. Guebels, Relation complète des travaux de la Commission Permanente pour la Protection des Indigènes, 1911–1951 (Brussels, 1952). An excellent introduction to the subject of the trade unions and social protest is found in the 1999 thematic issue of Brood en Rozen, "Sociale bewegingen in Belgisch-Congo." I also consulted André Corneille, Le syndicalisme au Katanga (Elisabethville, 1945), Arthur Doucy and Pierre Feldheim, Problèmes du travail et politique sociale au Congo belge (Brussels, 1952), and R. Poupart, Première esquisse de l'évolution du syndicalisme au Congo (Brussels, 1960).

I gained a better understanding of life in the colonial city from the books of Filip De Boeck and Marie-Françoise Plissart, Kinshasa: Tales of the Invisible City (Ghent, Belgium, 2004), and Johan Lagae, Kongo zoals het is: Drie architectuurverhalen uit de Belgische kolonisatiegeschiedenis (1920–1960) (Ghent, Belgium, 2002). The work of Suzanne Comhaire-Sylvian, Femmes de Kinshasa: hier et aujourd'hui (Paris, 1968), Valdo Pons, Stanleyville: An African Urban Community under Belgian Administration (Oxford, 1969), and W. C. Klein, De Congolese elite (Amsterdam, 1957), provided me with a vivid picture of the new urban culture. The operations and impact of the radio broadcasts for a Congolese audience are discussed by Greta Pauwels-Boon, *L'origine, l'évolution, et le fonctionnement de la radio-diffusion au Zaïre de 1937 à 1960* (Tervuren, Belgium, 1979), and Sara Boel, "Censuur in Belgisch Congo (1908–1960): Een onderzoek naar de controle op de pers, de film en de muziek door de koloniale overheid" (Vrije Universiteit Brussels, 2005). Concerning the association of Raphaël de la Kéthulle's alumni, see Charles Tshimanga, "*L'ADAPES et la formation d'une élite au Congo (1925–1945)*" in *Itinéraires croisés de la modernité: Congo belge (1920–1950)*, edited by J.-L. Vellut (Tervuren, Belgium, 2000).

The fate of the évolués has been discussed in many publications and by numerous authors such as Stengers, Young, and Ndaywel. The standard work is Jean-Marie Mutamba Makombo, *Du Congo Belge au Congo indépendant 1940–1960* (Kinshasa, 1998). A highly interesting study is by Mukala Kadima-Nzuji, who establishes a tic between social *ressentiment*, the press, and literature in *La littérature zaïroise de langue française (1945–1965)* (Paris, 1984). Concerning the establishment of the first Congolese university, see Ruben Mantels's interesting *Geleerd in de tropen: Leuven, Congo en de wetenschap, 1885–1960* (Leuven/Louvain, Belgium, 2007). King Baudouin's visit is colorfully depicted by Erik Raspoet, *Bwana Kitoko en de koning van de Bakuba: Een vorstelijke ontmoeting op de evenaar* (Antwerp, Belgium, 2005).

The lines at the end of this chapter were taken from the collection *Esanzo* by Antoine-Roger Bolamba, one of the loveliest works of Congolese poetry.

CHAPTER 6

The literature dealing with Congo's decolonization is abundant, but often also of inconsistent quality, dated and exaggeratedly "white" in its views. The very best book about the period remains *Politics in the Congo* by Crawford Young (Princeton, 1965). Almost half a century after its first publication, the reader is still amazed to see how—so soon after the events themselves—the author was able to lucidly analyze and document the major processes. In doing so, Young was undoubtedly helped by the fabulous work already done by the Centre de Recherche et d'Information Socio-Politiques (CRISP) in Brussels, an inspiring and conscientious documentation center where researchers such as Jean Van Lierde, Benoît Verhaegen, and Jules Gérard-Libois have done pioneering work. CRISP's yearbooks and studies of political movements remain to this day an indispensable source for historical research into the 1950s and 1960s in Congo. CRISP also published Young's standard work in French.

Another older, but still highly valuable study is that by Paule Bouvier, *L'accession du Congo belge à l'indépendance* (Brussels, 1965). More recently, Zana Aziza Etambala has collected a sizeable amount of new archive material in two highly readable volumes, *Congo 55/65: Van Koning Boudewijn tot president Mobutu* (Tielt, Belgium, 1999), and *De telo-organg van een modelkolonie: Belgisch Congo (1958–1960)* (Leuven/Louvain, Belgium, 2008). Of the many memoirs published concerning the turbulent decolonization, those by Jef Van Bilsen, a key figure in the process, are extremely worthwhile, *Kongo 1945–1965: Het einde van een kolonie* (Leuven/Louvain, Belgium, 1993).

Concerning the international context of the Congolese struggle for independence, I profited greatly from Pierre Queuille, *Histoire de l'afro-asiatisme jusqu'à Bandoung: la naissance du tiers-monde* (Paris, 1965), and Colin Legum, *Pan-Africanism: A Short Political Guide* (New York, 1965).

Kinshasa's youth cultures have been described by Didier Gondola in *Villes miroirs: Migrations et identités urbaines à Kinshasa et Brazzaville, 1930–1970* (Paris, 1997). The above-mentioned work by Filip De Boeck also gives attention to the phenomenon of the "bills" and the "*moziki.*" The political dimension of Congolese soccer is the subject of an excellent documentary by Jan Antonissen and Joeri Weyn: *F.C. Indépendance* (2007). The violent rioting in the capital in 1959 has attracted a great deal of attention. Jacques Marras and Pierre De Vos wrote the accessible *L'équinoxe de janvier: Les émeutes de Léopoldville* (Brussels, 1959), but *J'étais le général Janssens* (Brussels, 1961) by General Émile Janssens, who commanded the Force Publique and was therefore far from impartial, is also worth reading.

The first generation of political figures has been dealt with widely. Concerning Kasavubu, see Benoît Verhaegen and Charles Tshimanga, *L'Abako et l'indépendance du Congo belge: Dix ans de nationalisme kongo (1950–1960)* (Tervuren, Belgium, 2003). On Lumumba, see Jean Omasombo Tshonda and Benoît Verhaegen, *Patrice Lumumba: Jeunesse et apprentissage politique, 1925–1956* (Tervuren, Belgium, 1998), and their sequel, *Patrice Lumumba: De la prison aux portes du pouvoir, juillet 1956–février 1960* (Tervuren, Belgium, 2005). The best study of Lumumba was written by Jean-Claude Willame, *Patrice Lumumba: La crise congolaise revisitée* (Paris, 1990). Other works have come mostly from the outspokenly partisan, with all inherent advantages and disadvantages; what the reader gains in terms of *histoire vécue* (history as it was lived) is usually lost to a lack of nuance and perspective. Pierre De Vos wrote the very readable, but not always accurate *Vie et mort de Lumumba* (Paris, 1961); Francis Monheim seemed head-over-heels in love when he wrote *Mobutu, l'homme seul* (Brussels, 1962); and Jules Chomé avoided seeming enraged but was very much so when he published *Moïse Tshombe et l'escroquerie katangaise* (Brussels, 1966). In *La pensée politique de Patrice Lumumba* (Paris, 1963), Jean Van Lierde brought together Lumumba's most important speeches, articles, and correspondance. The foreword by Jean-Paul Sartre is, aside from its predictability, still impressive.

Studies examining the partisan squabbling with greater distance are also rare. P. Caprasse, however, with his *Leaders africains en milieu urbain (Elisabethville)* (Brussels, 1959), provided a magnificent sociological approach that went far beyond the local focus of his Katangan fieldwork. He devoted special attention to the rhetoric with which tribal awareness was exploited. Luc Fierlafyn went further in the same vein and submitted the political texts of that day to an interesting rhetorical analysis, *Le discours nationaliste au Congo belge durant la période 1955–1960* (Brussels, 1990).

CHAPTER 8

The whirlwind of events that combined around the formation of the First Republic have been subjected to examinations numerous enough to fill a bookcase. A recent and broad historical survey is lacking, but solid studies have appeared concerning all individual aspects. Walter Geerts's *Binza 10: De eerste tien onafhankelijkheidsjaren van de Democratische Republiek Congo* (Ghent, Belgium, 1970) still provides a clear introduction. Zana Aziza Etambala's *Congo 55/65: Van koning Boudewijn tot president Mobutu* (Leuven/Louvain, Belgium, 1999), and Jef Van Bilsens highly important *Kongo 1945–1965: Het einde van een kolonie* (Leuven/Louvain, Belgium, 1993) also provide accessible points of departure for the interested reader. The above-mentioned CRISP yearbooks are essential reading as well.

Concerning the mutiny within the national armed forces, Louis-François Vanderstraeten wrote the definitive study, *Histoire d'une mutinerie, juillet 1960: De la Force Publique à l'Armée nationale congolaise* (Paris, 1985). He gave a great deal of attention to the atmosphere of panic, the sudden exodus of the remaining Belgians, and the Belgian military intervention. For a vivid picture of those days, see two books by Peter Verlinden, *Weg uit Congo: Het drama van de kolonialen* (Leuven/Louvain, Belgium, 2002), and *Achterblijven in*

Congo: een drama voor de Congolezen? (Leuven/Louvain, Belgium, 2008). Marie-Bénédicte Dembour wrote an interesting anthropological study on the perspective of the former colonials, *Recalling the Belgian Congo* (New York, 2000).

How the Congo crisis drew Africa into the Cold War is the subject of a truly magnificent analysis in the epic documentary by Jihan El Tahri, *Cuba, une odyssée africaine* (Arte, 2007). The film not only includes interviews with Cuban veterans, but also with leading Congolese, Russian, and American figures of that day: it is a stunning portrait of the Cold War machinations within Africa. For the American perspective, see Stephen R. Weissman, *American Foreign Policy in the Congo 1960–1964* (Ithaca, NY, 1974), and Romain Yakem-tchouk, *Les relations entre les États-Unis et le Zaïre* (Brussels, 1986). For the communist perspective, see *Le monde communiste et la crise du Congo belge*, edited by Arthur Wauters (Brussels, 1961), and Edouard Mendiaux, *Moscou, Accra, et le Congo* (Brussels, 1960). Former CIA boss Larry Devlin recently published his strikingly frank memoires, *Chief of Station, Congo: A Memoir of 1960–67* (New York, 2007). More recently, Frank R. Villafaña has drawn attention to the confrontation between left- and right-wing Cubans in Congo in *Cold War in the Congo: The Confrontation of Cuban Military Forces, 1960–1967* (New Brunswick, NJ, 2009).

The UN operations have been the subject of commentary from various authors. Georges Abi-Saab analyzed the implications for international law in *The United Nations Operation in the Congo 1960–1964* (Oxford, 1978). Claude Leclercq granted a great deal of attention to the situation on the ground in *L'ONU et l'affaire du Congo* (Paris, 1964). Georges Martelli delivered a very negative verdict in *Experiment in World Government: An Account of the United Nations Operation in the Congo 1960–1964* (London, 1966). The United Nations played such a striking role that other forms of multilateralism have tended to be somewhat neglected. Concerning the establishment of the Organization of African Unity and its contribution to the conflict, see Catherine Hoskyns, *The Organization of African Unity and the Congo Crisis* (Dar es Salaam, Tanzania, 1969).

Lumumba's murder is know best of all from the oft-translated classic by Ludo De Witte, *De moord op Lumumba* (Leuven/Louvain, Belgium, 1999). In Belgium the book prompted the establishment of a parliamentary investigative subcommittee consisting of four historians charged with combing the available archives with a view to establishing the extent of Belgian culpability in the killing. Their report was bone dry but scrupulous: Luc De Vos et al., *Lumumba: De complotten? De moord* (Leuven/Louvain, Belgium, 2004). For the American involvement in the affair, see Madeleine Kalb, *The Congo Cables: The Cold War in Africa, from Eisenhower to Kennedy* (New York, 1982), and the recent article by Stephen R. Weissman, "An Extraordinary Rendition," *Intelligence and National Security* (2010). For the perspective of two Congolese politicians who had once been Lumumba's allies, see Cléophas Kamitatu, *La grande mystification du Congo-Kinshasa: Les crimes de Mobutu* (Paris, 1971), and Thomas Kanza, *Conflict in the Congo: The Rise and Fall of Lumumba* (Baltimore, 1972).

A thoroughgoing study of the Katangan succession was written surprisingly soon after the events themselves is by Jules Gérard-Libois, *Sécession au Katanga* (Brussels, 1963). For the historical roots of that secession, see Romain Yakemtchouk, *Aux origines du séparatisme katangais* (Brussels, 1988).

The uprisings in Kwilu and the east of the country have been dealt with exhaustively in the studies by Benoît Verhaegen, *Rébellions au Congo* (Brussels, 1966–1969), and the two-volume collection of abstracts, *Rébellions-révolution au Zaïre 1963–1965*, edited by Catherine Coquery-Vidrovitch et al. (Paris, 1987). See *Les rébellions dans l'est du Zaïre (1964–1967)*, edited by Herbert Weiss and Benoît Verhaegen (1986), an important thematic issue of *Les Cahiers du CEDAF,* a periodical publication by the Centre d'Etude et de Documentation Africaines. Ludo Martens wrote two sympathetic biographies about Pierre Mulele and his wife Léonie Abo, *Pierre Mulele ou la seconde vie de Patrice Lumumba* (Berchem, 1985), and *Une femme du Congo* (Berchem, 1991). An excellent journalistic account of the Congolese rebellion is by Jean Kestergat, *Congo Congo: de l'indépendance à la guerre civile* (Paris, 1965).

The social and economic conditions during the First Republic have received much less attention than the political and military infighting, yet there is a highly accurate picture of life in the big city by J. S. Lafontaine, *City Politics: A Study of Léopoldville, 1962–63* (Cambridge, UK, 1970). Concerning the complex question of the colonial stock portfolio and the negotiations dealing with its return to Congo, see Jean-Claude Willame, *Eléments pour une lecture du contentieux belgo-zaïrois* (Brussels, 1988).

CHAPTER 9

An outstanding, even formidable, introduction to Mobutu's life and work can be gleaned from the documentary by Thierry Michel, *Mobutu, roi du Zaïre* (Brussels, 1999). Readers wishing to dig more deeply into that period would do well to start with the highly illuminating chapter about the Second Republic by Jacques Vanderlinden in *Du Congo au Zaïre, 1960–1980,* edited by A. Huybrechts et al. (Brussels, 1980). To see how a political elite plundered the national economy, consult Fernard Bézy et al., *Accumulation et sous-développement au Zaïre 1960–1980* (Louvain-la-Neuve, Belgium, 1981), and David J. Gould, *Bureaucratic Corruption and Underdevelopment in the Third World: The Case of Zaire* (New York, 1980). But no one out to make a serious study of the era should omit the bulky study by Crawford Young and Thomas Turner, *The Rise and Decline of the Zairean State* (Madison, WI, 1985). That book focuses on the first half of the Mobutu regime, the period 1965–80, and provides a very convincing picture of how the state first became omnipresent and omnipotent, then fell into total disarray. Its style is sober, yet it contains a wealth of documentation. By far the most important book about this era.

Original Zairian sources from that period are numerous, but consistently fettered by fear of the regime. There is propaganda in abundance, without a drop of critical analysis. It was only outside the borders of the national territory that one could curse out loud. In Paris, Cléophas Kamitatu, cofounder of the Parti Solidaire Africain, wrote two well-documented works that also provide virulent critique of the regime, *La grande mystification du Congo-Kinshasa: Les crimes de Mobutu* (Paris, 1971), and *Zaïre: Le pouvoir à la portée du peuple* (Paris, 1977).

Two recent American books have provided a backstage glimpse. Mobutu's personal physician, the American William Close, father of actress Glenn Close, published his recollections of a turbulent period, *Beyond the Storm* (Marbleton, WY, 2007). Although his analysis is not always profound, the anecdotes are often highly revealing. For a better understanding of the ties of friendship between America and Zaïre, readers can best turn to Romain Yakemtchouk, *Les relations entre les États-Unis et le Zaïre* (Brussels, 1986), and the memoirs of CIA agent Larry Devlin, *Chief of Station,* mentioned above.

Kinshasa's staggeringly explosive growth has been described well by Marc Pain, *Kinshasa, la ville et la cité* (Paris, 1984), and René de Maximy, *Kinshasa, ville en suspens* (Paris, 1984). Both books devote attention not only to urban and demographic processes, but also to their social and cultural consequences.

In this burgeoning and youthful city, music played a major role. The Congolese music scene was probably never so vital as in the early 1970s, thanks in part to Mobutu's campaign of *authenticité.* Gary Stewart's exhaustive *Rumba on the River* (London,

2000), deals with this in detail. Also highly worthwhile is the recent *Rumba Rules: The Politics of Dance Music in Mobutu's Zaire* (Durham, NC, 2008), with as main theme the closely knit ties between politics and popular music. For the descriptions of the match between Muhammad Ali and George Foreman I made use not only of clips on YouTube, but also of Norman Mailer's classic *The Fight* (Boston, 1975), one of the best sports books ever written. In addition, I greatly enjoyed Leon Gast's Oscar-winning documentary *When We Were Kings* (1996), which also deals with the musical aspects of "The Rumble in the Jungle." Concerning the intertwining of the black emancipation struggle and boxing, I referred to several excellent essays in Gerard Early, *Speech and Power* (Hopewell, 1992).

CHAPTER 10

Accessible and well-documented works dealing with the madness of the Mobutu regime from 1975 exist in a number of languages. Jean-Claude Willame wrote the serene but shrewd *L'automne d'un despotisme* (Paris, 1992) and Colette Braeckman, journalist for *Le Soir*, the readable and in Congo highly influential *Le dinosaure* (Paris, 1991). In Flanders two journalists from the public broadcasting company wrote down their experiences and analyses in *Mobutu, de man van Kamanyola* by Walter Geerts (Leuven/Louvain, Belgium, 2005) and particularly *Mobutu, van mirakel tot malaise* by Walter Zinzen (Antwerp, 1995). The latter is extremely worthwhile, even if only for the chapter on the Shaba wars. The American historian Thomas Callaghy saw a parallel between the Mobutu regime and the ancien régime in France in *The State-Society Struggle: Zaire in Comparative Perspective* (New York, 1984). With her *In the Footsteps of Mr Kurtz: Living on the Brink of Disaster in the Congo* (London, 2000), the British journalist Michela Wrong has written a wonderful page-turner with a wealth of information about the 1990s. And more than twenty years after publication, the oft-translated *Terug naar Congo* by Lieve Joris (Amsterdam, 1987) still provides a very tangible and gripping picture of life under the dictatorship.

Jean-Claude Willame deals with the "white elephants," Mobutu's senseless building projects, in *Zaïre, l'épopée d'Inga: chronique d'une prédation industrielle* (Paris, 1986). Contrary to what the title seems to suggest, the book deals with more than the notorious hydroelectric station. Information about the German rocket program, I assembled piece by piece from the documentary *Mobutu, roi du Zaïre* by Thierry Michel (1999), the above-mentioned book by Walter Geerts, but above all from Otrag Rakete, the website of Bernd Leitenberger, http://www.bernd-leitenberger.de/otrag.shtml.

The standard work on the Shaba wars was written by Romain Yakemtchouk, *Les deux guerres du Shaba* (Brussels, 1988). He devoted a great deal of attention to the ties maintained by Belgium, France, and the United States with Mobutu's Zaïre. Before starting in on Sean Kelly's *Les relations entre les États-Unis et le Zaïre* (Brussels, 1986), I read his less technical work with the title-as-synopsis, *America's Tyrant: The CIA and Mobutu of Zaire: How the United States Put Mobutu in Power, Protected Him from His Enemies, Helped Him Become One of the Richest Men in the World, and Lived to Regret It* (Washington, DC, 1993).

Achieving an understanding of the economic and monetary policies of the period 1975–90 is no mean feat, especially in view of the absence of a good survey of the role of the IMF, the World Bank, and the Paris Club. Winsome J. Leslie focused on one of the key players in his *The World Bank and Structural Adjustment in Developing Countries: The Case of Zaire* (Boulder, CO, 1987). The work of Jean-Philippe Peemans, *Zaïre onder het Mobutu-regime* (Brussels, 1988), was lucid and interesting to read, not least of all because of his early warning for the undesired effects of the IMF measures. Kisangani Emizet further refined the arguments and provided important and convincing graph material in the first chapters of his *Zaire after Mobutu* (Helsinki, 1997). My verdict on the work of the IMF is greatly indebted to the bestseller *Globalization and Its Discontents* by Nobel-Prize laureate Joseph Stiglitz (London, 2002).

The dramatic consequences of the crisis and the rise of a "second," informal economy were examined by Janet MacGaffey and her team: *The Real Economy of Zaire* (London, 1991). For the role of women in that new economy, see Benoît Verhaegen, *Femmes zaïroises de Kisangani: Combats pour la survie* (Paris, 1990). Striking accounts are also found in *Manières de vivre: Économie de la "débrouille" dans les villes du Congo/Zaïre*, edited by G. de Villers et al. (Tervuren, Belgium, 2002).

For an understanding of the repressive state apparatus, the reader may turn to the bleak reports from Amnesty International and to the Sovereign National Conference's *Rapport sur les assassinats,* as reissued by Abdoulaye Yerodia (Kinshasa, 2004). A more academic approach is found in Michael Schatzberg's *The Dialectics of Oppression in Zaire* (Bloomington, IN, 1988). Urban legend, rumors, and news from the *radio-trottoir* were compiled by Cornelis Nlandu-Tsasa in *La rumeur au Zaïre de Mobutu: Radio-trottoir à Kinshasa* (Paris, 1997). Regarding popular painting, see *Art pictural zaïrois,* edited by Bogumil Jewsiewicki (Paris, 1992), and Johannes Fabian, *Remembering the Present: Painting and Popular History in Zaire* (Berkeley, CA, 1996).

The six thousand reports written after the 1990 "people's meetings" have never been released, but the best work dealing with the start of the process of democratization is that by A. Gbabendu Engunduka and E. Efolo Ngobaasu, *Volonté de changement au Zaïre: De la consultation populaire vers la conférence nationale* (Paris, 1992).

CHAPTER 11

A succinct but highly illuminating introduction to the turbulent period of transition between the Second and the Third Republics can be found in Flemish radio journalist Guy Poppe's *De tranen van de dictator: Van Mobutu tot Kabila* (Antwerp, 1998). Many of those actively involved have written about their vision of the political struggle and had it published by L'Harmattan in Paris. For years, that publisher has served as the major display case for the intellectual Francophone African diaspora; its noncritical publishing policy, however, sometimes makes it seem more like a glorified copy shop than any systematic distributor of knowledge. One of the more balanced works is that by Dieudonné Ilunga Mpunga, *Etienne Tshisekedi: Le sens d'un combat* (Paris, 2007), which chiefly examines the role of the UDPS. Loka-ne-Kongo wrote a critical retrospective about that chaotic period of democratization, *Lutte de libération et piège de l'illusion: Multipartisme intégral et dérive de l'opposition au Zaïre (1990–1997)* (Kinshasa, 2001). Axel Buyse summed up the major events of the initial years in *Democratie voor Zaïre: De bittere nasmaak van een troebel experiment* (Groot-Bijgaarden, Belgium, 1994). The most detailed work is that by Gauthier de Villers, *Zaïre: La transition manquée (1990–1997)* (Paris, 1997), the first volume of a highly valuable trilogy about the democratic transition.

The most complete study of the suppression of the student protest in Lubumbashi comes from Muela Ngalamulume Nkongolo, *Le campus martyr: Lubumbashi, 11–12 mai 1990* (Paris, 2000). Concerning the quashing of the big peace march in Kinshasa, see *Marche d'espoir, Kinshasa 16 février 1992: Non-violence pour la démocratie au Zaïre,* edited by Philippe de Dorlodot (Paris, 1994). There is, to the best of my knowledge, no standard work dealing with the Sovereign National Conference, but I supplemented the information I gained from talks with Régine Mutijima with the historical survey by

Georges Nzongola-Ntalaja, who was also a participant, *The Congo from Leopold to Kabila* (London, 2002).

It was my privilege on several occasions to talk with Baudouin Hamuli, the veritable godfather of Congo's *société civile*. He was the first chairman of the national council for NGOs and recorded his analyses in two interesting studies, *Donner sa chance au peuple congolais: Expériences de développement participatif (1985–2001)* (Paris, 2002), and, with two coauthors, *La société civile congolaise: État des lieux et perspectives* (Brussels, 2003).

The extremely precarious situation in which the common people lived has been dealt with in the compilations by *Manières de vivre: Économie de la "débrouille" dans les villes du Congo/Zaïre*, edited by Gautier de Villers et al. (Tervuren, Belgium, 2002), and *Chasse au diamant au Congo/Zaïre*, edited by L. Monnier et al. (Tervuren, Belgium, 2001). These books examine the rise of such phenomena as the *cambistes* in Kinshasa, the bicycle taxis in Kisangani, and diamond smuggling in Kasai. Concerning the opulence still enjoyed by Mobutu in the 1990s, one can learn a lot from the stories of his Belgian son-in-law, Pierre Janssen, *Aan het hof van Mobutu* (Paris, 1997). Concerning the rise of the new religiosity, see Isidore Ndaywel è Nziem, *La transition politique au Zaïre et son prophète Dominique Sakombi Inongo* (Québec, 1995). Anthropologist René Devisch wrote an important article about finding moral and social meaning in times of crisis, "Frenzy, Violence, and Renewal in Kinshasa," *Public Culture* (1995). Lieve Joris's *Dans van de lui-paard* (Amsterdam, 2001) is definitely the best-known work of literary journalism dealing with the end of the Mobutu era.

Entire libraries have been written about the Rwandan genocide. The standard work was and remains *Leave None to Tell the Story* by Human Rights Watch researcher Alison Des Forges (New York, 1999), who died far too young. In addition, I would recommend to the reader the classic by Gérard Prunier, *The Rwanda Crisis* (London, 1995). Several hefty tomes have recently appeared, dealing with the conflict in the area around the Great Lakes: Thomas Turner, *The Congo Wars: Conflict, Myth, and Reality* (London, 2007), René Lemarchand, *The Dynamics of Violence in Central-Africa* (Philadelphia, 2008), Filip Reyntjens, *De Grote Afrikaanse Oorlog: Congo in de regionale geopolitiek, 1996–2006* (Antwerp, Belgium, 2009), and Gérard Prunier, *Africa's World War: Congo, the Rwandan Genocide, and the Making of a Continental Catastrophe* (Oxford, 2009). While Turner is somewhat chaotic, Lemarchand tells a fascinating and well-organized story, Reyntjens provides an admirable overview, and Prunier a detailed analysis.

Dealing specifically with the advance of the AFDL is the excellent compilation by Colette Braeckman et al., *Kabila prend le pouvoir* (Brussels, 1998). Erik Kennes wrote a substantial biography of Kabila's life before his move to seize power, *Essai biographique sur Laurent Désiré Kabila* (Tervuren, Belgium, 2003). Unparalleled in its evocative power is once again a documentary by Egyptian filmmaker Jihan El-Tahri, *L'Afrique en morceaux: La tragédie des Grands Lacs* (2000), which can be viewed in its entirely on the Internet.

CHAPTER 12

The run-up to and course of the Second Congo War have been dealt with in detail in the above-mentioned surveys by Prunier and Reyntjens. An excellent introduction to the conflict has also been provided by Olivier Lanotte in his *Guerres sans frontières en République Démocratique du Congo* (Brussels, 2003). More analytical, but with a wealth of information, is the work by Gauthier de Villers, *Guerre et politique: Les trente derniers mois de L. D. Kabila* (Tervuren, Belgium, 2001). Concerning Kabila's regime before and after the invasion, see the critique by Wamu Oyatambwe, *De Mobutu à Kabila: Avatars d'une passation inopinée* (Paris, 1999). A great deal more hagiographic, almost to the point of being burlesque at times, is the compilation edited by Eddie Tambwe and Jean-Marie Dikanga Kazadi, *Laurent-Désiré Kabila: L'actualité d'un combat* (Paris, 2008). With regard to the motives of the countries taking part in that war, *The African Stakes of the Congo War* by John F. Clark (New York, 2002) appeared quite soon after the facts themselves. The toilsome peace negotiations leading to the agreements at Lusaka (1999) and Pretoria (2002) are discussed by Jean-Claude Willame, *Les "faiseurs de paix" au Congo* (Brussels, 2007). The book also grants a good deal of attention to the motives of combatants both domestic and foreign, and the role of the international UN peacekeeping force MONUC. The definitive study of the MONUC remains to be written, but Xavier Zeebroek recently wrote a useful report, *La Mission des Nations Unies au Congo: Le laboratoire de la paix introuvable* (Brussels, 2008), and Julie Reynaert produced a clear and concise master's thesis, "De balans na tien jaar Monuc in Congo" (Leuven/Louvain, Belgium, 2009).

The massive theft of raw materials has been proven irrefutably by consecutive reports from the United Nations panel of experts (www.un.org/News/dh/latest/dr congo.htm). An overall, quantitative analysis is lacking, but Stefaan Marysse and Catherine André carried out pioneering calculations for the years 1999 and 2000 in "Guerre et pillage en République Démocratique du Congo," *L'Afrique des Grands Lacs* (2001). The *L'Afrique des Grands Lacs* yearbooks, currently edited by Stefaan Marysse, Filip Reyntjens, and Stef Vandeginste, provide a wealth of information for all those wishing to study the more recent periods in Congolese (but also Rwandan and Burundian) history. Back issues can be downloaded in their entirety from the University of Antwerp website.

Marvelous work has also been done by a number of independent NGOs. Human Rights Watch documented the smuggling of gold by Uganda in two reports, *Uganda in Eastern DRC* (2001) and above all *The Curse of Gold* (2005). Global Witness investigated Rwanda's role in the smuggling of tin, *Under-Mining Peace: Tin, The Explosive Trade in Cassiterite in Eastern DRC* (2005). In a two-part study, IPIS looked at the international markets for coltan: *Supporting the War Economy in the DRC: European Companies and the Coltan Trade* (2002). Pole Institute, a Congolese studies center in Goma, published *The Coltan Phenomenon* (2002), with extensive interviews with mineworkers. All these reports are also available online.

Two studies in particular showed me that it is not sufficient to look only to the regimes of Rwanda and Uganda when it comes to the raw-materials robbery in the eastern Congo; there are other players as well, both "downstream" and "upstream." *Network War: An Introduction to Congo's Privatised War Economy* by Tim Raeymaekers (IPIS, 2002) pointed out the crucial role of private, "nonstate actors" in today's globalized world, while Koen Vlassenroot and Hans Romkema showed how normal Congolese citizens also profited: "The Emergence of a New Order? Resources and War in Eastern Congo," *Journal of Humanitarian Assistance* (2002).

Concerning the consequences of the war at the local level, social and otherwise, Koen Vlassenroot and Tim Raeymaekers edited the noteworthy compilation, *Conflict and Social Transformation in Eastern Congo* (Ghent, Belgium, 2004). Among others, the anthropological chapter by Luca Jourdan, "Being at War, Being Young: Violence and Youth in North Kivu," was one I read with great interest. As early as June 2002, Human Rights Watch published a report on sexual violence: *The War within the War*. For the ecological impact of the conflict, I consulted both the UNESCO report *Promoting and Preserving Congolese Heritage: Linking Biological and Cultural Diversity* (2005) and the ambitious survey *Forests in Post-Conflict Democratic Republic of Congo*, edited by L. Debroux et al. (2007).

CHAPTER 13

The political and military aspects of the transitional period are described clearly in the above-mentioned works by Reyntjens and Prunier. The most detailed study once again comes from Gauthier de Villers, *De la guerre aux élections* (Tervuren, Belgium, 2009), with which he completed his trilogy on Zaïre/Congo during the long transition from the Second to the Third Republic (Villers, 1997, 2001, 2009). The thoroughgoing character of these studies makes them a reference work for the period 1990–2008, like the yearbooks published by the CRISP for the period 1959–1967.

This chapter deals in some detail with the interplay between multinational concerns, pop music, Pentecostalism, and the mass media in urban Congolese society. Because these are recent phenomena, no integral studies have yet been written. Theodore Trefon's compilation, *Reinventing Order in the Congo: How People Respond to State Failure in Kinshasa* (London, 2004), contains a number of fine contributions. The standard work on life in the capital, however, is Filip De Boeck's masterful anthropological study, *Kinshasa, Tales of the Invisible City* (Ghent, Belgium, 2004), illustrated with photos by Marie-Françoise Plissart. Two of his doctoral students, Kristien Geenen and Katrien Pype, have in recent years published admirable studies of Kinshasa's street children, youth gangs, and religious soap operas. In 2010 De Boeck himself released the documentary *Cemetery State*, about youth and death in a city that eludes description.

Information about pop music I gleaned from the Internet and from countless conversations with Congolese people. In addition, my most important sources were *Rumba on the River* by Gary Stewart (London, 2000) and *Rumba Rules* by Bob White (Durham, NC, 2008). To the best of my knowledge, no systematic research has been carried out into Heineken's activities in Africa. In 2008 the Dutch television broadcaster RTL made the rather superficial and patriotic documentary *Een Hollands biertje in Afrika*. The documentary dealt solely with Bralima in Kinshasa, with a main role reserved for Dolf van den Brink, and can be seen on that broadcaster's website.

In addition to Katrien Pype's work dealing with religious broadcasters, I received insights into the workings of the Congolese media from Marie-Soleil Frère's *Afrique centrale, médias et conflits: Vecteurs de guerre ou acteurs de paix* (Brussels, 2005) and her more recent articles. For the impact of mobile telephony in Africa, see Mirjam de Bruijn et al., *Mobile Africa: Changing Patterns of Movement in Africa and Beyond* (Leiden, The Netherlands, 2001).

Concerning the rise of charismatic Christianity, I consulted, among other sources, Gerrie Ter Haar's *How God Became African: African Spirituality and Western Secular Thought* (Philadelphia, 2009). The interaction with the recent history of migration is described in Emma Wild-Wood's *Migration and Christian Identity in Congo* (Leiden, The Netherlands, 2008). For more about the rise of the Congolese diaspora in Europe, see Zana Etambala's *In het land van de Banoko* (Leuven/Louvain, Belgium, 1993) for Belgium, and Marc Tardieu's *Les Africains en France* (Monaco, 2006) for France. For the much more recent community in London, see the interviews collected by David Garbin and Wa Gamoka Pambu in *Roots and Routes: Congolese Diaspora in Multicultural Britain* (London, 2009).

A few articles in newspapers and magazines have described the interaction between popular culture and politics. In "La victoire en chantant" in *Jeune Afrique* (2006), Luc Olinga investigated the impact of Congolese pop music on the 2006 elections. Marie-Soleil Frère, in "Quand le pluralisme déraille" in *Africultures* (2007), looked at the influence of commercial and religious television on the electoral campaign.

In the field of cinematography, see *Congo River* by Thierry Michel (2005) for a lively

impression of Congo during the transitional years, and *Congo na biso* by Chuck de Liedekerke and Yannick Muller (2006) for a lucid political approach. Lieve Joris's *Het uur van
de rebellen* (Amsterdam, 2006) is a courageous book about the uphill battle to reform the
Congolese army.

CHAPTER 14

Few books have yet appeared, of course, about the most recent phase of Congolese history. A highly readable account of the first free elections in decades was written by the
Congolese Alphonse Muambi, who returned briefly to his former fatherland as an international observer, *Democratie kun je niet eten* (Amsterdam, 2009).

The early days of the Third Republic are described in two widely divergent works. In
Vers la deuxième indépendance du Congo (Brussels, 2009), *Le Soir* journalist Colette Braeckman presents a cautiously optimistic view, while the compilation edited by Theodore
Trefon, *Réforme au Congo (RDC): Attentes et désillusions* (Tervuren, 2009), strikes a much
more somber note. In addition to the regular printed media, I sought and found further
documentation in *Mo* magazine, *Le Monde Diplomatique,* and *Jeune Afrique.* The blogs
by Colette Braeckman (at lesoir.be) and Jason Stearns (congosiasa.blogspot.com) were a
great help in placing the recent developments in their proper perspective. I also profited
greatly from the razor-sharp analyses distributed by Kris Berwouts as director of EurAc,
the umbrella organization of European NGOs active in Central Africa.

The websites of the International Crisis Group (crisisgroup.org) and Human Rights
Watch (hrw.org) are without equal when it comes to conflict analysis and fieldwork
concerning human rights violations. The macroperspective provided by the former is
equaled only by the detailed, on-the-ground observations of the latter. For years, both
NGOs have been doing outstanding work that not only pleases historians, but above all
aims to save human lives.

The websites of *Le Potentiel* and Radio Okapi, the best newspaper and the best radio
broadcaster in Congo, respectively, allowed me to keep up to date on daily current
events in the country even from a distance. Rapper Alesh, who I interviewed in Kisangani, can also be heard on the Radio Okapi website. A number of brave Congolese NGOs
have recently started distributing reports on the Internet: special mention here goes to
Asadho (Association Africaine de Défense de Droits de l'Homme), Rodhecic (Réseau
d'Organisations des Droits Humains et d'Éducation Civique d'Inspiration Chrétienne),
and Journaliste en Danger.

Concerning the intricacies of the tumult in Katangan mining, Thierry Michel made
the interesting documentary *Katanga Business* (2009). I owe much to the reports from
IPIS, RAID, Global Witness, and Resource Consulting Services.

In recent years a few good studies have appeared dealing with the growing Chinese
presence in Africa. For an analytical approach, see Chris Alden, *China in Africa* (London, 2007), and Serge Michel and Michel Beuret's *La Chinafrique* (Paris, 2009) for a most
lively journalistic account. Outstanding by reason of its balanced approach is the study
by Martine Dahle Huse and Stephen L. Muyakwa, "China in Africa: Lending, Policy
Space, and Governance" (www.afrika.no, 2008). I found a fine analysis of the Congolese-
Chinese contract in Stefaan Marysse and Sara Geenen, "Les contrats chinois en RCD:
l'impérialisme rouge en marche?" *L'Afrique des Grands Lacs* (2007–2008).

CHAPTER 15

Little research has been carried out into Guangzhou's African community. The first
academic articles are now seeing the light of day, but are generally very descriptive in

nature. See Brigitte Bertoncelo and Sylvie Bredeloup, "The Emergence of New African 'Trading Posts' in Hong Kong and Guangzhou," *China Perspectives* (2007) and Li Zhang, "Ethnic Congregation in a Globalizing City: The Case of Guangzhou, China" (www.sciencedirect.com, 2008). See also Zhigang Li, Desheng Xue, Michael Lyons, Alison Brown "Ethnic Enclave of Transnational Migrants in Guangzhou" (asiandrivers. open.ac.uk, 2007), and Adams Bodomo, a Ghanian professor in Hong Kong, "The African Trading Community in Guangzhou," *China Quarterly* (2010). I learned a great deal from my conversations with the Belgian consul Frank Felix, with the Flemish economic attaché and sinologist Koen De Ridder, and with the China-based Congolese journalist Jaffar Mulassa; as stated, however, I learned the most from talking to those directly involved.

NOTES

INTRODUCTION

1. Booven 1913: 23–24.
2. http://www-odp.tamu.edu/ publications/175_SR/chap_11/c11_3. htm.
3. Julien 1953: 10.
4. Northrup 2002: 18–21; McLynn 1992: 321–22; Hilton 1985: 50.
5. Hilton 1985: 80.
6. Jadin 1968.
7. Hilton 1985: 69–84.
8. Vansina 1990: 86.
9. Harms 1981: 3–5.
10. Harms 1981: 21–9.
11. Harms 1981: 3.
12. Northrup 2002: 113–14.
13. Harms 1981: 54.
14. Vansina 1965: 146–52.
15. http://neveu01.chez-alice.fr/birasouf. htm.

CHAPTER 1

1. Makulo Akambu 1983: 15.
2. Makulo Akambu 1983: 15–16.
3. Bontinck 1974: 250.
4. Stanley 1899: 210, 212.
5. Jeal 2007: 199.
6. Jeal 2007: 469.
7. Vansina 1976: 30.
8. Wesseling 1991: 119.
9. Stengers 1997: 275.
10. Makulo Akambu 1983: 18.
11. Makulo Akambu 1983: 20–30.
12. Bontinck 1974: 269–71.
13. Jeal 2007: 274–76.
14. Stanley 1886: 2:147, 151–52.
15. Jeal 2007: 276.
16. Harms 1981: 33.
17. Makulo Akambu 1983: 32–34.
18. Fabian 2000: 103.
19. McLynn 1992: 322.
20. Johnston 1908: 222–24.
21. Bentley 1900: 81.
22. Bentley 1900: 126.
23. Johnston 1908: 328.
24. Alexander L. Bain, registry card, archival collection, Board of International Ministries (BIM), American Baptist Historical Society, Atlanta, GA.
25. Slade 1959: 154; Braekman 1961: 129–36, 351.
26. Etambala 1987: 237–85.
27. Ernest T. Welles, registry card, archival collection, Board of International Ministries (BIM), American Baptist Historical Society, Atlanta, GA.
28. Jeal 2007: 464–75.
29. Makulo Akambu 1983: 36.
30. Bailey 1894: 161–63.
31. Denuit-Somerhausen 1988: 77–146.
32. Wesseling 1991: 126.
33. Jeal 2007: 277–78.
34. Stengers 1989: 58–9.
35. Maquet-Tombu 1952: 56.

CHAPTER 2

1. Stengers and Vansina 1985: 351.
2. Ndaywel è Nziem 1998: 289–92.
3. Van der Smissen 1920: 425.
4. Vellut 2005b: 247.

5. Jeal 2007: 281.
6. Vansina and Stengers 1985: 351.
7. Jeal 2007: 294.
8. Makulo Akambu 1983: 36–37.
9. Etambala 1987.
10. Etambala 1993.
11. Maquet-Tombu 1952.
12. Meeuwis 1999.
13. Hawker 1909: 244.
14. Picard 1896: 161.
15. Bailey 1894: 246.
16. *New York Times*, April 16, 1899.
17. Lauro 2005: 78.
18. Makulo Akambu 1983: 38.
19. Makulo Akambu 1983: 38–39.
20. Makulo Akambu 1983: 45.
21. Hemmens 1949: 27.
22. Johnston 1908: 328.
23. Makulo Akambu 1983: 58.
24. Makulo Akambu 1983: 68–69.
25. Makulo Akambu 1983: 70.
26. Makulo Akambu 1983: 71.
27. Makulo Akambu 1983: 80–81.
28. Interview with Étienne Nkasi, Kinshasa, December 8, 2008.
29. Sadin 1918.
30. Sadin 1918: 16–17.
31. Sadin 1918: 20.
32. Travaux du Groupe d'Études Coloniales 1912: 7.
33. Travaux du Groupe d'Études Coloniales 1912: 7.
34. Sadin 1918.
35. Sadin 1918: 68.
36. Van Acker 1924: 164.
37. Interview with Victor Masunda Kukana, Boma, October 8, 2008.
38. Interview with Camille Mananga Nkanu, Boma, October 9, 2008.
39. Makulo Akambu 1983: 40–44.
40. Michaux 1913: 46, 52.
41. Flament 1952: 509, 516.
42. Flament 1952: 81–82.
43. Interview with Eugène Yoka Kinene, Kinshasa, November 11, 2008.
44. Joye and Lewin 1961: 18.
45. Stengers 1957: 32.
46. Stengers 1997: 277.
47. Stengers 1997: 240.
48. Catherine 1994: 126–30.
49. Interview with Martin Kabuya, Kinshasa, October 16, 2008.
50. Gann and Duignan 1979: 97; Poel 2006: 1, 26.
51. Van der Poel 2006: 1:8–30.
52. Interview with Etienne Nkasi, Kinshasa, November 6 and 10, 2008.
53. Goffin 1907: 79.
54. Interview with Étienne Nkasi, Kinshasa, November 10, 2008.
55. Vangroenweghe 2005: 376; Stengers 1989: 102.
56. Makulo Akambu 1983: 79–80.
57. FOD Buitenlandse Zaken, Africa archives, papers E. Janssens, D1366, 27/12/1904.
58. FOD Buitenlandse Zaken, Africa archives, papers E. Janssens, D1366, 12/12/1904.
59. FOD Buitenlandse Zaken, Africa archives, papers E. Janssens, D1366, 2/1/1905.
60. FOD Buitenlandse Zaken, Africa archives, papers E. Janssens, D1366, 12/12/1904.
61. FOD Buitenlandse Zaken, Africa archives, papers E. Janssens, D1366, 3/1/1905.
62. FOD Buitenlandse Zaken, Africa archives, papers E. Janssens, D1366, 22/11/1904.
63. FOD Buitenlandse Zaken, Africa archives, papers E. Janssens, D1366, 12/12/1904.
64. Bosschaerts 2007: 216.
65. FOD Buitenlandse Zaken, Africa archives, papers E. Janssens, D1366, 5/1/1905.
66. Vangroenweghe 1985: 64.
67. Vangroenweghe 1985: 62.
68. Marechal 2005: 45–46.
69. Johnston 1908: 378–79.
70. Stengers 1989: 109.
71. Johnston 1908: 380.
72. Singleton-Gates and Girodias 1959: 120–22.

73. Singleton-Gates and Girodias 1959: 114.
74. Janssens 1905: 197.
75. Cattier 1906: 341.

76. Cornevin 1963: 129; Gann and Duignan 1979: 79; Stengers and Vansina 1985: 346, 354.
77. Makulo Akambu 1983: 85.

CHAPTER 3

1. Thiel 1982: 20; Boelaert et al. 1995: 36–117.
2. Stanley 1886: 2:214.
3. Maquet-Tombu 1952 .
4. Cornevin 1963: 173–228; Stengers 1989.
5. Young 1968: 23.
6. De Meulder 2000: 50.
7. Cornevin 1963: 187.
8. Vanderkerken 1920: 235.
9. Vanderkerken 1920: 234.
10. Van Wing 1959: 128–29.
11. Carton de Wiart 1923: 70–71.
12. Cattier 1906: 321.
13. Cattier 1906: 322.
14. Couttenier 2005: 225.
15. Van Overbergh 1913: viii.
16. De Jonghe 1908: 304.
17. Van Overbergh 1913: 181.
18. Van Overbergh 1913: 183.
19. Depaepe et al. 2003: 233, 236.
20. Vinck 2002.
21. Depaepe et al. 2003: 191.
22. Frères Maristes 1927: 30–31.
23. Vinck 2002.
24. Kalundi Mango, interviewed by Johannes Fabian, Lubumbashi, June 1986, http://www2.fmg.uva.nl/lpca/aps/vol4/vocabulairekalundicomments.html.
25. Kaoze 1910.
26. Chalux 1925: 125.
27. Interview with Étienne Nkasi, Kinshasa, November 6, 2008.
28. Chalux 1925: 111–14.
29. Chalux 1925: 122–25.
30. Interview with Étienne Nkasi, Kinshasa, November 10, 2008.
31. Stengers 1989: 213–14.
32. Cornet 1944: 261.
33. Stengers 1989: 215.
34. Carton de Wiart 1923: 93.
35. Carton de Wiart 1923: 5.

36. Chalux 1925: 204.
37. Buelens 2007: 405.
38. Carton de Wiart 1923: 83.
39. Merlier 1962: 130.
40. Jewsiewicki 1988: 231–32.
41. Foire Internationale d'Elisabethville 1962: 71–73.
42. Yav 1965: 29.
43. Yav 1965: 5.
44. Brausch 1961: 21–22.
45. Higginson 1989: 33.
46. Higginson 1989: 35.
47. Chalux 1925: 79.
48. Yav 1965: 7.
49. Northrup 1988: 97–99.
50. Chalux 1925: 209.
51. Joye and Lewin 1961: 184.
52. Kimoni Iyay 1990: 155–82.
53. Vandewalle 1966: 45.
54. Banque Centrale du Congo 2007.
55. Chalux 1925: 147.
56. Boelaert et al. 1995.
57. Delcommune 1920: 26.
58. Cayen 1938: 58.
59. Cayen 1938: 47–54.
60. Cornevin 1963: 176–77.
61. Geernaert n.d.
62. Depaepe et al. 2003: 220.
63. Interview with Martin Kabuya, Kinshasa, October 16, 2008.
64. Kisobele Ndontoni 2008.
65. Interview with Hélène Nzimbu Diluzeyi and Leon Wasolua, Kinshasa, November 11, 2008.
66. Interview with Eugène Yoka Kinene, Kinshasa, November 11, 2008.
67. Vanthemsche 2010.
68. Brion and Moreau 2006: 95.
69. Jewsiewicki 1980.
70. Hulstaert 1990.
71. Ndaywel è Nziem 1998: 411.
72. Libotte n.d.

73. Delannoo 2006.
74. Etambala 1993: 33–37; Odette Kudjabo, press communiqué.
75. Dominiek Dendooven, press communiqué.
76. Habran 1925: 52–53.

CHAPTER 4

1. Geldof 1937: 131.
2. Interview with Marcel Wanzungasa, Nkamba, November 4, 2008.
3. Cited by Vellut 2005a: 10.
4. Mokoko Gampiot 2004: 60–63.
5. Sinda 1972: 73.
6. Munayi 1977.
7. Gérard 1969: 9–13.
8. Nelson 1994: 176–77.
9. Ndaywel è Nziem 1998: 411–12.
10. Kibari 1985.
11. Maquet-Tombu 1952: 135–36.
12. Thieffry 1926: 267.
13. Blanchart et al. 1999.
14. Guebels 1952: 262.
15. Vandewalle 1966: 13.
16. Nzula et al. 1979: 64.
17. Vanthemsche 1999: 17.
18. Vandewalle 1966: 45.
19. Davidson et al. 1987: 739.
20. FOD Buitenlandse Zaken, Africa archives, personnel dossier, Firmin Joseph Arthur Peigneux.
21. Interview with Pierre Diakanua, Kinshasa, December 8, 2008.
22. Interview with Étienne Nkasi, Kinshasa, November 10, 2008.
23. Vellut 1992: 201.
24. Vellut 1992: 175.
25. Nelson 1994: 155.
26. Sikitele 1973: 117–18.
27. Sikitele 1973: 109.
28. Cited by Nzula et al. 1979: 110.
29. Saint Moulin 2007: 42.
30. Fetter 1976: 74.
31. Brion and Moreau 2006: 115, 134.
32. Nelson 1994: 151; Northrup 1988: 206–209.
33. Fetter 1973: 23.
34. Higginson 1989: 56.
35. Northrup 1988: 208.
36. Joye and Lewin 1961: 160.
37. Joye and Lewin 1961: 159.
38. Fetter 1974: 216.
39. Grévisse 1951: 98.
40. Esgain 2000: 61.
41. Verhaegen 1999: 126.
42. Daye 1929: 207.
43. Davis 1933: 287–90.
44. Van Peel 2000: 152.
45. Daye 1929: 239.
46. Chalux 1925: 213.
47. Chalux 1925: 157–58.
48. Stewart 2000: 16.
49. Fetter 1973: 38.
50. Chalux 1925: 126.
51. Fabian 1986.
52. Cited by Depaepe et al. 2003: 164–65.
53. Boel 2005: 77, 83.
54. Boel 2005: 111, 139.
55. Boel 2005: 88.
56. Emongo Lomomba 1985: 136.
57. Fabian 1971: 60.
58. Young 1984: 700.
59. Emongo Lomomba 1985: 137–38.
60. Joye and Lewin 1961: 161–3; Fabian 1971: 55–60; Fetter 1973: 38.
61. Cited by Brion and Moreau 2006: 137.
62. Tilman 2000.
63. Van Peel 2000: 180.
64. Renson and Peeters 1994: 204.
65. Van Peel 2000: 185.
66. Interviews with Henri de la Kéthulle, Kinshasa, May 26, 2007, Kikwit, June 2, 2007, and September 20, 2008.
67. Fetter 1974.
68. Perrings 1979: 216.
69. Kalundi Mango, interviewed by Johannes Fabian, Lubumbashi, June 1986, http://www2.fmg.uva.nl/lpca/aps/vol4/vocabulairekalundicomments.html.
70. Yav 1965: 22.
71. Fetter 1974: 212–13.
72. Brausch 1961: 20.
73. Brausch 1961: 19–39.
74. Souchard 1983: 47.

75. Cited by Feuchaux 2000: 88–90.
76. Cited by Etambala 1993: 40.

77. Cited by Etambala 1993: 40.
78. Cited by Bontinck 1980: 608.

CHAPTER 5

1. Interview with André Kitadi, Kinshasa, October 16, 2008.
2. Bourla Errera 2000: 59.
3. McCrummen 2009.
4. Interview with André Kitadi, Kinshasa, November 11, 2008.
5. Interview with André Kitadi, Kinshasa, October 16, 2008.
6. Interview with Martin Kabuya, Kinshasa, October 16, 2008.
7. Ergo 2008: 132–34.
8. Interview with Libert Otenga, November 11, 2008.
9. Wrong 2000: 136–44.
10. Buelens 2007: 282.
11. Buelens 2007: 288.
12. Jewsiewicki et al. 1973: 160.
13. Tshibangu Kabet 1974: 297.
14. Etambala 1999a: 77–78.
15. Dibwe dia Mwembu 1999: 195.
16. Perrings 1979: 226.
17. Yav 1965: 24.
18. Emongo Lomomba 1985: 140.
19. Vellut 1983: 506–514.
20. Jewsiewicki et al. 1973.
21. Young 1984: 703.
22. Souchard 1983: 176.
23. Souchard 1983: 59.
24. Souchard 1983: 58.
25. Souchard 1983: 59–60.
26. Souchard 1983: 146–48.
27. Souchard 1983: 155.
28. Hulstaert 1983: 590.
29. Souchard 1983: 234.
30. Souchard 1983: 84–87.
31. Souchard 1983: 235.
32. Interview with Libert Otenga, November 11, 2008.
33. Vellut 1983: 505.
34. Interview with André Kitadi, Kinshasa, October 16, 2008.
35. Vanderlinden 1994: 604.
36. Van Bilsen 1993: 55.
37. Guebels 1952: 640.

38. Roussel 1949:49.
39. Wauters 1929: 142.
40. Dehoux 1950: 1:155.
41. Lefebvre 1952: 519.
42. Ceuppens 2009.
43. Young 1984: 704–707.
44. Stengers 1989: 226.
45. Drachoussoff 1954: 115–16.
46. Guebels 1952: 738.
47. Tempels 1944: 22.
48. Tempels 1945 [1946]: 105.
49. Guebels 1952: 659.
50. Saint Moulin 2007: 42.
51. Gourou 1955: 33.
52. Guebels 1952: 642.
53. Interviews with Longin Ngwadi, Kikwit, September 19–20, 2008.
54. Comhaire-Sylvain 1968: 54–56.
55. Interview with Sister Apolline Lemole Daringi, Kinshasa, September 29, 2008.
56. Interview with Victorine Ndjoli, Kinshasa, November 7, 2008.
57. Interview with François Ngombe, Kinshasa, November 9, 2009.
58. Comhaire-Sylvain 1968: 23.
59. Gondola 1997a.
60. Pauwels-Boon 1979: 137.
61. Jewsiewicki 1976: 69.
62. Guebels 1952: 664.
63. Martens 1999: 141.
64. Young 1968: 24–25.
65. Pons 1969: 147–50.
66. Young 1968: 150–53.
67. Pons 1969: 214.
68. Interview with Kipulu Sambo and Hery Mambo, Kinshasa, September 17, 2008.
69. Emongo Lomomba 1985: 139.
70. Leysen 1982: 35 min.04 sec.
71. Jewsiewicki 1976: 69.
72. Cited by Rubbens 1945: 128–29.
73. Mantels 2007: 206.
74. Kadima-Nzuji 1984: 55.
75. Interview with Camille Mananga Nkanu, Boma, October 9, 2008.

76. Interview with Victor Masunda
    Kukana, Boma, October 8, 2008.
77. Ndaywel è Nziem 1998: 462.
78. Interviews with Jean Lema, alias Jamais
    Kolonga, Kinshasa, October 3 and 14
    and November 6, 2008.

79. Interview with Paul Kasenge,
    Lubumbashi, June 29, 2007.
80. Interview with Zizi Kabongo, Kinshasa,
    April 21, 2008.
81. Tshitungu Kongolo 2003b: 62.

CHAPTER 6

1. Interview with Michel Lechat, Brussels,
   September 19, 2007.
2. De Backer 1959: 1:7.
3. Van Bilsen 1958: 164–202.
4. Labrique 1957: 253, 256.
5. Labrique 1957: 261.
6. Labrique 1957: 254.
7. Young 1968: 114–17.
8. Labrique 1957: 271.
9. Labrique 1957: 107–10.
10. Queuillle 1965: 315.
11. Klein 1957: 84; Pétillon 1985: 448.
12. Interview with Jean Cordy, Louvain-la-
    Neuve, September 5, 2009.
13. Verhaegen 1971: 419–21; Bouvier 1965:
    39–56.
14. Ghilain 1963: 90–91.
15. Archer 1971: 67.
16. Wolter et al. 1957: 55–58.
17. Sinatu Bolya 2003.
18. Interview with Victorine Ndjoli,
    Kinshasa, November 7, 2008.
19. Michel 1962: 72.
20. Verhaegen 1971: 419.
21. Gondola 1999.
22. Laude 1956: 29.
23. Sohier 1959: 236.
24. Etambala 1999b: 50.
25. Ganshof van der Meersch 1958: 40–54.
26. CRISP 1962: 136.
27. Interviews with Jean Lema, alias Jamais
    Kolonga, Kinshasa, October 3 and 14
    and November 6, 2008.
28. Monstelle 1962: 119.
29. Interviews with Longin Ngwadi,
    Kikwit, September 19–20, 2008.
30. Pétillon 1985: 446.
31. Pétillon 1985: 517.
32. De Backer 1959: 1:32.
33. Etambala 2008: 79–80.
34. Scott 1969: 21–22.

35. Etambala 2008: 82, 84.
36. De Vos 1961: 52.
37. De Backer 1959: 1:32.
38. Interview with Albert Tukeke Talulue,
    Kisangani, November 18, 2008.
39. Interview with Jean Mayani, Kisangani,
    November 19, 2008.
40. Interview with Raphaël Maindo,
    Kisangani, November 17, 2008.
41. Van Bilsen 1993: 124.
42. Interview with Jean Cordy, Louvain-la-
    Neuve, September 5, 2009.
43. Monheim 1961: 22–24.
44. Janssens 1961: 60.
45. Lumenganeso 2005: 108–109.
46. Janssens 1961: 59–61.
47. Interview with Jean Cordy, Louvain-la-
    Neuve, September 5, 2009.
48. Interview with Jean Cordy, Louvain-la-
    Neuve, September 5, 2009.
49. Interview with Jean Cordy, Louvain-la-
    Neuve, September 5, 2009.
50. CRISP 1960: 10.
51. Demunter 1975: 266.
52. Interview with Jean Cordy, Louvain-la-
    Neuve, September 5, 2009.
53. Young 1968: 159–60.
54. Caprasse 1959: 137–41.
55. CRISP 1960: 51–53.
56. De Backer 1959: 3:64.
57. Interview with Jean Mayani, Kisangani,
    November 19, 2008.
58. De Backer 1959: 2:20.
59. De Backer 1959: 3:158.
60. De Backer 1959: 2:83.
61. Interview with Jean Mayani, Kisangani,
    November 19, 2008.
62. Interview with Jean Mayani, Kisangani,
    November 19, 2008.
63. Ndaywel è Nziem 1998: 546.
64. Schöller 1982: 114.

65. Interview with Charly Henault, Méhaigne, August 28, 2008.
66. Ganshof van der Meersch 1960: 25; Scott 1969: 25.
67. Fierlafyn 1990: 200.
68. Demunter 1975: 276–77.
69. Archer 1971: 84.
70. Interview with Mario Cardoso, Kinshasa, October 1, 2008.
71. Buelens 2007: 327.
72. Merlier 1962: 292.
73. Eyskens 1994: 567.
74. Remilleux 1989: 46.
75. Joye and Lewin 1961: 290–95.
76. Young 1984: 712–13.
77. Weiss 1965: 2.
78. Inforcongo 1958; Verhaegen 1971: 421.
79. Kanza 1959: 39.

CHAPTER 7

1. Interviews with Jean Lema, alias Jamais Kolonga, Kinshasa, October 3 and 14 and November 6, 2008.
2. Interview with Victorine Ndjoli, Kinshasa, November 7, 2008.
3. Archer 1971: 11.
4. Etambala 1999b: 147.
5. Etambala 2008: 432–33.
6. CRISP 1961: 318–20.
7. Verlinden 2008: 140.
8. Interview with Jean Cordy, Louvain-la-Neuve, September 18, 2009.
9. De Vos 1961: 193–94.
10. CRISP1961: 323.
11. Interview with Victor Masunda Kukana, Boma, October 8, 2008.
12. Interview with Camille Mananga Nkanu, Boma, October 9, 2008.
13. Interview with Mario Cardoso, Kinshasa, October 1, 2008.
14. Schöller 1982: 178–83.
15. Ganshof van der Meersch 1958: 284.
16. Paulus 1962: 224.

CHAPTER 8

1. Janssens 1961: 12.
2. CRISP 1961: 353–54.
3. De Vos 1961: 202.
4. CRISP 1961: 381, 388; Geerts 1970: 79.
5. Verlinden 2002: 154.
6. CRISP 1961: 375.
7. Janssens 1961: 216.
8. Jorissen 2005: 115.
9. Verlinden 2002: 148–51.
10. De Craemer and Fox 1968: 3.
11. Interview with Jacques Courtejoie, Nsioni, October 5, 2008.
12. Souchard 1983: 254.
13. Kanza 1959: 40.
14. Souchard 1983: 248–49.
15. Geerts 1970: 11.
16. Interview with Bonyololo Lokombe, alias Papa Rovinscky, Kisangani, November 7, 2008.
17. Vesse 1961: 16.
18. Van den Bosch 1986: 57.
19. De Vos et al. 2004: 41.
20. De Vos et al. 2004: 40.
21. Boehme 2005.
22. De Vos et al. 2004: 521.
23. http://www.congo-1960.be/ WilfriedDeBrouwerFAF_Piloot.html.
24. Kestergat 1961.
25. Interview with Camille Mananga Nkanu, Boma, October 9, 2008.
26. Verlinden 2002: 151.
27. CRISP 1961: 544.
28. Abi-Saab 1978: 14.
29. CRISP 1961: 555.
30. Devlin 2007: 48.
31. Pardigon 1961: 89.
32. CRISP 1961: 555–56.
33. Devlin 2007.
34. El-Tahri 2007.
35. Interview with Jean Lema alias Jamais Kolonga, Kinshasa, October 3, 2008.
36. Brian Urquhart, cited by Meredith 2005: 104.
37. Meredith 2005: 104–105.
38. Eyskens 1994: 584.
39. CRISP 1961: 806.
40. CRISP 1961: 110.
41. Geerts 1970: 90–91.

42. Interview with Mario Cardoso, Kinshasa, October 1, 2008.
43. De Vos et al. 2004: 581.
44. Devlin 2007: 94–97.
45. Young 1984: 721.
46. De Vos et al. 2004: 255.
47. Interview with Mario Cardoso, Kinshasa, October 1, 2008.
48. Meredith 2005: 109.
49. De Vos et al. 2004: 363–422.
50. De Vos et al. 2004: 395.
51. Soete 1993: 98–101.
52. Walter Zinzen, press release, November 4, 2009.
53. Van Bilsen 1993: 161.
54. Interview with Anne Mutosh Amuteb, Lubumbashi, April 23, 2008.
55. Ziégler 1963.
56. Yakemthouck 1988a: 177.
57. Scholl-Latour 1986: 216.
58. Gérard-Libois 1963: 186–87.
59. Ziégler 1963: 38.
60. Interview with Walter and Alice Lumbeeck, Oostkamp, July 25, 2009.
61. Interview with Frans and Marja Vleeschouwers, Berchem, July 25, 2009.
62. Hammarskjöld 1964: 93.
63. Martelli 1966: 198.
64. http://www.congo-1960.be/huurlingencongo.htm.
65. Interview with Walter en Alice Lumbeeck, Oostkamp, July 25, 2009.
66. Interview with Frans and Marja Vleeschouwers, Berchem, July 25, 2009.
67. Interview with Walter en Alice Lumbeeck, Oostkamp, July 25, 2009.
68. CRISP 1964: 99.
69. Kamitatu 1971: 97.
70. Young 1968: 319–48.

71. CRISP 1965: 104.
72. CRISP 1964: 105–107.
73. Verhaegen 1966–1969: 1:122.
74. Fox et al. 1965: 22.
75. Makulo Akambu 1983: 91–94.
76. Manya K'Omalowete 1986.
77. Verhaegen 1986: 7, 12.
78. Takizala 1964: 69.
79. Manya K'Omalowete 1986: 102.
80. Zinzen 2004: 101.
81. Geerts 1970: 189.
82. Etambala 1999b: 266.
83. Etambala 1999b: 258.
84. Tielemans 1966; Esposito 1978.
85. Ndaywel è Nziem 1998: 638–39.
86. CRISP 1965: 141.
87. Ndaywel è Nziem 1998: 639.
88. Verbeken 2005: 36.
89. Devlin 2007: 225.
90. Ndaywel è Nziem 1998: 623.
91. Brion 1986: 63.
92. Makulo Akambu 1983: 92.
93. Makulo Akambu 1983: 93–94.
94. El-Tahri 2007.
95. Guevara 2001: 298.
96. Guevara 2001: 83.
97. Guevara 2001: 313.
98. Guevara 2001: 281.
99. Zinzen 1995: 19–20.
100. CRISP 1966: 441.
101. CRISP 1966: 257.
102. Houyoux 1973: 30.
103. La Fontaine 1970: 64.
104. Close 2007: 164.
105. CRISP 1966: 6.
106. Interview with Jean Lema, alias Jamais Kolonga, Kinshasa, October 3, 2008.
107. Ilosono 1985: 67–72.

CHAPTER 9

1. Interviews with Zizi Kabongo, Kinshasa, May 31, November 14 and 16, 2007, April 21 and September 16, 2008.
2. Buana Kabue 1975: 105.
3. CRISP 1966: 438–44.
4. Joris 2001: 33.
5. Interview with Vincent Lombume Kalimasi, Kinshasa, June 14, 2007.

6. Diallo 1977: 88.
7. CRISP 1967: 441.
8. CRISP 1967: 442–43.
9. Saint Moulin 2007: 42.
10. Houyoux 1973: 30–31.
11. Geerts 1970: 358.
12. CRISP 1966: 415–16.
13. CRISP 1967: 102.

14. Geerts 1970: 286–95.
15. CRISP 1967: 120.
16. CRISP 1967: 179.
17. Geerts 1970: 274.
18. Interview with Alphonsine Mosolo Mpiaka, Kinshasa, November 7, 2008.
19. Zinzen 1995: 29–31.
20. Verhaegen 1970: 23.
21. Braeckman 1992: 38–40.
22. Geerts 1970: 255.
23. Huybrechts et al. 1980: 152–63.
24. Ikembana 2007: 31.
25. Bureau du Président de la République 1972: 384–85.
26. Sakombi Inongo 1974b: 409.
27. Close 2007: 235.
28. Close 2007: 190.
29. Close 2007: 251.
30. Interview with François Ngombe, alias Maître Taureau, Kinshasa, November 9, 2009.
31. Lubabu Mpasi-A-Mbongo and Musangi Ntemo 1987: 56.
32. Mwabila Malela 1979: 128.
33. Huybrechts et al. 1980: 239.
34. Interview with Paul Kasenge, Lubumbashi, June 29, 2007.
35. Huybrechts et al. 1980: 170.
36. Interview with André Kitadi, Kinshasa, October 16, 2008.
37. Gast 1996.
38. Lubabu Mpasi-A-Mbongo and Musangi Ntemo 1987: 76.
39. Verhaegen 1978: 126–30.
40. Interview with Adolphine Ngoy, Bukavu, June 19, 2007.
41. Interview with Bertrand Bisengimana, Kivumeer, April 25, 2008.
42. Young and Turner 1985: 167–68.
43. Ndaywel è Nziem 1995.
44. Sakombi Inongo 1974b: 409.
45. Sakombi Inongo 1974a: 334.
46. Ekanga Botombele 1975.
47. Sakombi Inongo 1974b: 318.
48. Interview with Joseph Ibongo, Kinshasa, June 1, 2007.
49. with Jean-Pierre Mukoko, Kinshasa, March 19, 2005.
50. White 2008: 73–79.
51. Huybrechts et al. 1980: 155.
52. Schatzberg 1988: 122–25.
53. Nzongola-Ntalaja 2002: 148.
54. Bézy et al. 1981: 57–68.
55. Van den Bosch 1992: 90.
56. Bézy et al. 1981: 61.
57. Mobutu 1973: 233, 243.
58. Chomé 1975: 28.
59. Chomé 1978: 142–43.
60. Stewart 2000: 199.
61. Mailer 1975.
62. Young and Turner 1985: 326–62.
63. Houyoux 1972 ; Houyoux 1973.
64. Pain 1984: 114.

CHAPTER 10

1. Braeckman 1992: 298.
2. Remilleux 1989: 91.
3. Remilleux 1989: 92.
4. http://www.bernd-leitenberger.de/otrag.shtml.
5. Kalonga 1978: 32–47.
6. Mende Omalanga and Tshilenge wa Kabamb 1992: 75–77.
7. Willame 1986.
8. Ndaywiel è Nziem 1998: 737.
9. Willame 1986: 132.
10. Willame 1986: 80–81.
11. Interviews with Zizi Kabongo, Kinshasa, May 31and November 14 and 16, 2007, April 21 and September 16, 2008.
12. Geerts 2005: 173–76.
13. Yambuya 1991: 34–36.
14. Yambuya 1991: 33.
15. Yambuya 1991: 28.
16. Cabinet du Département de la Défense Nationale 1974: 40.
17. Kamitatu-Massamba 1977: 103.
18. Yambuya 1991: 69.
19. Yakemtchouk 1988b: 399–402.
20. Young and Turner 1985: 375.
21. Wrong 2000: 190.
22. Interview with Eugène Yoka Kinene, Kinshasa, November 11, 2008.
23. Interview with Alphonsine Mosolo Mpiaka, Kinshasa, November 7, 2008.

24. Young and Turner 1985: 324.
25. Ndaywel è Nziem 1998: 732.
26. Young and Turner 1985: 324.
27. Ndaywel è Nziem 1998: 732.
28. Willame 1992: 96.
29. Peemans 1988: 23.
30. Young and Turner 1985: 379.
31. Mende Omalanga and Tshilenge wa Kabamb 1992: 28–29.
32. Blumenthal 1982: 8.
33. Blumenthal 1982: 15.
34. Nguza Karl I Bond 1982.
35. Released CIA document, "Zaire's IMF Problem," September 15, 1986, http://www.foia.cia.gov/.
36. Stiglitz 2002: 12.
37. Kabuya Kalala et al. 1980.
38. Interview with Didace Kawang, Kinshasa, September 30, 2008.
39. Stiglitz 2002: 18.
40. Young and Turner 1985: 323.
41. Emizet 1997: 22, 26.
42. Nzongola-Ntalaja 1986: 4.
43. Peemans 1988: 39.
44. Emizet 1997: 25.
45. Young and Turner 1985: 311.
46. Janssen 1997: 78–83.
47. Janssen 1997: 73–78.
48. Interview with Kibambi Shintwa, Kinshasa, September 25, 2008.
49. Interviews with Jean Lema, alias Jamais Kolonga, Kinshasa, October 3, 14, and November 6, 2008.

50. Bender 2006: 304.
51. Interview with Alfons Mertens, Puurs, November 5, 2007.
52. Willame 1992: 132.
53. Interview with Raymond Mukoka, Kinshasa, September 29, 2008.
54. Wrong 2000: 97.
55. Braeckman 1992: 72.
56. Interview with Mrs.A., Kinshasa, March 17, 2005; cf. Nlandu-Tsasa 1997: 16.
57. Vandommele 1983: 78–79; Batumike 1986: 52–53.
58. Amnesty International 1980; Amnesty International 1983.
59. Yambuya 1991: 91.
60. Nlandu-Tsasa 1997: 98.
61. Mvumbi Ngolu Tsasa 1986: 68–69.
62. Interview with Papy Mbwiti, Kinshasa, September 30, 2008.
63. MacGaffey 1991: 17–19.
64. Nlandu-Tsasa 1997: 64.
65. Renton et al. 2007: 136.
66. Interview with Papy Mbwiti, Kinshasa, September 30, 2008.
67. Interview with Rufin Kibari Nsanga, Kikwit, September 21, 2008.
68. Newbury 1984.
69. Interview with Régine Mutijima, Kinshasa, December 10, 2008.
70. Interview with Thérèse Pakasa, Kinshasa, November 8, 2008.
71. Michel 1999.
72. Braeckman 1992: 303.

CHAPTER 11

1. Interview with Régine Mutijima, Kinshasa, December 10, 2008.
2. Interview with Modeste Mutinga, Kinshasa, October 3, 2008.
3. Interview with Baudouin Hamuli, Kinshasa, September 16 and 24, 2008.
4. Hamuli Kabarhuza et al. 2003: 27.
5. Braeckman 1992: 205.
6. Interview with Baudouin Hamuli, Kinshasa, September 16 and 24, 2008.
7. Mende Omalanga and Tshilenge wa Kabamb 1992: 45–47.
8. Buyse 1994: 29.

9. Interview with Kibambi Shintwa, Kinshasa, September 25, 2008.
10. Interview with Baudouin Hamuli, Kinshasa, September 16 and 24, 2008.
11. Interview with José Mpundu, Kinshasa, September 23, 2008.
12. De Dorlodot 1994: 90–94.
13. Interviews with Zizi Kabongo, Kinshasa, May 31, November 14 and 16, 2007, April 21 and September 16, 2008.
14. Nzongola-Ntalaja 2002: 193.
15. Nzongola-Ntalaja 2002: 199.
16. Villers and Omasomba 2002: 403.

17. Nzongola-Ntalaja 2002: 196–98.
18. De Herdt and Marysse 2002: 175.
19. Banque Centrale du Congo 2007: 92.
20. Michel 1992.
21. Smith 2005: 234.
22. Devisch 1995: 608.
23. Hamuli Kabarhuza et al. 2003: 18.
24. Devisch 1995.
25. Ndaywel è Nziem 1995: 78, 116.
26. Ndaywel è Nziem 1995: 65–66.
27. Banque Centrale du Congo 2007: 117.
28. Beaugrand 1997: 9.
29. Banque Centrale du Congo 2007: 119;
    De Boeck and Plissart 2004: 187.
30. Beaugrand 2003.
31. Interview with Alexandre Kasumba
    Dunia, zoo manager, Kinshasa,
    December 29, 2003.
32. Kabuya Kalala and Matata Ponyo 1999.
33. Interview with Sekombi Katondolo,
    Goma, November 26, 2008.
34. Mamdani 2001: 253.
35. Interview with anonymous Mai-mai,
    Goma, December 4, 2008.
36. Interview with Pierrot Bushala, Goma,
    November 26, 2008.
37. Vlassenroot 2000.
38. Prunier 2009: 25.
39. Interview with Ruffin Luliba, Kinshasa,
    October 1, 2008.
40. Reyntjens 2009: 88.
41. Prunier 2009: 122.
42. Interview with Katelijne Hermans,
    Brussels, September 12, 2007.
43. Interview with Dr. Soki, Kisangani,
    November 16, 2008.
44. Interview with Sekombi Katondolo,
    Goma, November 26, 2008.
45. Efinda 2009: 49.
46. Efinda 2009: 73.
47. Human Rights Watch 1997a: 23.
48. Interview with Papy Bulaya, Kinshasa,
    December 11, 2008.
49. Van Dijck 1997: 9.
50. Prunier 2009: 148.
51. Interview with Rufin Kibari Nsanga,
    Kikwit, September 21, 2008.
52. Interview with Katelijne Hermans,
    Brussels, September 12, 2007.
53. Interview with Rufin Kibari Nsanga,
    Kikwit, September 21, 2008.
54. Interview with Rufin Kibari Nsanga,
    Kikwit, September 21, 2008.
55. Willame 1998.
56. Reyntjens 2009: 67–77.
57. Prunier 2009: 137–43.
58. El-Tahri 2000.
59. Tokwaulu Aena n.d.: 187.
60. El-Tahri 2000.

CHAPTER 12

1. Reyntjens 2009: 152.
2. Reyntjens 2009: 152.
3. Banque Centrale du Congo 2007: 95.
4. Human Rights Watch 1997b: 16.
5. Rights Watch 1997b: 48.
6. Reyntjens 2009: 152.
7. Interview with Bertin Punga, Kinshasa,
   September 29, 2008.
8. Frère 2005: 100.
9. Interview with Zizi Kabongo, Kinshasa,
   November 14, 2007.
10. Reyntjens 2009: 154.
11. Prunier 2009: 150.
12. Reyntjens 2009: 57.
13. Ngbanda 2004: 41.
14. Braeckman et al. 1998: 132.
15. Prunier 2009: 178.
16. International Rescue Committee 2007.
17. Interview with Simba Regis, Kigali,
    June 25, 2007.
18. Interview with Ruffin Luliba, Kinshasa,
    December 11, 2008.
19. Interview with Rufin Kibari Nsanga,
    Kikwit, September 21, 2008.
20. Reyntjens 2009: 186.
21. Lanotte 2003: 109.
22. Interview with Jeanine Mukanirwa,
    Kinshasa, November 12, 2007.
23. Interview with Tharcisse Kayira,
    Bukavu, June 20, 2007.
24. Interview with Pierrot Bushala, Goma,
    November 26 2008.

25. Human Rights Watch 2000.
26. Interview with Muhindo, Goma, December 4, 2008.
27. Interview in *De Morgen*, June 23, 2007.
28. Interview with Yves Van Winden, Goma, December 4, 2008.
29. Villers 2001: 12.
30. Zinzen 2004: 180–5.
31. Interview with Agustin Utshudi, Goma, December 1, 2008.
32. Lanotte 2003: 111–15.
33. Interview with Agustin Utshudi, Goma, December 1, 2008.
34. Interview with Dr. Soki, Kisangani, November 16, 2008.
35. Prunier 2009: 229.
36. Lanotte 2003: 122.
37. Villers 2001: 159–60.
38. Human Rights Watch 2005: 18.
39. Marysse and André 2001.
40. Leclercq 2001: 69.
41. Marysse and André 2001.
42. Global Witness 2005: 30.
43. Wamu Oyatambwe 1999: 122.
44. Harden 2001.
45. Marysse and André 2001.
46. UNESCO 2005: 34.
47. Debroux et al. 2007: 7, 10.
48. Interview with Papy Bulaya, Kinshasa, December 11, 2008.
49. Pole Institute 2002: 11, 13.
50. Vlassenroot and Romkema 2002.
51. Human Rights Watch 2001.
52. Human Rights Watch 2005: 27, 29, 45.
53. Interview with UPC veteran, Kasenyi, November 23, 2008.
54. Jourdan 2004.
55. Interview with Baptiste Uzele-Uparpiu, Kasenyi, November 23, 2008.
56. Interview with Masika Katsuva, Goma, December 2, 2008.

CHAPTER 13

1. Prunier 2009: 252.
2. Braeckman 2003: 97–125.
3. Villers 2009: 15–20.
4. Interview with Antoine Vumilia, Kinshasa, December 12, 2008.
5. Zeebroek 2008: 8–9.
6. Villers 2009: 222.
7. Interview with Johan Swinnen, Kinshasa, November 17, 2007.
8. Reyntjens 2009: 239.
9. Reyntjens 2009: 243–44.
10. NIZA 2006: 20.
11. Interview with Rufin Kibari Nsanga, Kikwit, September 21, 2008.
12. Nzeza 2004: 29.
13. NIZA 2006: 20–21.
14. NIZA 2006: 9.
15. Prunier 2009: 306.
16. Reyntjens 2009: 243.
17. Interview with Frank Werbrouck, Kinshasa, November 11, 2008.
18. Vlassenroot and Raeymaekers 2004b.
19. Prunier 2009: 325–26.
20. Interviews with Marie Djoza, Bunia, November 28, 2008; interview with Marie Pacuriema and Jacqueline Dz'ju Malosi, Bunia, November 22, 2008.
21. Human Rights Watch 2003: 6.
22. Interview with Johan Swinnen, Kinshasa, November 17, 2007.
23. Human Rights Watch 2004: 5.
24. Market Volume DRC, internal document Bralima, September 2008.
25. Interview with Jo De Neckere, Brussels, January 23, 2010.
26. Bruijn et al. 2001.
27. Giovannoni et al. 2004: 102–105.
28. Interview with Riza Labwe, Kikwit, September 21, 2008.
29. Interview with Justine Masika, Goma, December 1, 2008.
30. Interview with Dolf van den Brink, September 17, 2008.
31. Yoka 2005: 60.
32. Interview with Papy Mbwiti, Kinshasa, September 16, 2008.
33. De Boeck and Plissart 2004: 41.
34. "Werrason," Wikipedia, January 30, 2010.

35. Interviews with seven Werrason fans, Kikwit, September 21, 2008.
36. Werrason," Wikipedia, January 30, 2010.
37. Interview with Flamme Kapaya, Kisangani, November 13 and 16, 2008.
38. Interview with Dolf van den Brink, October 4, 2008.
39. *Le Potentiel,* September 25, 2008.
40. Interview with Beko, May 28, 2007.
41. Frère 2005: 124.
42. Tigo 2007: *Annual Report and Accounts.*
43. Frère 2009: 210.
44. Frère 2009: 209.
45. Pype 2009a: 103.
46. Pype 2009a: 103.
47. Ayad 2001.
48. Pype 2007.
49. Kalulambi Pongo 2004: 59.
50. Pype, 2009c.
51. *Le Potentiel,* February 4, 2006.
52. Muambi 2009: 96.

53. Interview with Martin Kayembe, Kinshasa, April 22, 2008.
54. *Le Potentiel,* May 31, 2007.
55. De Boeck 2004: 162–64.
56. Ndaywel è Nziem 2002.
57. Interview with Dominique Khonde Mpolo, Kinshasa, November 10, 2008.
58. Interview with José Mpundu, Kinshasa, September 23, 2008.
59. Kennes 2003: 298.
60. Amnesty International 2006.
61. Olinga 2006.
62. Reyntjens 2009: 237.
63. White 2008: 184, 187.
64. White 2008: 173; Olinga 2006.
65. Interview with Israël Tshipamba, Kinshasa, May 29, 2007.
66. Frère 2007.
67. Villers 2009: 372.
68. www.deboutcongolais.info/actualite5/art_330.htm.

CHAPTER 14

1. Interview with Pascal Rukengwa, Kinshasa, September 26, 2008.
2. Muambi 2009: 78.
3. Villers 2009: 366.
4. Braeckman 2009: 123.
5. Villers 2009: 369.
6. Saint Moulin 2009: 54.
7. Telephone interview with Pascal Rukengwa, Brussels-Kinshasa, February 9, 2010.
8. Vandaele 2008: 30.
9. De Valon 2006: 23.
10. Interview with Carl De Keyzer, Kinshasa, May 27, 2007.
11. Braeckman 2009: 132.
12. International Crisis Group 2007: 10.
13. Human Rights Watch 2008a: 16–65.
14. Soudan 2007: 28.
15. Quiproquo, 20 juni 2007.
16. Interview with Baudouin Waterkeyn, hospital chaplain, Lubumbashi, June 29, 2007.

17. Villers 2009: 430.
18. *The Africa Report,* accessed December 2009: 197.
19. Muzito 2010: 4–8.
20. hrd.undp.org.
21. www.foreignpolicy.com.
22. www.doingbusiness.org.
23. www.unicef.org/infobycountry/drcongo_statistics.html.
24. Villers 2009: 426.
25. IPIS 2008a.
26. *Africa Confidential,* October 8, 2009: 2.
27. Interview with Alain Chirwisa, alias Alesh, Kisangani, November 16, 2008.
28. Stiglitz 2002: 18.
29. Radio France International, September 22, 2009.
30. Amnesty International 2010.
31. Baldauf, 2008.
32. Scott 2008: 170.
33. Lemarchand 2009: 270.
34. IPIS 2008b.
35. Interview in *De Morgen,* May 6, 2008.
36. Human Rights Watch 2008b.

37. Interview with Major Antoine, Rwanguba, November 27, 2008.
38. Human Rights Watch 2002b: 18–20.
39. Interview with Laurent Nkunda, Rwanguba, November 27, 2008.
40. United Nations Security Council 2008; Human Rights Watch 2008b.
41. Interview with Grâce Nirahabimana, Mugunga, December 2, 2008.
42. International Crisis Group 2009a, b.
43. Hoebeke et al. 2009: 136.
44. Berwouts 2010.
45. De Keyzer 2009: 14–15.
46. Campbell 2007.
47. Braeckman 2009: 158–59.
48. RAID 2009: 8–26.
49. IPIS 2009.
50. Braeckman 2010: 52.
51. Braeckman 2009: 175.
52. Soudan 2007: 28.
53. Braeckman 2009: 177.
54. IMF, December 11, 2009: press release 09/445.
55. Minani 2010.
56. Vandaele 2005: 166; www.imf.org.
57. Reuters, December 12, 2009.
58. ASADHO 2010.
59. Stearns 2010.
60. Interview with Dadine Musitu, Kinshasa, September 16, 2008.
61. Interview with Rosemonde N., Kinshasa, September 25, 2008.

CHAPTER 15

1. *China Daily,* October 28, 2008.
2. Interview with Jules Bitulu, Guangzhou, October 27, 2008.
3. Interview with Lukisu Fule, Guangzhou, October 25, 2008.
4. Interview with Patou Lelo, Guangzhou, October 26, 2008.
5. Interviews with Commandant "César" and "Timothée," Guangzhou, October 26, 2008.

REFERENCES

Abi-Saab, G. 1978: *The United Nations Operation in the Congo 1960–1964*. Oxford.

Africa Confidential 2009: "Congo-Kinshasa: Mines, Dollars and Dams." *Africa Confidential,* October 9, 2009, 50, 20, 2–3.

Alden, C. 2007: *China in Africa*. London.

Alpers, E. A. 1975: *Ivory and Slaves in East Central Africa*. London.

Amnesty International. 1980: *Les violations des droits de l'homme au Zaïre*. Brussels.

———. 1983: *Zaïre: dossier sur l'emprisonnement politique et commentaires des autorités.* Paris.

———. 2006: "Democratic Republic of Congo: Acts of Political Repression on the Increase." AI Index: AFR 62/014/2006, July 4.

———. 2010: "Human Rights Defenders under Attack in the Democratic Republic of Congo." February 2010. www.amnesty.org.

Antipas, G. 2007: *Pionniers méconnus du Congo Belge*. Brussels.

Antonissen, J., and J. Weyn 2007: *F.C. Indépendance*. Canvas documentary.

Archer, J. 1971: *Congo: The Birth of a New Nation*. Folkestone, UK.

Asadho (Association Africaine de Défense de Droits de l'Homme). 2010: "Les conditions de travail des congolais au sein de l'entreprise chinoise CREC sont inacceptables!" January 2010, www.asadho-rdc.org.

Asch, S. 1982: *L'église du prophète Simon Kimbangu: De ses origines à son rôle actuel au Zaïre.* Paris.

Ayad, C. 2001: "Les sectes, sauve-qui-peut au Congo-Kinshasa." *Libération,* January 31.

Bailey, H. 1894: *Travel and Adventures in the Congo Free State and Its Big Game Shooting.* London.

Baldauf, S. 2008: "Congo's Riches Fuel Its War." *The Christian Science Monitor,* November 4, 2008.

Banque Centrale du Congo. 2007: *La Banque Centrale du Congo: Une rétrospective historique*. Kinshasa.

Barham, L., and P. Mitchell. 2008: *The First Africans: African Archaeology from the Earliest Toolmakers to the Most Recent Foragers*. Cambridge, UK.

Batumike, C. 1986: *Une liberté de moins: Témoignage de prison et autres rubriques*. Langenthal, Switzerland.

Beaugrand, P. 1997: "Zaïre's hyperinflation, 1990–96." IMF Working Paper 97/50.

———. 2003: "Overshooting and Dollarization in the Democratic Republic of the Congo." IMF Working Paper 03/105.

Bender, K. W. 2006: *Moneymakers: The Secret World of Banknote Printing*. Weinheim, Germany.

Bentley, W. H. 1900: *Pioneering on the Congo*. London.

Bertoncelo, B., and S. Bredeloup. 2007: "The Emergence of New African "Trading Posts" in Hong Kong and Guangzhou." *China Perspectives* 1, chinaperspectives.revues.org.

Berwouts, K. 2010: "Un semblant d'état en état de ruine." Internal document EurAc, January 27, 2010.

Bézy, F., J.-P. Peemans, and J.-M. Wautelet. 1981: *Accumulation et sous-développement au Zaïre 1960–1980.* Louvain-la-Neuve, Belgium.

Blanchart, C., J. de Deurwaerder, G. Nève, M. Robeyns, and P. van Bost. 1993: *Le rail au Congo Belge,* vol. 1: 1890–1920. Brussels.

———. 1999: *Le rail au Congo Belge,* vol. 2: 1920–1945. Brussels.

Blumenthal, E. 1982: "Zaïre: rapport over zijn internationale financiële credibiliteit." *Info Zaïre* 36: 3–15.

Bodomo, A. 2010: "The African Trading Community in Guangzhou: An Emerging Bridge for Africa-China Relations," *China Quarterly* (in press).

Boehme, O. 2005: "The Involvement of the Belgian Central Bank in the Katanga Secession, 1960–1963," *African Economic History* 33: 1–29.

Boel, S. 2005: "Censuur in Belgisch Congo (1908–1960): Een onderzoek naar de controle op de pers, de film en de muziek door de koloniale overheid." Unpublished thesis, Vrije Universiteit Brussels.

Boelaert, E., H. Vinck, and C. Lonkama. 1995: "Témoignages africains de l'arrivée des premiers blancs aux bords des rivières de l'Equateur," *Annales Aequatoria* 16: 36–117.

Bontinck, F. 1974: *L'autobiographie de Hamed ben Mohammed el-Murjebi: Tippo Tip (ca. 1840–1905).* Brussels.

———. 1980: "Mfumu Paul Panda Farnana, 1888–1930: premier (?) nationaliste congolais," In *La dépendance de l'Afrique et les moyens d'y remédier.* Edited by V. Y. Mudimbe. Paris. 591–610.

Boone, O. 1951: *Les tambours du Congo-belge et du Ruanda-Urundi.* Tervuren, Belgium.

Booven, H. van. 1913: *Tropenwee.* Amsterdam (orig. pub.1904).

Bosschaerts, D. 2007: *Herinneringen aan Congo: ambtenaar in Boma (1904–1907).* Antwerp, Belgium.

Bourla Errera, M. 2000: *Moïse Levy: Un rabbin au Congo (1937–1991).* Brussels.

Bouvier, P. 1965: *L'accesion du Congo belge à l'indépendance: Essai d'analyse sociologique.* Brussels.

Braeckman, C. 1992: *De dinosaurus: Het Zaïre van Mobutu.* Berchem, Belgium.

———. 2003: *Les nouveaux prédateurs: politique des puissances en Afrique.* Paris.

———. 2009: *Vers la deuxième indépendance du Congo.* Brussels.

———. 2010:"Les amis chinois du Congo," *Manière de Voir* (December 2009–January 2010): 52–54.

Braeckman, C., M.-F. Cros, G. de Villers, F. François, F. Reyntjens, F. Ryckmans, and J.-C. Willame. 1998: *Kabila prend le pouvoir.* Brussels.

Braekman, E. M. 1961: *Histoire du protestantisme au Congo.* Brussels.

Brausch, G. 1961: *Belgian Administration in the Congo.* London.

Brion, E. 1986: "L'Eglise catholique et la rébellion au Zaïre (1964–1967)," *Les Cahiers du CEDAF* 7–8: 61–78.

Brion, R., and J.-L. Moreau. 2006: *Van mijnbouw tot Mars: de ontstaansgeschiedenis van Umicore.* Tielt, Belgium.

Brousmiche, P. 1987: *Bortaï: Faradje, Asosa, Gambela, Saio. Journal de campagne.* Doornik/Tournai, Belgium.

Brower, A., ed. 2005: *African Archaeology: A Critical Introduction.* Oxford.

Bruijn, M. de, R. van Dijk, and D. Foeken, eds. 2001: *Mobile Africa: Changing Patterns of Movement in Africa and Beyond*. Leiden, The Netherlands.

Buana Kabue. 1975: *L'expérience zaïroise: Du casque colonial à la toque de léopard*. Paris.

Buch, P., and J. Vanderlinden. 1995: *L'uranium, la Belgique, et les puissances*. Brussels.

Buelens, F. 2007: *Congo 1885–1960: Een financieel-economische geschiedenis*. Berchem, Belgium.

Bureau du Président de la République. [1972]: *Profils du Zaïre*. Kinshasa.

Buyse, A. 1994: *Democratie voor Zaïre: De bittere nasmaak van een troebel experiment*. Groot-Bijgaarden, Belgium.

Cabinet du Département de la Défense Nationale. 1974: *Forces Armées Zaïroises: Mémorandum de Réflexion, d'Action et d'Information*. Kinshasa.

Callaghy, T. 1984: *The State-Society Struggle: Zaire in Comparative Perspective*. New York.

Campbell, K. 2007: "800 Chinese State-Owned Enterprises Active in Africa, Covering Every Country," *Mining Weekly* (September 28, 2007).

Caprasse, P. 1959: *Leaders africains en milieu urbain (Elisabethville)*. Brussels.

Carrington, J. F. 1974: *La voix des tambours*. Kinshasa.

Carton de Wiart, H. 1923: *Mes vacances au Congo*. Brussels.

Catherine, L. 1994: *Manyiema, de enige oorlog die België won*. Antwerp.

Catteeuw, K. 1999: "Cardijn in Congo: De ontwikkeling en betekenis van de Katholieke Arbeidersjeugd in Belgisch-Congo." *Brood en Rozen* 4, no. 2: 153–69.

Cattier, F. 1906: *Étude sur la situation de l'État Indépendant du Congo*. Brussels.

Cayen, A. 1938: *Au service de la colonie*. Brussels.

Ceuppens, B. 2003: *Congo Made in Flanders? Koloniale Vlaamse visies op "bank" en "zwart" in Belgisch-Congo*. Ghent, Belgium.

———. 2009: "Een Congolese kolonie in Brussels." In *Congo in België: koloniale cultuur in de metropool*. Edited by V. Viaene, D. Van Reybrouck, and B. Ceuppens. Louvain, Belgium. 231–50.

Chalux 1925: *Un an au Congo Belge*. Brussels.

Chomé, J. 1959: *La passion de Simon Kimbangu, 1921–1951*. Brussels.

———. 1966: *Moïse Tshombe et l'escroquerie katangaise*. Brussels.

———. 1975: *Mobutu, guide suprême*. Brussels.

———. 1978: *Mobutu of de opgang van een sergeant-hulpboekhouder tot Opperste Leider van Zaïre*. Antwerp.

Clark, J. F. 2002: *The African Stakes of the Congo War*. New York.

Close, W. T. 2007: *Beyond the Storm: Treating the Powerless and the Powerful in Mobutu's Congo/Zaire*. Marbleton, WY.

Comhaire-Sylvain, S. 1968: *Femmes de Kinshasa: Hier et aujourd'hui*. Paris.

Connah, G. 2004: *Forgotten Africa: An Introduction to Its Archaeology*. London.

Convents, G. 2006: *Images et démocratie: Les Congolais face au cinéma et à l'audiovisuel*. Louvain, Belgium.

Coquery-Vidrovitch, C., A. Forest, and H. Weiss, eds. 1987: *Rébellions-révolution au Zaïre 1963–1965*. 2 vols. Paris.

Corneille, A. 1945: *Le syndicalisme au Katanga*. Elisabethville.

Cornet, R. J. 1944: *Katanga: le Katanga avant les Belges*. Brussels.

———. 1947: *La bataille du rail: La construction du chemin de fer de Matadi au Stanley Pool*. Brussels.

Cornevin, R. 1963: *Histoire du Congo (Léopoldville)*. Paris.

Couttenier, M. 2005: *Congo tentoongesteld: Een geschiedenis van de Belgische antropologie en het museum van Tervuren (1882–1925)*. Louvain, Belgium.

CRISP (Centre de Recherche et d'Information Socio-Politiques). 1960: *Congo 1959: Documents belges et africains.* Brussels.

———. 1961: *Congo 1960.* 2 vols. and supplement. Brussels.

———. 1962: *Abako 1950–1960: documents.* Brussels.

———. 1963: *Congo 1962.* Brussels.

———. 1964: *Congo 1963.* Brussels.

———. 1965: *Congo 1964.* Brussels.

———. 1966: *Congo 1965.* Brussels.

———. 1967: *Congo 1966.* Brussels.

———. 1969: *Congo 1967.* Brussels.

Cros, M.-F., and F. Misser. 2006: *Géopolitique du Congo (RDC).* Brussels.

Crowder, M. 1984: "The Second World War: Prelude to Decolonization in Africa." In *The Cambridge History of Africa,* vol 8: From c. 1940 to c. 1975. Edited by M. Crowder. Cambridge, UK. 8–51.

Dahle Huse, M., and S. L. Muyakwa. 2008: "China in Africa: Lending, Policy Space, and Governance." www.afrika.no.

Danniau, F. 2005: "'Il s'agit d'un peuple': Het antropologisch onderzoek van het Bureau international d'ethnographie (1905–1913)." Unpublished thesis, University of Ghent.

Davidson, A. B., A. F. Isaacman, and R. Pélissier. 1987: "La politique et le nationalisme en Afrique centrale et méridionale, 1919–1935." In: *Histoire générale de l'Afrique,* vol. 7: *L'Afrique sous la domination coloniale.* Edited by A. Adu Boahen. Paris. 721–60.

Davis, J. M. 1933: *Modern Industry and the African: An Inquiry into the Effect of the Copper Mines of Central Africa upon Native Society and the Work of the Christian Missions.* London.

Daye, P. 1929: *Congo et Angola.* Brussels.

De Backer, M. C. C. 1959: "Notes pour servir à l'étude des 'groupements politiques' à Léopoldville." 3 parts, typescript. Brussels.

De Boeck, F. 2004: "On Being Shege in Kinshasa: Children, the Occult, and the Street. In *Reinventing Order in the Congo: How People Respond to State Failure in Kinshasa.* Edited by T. Trefon. London. 155–73.

———. 2005: "The Apocalyptic Interlude: Revealing Death in Kinshasa." *African Studies Review* 48, no. 2: 11–32.

De Boeck, F., and M.-F. Plissart. 2004: *Kinshasa: Tales of the Invisible City.* Ghent, Belgium.

Debroux, L., T. Hart, D. Kalmowitz, A. Karsenty, and G. Topa, eds. 2007: "Forests in Post-Conflict Democratic Republic of Congo: Analysis of a Priority Agenda." www.cifor.cgiar.org.

De Craemer, W., and R. C. Fox. 1968: *The Emerging Physician: A Sociological Approach to the Development of a Congolese Medical Profession.* Stanford, CA.

De Dorlodot, P. 1994: *Marche d'espoir, Kinshasa 16 février 1992: Non-violence pour la démocratie au Zaïre.* Paris.

De Herdt, T., and S. Marysse. 2002: "La réinvention du marché par le bas: circuits monétaires et personnes de confiance dans les rues de Kinshasa." In *Manières de vivre: Économie de la "débrouille" dans les villes du Congo/Zaïre.* Edited by G. de Villers, B. Jewsiewicki, and L. Monnier. Tervuren, Belgium.

Dehoux, E. 1950: *L'Afrique centrale à la croisée des chemins: un reportage critique.* 2 vols. Brussels.

De Jonghe, E. 1908: "L'activité ethnographique des Belges au Congo." *Bulletin de la Société d'Études Coloniales* 15, no. 4: 283–308.

De Keyzer, C. 2009: *Congo (Belge)*. Tielt, Belgium.

De Keyzer, C., and J. Lagae 2010: *Congo Belge en images*. Tielt, Belgium.

Delannoo, E. 2006: "Het kortstondige verhaal van het Kongolese Vrijwilligerskorps." *Shrapnel*, June 2006: 49–52.

Delathuy A. M. 1986: *Jezuïeten in Kongo met zwaard en kruis*. Berchem, Belgium.

———. 1989: *De Kongostaat van Leopold II*. Antwerp.

———. 1992–1994: *Missie en staat in Oud-Kongo*. 2 vols. Berchem, Belgium.

Delcommune, A. 1920: *Le Congo, la plus belle colonie du monde*. Brussels.

De Liedekerke, C., and Y. Muller. 2006: *Congo na biso*. Documentary, Paris.

Delpierre, G. 2002: "Tabora 1916: de la symbolique d'une victoire." *Belgisch Tijdschrift voor Nieuwste Geschiedenis* 32, nos. 3–4: 351–81.

Delval, N. 1966: *Schuld in Kongo?* Louvain, Belgium.

De Meulder, B. 1996: *De kampen van Kongo: Arbeid, kapitaal, en rasveredeling in de koloniale planning*. Amsterdam.

———. 2000: *Kuvuande Mbote: een eeuw koloniale architectuur en stedenbouw in Kongo*. Antwerp.

Dembour, M.-B. 2000: *Recalling the Belgian Congo*. New York.

Demunter, P. 1975: *Luttes politiques au Zaïre: le processus de politisation des masses rurales du Bas-Zaïre*. Paris.

Dendooven, D., and P. Chielens. 2008: *Wereldoorlog I: Vijf continenten in Vlaanderen*. Tielt, Belgium.

Denuit-Somerhausen, C. 1988: "Les traités de Stanley et de ses collaborateurs avec les chefs africains, 1880–1885." In Koninklijke Academie voor Overzeese Wetenschappen, *Bijdragen over de honderdste verjaring van de Onafhankelijke Kongostaat*. Brussels. 77–146.

Denuit-Somerhausen, C., and F. Balace. 1992: "Abyssinie 41: Du mirage à la victoire." In *Jours de lutte*. Edited by F. Balace. Brussels. 15–49.

Depaepe, M., J. Briffaerts, P. Kita Kyankenge Masandi, and H. Vinck 2003: *Manuels et chansons scolaires au Congo Belge*. Louvain, Belgium.

Derrick, J. 2008: *Africa's "Agitators": Militant Anti-Colonialism in Africa and the West, 1918–1939*. London.

Devisch, R. 1995: "Frenzy, Violence, and Ethical Renewal in Kinshasa." *Public Culture* 7, 593–629.

Devlin, L. 2007: *Chief of Station, Congo: A Memoir of 1960–67*. New York.

De Vos, L., E. Gerard, P. Raxhon, and J. Gérard-Libois. 2004: *Lumumba: De complotten? De moord*. Louvain, Belgium.

De Vos, P. 1961: *Vie et mort de Lumumba*. Paris.

De Waele, J. 2007–2008: "Voor vorst en vaderland: zwarte soldaten en dragers tijdens de Eerste Wereldoorlog in Congo." *Militaria Belgica*, 107–26.

De Witte, L. 1999: *De moord op Lumumba*. Louvain, Belgium.

Diallo, S. 1977: *Zaire Today*. Paris.

Diangienda Kuntima, J. 2007: *L'histoire du Kimbanguisme*. Châtenay-Malabry, France.

Dibwe dia Mwembu, D. 1999: "De la surpolitisation à l'antipolitique, quelques remarques en marge de l'histoire du mouvement ouvrier à l'Unoin minière du Haut-Katanga (UMHK) et à la Gécamines, 1920–1960." *Brood en Rozen* 4, no. 2: 184–199.

Dibwe dia Mwembu, D. 2001a: *Histoire des conditions de vie des travailleurs de l'Union Minière du Haut-Katanga/Gécamines (1910–1999)*. Lubumbashi.

———. 2001b: *Bana Shaba abandonnés par leur père: Structure de l'autorité et histoire sociale de la famille ouvrière au Katanga, 1910–1997*. Paris.

————. 2004: *Le travail hier et aujourd'hui: mémoires de Lubumbashi.* Paris.

Doucy, A., and P. Feldheim. 1952: *Problèmes du travail et politique sociale au Congo belge.* Brussels.

Drachoussoff, V. 1954: *L'évolution de l'agriculture indigène dans la zone de Léopoldville.* Brussels.

Dujardin, V., V. Rosoux, and T. de Wilde d'Estmael, eds. 2009: *Leopold II: Schaamteloos genie?* Tielt, Belgium.

Early, G., ed. 1992: *Speech and Power: The African-American Essay and Its Cultural Content from Polemics to Pulpit.* 2 vols. Hopewell.

Efinda, E. 2009: *Grand Lacs: Sur les routes malgré nous!* Paris.

Ekanga Botombele, B. 1975: *La politique culturelle en République du Zaïre.* Paris.

El-Tahri, J. 2000: *L'Afrique en morceaux: la tragédie des Grands Lacs.* Canal Plus documentary.

————. 2007: *Cuba, une odyssée africaine.* ARTE documentary.

Emizet, K. N. F. 1997: *Zaire after Mobutu: A Case of Humanitarian Emergency.* Helsinki.

Emongo Lomomba. 1985: "Le 'Blanc-belge' au Congo: Entretien avec Lomami Tshibamba." In: *Zaïre 1885–1985: Cent ans de regards belges.* Brussels. 135–47.

Ergo, A.-B. 2008: *Congo belge: La colonie assassinée.* Paris.

Esgain, N. 2000: "Scènes de la vie quotidienne à Elisabethville dans les années vingt." In *Itinéraires croisés de la modernité: Congo belge (1920–1950).* Edited by J.-L. Vellut. Tervuren, Belgium. 57–60.

Esposito, R. F. 1978: *Anuarite, vierge et martyre zaïroise.* Kinshasa.

Etambala, Zana Aziza. 1987: "Congolese Children at the Congo House in Colwyn Bay (North Wales, Great-Britain), at the End of the 19th Century." *Afrika Focus* 3: 237–85.

————. 1993: *In het land van de Banoko: De geschiedenis van de Kongolese/Zaïrese aanwezigheid in België van 1885 tot heden.* Louvain, Belgium.

————. 1999a: "Arbeidersopstanden en het ontstaan van inlandse syndicaten: de houding van de Katholiek Kerk (1940–1947)." *Brood en Rozen* 4, no. 2: 67–111.

————. 1999b: *Congo 55|65: Van koning Boudewijn tot president Mobutu.* Tielt, Belgium.

————. 2008: *De teloorgang van een modelkolonie: Belgisch Congo (1958–1960).* Louvain, Belgium.

Eyskens, G. 1994: *De memoires.* Tielt, Belgium.

Fabian, J. 1971: *Jamaa: A Charismatic Movement in Katanga.* Evanston, IL.

————. 1986: *Language and Colonial Power: The Appropriation of Swahili in the Former Belgian Congo, 1880–1938.* Cambridge, UK.

————. 1996: *Remembering the Present: Painting and Popular History in Zaire.* Berkeley, CA.

————. 2000: *Out of Our Minds: Reason and Madness in the Exploration of Central Africa.* Berkeley, CA.

Fetter, B. 1973: *L'Union Minière du Haut-Katanga, 1920–1940: La naissance d'une sous-culture totalitaire.* Brussels.

————. 1974: "African associations in Elisabethville, 1910–1935: their origins and development." *Études d'Histoire Africaine* 6: 205–23.

————. 1976: *The Creation of Elisabethville, 1910–1940.* Stanford, CA.

Feuchaux, L. 2000: "Vie coloniale et faits divers à Léopoldville (1920–1940)." In *Itinéraires croisés de la modernité: Congo belge (1920–1950).* Edited by J.-L. Vellut. Tervuren, Belgium. 71–101.

Fierlafyn, L. 1990: *Le discours nationaliste au Congo belge durant la période 1955–1960.* Brussels.

Flament, F. 1952: *La Force Publique de sa naissance à 1914: participation des militaires à l'histoire des premières années du Congo.* Brussels.

Foden, G. 2004: *Mimi and Toutou Go Forth: The Bizarre Battle of Lake Tanganyika.* London.

Foire Internationale d'Elisabethville. 1962: *Elisabethville 1911–1961.* Brussels.

Forbath, P. 1977: *The River Congo: The Discovery, Exploration and Exploitation of the World's Most Dramatic River.* New York.

Fox, R. C., W. De Craemer, and J.-M. Ribeaucourt. 1965: "La deuxième indépendance: Étude d'un cas, la rébellion au Kwilu." *Études Congolaises* 8, no. 1: 1–35.

Frère, M.-S. 2005: *Afrique centrale, médias et conflits: Vecteurs de guerre ou acteurs de paix.* Brussels.

———. 2007: "Quand le pluralisme déraille: Images et manipulations télévisuelles à Kinshasa." *Africultures*, November 20, 2007, www.africultures.com.

———. 2009: "Appui au secteur des médias: Quel bilan pour quel avenir?" In *Réforme au Congo (RDC): Attentes et désillusions.* Edited by T. Trefon. Tervuren, Belgium. 191–210.

Frères Maristes. 1927: *Buku na kutanga o lingala (Livre de lecture en lingala).* Liège, Belgium. www.abbol.com.

Gann, L. H., and P. Duignan. 1979: *The Rulers of Belgian Congo, 1884–1914.* Princeton, NJ.

Ganshof van der Meersch, W. J. 1958: *Le droit électoral au Congo belge: Status des villes et des communes.* Brussels.

———. 1960: *Congo, mei–juni 1960.* section l. N.p.

Garbin, D., and Wa Gamoka Pandu 2009: *Roots and Routes: Congolese Diaspora in Multicultural Britain.* London.

Gast, L. 1996: *When We Were Kings.* Documentary. Los Angeles.

Gbabendu Engunduka, A., and E. Efolo Ngobaasu. 1991: *Volonté de changement au Zaïre: De la consultation populaire vers la conférence nationale.* 2 vols. Paris.

Geenen, K. 2009: "'Sleep Occupies No Space': The Use of Public Space by Street Gangs in Kinshasa." *Journal of the African International Institute* 79, no. 3: 347–68.

Geernaert, J. no date: *Congophilie: Solution de la question coloniale belge.* Brussels.

Geerts, W. 1970: *Binza 10: De eerste tien onafhankelijkheidsjaren van de Democratische Republiek Congo.* Ghent, Belgium.

———. 2005: *Mobutu: de man van Kamanyola.* Louvain, Belgium.

Geldof, J. 1937: *Belgisch-Congo.* 2nd ed. Bruges, Belgium.

Gérard, J. E. 1969: *Les fondements syncrétiques du Kitawala.* Brussels.

Gérard-Libois, J. 1963: *Sécession au Katanga.* Brussels.

Ghilain, J. 1963: *Le revenu des populations indigènes du Congo-Léopoldville.* Brussels.

Giovannoni, M., T. Trefon, J. Kasongo Banga, and C. Mwema. 2004: "Acting on Behalf (and in Spite) of the State: NGOs and Civil Society Associations in Kinshasa." In *Reinventing Order in the Congo: How People Respond to State Failure in Kinshasa.* Edited by T. Trefon. London. 99–115.

Global Witness. 2005: "Under-Mining Peace: Tin, The Explosive Trade in Cassiterite in Eastern DRC." June 2005, www.globalwitness.org.

Goffin, L. 1907: *Le chemin de fer du Congo (Matadi-Stanley Pool).* Brussels.

Gondola, C. D. 1997a: "Unies pour le meilleur et le pire: Femmes africaines et villes coloniales: une histoire du métissage." *Clio: Histoire, Femmes, et Sociétés* 6, http://clio.revues.org/index377.html.

———. 1997b: *Villes miroirs: Migrations et identités urbaines à Kinshasa et Brazzaville, 1930–1970.* Paris.

————. 1999: "La contestation politique des jeunes à Kinshasa à travers l'exemple du mouvement 'kindoubill' (1950–1959)." *Brood en Rozen* 4, no. 2: 171–83.

Gould, D. J. 1980: *Bureaucratic Corruption and Underdevelopment in the Third World: The Case of Zaire.* New York.

Gourou, P. 1955: *La densité de la population rurale au Congo belge.* Brussels.

Govaerts, B. 2009: "De strop of de kogel? Over de toepassing van de doodstraf in Kongo en Ruanda-Urundi (1885–1962)." *Brood en Rozen* 1: 59–77.

Grévisse, F. 1951: *Le centre extra-coutumier d'Elisabethville: Quelques aspects de la politique indigène du Haut-Katanga industriel.* Brussels.

Guebels, L. 1952: *Relation complète des travaux de la Commission Permanente pour la Protection des Indigènes, 1911–1951.* Gembloux, Belgium.

Guevara, E. 2001: *De Afrikaanse droom: De revolutionaire dagboeken uit de Kongo 1965–1966.* Amsterdam.

Habran, L. 1925: *Coup d'œil sur le problème politique et militaire du Congo Belge.* Brussels.

Hammarskjöld, D. 1964: *Markings.* London.

Hamuli Kabarhuza, B. 2002: *Donner sa chance au peuple congolais: Expériences de développement participatif (1985–2001).* Paris.

Hamuli Kabarhuza, B., F. Mushi Mugumo, and N. Yambayamba Shuku. 2003: *La société civile congolaise: État des lieux et perspectives.* Brussels.

Harden, B. 2001: "A Black Mud from Africa Helps Power the New Economy." *New York Times,* August 12.

Harms, R. W. 1981: *River of Wealth, River of Sorrow: The Central Zaire Basin in the Era of the Slave and Ivory Trade, 1500–1891.* New Haven, CT.

Hawker, G. 1909: *The Life of George Grenfell: Congo Missionary and Explorer.* London.

Helmreich, J. E. 1983: "The Uranium Nnegotiations of 1944." In *Le Congo belge durant la Seconde Guerre Mondiale.* Brussels. 253–84.

————. 1986: *Gathering Rare Ores: The Diplomacy of Uranium Acquisition, 1943–1954.* Princeton, NJ.

Hemmens, H. L. 1949: *George Grenfell, Master Builder of Foundations.* London.

Hibbert, C. 1982: *Africa Explored: Europeans in the Dark Continent, 1769–1889.* London.

Higginson, J. 1989: *A Working Class in the Making: Belgian Colonial Labor Policy, Private Enterprise, and the African Mineworker, 1907–1951.* Madison, WI.

Hilton, A. 1985: *The Kingdom of Kongo.* Oxford.

Hochschild, A. 1998: *De geest van koning Leopold II en de plundering van de Congo.* Amsterdam.

Hoebeke, H., H. Boshoff, and K. Vlassenroot. 2009: " 'Monsieur le Président, vous n'avez pas d'armée . . . ': La réforme du secteur de sécurité vue du Kivu." In *Réforme au Congo (RDC): Attentes et désillusions.* Edited by T. Trefon. Tervuren, Belgium. 119–37.

Hoskyns, C. 1969: *The Organization of African Unity and the Congo Crisis.* Case Studies in African Diplomacy 1. Dar es Salaam.

Houyoux, J. 1972: *Budgets menagers à Kisangani, juin-juillet-août 1972.* section 1.

————. 1973: *Budgets menagers, nutrition, et mode de vie à Kinshasa.* Kinshasa.

Hulstaert, G. 1983: "Herinneringen aan de oorlog." In *Le Congo belge durant la Seconde Guerre Mondiale.* Brussels. 587–95.

————. 1990: "Marie aux Léopards: Quelques souvenirs historiques." *Annales Aequatoria* 11: 433–35.

Human Rights Watch. 1997a: "What Kabila Is Hiding: Civilian Killings and Impunity in Congo." October 1997, www.hrw.org.

————. 1997b: "Uncertain Course: Transition and Human Rights Violations in the Congo." December 1997, www.hrw.org.

————. 2000: "Eastern Congo Ravaged: Killing Civilians and Silencing Protest." May 2000, www.hrw.org.

————. 2001: "Uganda in Eastern DRC: Fuelling Political and Ethnic Strife." March 2001, www.hrw.org.

————. 2002a: "The War within the War: Sexual Violence against Women and Girls in Eastern Congo." June 2002, www.hrw.org.

————. 2002b: "War Crimes in Kisangani: The Response of Rwandan-backed Rebels to the May 2002 Mutiny." August 2002, www.hrw.org.

————. 2003: "Ituri 'Covered in Blood': Ethnically Targeted Violence in Northeastern DR Congo." July 2003, www.hrw.org.

————. 2004: "D.R. Congo: War Crimes in Bukavu." June 2004, www.hrw.org.

————. 2005: "The Curse of Gold." June 2005, www.hrw.org.

————. 2008a: "'We Will Crush You': The Restriction of Political Space in the Democratic Republic of Congo." November 2008, www.hrw.org.

————. 2008b: "Killings in Kiwanja." December 2008, www.hrw.org.

Hunt, N. R. 1999: *A Colonial Lexicon: Of Birth Ritual, Medicalization, and Mobility in the Congo*. Durham, NC.

Huybrechts, A., V. Y. Mudimbe, L. Peeters, J. Vanderlinden, D. Van Der Steen, and B. Verhaegen. [1980]: *Du Congo au Zaïre, 1960–1980: Essai de bilan*. Brussels.

Ikembana, P. 2007: *Mobutu's Totalitarian Political System: An Afrocentric Analysis*. London.

Ilosono Bekili B'Inkonkoy. 1985: *L'épopée du 24 novembre: Témoignage*. Kinshasa.

Ilunga Mpunga, D. 2007: *Étienne Tshisekedi: Le sens d'un combat*. Paris.

Inforcongo. 1958: *Belgisch-Congo en Ruandi-Urundi: Reisgids*. Brussels.

International Crisis Group. 2007: "Congo: Consolidating the Peace." July 2007, www.crisisgroup.org.

————. 2009a: "Congo: Five Priorities for a Peacebuilding Strategy." May 2009, www.crisisgroup.org.

————. 2009b: "A Comprehensive Strategy to Disarm the FDLR." July 2009, www.crisisgroup.org.

International Rescue Committee 2007: "Mortality in the Democratic Republic of Congo: An Ongoing Crisis." January 2007, www.theirc.org.

IPIS (International Peace Information Service). 2002: "Supporting the War Economy in the DRC: European Companies and the Coltan Trade." www.ipisresearch.be.

————. 2002: "European Companies and the Coltan Trade: An Update." www.ipisresearch.be.

————. 2002: "Network War: An Introduction to Congo's Privatised War Economy." www.ipisresearch.be.

————. 2008a: "The Congo Wants to Raise the Profits of Its Mining Sector." www.ipisresearch.be.

————. 2008b: "Mapping Conflict Motives: Eastern DRC." www.ipisresearch.be.

————. 2009: "The Impact of the Global Financial Crisis in Katanga." www.ipisresearch.be.

Isaacman, A., and J. Vansina 1987: "Initiatives et résistances africaines en Afrique centrale de 1880 et 1914." In *Histoire générale de l'Afrique*, vol. 7: *L'Afrique sous domination coloniale*. Edited by A. Adu Boahen. Paris. 191–216.

Jadin, L. 1968: "Les sectes réligieuses sécrètes des Antoniens au Congo, 1703–1709." *Cahiers des religions africaines* 2: 113–20.

Janssen, P. 1997: *Aan het hof van Mobutu*. Paris.

Janssens, E. 1905: "Rapport de la Commission d'Enquête." *Bulletin Officiel de l'État Indépendent du Congo* 21, nos. 9–10: 135–287.

———. 1961: *J'étais le général Janssens*. Brussels.

———. 1979: *Histoire de la Force Publique*. Brussels.

———. 1982–84: *Contribution à l'histoire militaire du Congo belge pendant la Seconde Guerre mondiale, 1940–45*. Brussels.

Jeal, T. 2007: *The Impossible Life of Africa's Greatest Explorer*. London.

Jewsiewicki, B. 1976: "La contestation sociale et la naissance du prolétariat au Zaïre au cours de la première moitié du XXe siècle." *Revue canadienne des études africaines* 10, no. 1: 47–71.

———. 1980: "Political Consciousness among African Peasants in the Belgian Congo." *Review of African Political Economy* 7, no. 19: 23–32.

———. 1988: "Mémoire collective et passé récent dans les discours historiques populaires zaïrois." In *Dialoguer avec le léopard*. Edited by B. Jewsiewicki and H. Moniot. Paris. 218–68.

Jewsiewicki, B., ed. 1992: *Art pictural zaïrois*. Paris.

Jewsiewicki, B., Kilola Lema, and J.-L. Vellut. 1973: "Documents pour servir à l'histoire sociale du Zaïre: Grèves dans le Bas-Congo (Bas-Zaïre) en 1945." *Études d'Histoire Africaine* 5: 155–88.

Johnston, H. 1908: *George Grenfell and the Congo*. London.

Joris, L. 1987: *Terug naar Congo*. Amsterdam.

———. 2001: *Dans van de luipaard*. Amsterdam.

———. 2006: *Het uur van de rebellen*. Amsterdam.

Jorissen, F. 2005: *Dagboek van een koloniaal: herinneringen van Belgisch Kongo 1953–1960*. Hasselt, Belgium.

Jourdan, L. 2004: "Being at War, Being Young: Violence and Youth in North Kivu." In *Conflict and Social Transformation in Eastern DRC*. Edited by K. Vlassenroot and T. Raeymaekers. Ghent, Belgium. 157–76.

Joye, P., and R. Lewin 1961: *Les trusts au Congo*. Brussels.

Julien, P. 1953: *Pygmeeën: Vijfentwintig jaar dwergen-onderzoek in Equatoriaal Afrika*.

Kabuya Kalala, Kalonji Nsenga, and Itimelongo Titi. 1980: "Les mésures de démonétisation du 25 décembre 1979 au Zaïre: Impacts et conséquences probables." *Zaïre-Afrique* 20: 144, 197–214.

Kabuya Kalala F., and Matata Ponyo. 1999: *L'espace monétarie kasaïen: Crise de légitimité et de souveraineté monétaire en période d'hyperinflation au Congo (1993–1997)*. Paris.

Kadima-Nzuji, M. 1984: *La littérature zaïroise de langue française (1945–1965)*. Paris.

Kalb, M. 1982: *The Congo Cables: The Cold War in Africa, from Eisenhower to Kennedy*. New York.

Kalonga, A. 1978: *Le mal zaïrois*. Brussels.

Kalulambi Pongo, M. 2004: "Le ndombolo du Seigneur: itinéraires et logiques des musiques religieuses en Afrique centrale." *Rupture-Solidarité* 5: 47–67.

Kamitatu, C. 1971: *La grande mystification du Congo-Kinshasa: les crimes de Mobutu*. Paris.

Kamitatu-Massamba, C. 1977: *Zaïre: le pouvoir à la portée du peuple*. Paris.

Kanza, T. R. 1959: *Propos d'un congolais naïf*. Brussels.

Kanza, T. 1972: *Conflict in the Congo: The Rise and Fall of Lumumba*. Baltimore.

Kaoze, S. 1910: "La psychologie des Bantu." *La Revue Congolaise* 1: 406–37.

Kelly, S.1993: *America's Tyrant: The CIA and Mobutu of Zaire: How the United States Put*

Mobutu in Power, Protected Him from His Enemies, Helped Him Become One of the Richest Men in the World, and Lived to Regret It. Washington, DC.

Kennes, E. 2003: *Essai biographique sur Laurent Désiré Kabila.* Tervuren, Belgium.

Kestergat, J. 1961: *André Ryckmans.* Brussels.

———. 1965: *Congo Congo: De l'indépendance à la guerre civile.* Paris.

Kibari Nsanga, R. 1985: "Mouvements 'anti-sorciers' dans les Provinces de Leopolville [sic] et du Kasaï, à l'époque coloniale." Unpublished typescript, Kikwit.

Kimoni Iyay. 1990: "Kikwit et son destin: Aperçu historique et sociologique." *Pistes et Recherches: Revue Scientifique* 5: 155–82.

Kisangani, E. F., and F. S. Bobb. 2010: *Historical Dictionary of the Democratic Republic of the Congo.* Lanham, MD.

Kisobele Ndontoni, N. 2008: "Mot de Circonstance des Anciens Combattants 40–45." Unpublished speech, Kinshasa, November 11, 2008.

Klein, W. C. 1957: *De Congolese elite.* Amsterdam.

Koninklijke Academie voor Overzeese Wetenschappen. 1976: *Bijdragen over de Aardrijkskundige Conferentie van 1876.* Brussels.

———. 1988: *Bijdragen over de honderdste verjaring van de Onafhankelijke Kongostaat.* Brussels.

Labrique, J. 1957: *Congo politique.* Léopoldville.

La Fontaine, J. S. 1970: *City Politics: A Study of Léopoldville, 1962–63.* Cambridge, UK.

Lagae, J. 2002: *Kongo zoals het is: Drie architectuurverhalen uit de Belgische kolonisatiegeschiedenis (1920–1960).* Ghent, Belgium.

Lagae, J., T. de Keyser, and J. Vervoort. 2006: *Boma 1880–1920: Koloniale hoofdstad of kosmopolitische handelspost.* cd-rom, Ghent, Belgium.

Lanotte, O. 2003: *Guerres sans frontières en République Démocratique du Congo.* Brussels.

Laude, N. 1956: *La délinquance juvénile au Congo belge et au Ruanda-Urundi.* Brussels.

Lauro, A. 2005: *Coloniaux, ménagères et prostituées au Congo belge (1885–1930).* Loverval.

Leclercq, C. 1964: *L'ONU et l'affaire du Congo.* Paris.

Leclercq, H. 2001: "Le rôle économique du diamant dans le conflit congolais." In *Chasse au diamant au Congo/Zaïre.* Edited by L. Monnier, B. Jewsiewicki, and G. de Villers. Tervuren, Belgium. 47–78.

Lefebvre, V. 1952: *La Belgique et le Congo au milieu du XXᵉ siècle.* Charleroi, Belgium.

Legum, C. 1965: *Pan-Africanism: A Short Political Guide.* New York.

Lemarchand, R. 2009: *The Dynamics of Violence in Central-Africa.* Philadelphia.

Leslie, W. J. 1987: *The World Bank and Structural Adjustment in Developing Countries: The Case of Zaire.* Boulder, CO.

Leysen, L. 1982: *Heimweh nach den Tropen.* ARD documentary.

Li, Z., D. Xue, M. Lyons, and A. Brown 2007: "Ethnic Enclave of Transnational Migrants in Guangzhou: A Case Study of Xiaobei." asiandrivers.open.ac.uk.

Libotte, B. no date.: "Droeven J.: De eerste kleurling in het Belgische leger," http://www.forumeerstewereldoorlog.nl/viewtopic.php?p=117626.

Loka-ne-Kongo 2001: *Lutte de libération et piège de l'illusion: Multipartisme intégral et dérive de l'opposition au Zaïre (1990–1997).* Kinshasa.

Lubabu Mpasi-A-Mbongo and Musangi Ntemo 1987: "Histoire du MPR." In *Mélanges pour une révolution.* Edited by Sakombi Inongo. Kinshasa. 35–126.

Lumenganeso, A. 2005: "Stedelijk vervoer in Leopoldstad: De gyrobus." In *Het geheugen van Congo: De koloniale tijd.* Edited by J.-L. Vellut. Tervuren, Belgium. 108–9.

Lyons, M. 1992: *The Colonial Disease: A Social History of Sleeping Sickness in Northern Zaire, 1900–1940.* Cambridge, UK.

MacGaffey, W. 1986: *Religion and Society in Central Africa*. 1986.

MacGaffey, J. 1991: *The Real Economy of Zaire: The Contribution of Smuggling and Other Unofficial Activities to National Wealth*. London.

Mailer, N. 1975: *The Fight*. Boston.

Makulo Akambu. 1983: *La vie de Disasi Makulo, ancien esclave de Tippo Tip et catéchiste de Grenfell, par son fils Makulo Akambu*. Kinshasa.

Malu-Malu, J.-J. 2002: *Le Congo Kinshasa*. Paris.

Mamdani, M. 2001: *When Victims Become Killers: Colonialism, Nativism and the Genocide in Rwanda*. Princeton, NJ.

Mantels, R. 2007: *Geleerd in de tropen: Leuven, Congo en de wetenschap, 1885–1960*. Louvain, Belgium.

Manya K'Omalowete. 1986: "Utilisation des procédés d'initiation et d'immunisation à caractère magique par le mouvement Simba." *Les cahiers du CEDAF* 7–8: 87–112.

Maquet-Tombu, J. 1952: *Le Siècle marche . . . Vie du chef congolais Lutunu*. Brussels.

Marchal, J. 1985: *E. D. Morel tegen Leopold II en de Kongostaat*. Berchem, Belgium.

———. 2001: *L'histoire du Congo 1910–1945*, vol. 3: *Travail forcé pour l'huile de palme de Lord Leverhulme*. Borgloon, Belgium.

Marechal, P. 1992: *De "Arabische" campagne in het Maniema-gebied (1892–1894)*. Tervuren, Belgium, 1992.

———. 2005: "Kritische bedenkingen bij de controverses over Leopold II en Congo in de literatuur en de media." In *Het geheugen van Congo: De koloniale tijd*. Edited by J.-L. Vellut. Tervuren, Belgium. 45–46.

Marres, J., and P. De Vos. 1959: *L'équinoxe de janvier: Les émeutes de Léopoldville*. Brussels.

Martelli, G. 1962: *Leopold to Lumumba: A History of the Belgian Congo 1877–1960*. London.

———. 1966: *Experiment in World Government: An Account of the United Nations Operation in the Congo 1960–1964*. London.

Martens, G. 1999: "Congolese Trade Unionism: The Colonial Heritage." *Brood en Rozen* 4, no. 2: 129–49.

Martens, L. 1985: *Pierre Mulele ou la seconde vie de Patrice Lumumba*. Berchem, Belgium.

———. 1991: *Une femme du Congo*. Berchem, Belgium.

Martin, M.-L. 1981: *Simon Kimbangu: Un prophète et son église*. Lausanne, Switzerland.

Marysse, S., and C. André. 2001: "Guerre et pillage en République Démocratique du Congo." *L'Afrique des Grands Lacs, annuaire 2000–2001*. 307–32.

Marysse, S., and S. Geenen. 2008: "Les contrats chinois en rdc: l'impérialisme rouge en marche?" *L'Afrique des Grands Lacs, 2007–2008*. 287–313.

Maximy, R. de 1984: *Kinshasa, ville en suspens: Dynamique de la croissance et problèmes d'urbanisme: étude socio-politique*. Paris.

McBrearty, S., and A. S. Brooks. 2000: "The Revolution that Wasn't: A New Interpretation of the Origin of Modern Behavior." *Journal of Human Evolution* 39: 453–563.

McCrummen, S. 2009: "Nearly Forgotten Forces of WWII." *Washington Post*, August 4.

McLynn, F. 1992: *Hearts of Darkness: The European Exploration of Africa*. London.

Meeuwis, M. 1999: "Buntungu's 'Mokingi mwa Mputu': A Boloki Perception of Europe at the End of the 19th Century." LPCA Text Archives 1, www2.fmg.uva.nl/lpca/textarchives/buntungu.html.

Mende Omalanga, L., and Tshilenge wa Kabamb. 1992: *Rapport sur les Biens Mal Acquis*. Kinshasa.

Mendiaux, E. 1960: *Moscou, Accra, et le Congo*. Brussels.

Mercader, J. 2003: "Foragers of the Congo: The Early Settlement of the Ituri Forest." In

Under the Canopy: The Archaeology of Tropical Rain Forests. Edited by J. Mercader. New Brunswick, NJ. 93–116.

Meredith, M. 2005: *The State of Africa: A History of Fifty Years of Independence.* London.

Merlier, M. 1962: *Le Congo: De la colonisation belge à l'indépendance.* Paris.

Mertens, M. 2009: "Chemical Compounds in the Congo: A Belgian Colony's Role in the Chemotherapeutic Knowledge Production during the 1920s." Unpublished lecture, Leipzig, Germany.

Meyers, J. 1964: *Le prix d'un empire.* Brussels.

Michaux, O. 1913: *Au Congo: Carnet de campagne.* Namur, Belgium.

Michel, S. 1962: *Uhuru Lumumba.* Paris.

Michel, S., and M. Beuret. 2009: *La Chinafrique: Pékin à la conquête du continent noir.* Paris.

Michel, T. 1992: *Zaïre, le cycle du serpent.* Documentary, Brussels.

———. 1999: *Mobutu, roi du Zaïre.* Documentary, Brussels.

———. 2005: *Congo River.* Documentary, Brussels.

———. 2009: *Katanga Business.* Documentary, Brussels.

Minani, R. 2010: "2010, année charnière: Bref aperçu de la situation sociopolitique." February 19, 2010, www.rodhecic.org.

Mobutu Sese Seko. 1973: "Discours du 30 novembre 1973 devant le Conseil Législatif National." In Cabinet du Département de la Défense Nationale, Forces Armées Zaïroises. *Mémorandum de Réflexion, d'Action et d'Information.* Kinshasa. 229–69.

Mokoko Grampiot, A. 2004: *Kimbanguisme et identité noire.* Paris.

Monheim, F. 1961: *Réponse à Pierre De Vos au sujet de "Vie et mort de Lumumba."* Antwerp, Belgium.

———. 1962: *Mobutu, l'homme seul.* Brussels.

Monnier, L., B. Jewsiewicki, and G. de Villers, eds. 2001: *Chasse au diamant au Congo/Zaïre.* Tervuren, Belgium.

Monstelle, A. de 1965: *La débâcle du Congo Belge.* Brussels.

Muambi, A. 2009: *Democratie kan je niet eten: Reisverslag van een verkiezingswaarnemer.* Amsterdam.

Mumengi, D. 2005: *Panda Farnana, premier universitaire congolais, 1888–1930.* Paris.

Munayi Muntu-Monji. 1977: "La déportation et le séjour des Kimbanguistes dans le Kasaï-Lukenié (1921–1960)." *Zaïre-Afrique* 119: 555–73.

Mutamba Makombo Kitashima, J.-M. 1998: *Du Congo Belge au Congo indépendant 1940–1960.* Kinshasa.

Muzito, A. 2010: "Les années des nationalistes au pouvoir en chiffres." Unpublished prime ministerial PowerPoint presentation, Kinshasa.

Mvumbi Ngolu Tsasa. 1986: "'Révolution' sexuelle, intention éthique et ordre politique." In *Crise morale et vie économique au Zaïre.* Edited by Association des Moralistes Zaïrois. Kinshasa. 65–72.

Mwabila Malela. 1979: *Travail et travailleurs au Zaïre: Essai sur la conscience ouvrière du prolétariat urbain de Lubumbashi.* Kinshasa.

Ndaywel è Nziem, I. 1995: *La transition politique au Zaïre et son prophète Dominique Sakombi Inongo.* Québec.

———. 1998: *L'histoire générale du Congo: de l'héritage ancien à la République Démocratique.* Paris.

———. 2002: "Identité congolaise contemporaine du prénom écrit au prénom oral." In *Figures et paradoxes de l'histoire au Burundi, au Congo, et au Rwanda.* 2 vols. Edited by M. Quaghebeur. Paris. 766–79.

Nelson, S. 1994: *Colonialism in the Congo Basin, 1880–1940*. Athens, OH.

Newbury, M. C. 1984: "Ebutumwa Bw'Emiogo, the Tyranny of Cassava: A Women's Tax Revolt in Eastern Zaïre." *Canadian Journal of African Studies* 18, no. 1: 35–54.

Ngalamulume Nkongolo, M. 2000: *Le campus martyr: Lubumbashi, May 11–12, 1990*. Paris.

Ngbanda Nzambo, H. 2004: *Crimes organisés en Afrique Centrale: Révélations sur les réseaux rwandais et occidentaux*. Paris.

Nguza Karl I Bond. 1982: *Mobutu ou l'incarnation du Mal Zaïrois*. London.

NIZA. 2006: *The State vs. the People: Governance, Mining and the Transitional Regime in the Democratic Republic of Congo*. Amsterdam.

Nlandu-Tsasa, C. 1997: *La rumeur au Zaïre de Mobutu: Radio-trottoir à Kinshasa*. Paris.

Northrup, D. 1988: *Beyond the Bend in the River: African Labor in Eastern Zaire, 1865–1940*. Athens, OH.

———. 2002: *Africa's Discovery of Europe*. New York.

———. 2007: "Slavery and Forced Labour in the Eastern Congo, 1850–1910." In *Slavery in the Great Lakes Region of East Africa*. Edited by H. Médard and S. Doyle. Oxford. 111–23.

Nzeza Bilakila, A. 2004: "The Kinshasa Bargain." In *Reinventing Order in the Congo: How People Respond to State Failure in Kinshasa*. Edited by T. Trefon. London. 20–32.

Nzongola-Ntalaja, G., ed. 1986: *The Crisis in Zaire: Myths and Realities*. Trenton, NJ.

Nzongola-Ntalaja, G. 2002: *The Congo from Leopold to Kabila: A People's History*. London.

Nzula, A. T., I. I. Potekhin, and A. Z. Zusmanovich. 1979: *Forced Labour in Colonial Africa*. London. Original Russian edition, 1933.

Olinga, L. 2006: "La victoire en chantant." *Jeune Afrique*, March 7, 2006, www.jeunea frique.com.

Omasombo Tshonda, J., and Benoît Verhaegen. 1998: *Patrice Lumumba: Jeunesse et apprentissage politique, 1925–1956*. Tervuren, Belgium.

———. 2005: *Patrice Lumumba: de la prison aux portes du pouvoir, juillet 1956–février 1960*. Tervuren, Belgium.

Paice, E. 2007: *Tip and Run: The Untold Tragedy of the Great War in Africa*. London.

Pain, M. 1984: *Kinshasa, la ville et la cité*. Paris.

Pakenham, T. 1991: *The Scramble for Africa, 1876–1912*. London.

Pardigon, V. 1961: "L'U.R.S.S." In *Le monde communiste et la crise du Congo belge*. Edited by A. Wauters. Brussels. 59–92.

Paulus, J.-P. 1962: *Congo 1956–1960*. Brussels.

Pauwels-Boon, G. 1979: *L'origine, l'évolution, et le fonctionnement de la radiodiffusion au Zaïre de 1937 à 1960*. Tervuren, Belgium.

Peemans, J.-P. 1988: "Zaïre onder het Mobutu-regime: Grote stappen in de economische en sociale ontwikkeling." In *Wederzijds: De toekomst van de Belgisch-Zaïrese samenwerking*. Edited by J. Devos, J.-P. Peemans, R. Renard, E. Vervliet, and J.-C. Willame. Brussels. 16–49.

Perrings, C. 1979: *Black Mineworkers in Central Africa: Industrial Strategies and the Evolution of an African Proletariat in the Copperbelt 1911–41*. London.

Pétillon, L. A. 1985: *Récit: Congo 1929–1958*. Brussels.

Picard, E. 1896: *En Congolie*. Brussels.

Poddar, P., R. S. Patke, and L. Jensen, eds. 2008: *A Historical Companion to Postcolonial Literatures: Continental Europe and Its Empires*. Edinburgh.

Poel, I. van der. 2006: *Congo-Océan: Un chemin de fer colonial controversé*. 2 vols. Paris.

Pole Institute. 2002: *The Coltan Phenomenon*. Goma.

Poncelet, M. 2008: *L'invention des sciences coloniales belges*. Paris.

Pons, V. 1969: *Stanleyville: An African Urban Community under Belgian Administration*. Oxford.

Popovitch, M. D., and F. De Moor 2007: *Eza Congo: Photographes de RDC*. Roeselare.

Poppe, G. 1998: *De tranen van de dictator: Van Mobutu tot Kabila*. Antwerp, Belgium.

Poupart, R. 1960: *Première esquisse de l'évolution du syndicalisme au Congo*. Brussels.

Prunier, G. 1995: *The Rwanda Crisis: History of a Genocide*. London.

———. 2009: *Africa's World War: Congo, the Rwandan Genocide, and the Making of a Continental Catastrophe*. Oxford.

Pype, K. 2006: "Dancing for God or the Devil: Pentecostal Discourse on Popular Dance in Kinshasa." *Journal of Religion in Africa* 36, nos. 3–4: 296–318.

———. 2007: "Fighting Boys, Strong Men and Gorillas: Notes on the Imagination of Masculinities in Kinshasa." *Journal of the International African Institute* 77, no. 2: 250–71.

———. 2009a: " 'We Need to Open Up the Country': Development and the Christian Key Scenario in the Social Space of Kinshasa's Tele-serials." *Journal of African Media Studies* 1, no. 1: 101–16.

———. 2009b: "Media Celebrity, Charisma, and Morality in Post-Mobutu Kinshasa." *Journal of Southern African Studies* 35, no. 3: 541–55.

———. 2009c: "A Historical Analysis of Christian Visual Media in Postcolonial Kinshasa." *Studies in World Christianity* 15, no. 2: 131–48.

Queuille, P. 1965: *Histoire de l'afro-asiatisme jusqu'à Bandoung: La naissance du tiers-monde*. Paris.

RAID. 2009: "Chinese Mining Operations in Katanga." September 2009, www.raid-uk.org.

Raspoet, E. 2005: *Bwana Kitoko en de koning van de Bakuba: een vorstelijke ontmoeting op de evenaar*. Antwerp, Belgium.

Raymaekers, P., and H. Desroche. 1983: *L'administration et le sacré (1921–1957)*. Brussels.

Remilleux, J.-L. 1989: *Mobutu: Dignité pour l'Afrique*. Paris.

Renson, R., and C. Peeters. 1994: "Sport als missie: Raphaël de la Kéthulle de Ryhove (1890–1956)." In *Voor lichaam en geest: Katholieken, lichamelijke opvoeding en sport in de 19de en 20ste eeuw*. Edited by M. D'hoker, R. Renson, and J. Tolleneer. Louvain, Belgium, 200–15.

Renton, D., D. Seddon, and L. Zeilig. 2007: *The Congo: Plunder and Resistance*. London.

Reynaert, J. 2009: *De balans na tien jaar MONUC in Congo: hoe effectief is de VN-vredesmissie op het terrein?* Louvain, Belgium.

Reyntjens, F. 2009: *De grote Afrikaanse oorlog: Congo in de regionale geopolitiek, 1996–2006*. Antwerp, Belgium.

Riva, S. 2000: *Nouvelle histoire de la littérature du Congo-Kinshasa*. Paris.

Roberts, A. F. 1989: "History, Ethnicity, and Change in the 'Christian Kingdom' of Southeastern Zaire": In *The Creation of Tribalism in Southern Africa*. Edited by L. Vail. Berkeley, CA. 176–207.

Roes, A. 2008: "Thinking with and beyond the State: The Sub- and Supranational Perspectives on the Exploitation of Congolese Natural Resources, 1885–1914." Unpublished lecture, Tervuren, Belgium.

Roussel, J. 1949: *Déontologie coloniale: Consignes de vie et d'action coloniales pour l'élite des blancs et l'élite des noirs*. Louvain, Belgium.

Rubbens, A. 1945: *Dettes de guerre*. Elisabethville.

————. 1983: *De naweeën van de oorlogsinspanning. In: Le Congo belge durant la Seconde Guerre Mondiale*. Brussels, 579–85.

Ryckmans, A. 1970: *Les mouvements prophétiques kongo en 1958: contribution à l'étude de l'Histoire du Congo*. Kinshasa.

Ryckmans, P. 1948: *Dominer pour servir*. Brussels.

Sadin, F. 1918: *La mission des jésuites au Kwango: Notice historique*. Kisantu.

Saint Moulin, L. de. 1983: "La population du Congo pendant la Seconde Guerre Mondiale." In *Le Congo belge durant la Seconde Guerre Mondiale*. Brussels. 15–49.

————. 2007: "Croissance de Kinshasa et transformations du réseau urbain de la République du Congo depuis l'indépendance." In *Villes d'Afrique: Explorations en histoire urbaine*. Edited by J.-L. Vellut. Paris. 41–65.

————. 2009: "Analyse du paysage sociopolitique à partir du résultat des élections de 2006." In *Réforme au Congo (RDC): Attentes et désillusions*. Edited by T. Trefon. Tervuren, Belgium. 49–65.

Sakombi Inongo. 1974a: "L'authenticité à Dakar." InCabinet du Département de la Défense Nationale, Forces Armées Zaïroises, *Mémorandum de Réflexion, d'Action, et d'Information*. Kinshasa. 339–63.

————. 1974b: "L'authenticité à Paris." In Cabinet du Département de la Défense Nationale, Forces Armées Zaïroises, *Mémorandum de Réflexion, d'Action, et d'Information*. Kinshasa. 365–93.

————. 1987: *Mélanges pour une révolution*. Kinshasa.

Salmon, P. 1977: *La révolte des Batetela de l'expédition du Haut-Ituri (1897)*. Brussels.

Schatzberg, M. G. 1988: *The Dialectics of Oppression in Zaire*. Bloomington, IN.

Schöller, A. 1982: *Congo 1959–1960: Mission au Katanga, intérim à Léopoldville*. Paris.

Scholl-Latour, P. 1986: *Mort sur le grand fleuve: Du Congo au Zaïre, chronique d'une indépendance*. Paris.

Scott, I. 1969: *Tumbled House: The Congo at Independence*. London.

Scott, S. A. 2008: *Laurent Nkunda et la rébellion du Kivu: Au cœur de la guerre congolaise*. Paris.

Segal, R. 2001: *Islam's Black Slaves: The Other Black Diaspora*. New York.

Serufuri Hakiza, P. 1984: *Les auxiliaires autochtones des missions protestantes au Congo, 1878–1960: Etude de cinq Sociétés missionaires*. Louvain-la-Neuve, Belgium.

————. 2004: *L'évangélisation de l'ancien royaume Kongo, 1491–1835*. Kinshasa.

Sheriff, A. 1987: *Slaves, Spices, and Ivory in Zanzibar*. London.

Sikitele Gize 1973: "Les racines de la révolte Pende de 1931." *Études d'Histoire Africaine* 5: 99–153.

Sinatu Bolya, C. 2003: "Des sociétés d'élégance aux mouvements d'émancipation féministes." http://www.mvca.be/_realisations/realisations_11.html.

Sinda, M. 1972: *Le messianisme congolais et ses incidences politiques: Kimbanguisme—Matsouanisme—Autres mouvements*. Paris.

Singleton-Gates, P., and M. Girodias. 1959: *The Black Diaries: An Account of Roger Casement's Life and Times with a Collection of His Diaries and Public Writings*. N.p.

Slade, R. 1959: *English-Speaking Missions in the Congo Independent State, 1878–1908*. Brussels.

————. 1962: *King Leopold's Congo: Aspects of the Development of Race Relations in the Congo Independent State*. London.

Smith, J. 2005: *Dinner with Mobutu: A Chronicle of My Life and Times*. N.p.

Soete, G. 1993: *Het einde van de grijshemden: Onze koloniale politie*. Zedelgem.

Sohier, J. 1959: *Essai sur la criminalité dans la province de Léopoldville: Meurtres et infractions apparentées*. Brussels.

Souchard, V. 1983: *Jours de brousse: Congo 1940–1945*. Brussels.

Soudan, F. 2007: "Kabila, l'heure des choix." *Jeune Afrique*, December 16, 2007, 26–29.

Stanley, H. M. 1886: *Zes jaren aan den Congo en de stichting van een nieuwen vrijen staat*. 2 vols. Amsterdam.

————. 1899: *Through the Dark Continent*. 2 vols. London. Original publication 1878.

Stearns, J. 2010: "What Is Obama Doing for Congo?" congosiasa.blogspot.com, January 10.

Stengers, J. 1957: *Combien le Congo a-t-il couté à la Belgique*. Brussels.

————. 1963: *Belgique et Congo: L'élaboration de la Charte coloniale* (Brussels).

————. 1989: *Congo: Mythes et réalités. Cent ans d'histoire*. Paris.

————. 1997: "De uitbreiding van België: Tussen droom en werkelijkheid." In: G. Janssens and J. Stengers (ed.), *Nieuw licht op Leopold I & Leopold II: het archief Goffinet*. Brussels, 237–85.

Stengers J. & J. Vansina 1985: King Leopold's Congo, 1886-1908. In: R. Oliver & G.N. Sanderson (ed.), *The Cambridge History of Africa*, volume 6: *from 1870 to 1905*. Cambridge, 315-58.

Stewart, G. 2000: *Rumba on the River: A History of the Popular Music of the Two Congos*. London.

Stiglitz, J. E. 2002: *Globalization and Its Discontents*. London.

Storme, M. 1970: *La mutinerie militaire au Kasai en 1895*. Brussels.

Strachan, H. 2004: *The First World War in Africa*. Oxford.

Strizek, H. 2006: *Geschenkte Kolonien: Ruanda und Burundi unter deutscher Herrschaft*. Berlin.

Takizala, H. D. 1964: "Situation de l'enseignement durant la première législature." *Études Congolaises* 7, no. 8: 61–79.

Tambwe, E., and J.-M. Dikanga Kazadi, eds. 2008: *Laurent-Désiré Kabila: L'actualité d'un combat*. Paris.

Tardieu, M. 2006: *Les Africains en France: De 1914 à nos jours*. Monaco.

Tempels, P. 1944: "La philosophie de la rébellion." *L'Essor du Congo*, August 31. Included in A. Rubbens 1945: *Dettes de guerre*. Elisabethville. 17–23.

————. 1946 [1945]: *Bantoe-Filosofie*. Antwerp.

Ter Haar, G. 2009: *How God became African: African Spirituality and Western Secular Thought*. Philadelphia.

Thieffry, E. 1926: *En avion de Bruxelles au Congo Belge: Histoire de la première liaison aérienne entre la Belgique et sa colonie*. Brussels.

Thiel, H. van. 1982: *Wij Ngombe: Volk in Zaïre*. Deurne.

Tielemans, H. 1966: *Gijzelaars in Congo: Overzicht van de dramatische gebeurtenissen in het missiegebied Isangi tijdens de Congolese rebellie, 4 augustus 1964–27 februari 1965*. N.p.

Tilman, S. 2000: "L'implantation du scoutisme au Congo belge." In *Itinéraires croisés de la modernité: Congo belge (1920–1950)*. Edited by J.-L. Vellut. Tervuren, Belgium. 103–40.

Tokwaula Aena, B. n.d.: *Tant que je vivrai, tu vivras*. Kinshasa.

Travaux du Groupe d'Études Coloniales 1912: "Les fermes-chapelles au point de vue économique et civilisateur." *Bulletin de la Société Belge d'Etudes Coloniales* 5.

Trefon, T., ed. 2004: *Reinventing Order in the Congo: How People Respond to State Failure in Kinshasa*. London.

————. 2009: *Réforme au Congo (RDC): Attentes et désillusions.* Tervuren, Belgium.

Tshibangu Kabet Musas. 1974: "La situation sociale dans le ressort administratif de Likasi (ex-Territoire de Jadotville) pendant la Guerre 1940–1945." Études d'Histoire Africaine 6: 275–311.

Tshimanga, C. 2000: "L'adaptes et la formation d'une élite au Congo (1925–1945)." In *Itinéraires croisés de la modernité: Congo belge (1920–1950).* Edited by J.-L. Vellut. Tervuren, Belgium. 189–204.

Tshitungu Kongolo, A. 2003a: "Paul Panda Farnana (1888–1930), panafricaniste, nationaliste, intellectuel engagé: Une contribution à l'étude de sa pensée et de son action." *L'Africain* 211 (October–November 2003): 1–7.

————. 2003b: *Poète ton silence est crime.* Paris.

Turine, Roger Pierre. 2007: *Les arts du Congo, d'hier à nos jours.* Brussels.

Turner, T. 2007: *The Congo War: Conflict, Myth and Reality.* London.

UNESCO. 2005: *Promoting and Preserving Congolese Heritage: Linking Biological and Cultural Diversity.* Paris.

United Nations Security Council. 2008: *Final Report of the Group of Experts on the Democratic Republic of the Congo,* S/2008/773.

Valon, A. de 2006: "Mission de renforcement des capacités du Commissariat de District de l'Ituri." Unpublished typescript. Bunia.

Van Acker, G. 1924: *Een Vlaamsch geloofszendeling bij de Baloeba's in Congoland.* N.p.

Van Bilsen, A. A. J. 1962: *L'indépendance du Congo.* Doornik/Tournai, Belgium.

Van Bilsen, J. 1958: *Vers l'indépendance du Congo et du Ruanda-Urundi.* Kraainem, Belgium.

————. 1993: *Kongo 1945–1965: Het einde van een kolonie.* Louvain, Belgium.

Vandaele, J. 2005: *Het recht van de rijkste: Hebben andersglobalisten gelijk?* Antwerp, Belgium.

————. 2008: "Het roofdier, Mozes en de Chinezen." Mo, February 2008, 28–33.

Van den Bosch, J. 1986: *Pré-Zaïre: Le cordon mal coupé.* Brussels.

Van den Bosch, P. 1992: *Vijf en twintig jaren in de branding: Congo-Zaïre, november 1949—januari 1975.* N.p.

Vanderkerken, G. 1920: *Les sociétés bantoues du Congo Belge et les problèmes de la politique indigène.* Brussels.

Vanderlinden, J. 1991: *À propos de l'uranium congolais.* Brussels.

————. 1994: *Pierre Ryckmans, 1891–1959: Coloniser dans l'honneur.* Brussels.

Van der Smissen, E. 1920: *Léopold ii et Beernaert, d'après leur correspondance inédite de 1884 à 1894.* 2 vols. Brussels.

Vandersmissen, J. 2008: *Koningen van de wereld: de aardrijkskundige beweging en de ontwikkeling van de koloniale doctrine van Leopold II.* Ghent, Belgium.

Vanderstraeten, L.-F. 1985: *Histoire d'une mutinerie, juillet 1960: De la Force Publique à l'Armée nationale congolaise.* Paris.

————. 2001: *La répression de la révolte des Pende du Kwango en 1931.* Brussels.

Vandewalle, G. 1966: *De conjuncturele evolutie in Kongo en Ruanda-Urundi van 1920 tot 1939 en van 1949 tot 1958.* Ghent, Belgium.

Van Dijck, H. 1997: "Rapport sur les violations des Droits de l'Homme dans le Sud-Equateur du 15 Mars 1997 au 15 Septembre 1997." Unpublished report.

Vandommele, M. 1983: *Zaïre: Buitenlandse belangen, binnenlandse pijnbank.* Brussels.

Vangansbeke, J. 2006: "Afrikaanse verdedigers van het Belgisch grondgebied, 1914–1918." *Belgische Bijdragen tot de Militaire Geschiedenis* 4: 123–34.

Vangroenweghe, D. 1985: *Rood rubber: Leopold II en zijn Kongo.* Brussels.

————. 2005: *Voor rubber en ivoor: Leopold II en de ophanging van Stokes.* Louvain, Belgium.

Van Lierde, J. 1963: *La pensée politique de Patrice Lumumba.* Paris.

Van Overbergh, C. 1913: *Les nègres d'Afrique.* Brussels.

Van Peel, B. 2000: "Aux débuts du football congolais." In *Itinéraires croisés de la modernité: Congo belge (1920–1950).* Edited by J.-L. Vellut. Tervuren, Belgium. 141–87.

Van Reybrouck, D. 2009: "Congo in de populaire cultuur." In *Congo in België: Koloniale cultuur in de metropool.* Edited by V. Viaene, D. Van Reybrouck, and B. Ceuppens. Louvain, Belgium. 169–81.

Vansina, J. 1965: *Les anciens royaumes de la savane: les états des savanes méridionales de l'Afrique centrale des origines à l'occupation coloniale.* Léopoldville.

————. 1976: "L'Afrique centrale vers 1875." In Koninklijke Academie voor Overzeese Wetenschappen, *Bijdragen over de Aardrijkskundige Conferentie van 1876.* Brussels. 1–31.

————. 1990: *Paths in the Rainforest: Toward a History of Political Tradition in Equatorial Africa.* Madison, WI.

————. 2004: *How Societies Are Born: Governance in West Central Africa before 1600.* Charlottesville, VA.

Vanthemsche, G. 1999: "Radioscopie van een kolonie: Belgisch-Congo 1908–1960." *Brood en Rozen* 4, no. 2: 9–29.

————. 2007: *Congo: de impact van de kolonie op België.* Tielt, Belgium.

————. 2009: *Le Congo belge pendant la Première Guerre mondiale: les rapports du ministre des Colonies Jules Renkin au roi Albert Iᵉʳ, 1914–1918.* Brussels.

————. 2010: *Belgian Congo during the First World War, as seen through the reports of Jules Renkin, minister of Colonies, to King Albert I, 1914–1918.* Mededelingen der zittingen van de Koninklijke Academie voor Overzeese Wetenschappen.

Van Wing, J. 1959: *Études Bakongo: Sociologie, réligion, et magie.* Bruges, Belgium.

Vellut, J.-L. 1981: *Les bassins miniers de l'ancien Congo Belge: Essai d'histoire économique et sociale (1900–1960).* Brussels.

————. 1983: "Le Katanga industriel en 1944: Malaises et anxiétés dans la société coloniale." In *Le Congo belge durant la Seconde Guerre Mondiale.* Brussels. 495–523.

————. 1984: "La violence armée dans l'Etat Indépendant du Congo." *Cultures et Développement* 16, nos. 3–4: 671–707.

————. 1992: "Une exécution publique à Elisabethville (20 septembre 1922): Notes sur la pratique de la peine capitale dans l'histoire coloniale du Congo." In *Art pictural zaïrois.* Edited by B. Jewsiewicki. Paris. 171–222.

————. 1996: "Le bassin du Congo et l'Angola." In *Histoire générale de l'Afrique,* vol. 6: *L'Afrique au XIXᵉ siècle jusque vers les années 1880.* Edited by J. F. Ade Ajayi. Paris. 331–61.

————. 2000: *Itinéraires croisés de la modernité: Congo belge (1920–1950).* Tervuren, Belgium.

————. 2005a: *Simon Kimbangu. 1921: de la prédication à la déportation. Les sources.* Vol. 1: *Fonds missionaires protestants (1). Alliance missionnaire suédoise.* Brussels.

————. 2005b: *Het geheugen van Congo: De koloniale tijd.* Tervuren, Belgium.

————. 2009: "Contextes africains du projet colonial de Léopold II." Unpublished lecture. Louvain-la-Neuve, Belgium.

Verbeken, A. 1956: *Msiri, roi du Garenganze: "l'homme rouge" du Katanga.* Brussels.

————. 1958: *La révolte des Batetela en 1895.* Brussels.

Verbeken, P. 2005: "Ik zeg het eerlijk: het was een prachtjob." *Humo,* July 26, 32–37.

Verhaegen, B. 1966–1969: *Rébellions au Congo.* 2 vols. Brussels.

―――. 1970: "Dix ans d'indépendance." *Revue Française d'Études Politiques Africaines* 57: 17–25.

―――. 1971: "Étude sur la rébellion." In*Mouvements nationaux d'indépendance et classes populaires*. Paris. 418–43.

―――. 1978: *L'enseignement universitaire au Zaïre: de Lovanium à l'Unaza 1958–1978*. Paris.

―――. 1983: "La guerre vécue au Centre Extra-Coutumier de Stanleyville." In *Le Congo belge durant la Seconde Guerre Mondiale*. Brussels. 439–93.

―――. 1986: "Conditions politiques et participation sociale à la rébellion dans l'Est du Zaïre." *Les cahiers du CEDAF* 7–8: 1–14.

―――. 1990: *Femmes zaïroises de Kisangani: Combats pour la survie*. Paris.

―――. 1999: "Communisme et anticommunisme au Congo (1920–1960)." *Brood en Rozen* 4, no. 2: 113–27.

Verhaegen B., and C. Tshimanga. 2003: *L'Abako et l'indépendance du Congo belge: dix ans de nationalisme kongo (1950–1960)*. Tervuren, Belgium.

Verlinden, P. 2002: *Weg uit Congo: Het drama van de kolonialen*. Louvain, Belgium.

―――. 2008: *Achterblijven in Congo: Een drama voor de Congolezen?* Louvain, Belgium.

Vesse, A. 1961: *Note sur l'évolution de l'économie congolaise après l'indépendance du pays*. N.p.

Viaene, V. 2008: "King Leopold's Imperialism and the Origins of the Belgian Colonial Party, 1860–1905." *Journal of Modern History* 80: 741–90.

―――. 2009: ""Reprise-remise: de Congolese identiteitscrisis van België rond 1908. In V. Viaene, D. Van Reybrouck, and B. Ceuppens (ed.), De overname van België door Congo: aspecten van de Congolese 'aanwezigheid' in de Belgische samenleving, 1908–1958. Louvain, Belgium, 43–62.

―――. 2009: "De religie van de prins: Leopold II, de Heilige Stoel, België en Congo (1855–1909)." In *Leopold II, ongegeneerd genie? Buitenlandse politiek en kolonisatie*. Edited by V. Dujardin, V. Rosoux, and T. de Wilde. Tielt, Belgium. 143–64.

Viaene, V., D. Van Reybrouck, and B. Ceuppens, ed. 2009: *Congo in België: Koloniale cultuur in de metropool*. Louvain, Belgium.

Vijgen, I. 2005: *Tussen mandaat en kolonie: Rwanda, Burundi en het Belgische bestuur in opdracht van de Volkenbond (1916–1932)*. Louvain, Belgium.

Villafaña, F. R. 2009: *Cold War in the Congo: The Confrontation of Cuban Military Forces, 1960–1967*. New Brunswick, NJ.

Villers, G. de. 1997: *Zaïre: La transition manquée (1990–1997)*. Paris.

―――. 2001: *Guerre et politique: Les trente derniers mois de L. D. Kabila (août 1998–janvier 2001)*. Tervuren, Belgium.

―――. 2009: *De la guerre aux élections: L'ascension de Joseph Kabila et la naissance de la Troisième République (janvier 2001–août 2008)*. Tervuren, Belgium.

Villers, G. de, B. Jewsiewicki, and L. Monnier, eds. 2002: *Manières de vivre: économie de la "débrouille" dans les villes du Congo/Zaïre*. Tervuren, Belgium.

Villers, G. de, and J. Omasombo Tshonda. 2002: "An Intransitive Transition." *Review of African Political Economy* 93–94: 399–410.

Vinck, H. 2002: *Colonial Schoolbooks (Belgian Congo): Anthology*. www.abbol.com.

Vlassenroot, K. 2000: "The Promise of Ethnic Conflict: Militarisation and Enclave-Formation in South Kivu." In *Conflict and Ethnicity in Central Africa*. Edited by D. Goyvaerts. Tokyo. 59–104.

Vlassenroot, K., and H. Romkema. 2002: "The Emergence of a New Order? Resources and War in Eastern Congo." *Journal of Humanitarian Assistance*, http://www.jha.ac/articles/a111.htm.

Vlassenroot, K., and T. Raeymaekers, eds. 2004a: *Conflict and Social Transformation in Eastern DRC*. Ghent, Belgium.

Vlassenroot, K., and T. Raeymaekers. 2004b: "Le conflit en Ituri." *L'Afrique des Grands Lacs, annuaire 2003–2004*, 207–33.

Wamu Oyatambwe. 1999: *De Mobutu à Kabila: Avatars d'une passation inopinée*. Paris.

Wauters, A. 1929: *D'Anvers à Bruxelles via le Lac Kivu: Le Congo vu par un socialiste*. Brussels.

Wauters, A., ed. 1961: *Le monde communiste et la crise du Congo belge*. Brussels.

Weiss, H. 1965: "L'évolution des élites." *Études Congolaises* 8, no. 5: 1–14.

Weiss, H., and B. Verhaegen, eds. 1986: *Les rébellions dans l'est du Zaïre (1964–1967)*. Thematic issue of *Les Cahiers du CEDAF*, 7–8.

Weissman, S. R. 1974: *American Foreign Policy in the Congo 1960–1964*. Ithaca, NY.

———. 2010: "An Extraordinary Rendition." *Intelligence and National Security* 25, no. 2: 198–222.

Werbrouck, R. 1945: *La campagne des troupes coloniales belges en Abyssinie*. Léopoldville.

Wesseling, H. L. 1991: *Verdeel en heers: De deling van Afrika, 1880–1914*. Amsterdam.

White, B. W. 2008: *Rumba Rules: The Politics of Dance Music in Mobutu's Zaire*. Durham.

Wildwood, E. 2008: *Migration and Christian Identity in Congo*. Leiden, The Netherlands.

Willame, J.-C. 1986: *Zaïre, l'épopée d'Inga: Chronique d'une prédation industrielle*. Paris.

———. 1988: *Eléments pour une lecture du contentieux belgo-zaïrois*. Brussels.

———. 1990: *Patrice Lumumba: La crise congolaise revisitée*. Paris

———. 1992: *L'automne d'un despotisme: Pouvoir et argent et obéissance dans le Zaïre des années quatre-vingt*. Paris.

———. 1998: "La 'nouvelle' politique américaine en Afrique centrale." In *Kabila prend le pouvoir*. Contributed volume by C. Braeckman, M.-F. Cros, G. de Villers, F. François, F. Reyntjens, F. Ryckmans, and J.-C. Willame. Brussels. 134–44.

———. 2007: *Les "faiseurs de paix" au Congo: gestion d'une crise internationale dans un Etat sous tutelle*. Brussels.

Wolf, E. 1982: *Europe and the People without History*. Berkeley.

Wolter, R., L. Davreux, and R. Regnier 1957: *Le chômage au Congo belge*. Brussels.

Wotzka, H.-P. 1995: *Studien zur Archäologie des zentral-afrikanische Regenwaldes: die Keramik des inneren Zaïre-Beckens und ihre Stellung im Kontext der Bantu-Expansion*. Cologne.

Wrong, M. 2000: *In the Footsteps of Mr. Kurtz: Living on the Brink of Disaster in the Congo*. London.

Wynants, M. 1997: *Van hertogen en Kongolezen: Tervuren en de koloniale tentoonstelling 1897*. Tervuren.

Yakemtchouk, R. 1986: "Les relations entre les Etats-Unis et le Zaïre." *Studia Diplomatica* 39, no. 1: 5–127.

———. 1988a: *Aux origines du séparatisme katangais*. Brussels.

———. 1988b: "Les deux guerres du Shaba: Les relations entre la Belgique, la France et le Zaïre." *Studia Diplomatica* 41, nos. 4–6: 375–742.

Yambuya, P. 1991: *Zaïre, het abattoir: Over gruweldaden van het leger van Mobutu*. Antwerp.

Yav, A. 1965: "Vocabulaire de la ville de Elisabethville: A History of Elisabethville from Its Beginnings to 1965." Compiled and written by André Yav, edited, translated, and commented by Johannes Fabian with assistance from Kalundi Mango. *Archives of Popular Swahili* 4 (2001), http://www2.fmg.uva.nl/lpca/aps/vol4/vocabulairesha baswahili.html.

Yellen, J. E. 1996: "Behavioral and Taphonomic Patterning at Katanda 9: A Middle Stone Age Site, Kivu Province, Zaïre." *Journal of Archaeological Science* 23: 915–32.

Yerodia, A. 2004: *Rapport sur les assassinats et violations des droits de l'homme.* Kinshasa.

Yoka, L. M. 2005: *Kinshasa, carnets de guerre.* Kinshasa.

Young, C. M. 1968: *Introduction à la politique congolaise.* Brussels.

———. 1984: "Zaire, Rwanda, and Burundi." In *The Cambridge History of Africa,* vol. 8: *From c. 1940 to c. 1975.* Edited by M. Crowder. Cambridge, UK. 698–754.

Young, C., and T. Turner. 1985: *The Rise and Decline of the Zairean State.* Madison, WI.

Zeebroek, X. 2008: *La Mission des Nations Unies au Congo: Le laboratoire de la paix introuvable.* Brussels.

Zhang, L. 2008: "Ethnic Congregation in a Globalizing City: The Case of Guangzhou, China." www.sciencedirect.com.

Ziégler, J. 1963: *La contre-révolution en Afrique.* Paris.

Zinzen, W. 1995: *Mobutu: Van mirakel tot malaise.* Antwerp.

———. 2004: *Kisangani, verloren stad.* Louvain, Belgium.

INDEX